# VW/AUDI

# PASSAT 1998-
# A4 1996-01 REPAIR MANUAL

T0260314

## Covers U.S. and Canadian models of Volkswagen Passat and Audi A4; 1.8L four-cylinder turbo and 2.8L V6 engines

*Does not include diesel engine, W8 engine or S4 model information*

### by Eric Godfrey

## CHILTON *Automotive Books*

PUBLISHED BY **HAYNES NORTH AMERICA, Inc.**

©2008 Haynes North America, Inc.
ISBN-13: 978-1-56392-748-5
ISBN-10: 1-56392-748-9
Library of Congress Control Number 2008937128

**Haynes UK**
Sparkford Nr Yeovil
Somerset BA22 7JJ England

**Haynes North America, Inc**
859 Lawrence Drive
Newbury Park
California 91320 USA

10U11

Chilton is a registered trademark of W.G. Nichols, Inc., and has been licensed to Haynes North America, Inc.

# Contents

**Photographer, author and mechanic with a 1999 Volkswagen Passat**

## ACKNOWLEDGEMENTS

Technical writers who contributed to this project include Jeff Kibler, Tim Imhoff, Robert Maddox and Bob Henderson.

All rights reserved. No part of this book may be reproduced or transmitted in any form or by any means, electronic or mechanical, including photocopying, recording or by any information storage or retrieval system, without permission in writing from the copyright holder.

## DISCLAIMER

There are risks associated with automotive repairs. The ability to make repairs depends on the individual's skill, experience and proper tools. Individuals should act with due care and acknowledge and assume the risk of performing automotive repairs.

The purpose of this manual is to provide comprehensive, useful and accessible automotive repair information, to help you get the best value from your vehicle. However, this manual is not a substitute for a professional certified technician or mechanic.

This repair manual is produced by a third party and is not associated with an individual vehicle manufacturer. If there is any doubt or discrepancy between this manual and the owner's manual or the factory service manual, please refer to the factory service manual or seek assistance from a professional certified technician or mechanic.

Even though we have prepared this manual with extreme care and every attempt is made to ensure that the information in this manual is correct, neither the publisher nor the author can accept responsibility for loss, damage or injury caused by any errors in, or omissions from, the information given.

## About this manual

### ITS PURPOSE

The purpose of this manual is to help you get the best value from your vehicle. It can do so in several ways. It can help you decide what work must be done, even if you choose to have it done by a dealer service department or a repair shop; it provides information and procedures for routine maintenance and servicing; and it offers diagnostic and repair procedures to follow when trouble occurs.

We hope you use the manual to tackle the work yourself. For many simpler jobs, doing it yourself may be quicker than arranging an appointment to get the vehicle into a shop and making the trips to leave it and pick it up. More importantly, a lot of money can be saved by avoiding the expense the shop must pass on to you to cover its labor and overhead costs. An added benefit is the sense of satisfaction and accomplishment that you feel after doing the job yourself.

### USING THE MANUAL

The manual is divided into Chapters. Each Chapter is divided into numbered Sections, which are headed in bold type between horizontal lines. Each Section consists of consecutively numbered paragraphs.

At the beginning of each numbered Section you will be referred to any illustrations which apply to the procedures in that Section. The reference numbers used in illustration captions pinpoint the pertinent Section and the Step within that Section. That is, illustration 3.2 means the illustration refers to Section 3 and Step (or paragraph) 2 within that Section.

Procedures, once described in the text, are not normally repeated. When it's necessary to refer to another Chapter, the reference will be given as Chapter and Section number. Cross references given without use of the word "Chapter" apply to Sections and/or paragraphs in the same Chapter. For example, "see Section 8" means in the same Chapter.

References to the left or right side of the vehicle assume you are sitting in the driver's seat, facing forward.

Even though we have prepared this manual with extreme care, neither the publisher nor the author can accept responsibility for any errors in, or omissions from, the information given.

### ➡NOTE

A *Note* provides information necessary to properly complete a procedure or information which will make the procedure easier to understand.

### ✳✳ CAUTION

A *Caution* provides a special procedure or special steps which must be taken while completing the procedure where the Caution is found. Not heeding a Caution can result in damage to the assembly being worked on.

### ✳✳ WARNING

A *Warning* provides a special procedure or special steps which must be taken while completing the procedure where the Warning is found. Not heeding a Warning can result in personal injury.

## Introduction

The models covered by this manual are offered in four-door sedan and five-door wagon body styles.

The front suspension on all models is independent, featuring four transverse links (control arms) per side, coil-over shock absorber assemblies and a stabilizer bar. The rear suspension on front-wheel drive models uses a semi-independent torsion beam axle with integral trailing arms, with coil-over shock absorber assemblies on the Audi A4 and individual coil springs and shock absorbers on the VW Passat. The rear suspension on all-wheel drive models uses upper and lower control arms, track rods, coil-over shock absorber assemblies and a stabilizer bar.

Power comes from either a turbocharged 1.8L four-cylinder engine or a SOHC or DOHC V6 engine. All engines are fuel-injected. The engine on all models is mounted longitudinally and transfers power to the front wheels (front-wheel drive models) or to all of the wheels (all-wheel drive models) through either a five-speed manual transaxle or a four- or five-speed automatic transaxle, and driveaxles with constant velocity joints.

The power-assisted steering is rack-and-pinion, mounted behind the engine.

All models have a power assisted brake system with disc brakes at the front and rear. An Anti-lock Brake System (ABS) is standard equipment.

## VEHICLE IDENTIFICATION NUMBERS

Modifications are a continuing and unpublicized process in vehicle manufacturing. Since spare parts manuals and lists are compiled on a numerical basis, the individual vehicle numbers are essential to correctly identify the component required.

## VEHICLE IDENTIFICATION NUMBER (VIN)

The Vehicle Identification Number (VIN), which appears on the Vehicle Certificate of Title and Registration, is also embossed on a plate located on the upper left (driver's side) corner of the dashboard, near the windshield (see illustration). The VIN tells you when and where a vehicle was manufactured, its country of origin, make, type, passenger safety system, line, series, body style, engine and assembly plant.

## VIN ENGINE AND MODEL YEAR CODES

Two particularly important pieces of information found in the VIN are the engine code and the model year code. Counting from the left, the engine code is the 5th character; the model year code is the 10th character.

**On the models covered by this manual the model year codes are:**

| | |
|---|---|
| T | 1996 |
| V | 1997 |
| W | 1998 |
| X | 1999 |
| Y | 2000 |
| 1 | 2001 |
| 2 | 2002 |
| 3 | 2003 |
| 4 | 2004 |
| 5 | 2005 |

**The engine codes are:**

1996 (Audi A4)
    D .......... 2.8L V6 (SOHC)
1997 (Audi A4)
    A .......... 2.8L V6 (SOHC)
    B .......... 1.8L four-cylinder turbo
1998 - 2000 Audi A4
    B .......... 1.8L four-cylinder turbo
    D .......... 2.8L V6 (DOHC)
2001 Audi A4
    C .......... 1.8L four-cylinder turbo
    H .......... 2.8L V6 (DOHC)
1998 - 2002 VW Passat
    A .......... 1.8L four-cylinder turbo
    D .......... 2.8L V6 (DOHC)
2003 - 2004 VW Passat
    D .......... 1.8L four-cylinder turbo
    H .......... 2.8L V6 (DOHC)
2005 VW Passat
    D .......... 1.8L four-cylinder turbo
    U .......... 2.8L V6 (DOHC)

## VEHICLE SAFETY CERTIFICATION LABEL

The Vehicle Safety Certification label is attached to the front of the driver's door post (see illustration). The label contains the name of the manufacturer, the month and year of production, the Gross Vehicle Weight Rating (GVWR), the Gross Axle Weight Rating (GAWR) and the certification statement. On most models, the label also includes the OEM tire sizes and pressures.

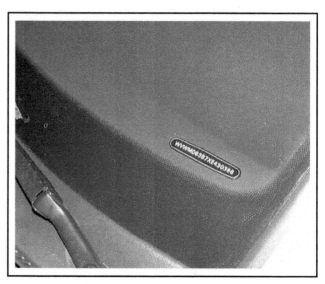

**The VIN plate is visible from the outside of the vehicle, through the driver's side of the windshield**

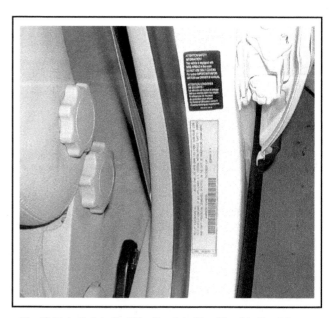

**The Vehicle Safety Certification label is affixed to the drivers side door post**

## SERVICE PARTS IDENTIFICATION LABEL

Located on the inside of the spare tire well in the trunk, this label contains information about the options on your vehicle and the paint and trim codes (see illustration). This information is important when ordering parts or when bodywork and repainting is done.

## ENGINE IDENTIFICATION CHARACTERS

The engine identification characters are stamped into a machined pad on the left side of the engine block just above the oil filter (four-cylinder engines), on the right side of the cylinder block between the cylinder head and the power steering pump (SOHC V6 engines) or at the front of the right-side cylinder head (DOHC V6 engines).

## TRANSAXLE IDENTIFICATION NUMBER (TIN)

The Transaxle Identification Number (TIN) is stamped into a plate fastened to the bottom of the transaxle (see illustration).

**The Service Parts Identification label (arrow) contains information on options and trim/paint codes - label is in the spare tire well in the trunk**

**Typical transaxle identification number location (arrow)**

## Anti-theft audio system

## GENERAL INFORMATION

1   All models are equipped with a stereo that has an anti-theft feature that will render the stereo inoperative if stolen or if the battery is disconnected. If the power source to the stereo is cut, the stereo will be inoperative. Even if the power source is immediately re-connected, the stereo will not function.

2   Do not disconnect the battery, remove the stereo or disconnect related components unless you have the individual ID (code) number for the stereo.

### Unlocking the stereo after a power loss

3   When the power is restored to the stereo, the stereo won't operate. Enter your ID code to reactivate it, using the following Steps.

4   Turn the radio ON. The lower display window (where radio stations are displayed) should show "SAFE". After a few seconds, the SAFE message should go out and the numerals "1000" should display.

5   At the bottom of the radio, the numbered buttons used to preset stations are used to enter your anti-theft code.

6   Press the number 1 preset button repeatedly until the first number of your four-digit code is displayed, then enter the second number of your code on button #2, and so on until all four code numbers are displayed.

7   Press the SEEK button until you hear a sound from the radio. The display will read "LSM" if you've entered the correct code.

8   If you have entered the wrong code during the above procedure, the display will read "SAFE," but you have one more try. If the second code is incorrect, the radio will stay locked for one hour, at which time two more code attempts can be made. Leave the radio on for an hour (while it is locked) and the key in the ignition switch, then try again.

9   You should have the code written down in a secure place, for use in unlocking the anti-theft feature.

## Buying parts

Replacement parts are available from many sources, which generally fall into one of two categories - authorized dealer parts departments and independent retail auto parts stores. Our advice concerning these parts is as follows:

**Retail auto parts stores:** Good auto parts stores will stock frequently needed components which wear out relatively fast, such as clutch components, exhaust systems, brake parts, tune-up parts, etc. These stores often supply new or reconditioned parts on an exchange basis, which can save a considerable amount of money. Discount auto parts stores are often very good places to buy materials and parts needed for general vehicle maintenance such as oil, grease, filters, spark plugs, belts, touch-up paint, bulbs, etc. They also usually sell tools and general accessories, have convenient hours, charge lower prices and can often be found not far from home.

**Authorized dealer parts department:** This is the best source for parts which are unique to the vehicle and not generally available elsewhere (such as major engine parts, transmission parts, trim pieces, etc.).

**Warranty information:** If the vehicle is still covered under warranty, be sure that any replacement parts purchased - regardless of the source - do not invalidate the warranty!

To be sure of obtaining the correct parts, have engine and chassis numbers available and, if possible, take the old parts along for positive identification.

**Maintenance techniques, tools and working facilities**

## MAINTENANCE TECHNIQUES

There are a number of techniques involved in maintenance and repair that will be referred to throughout this manual. Application of these techniques will enable the home mechanic to be more efficient, better organized and capable of performing the various tasks properly, which will ensure that the repair job is thorough and complete.

### Fasteners

Fasteners are nuts, bolts, studs and screws used to hold two or more parts together. There are a few things to keep in mind when working with fasteners. Almost all of them use a locking device of some type, either a lockwasher, locknut, locking tab or thread adhesive. All threaded fasteners should be clean and straight, with undamaged threads and undamaged corners on the hex head where the wrench fits. Develop the habit of replacing all damaged nuts and bolts with new ones. Special locknuts with nylon or fiber inserts can only be used once. If they are removed, they lose their locking ability and must be replaced with new ones.

Rusted nuts and bolts should be treated with a penetrating fluid to ease removal and prevent breakage. Some mechanics use turpentine in a spout-type oil can, which works quite well. After applying the rust penetrant, let it work for a few minutes before trying to loosen the nut or bolt. Badly rusted fasteners may have to be chiseled or sawed off or removed with a special nut breaker, available at tool stores.

If a bolt or stud breaks off in an assembly, it can be drilled and removed with a special tool commonly available for this purpose. Most automotive machine shops can perform this task, as well as other repair procedures, such as the repair of threaded holes that have been stripped out.

Flat washers and lockwashers, when removed from an assembly, should always be replaced exactly as removed. Replace any damaged washers with new ones. Never use a lockwasher on any soft metal surface (such as aluminum), thin sheet metal or plastic.

### Fastener sizes

For a number of reasons, automobile manufacturers are making wider and wider use of metric fasteners. Therefore, it is important to be able to tell the difference between standard (sometimes called U.S.

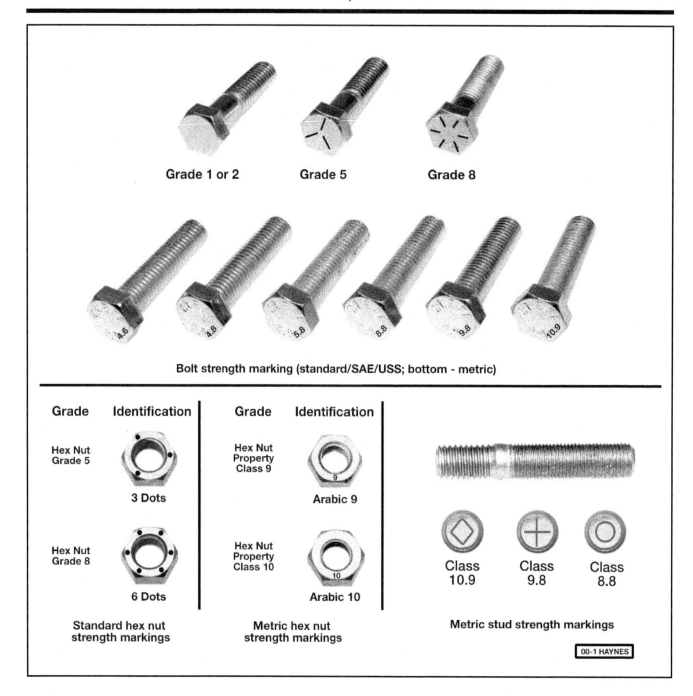

Grade 1 or 2    Grade 5    Grade 8

Bolt strength marking (standard/SAE/USS; bottom - metric)

| Grade | Identification | Grade | Identification |
|---|---|---|---|

Hex Nut Grade 5 — 3 Dots

Hex Nut Grade 8 — 6 Dots

Standard hex nut strength markings

Hex Nut Property Class 9 — Arabic 9

Hex Nut Property Class 10 — Arabic 10

Metric hex nut strength markings

Class 10.9    Class 9.8    Class 8.8

Metric stud strength markings

00-1 HAYNES

or SAE) and metric hardware, since they cannot be interchanged.

All bolts, whether standard or metric, are sized according to diameter, thread pitch and length. For example, a standard 1/2 - 13 x 1 bolt is 1/2 inch in diameter, has 13 threads per inch and is 1 inch long. An M12 - 1.75 x 25 metric bolt is 12 mm in diameter, has a thread pitch of 1.75 mm (the distance between threads) and is 25 mm long. The two bolts are nearly identical, and easily confused, but they are not interchangeable.

In addition to the differences in diameter, thread pitch and length, metric and standard bolts can also be distinguished by examining the bolt heads. To begin with, the distance across the flats on a standard bolt head is measured in inches, while the same dimension on a metric bolt is sized in millimeters (the same is true for nuts). As a result, a standard wrench should not be used on a metric bolt and a metric wrench should not be used on a standard bolt. Also, most standard bolts have slashes

radiating out from the center of the head to denote the grade or strength of the bolt, which is an indication of the amount of torque that can be applied to it. The greater the number of slashes, the greater the strength of the bolt. Grades 0 through 5 are commonly used on automobiles. Metric bolts have a property class (grade) number, rather than a slash, molded into their heads to indicate bolt strength. In this case, the higher the number, the stronger the bolt. Property class numbers 8.8, 9.8 and 10.9 are commonly used on automobiles.

Strength markings can also be used to distinguish standard hex nuts from metric hex nuts. Many standard nuts have dots stamped into one side, while metric nuts are marked with a number. The greater the number of dots, or the higher the number, the greater the strength of the nut.

Metric studs are also marked on their ends according to property class (grade). Larger studs are numbered (the same as metric bolts), while smaller studs carry a geometric code to denote grade.

| Metric thread sizes | Ft-lbs | Nm |
|---|---|---|
| M-6 | 6 to 9 | 9 to 12 |
| M-8 | 14 to 21 | 19 to 28 |
| M-10 | 28 to 40 | 38 to 54 |
| M-12 | 50 to 71 | 68 to 96 |
| M-14 | 80 to 140 | 109 to 154 |

| Pipe thread sizes | | |
|---|---|---|
| 1/8 | 5 to 8 | 7 to 10 |
| 1/4 | 12 to 18 | 17 to 24 |
| 3/8 | 22 to 33 | 30 to 44 |
| 1/2 | 25 to 35 | 34 to 47 |

| U.S. thread sizes | | |
|---|---|---|
| 1/4 - 20 | 6 to 9 | 9 to 12 |
| 5/16 - 18 | 12 to 18 | 17 to 24 |
| 5/16 - 24 | 14 to 20 | 19 to 27 |
| 3/8 - 16 | 22 to 32 | 30 to 43 |
| 3/8 - 24 | 27 to 38 | 37 to 51 |
| 7/16 - 14 | 40 to 55 | 55 to 74 |
| 7/16 - 20 | 40 to 60 | 55 to 81 |
| 1/2 - 13 | 55 to 80 | 75 to 108 |

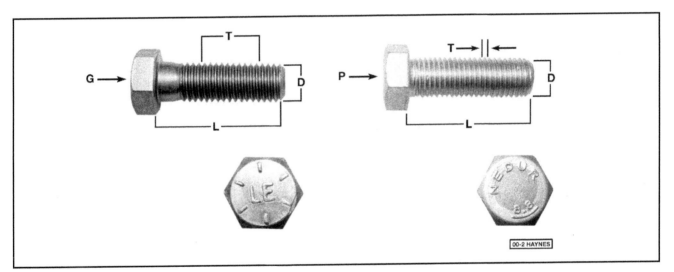

00-2 HAYNES

**Standard (SAE and USS) bolt dimensions/grade marks**

G   Grade marks (bolt strength)
L   Length (in inches)
T   Thread pitch (number of threads per inch)
D   Nominal diameter (in inches)

**Metric bolt dimensions/grade marks**

P   Property class (bolt strength)
L   Length (in millimeters)
T   Thread pitch (distance between threads in millimeters)
D   Diameter

It should be noted that many fasteners, especially Grades 0 through 2, have no distinguishing marks on them. When such is the case, the only way to determine whether it is standard or metric is to measure the thread pitch or compare it to a known fastener of the same size.

Standard fasteners are often referred to as SAE, as opposed to metric. However, it should be noted that SAE technically refers to a non-metric fine thread fastener only. Coarse thread non-metric fasteners are referred to as USS sizes.

Since fasteners of the same size (both standard and metric) may have different strength ratings, be sure to reinstall any bolts, studs or nuts removed from your vehicle in their original locations. Also, when replacing a fastener with a new one, make sure that the new one has a strength rating equal to or greater than the original.

### Tightening sequences and procedures

Most threaded fasteners should be tightened to a specific torque value (torque is the twisting force applied to a threaded component such as a nut or bolt). Overtightening the fastener can weaken it and cause it to break, while undertightening can cause it to eventually come loose. Bolts, screws and studs, depending on the material they are made of and their thread diameters, have specific torque values, many of which are noted in the Specifications at the end of each Chapter. Be sure to follow the torque recommendations closely. For fasteners not assigned a specific torque, a general torque value chart is presented here as a guide. These torque values are for dry (unlubricated) fasteners threaded into steel or cast iron (not aluminum). As was previously mentioned, the size and grade of a fastener determine the amount of torque that can

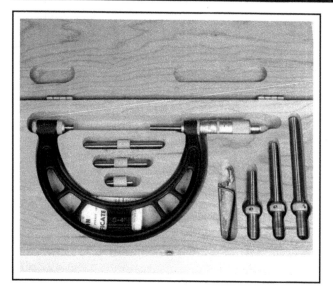

**Micrometer set**

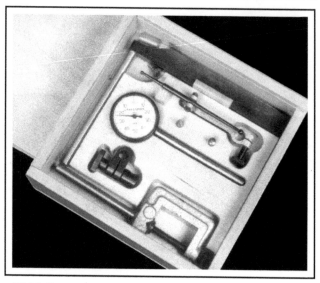

**Dial indicator set**

safely be applied to it. The figures listed here are approximate for Grade 2 and Grade 3 fasteners. Higher grades can tolerate higher torque values.

Fasteners laid out in a pattern, such as cylinder head bolts, oil pan bolts, differential cover bolts, etc., must be loosened or tightened in sequence to avoid warping the component. This sequence will normally be shown in the appropriate Chapter. If a specific pattern is not given, the following procedures can be used to prevent warping.

Initially, the bolts or nuts should be assembled finger-tight only. Next, they should be tightened one full turn each, in a criss-cross or diagonal pattern. After each one has been tightened one full turn, return to the first one and tighten them all one-half turn, following the same pattern. Finally, tighten each of them one-quarter turn at a time until each fastener has been tightened to the proper torque. To loosen and remove the fasteners, the procedure would be reversed.

## Component disassembly

Component disassembly should be done with care and purpose to help ensure that the parts go back together properly. Always keep track of the sequence in which parts are removed. Make note of special characteristics or marks on parts that can be installed more than one way, such as a grooved thrust washer on a shaft. It is a good idea to lay the disassembled parts out on a clean surface in the order that they were removed. It may also be helpful to make sketches or take instant photos of components before removal.

When removing fasteners from a component, keep track of their locations. Sometimes threading a bolt back in a part, or putting the washers and nut back on a stud, can prevent mix-ups later. If nuts and bolts cannot be returned to their original locations, they should be kept in a compartmented box or a series of small boxes. A cupcake or muffin tin is ideal for this purpose, since each cavity can hold the bolts and nuts from a particular area (i.e. oil pan bolts, valve cover bolts, engine mount bolts, etc.). A pan of this type is especially helpful when working on assemblies with very small parts, such as the carburetor, alternator, valve train or interior dash and trim pieces. The cavities can be marked with paint or tape to identify the contents.

Whenever wiring looms, harnesses or connectors are separated, it is a good idea to identify the two halves with numbered pieces of masking tape so they can be easily reconnected.

## Gasket sealing surfaces

Throughout any vehicle, gaskets are used to seal the mating surfaces between two parts and keep lubricants, fluids, vacuum or pressure contained in an assembly.

Many times these gaskets are coated with a liquid or paste-type gasket sealing compound before assembly. Age, heat and pressure can sometimes cause the two parts to stick together so tightly that they are very difficult to separate. Often, the assembly can be loosened by striking it with a soft-face hammer near the mating surfaces. A regular hammer can be used if a block of wood is placed between the hammer and the part. Do not hammer on cast parts or parts that could be easily damaged. With any particularly stubborn part, always recheck to make sure that every fastener has been removed.

Avoid using a screwdriver or bar to pry apart an assembly, as they can easily mar the gasket sealing surfaces of the parts, which must remain smooth. If prying is absolutely necessary, use an old broom handle, but keep in mind that extra clean up will be necessary if the wood splinters.

After the parts are separated, the old gasket must be carefully scraped off and the gasket surfaces cleaned. Stubborn gasket material can be soaked with rust penetrant or treated with a special chemical to soften it so it can be easily scraped off.

---

**❊❊ CAUTION:**

**Never use gasket removal solutions or caustic chemicals on plastic or other composite components.**

---

A scraper can be fashioned from a piece of copper tubing by flattening and sharpening one end. Copper is recommended because it is usually softer than the surfaces to be scraped, which reduces the chance of gouging the part. Some gaskets can be removed with a wire brush, but regardless of the method used, the mating surfaces must be left clean and smooth. If for some reason the gasket surface is gouged, then a gasket sealer thick enough to fill scratches will have to be used during reassembly of the components. For most applications, a non-drying (or semi-drying) gasket sealer should be used.

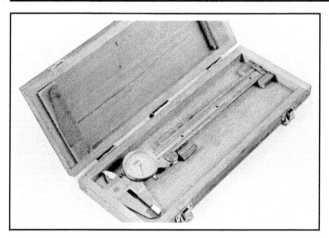

**Dial caliper**

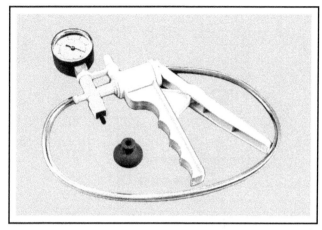

**Hand-operated vacuum pump**

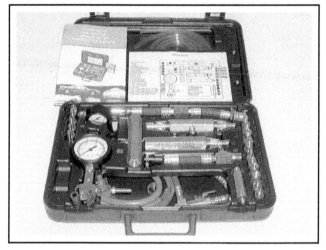

**Fuel pressure gauge set**

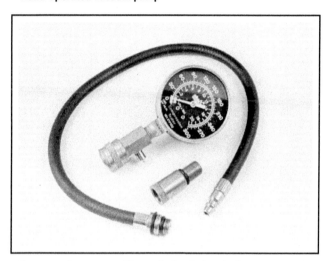

**Compression gauge with spark plug hole adapter**

## Hose removal tips

**✳✳ WARNING:**

**If the vehicle is equipped with air conditioning, do not disconnect any of the A/C hoses without first having the system depressurized by a dealer service department or a service station.**

Hose removal precautions closely parallel gasket removal precautions. Avoid scratching or gouging the surface that the hose mates against or the connection may leak. This is especially true for radiator hoses. Because of various chemical reactions, the rubber in hoses can bond itself to the metal spigot that the hose fits over. To remove a hose, first loosen the hose clamps that secure it to the spigot. Then, with slip-joint pliers, grab the hose at the clamp and rotate it around the spigot. Work it back and forth until it is completely free, then pull it off. Silicone or other lubricants will ease removal if they can be applied between the hose and the outside of the spigot. Apply the same lubricant to the inside of the hose and the outside of the spigot to simplify installation.

As a last resort (and if the hose is to be replaced with a new one anyway), the rubber can be slit with a knife and the hose peeled from the spigot. If this must be done, be careful that the metal connection is not damaged.

If a hose clamp is broken or damaged, do not reuse it. Wire-type clamps usually weaken with age, so it is a good idea to replace them with screw-type clamps whenever a hose is removed.

## TOOLS

A selection of good tools is a basic requirement for anyone who plans to maintain and repair his or her own vehicle. For the owner who has few tools, the initial investment might seem high, but when compared to the spiraling costs of professional auto maintenance and repair, it is a wise one.

To help the owner decide which tools are needed to perform the tasks detailed in this manual, the following tool lists are offered: *Maintenance and minor repair, Repair/overhaul and Special.*

The newcomer to practical mechanics should start off with the *maintenance and minor repair* tool kit, which is adequate for the simpler jobs performed on a vehicle. Then, as confidence and experience grow, the owner can tackle more difficult tasks, buying additional tools as they are needed. Eventually the basic kit will be expanded into the *repair and overhaul* tool set. Over a period of time, the experienced do-it-yourselfer will assemble a tool set complete enough for most repair and overhaul procedures and will add tools from the special category when it is felt that the expense is justified by the frequency of use.

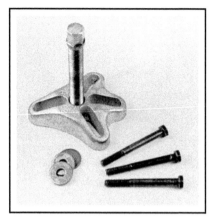

**Damper/steering wheel puller**

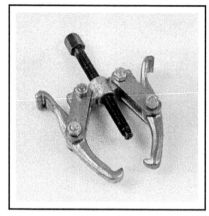

**General purpose puller**

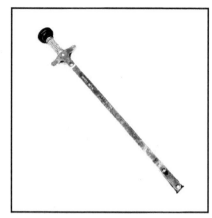

**Hydraulic lifter removal tool**

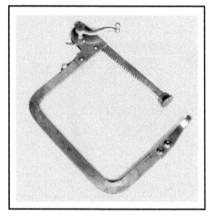

**Valve spring compressor**

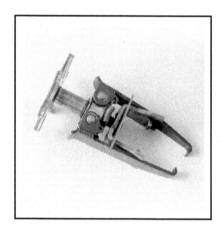

**Valve spring compressor**

**Ridge reamer**

## Maintenance and minor repair tool kit

The tools in this list should be considered the minimum required for performance of routine maintenance, servicing and minor repair work. We recommend the purchase of combination wrenches (box-end and open-end combined in one wrench). While more expensive than open end wrenches, they offer the advantages of both types of wrench.

*Combination wrench set (1/4-inch to 1 inch or 6 mm to 19 mm)*
*Adjustable wrench, 8 inch*
*Spark plug wrench with rubber insert*
*Spark plug gap adjusting tool*
*Feeler gauge set*
*Brake bleeder wrench*
*Standard screwdriver (5/16-inch x 6 inch)*
*Phillips screwdriver (No. 2 x 6 inch)*
*Combination pliers - 6 inch*
*Hacksaw and assortment of blades*
*Tire pressure gauge*
*Grease gun*
*Oil can*
*Fine emery cloth*
*Wire brush*
*Battery post and cable cleaning tool*
*Oil filter wrench*
*Funnel (medium size)*
*Safety goggles*
*Jackstands (2)*
*Drain pan*

➡️**Note: If basic tune-ups are going to be part of routine maintenance, it will be necessary to purchase a good quality stroboscopic timing light and combination tachometer/dwell meter. Although they are included in the list of special tools, it is mentioned here because they are absolutely necessary for tuning most vehicles properly.**

## Repair and overhaul tool set

These tools are essential for anyone who plans to perform major repairs and are in addition to those in the maintenance and minor repair tool kit. Included is a comprehensive set of sockets which, though expensive, are invaluable because of their versatility, especially when various extensions and drives are available. We recommend the 1/2-inch drive over the 3/8-inch drive. Although the larger drive is bulky and more expensive, it has the capacity of accepting a very wide range of large sockets. Ideally, however, the mechanic should have a 3/8-inch drive set and a 1/2-inch drive set.

*Socket set(s)*
*Reversible ratchet*
*Extension - 10 inch*
*Universal joint*
*Torque wrench (same size drive as sockets)*
*Ball peen hammer - 8 ounce*
*Soft-face hammer (plastic/rubber)*
*Standard screwdriver (1/4-inch x 6 inch)*
*Standard screwdriver (stubby - 5/16-inch)*
*Phillips screwdriver (No. 3 x 8 inch)*
*Phillips screwdriver (stubby - No. 2)*
*Pliers - vise grip*

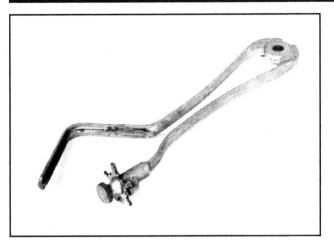

**Piston ring groove cleaning tool**

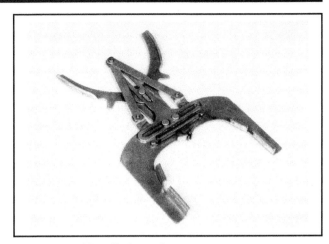

**Ring removal/installation tool**

**Ring compressor**

**Cylinder hone**

**Brake hold-down spring tool**

*Pliers - lineman's*
*Pliers - needle nose*
*Pliers - snap-ring (internal and external)*
*Cold chisel - 1/2-inch*
*Scribe*
*Scraper (made from flattened copper tubing)*
*Centerpunch*
*Pin punches (1/16, 1/8, 3/16-inch)*
*Steel rule/straightedge - 12 inch*
*Allen wrench set (1/8 to 3/8-inch or 4 mm to 10 mm)*
*A selection of files*
*Wire brush (large)*
*Jackstands (second set)*
*Jack (scissor or hydraulic type)*

➡**Note: Another tool which is often useful is an electric drill with a chuck capacity of 3/8-inch and a set of good quality drill bits.**

## Special tools

The tools in this list include those which are not used regularly, are expensive to buy, or which need to be used in accordance with their manufacturer's instructions. Unless these tools will be used frequently, it is not very economical to purchase many of them. A consideration would be to split the cost and use between yourself and a friend or friends. In addition, most of these tools can be obtained from a tool rental shop on a temporary basis.

This list primarily contains only those tools and instruments widely available to the public, and not those special tools produced by the vehicle manufacturer for distribution to dealer service depart-

ments. Occasionally, references to the manufacturer's special tools are included in the text of this manual. Generally, an alternative method of doing the job without the special tool is offered. However, sometimes there is no alternative to their use. Where this is the case, and the tool cannot be purchased or borrowed, the work should be turned over to the dealer service department or an automotive repair shop.

*Valve spring compressor*
*Piston ring groove cleaning tool*
*Piston ring compressor*
*Piston ring installation tool*
*Cylinder compression gauge*
*Cylinder ridge reamer*
*Cylinder surfacing hone*
*Cylinder bore gauge*
*Micrometers and/or dial calipers*
*Hydraulic lifter removal tool*
*Balljoint separator*
*Universal-type puller*
*Impact screwdriver*
*Dial indicator set*
*Stroboscopic timing light (inductive pick-up)*
*Hand operated vacuum/pressure pump*
*Tachometer/dwell meter*
*Universal electrical multimeter*
*Cable hoist*
*Brake spring removal and installation tools*
*Floor jack*

**Torque angle gauge**

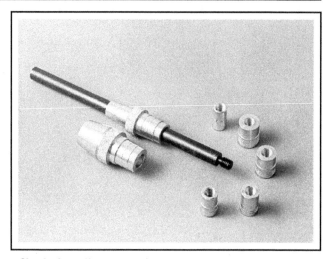

**Clutch plate alignment tool**

### Buying tools

For the do-it-yourselfer who is just starting to get involved in vehicle maintenance and repair, there are a number of options available when purchasing tools. If maintenance and minor repair is the extent of the work to be done, the purchase of individual tools is satisfactory. If, on the other hand, extensive work is planned, it would be a good idea to purchase a modest tool set from one of the large retail chain stores. A set can usually be bought at a substantial savings over the individual tool prices, and they often come with a tool box. As additional tools are needed, add-on sets, individual tools and a larger tool box can be purchased to expand the tool selection. Building a tool set gradually allows the cost of the tools to be spread over a longer period of time and gives the mechanic the freedom to choose only those tools that will actually be used.

Tool stores will often be the only source of some of the special tools that are needed, but regardless of where tools are bought, try to avoid cheap ones, especially when buying screwdrivers and sockets, because they won't last very long. The expense involved in replacing cheap tools will eventually be greater than the initial cost of quality tools.

### Care and maintenance of tools

Good tools are expensive, so it makes sense to treat them with respect. Keep them clean and in usable condition and store them properly when not in use. Always wipe off any dirt, grease or metal chips before putting them away. Never leave tools lying around in the work area. Upon completion of a job, always check closely under the hood for tools that may have been left there so they won't get lost during a test drive.

Some tools, such as screwdrivers, pliers, wrenches and sockets, can be hung on a panel mounted on the garage or workshop wall, while others should be kept in a tool box or tray. Measuring instruments, gauges, meters, etc. must be carefully stored where they cannot be damaged by weather or impact from other tools.

When tools are used with care and stored properly, they will last a very long time. Even with the best of care, though, tools will wear out if used frequently. When a tool is damaged or worn out, replace it. Subsequent jobs will be safer and more enjoyable if you do.

## HOW TO REPAIR DAMAGED THREADS

Sometimes, the internal threads of a nut or bolt hole can become stripped, usually from overtightening. Stripping threads is an all-too-

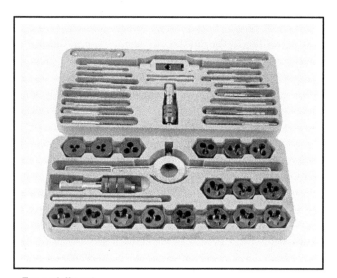

**Tap and die set**

common occurrence, especially when working with aluminum parts, because aluminum is so soft that it easily strips out.

Usually, external or internal threads are only partially stripped. After they've been cleaned up with a tap or die, they'll still work. Sometimes, however, threads are badly damaged. When this happens, you've got three choices:

1) *Drill and tap the hole to the next suitable oversize and install a larger diameter bolt, screw or stud.*
2) *Drill and tap the hole to accept a threaded plug, then drill and tap the plug to the original screw size. You can also buy a plug already threaded to the original size. Then you simply drill a hole to the specified size, then run the threaded plug into the hole with a bolt and jam nut. Once the plug is fully seated, remove the jam nut and bolt.*
3) *The third method uses a patented thread repair kit like Heli-Coil or Slimsert. These easy-to-use kits are designed to repair damaged threads in straight-through holes and blind holes. Both are available as kits which can handle a variety of sizes and thread patterns. Drill the hole, then tap it with the special included tap. Install the Heli-Coil and the hole is back to its original diameter and thread pitch.*

Regardless of which method you use, be sure to proceed calmly and carefully. A little impatience or carelessness during one of these relatively simple procedures can ruin your whole day's work and cost you a bundle if you wreck an expensive part.

## WORKING FACILITIES

Not to be overlooked when discussing tools is the workshop. If anything more than routine maintenance is to be carried out, some sort of suitable work area is essential.

It is understood, and appreciated, that many home mechanics do not have a good workshop or garage available, and end up removing an engine or doing major repairs outside. It is recommended, however, that the overhaul or repair be completed under the cover of a roof.

A clean, flat workbench or table of comfortable working height is an absolute necessity. The workbench should be equipped with a vise that has a jaw opening of at least four inches.

As mentioned previously, some clean, dry storage space is also required for tools, as well as the lubricants, fluids, cleaning solvents, etc. which soon become necessary.

Sometimes waste oil and fluids, drained from the engine or cooling system during normal maintenance or repairs, present a disposal problem. To avoid pouring them on the ground or into a sewage system, pour the used fluids into large containers, seal them with caps and take them to an authorized disposal site or recycling center. Plastic jugs, such as old antifreeze containers, are ideal for this purpose.

Always keep a supply of old newspapers and clean rags available. Old towels are excellent for mopping up spills. Many mechanics use rolls of paper towels for most work because they are readily available and disposable. To help keep the area under the vehicle clean, a large cardboard box can be cut open and flattened to protect the garage or shop floor.

Whenever working over a painted surface, such as when leaning over a fender to service something under the hood, always cover it with an old blanket or bedspread to protect the finish. Vinyl covered pads, made especially for this purpose, are available at auto parts stores.

## Booster battery (jump) starting

Observe these precautions when using a booster battery to start a vehicle:

*a) Before connecting the booster battery, make sure the ignition switch is in the Off position.*

*b) Turn off the lights, heater and other electrical loads.*

*c) Your eyes should be shielded. Safety goggles are a good idea.*

*d) Make sure the booster battery is the same voltage as the dead one in the vehicle.*

*e) The two vehicles MUST NOT TOUCH each other!*

*f) Make sure the transaxle is in Neutral (manual) or Park (automatic).*

*g) If the booster battery is not a maintenance-free type, remove the vent caps and lay a cloth over the vent holes.*

The battery on these vehicles is located in the center of the cowl chamber, just below the windshield. On some models a cover may have to be removed.

Connect the red-colored jumper cable to the positive (+) terminal of the booster battery and the other end to the positive (+) terminal of the dead battery. Then connect one end of the black jumper cable to the negative (-) terminal of the booster battery, and the other end of that cable to a good ground point on the engine of the disabled vehicle, preferably not too near the battery.

Start the engine using the booster battery and let the booster vehicle run at 2000 rpm for a few minutes to put some charge into the weak battery, then, with the engine running at idle speed, disconnect the jumper cables in the reverse order of connection.

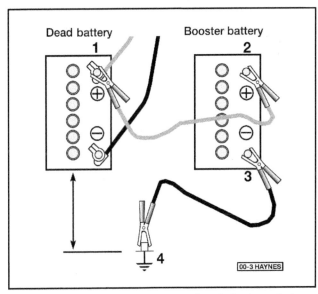

**Make the booster battery cable connections in the numerical order shown (note that the negative cable of the booster battery is NOT attached to the negative terminal of the dead battery)**

## Jacking and towing

### JACKING

✳✳ **WARNING:**

The jack supplied with the vehicle should only be used for changing a tire or placing jackstands under the frame. Never work under the vehicle or start the engine while this jack is being used as the only means of support.

The vehicle should be parked on level ground with the wheels blocked, the parking brake applied and the transmission in Park (automatic) or Reverse (manual). If the vehicle is parked alongside the roadway, or in any other hazardous situation, turn on the emergency hazard flashers. If a tire is to be changed, loosen the wheel bolts one-half turn before raising the vehicle off the ground.

Place the jack under the vehicle in the indicated position (see illustrations). Operate the jack with a slow, smooth motion until the wheel is raised off the ground. Remove the wheel bolts, pull off the wheel, install the spare and thread the wheel bolts back on. Tighten the wheel bolts snugly, lower the vehicle until some weight is on the wheel, tighten the wheel bolts completely in a criss-cross pattern and remove the jack.

### TOWING

#### Front-wheel drive models

As a general rule, the vehicle should be towed with the front (drive) wheels off the ground.

Vehicles equipped with an automatic transaxle can be towed from the front only with all four wheels on the ground, provided that speeds don't exceed 50 mph and the distance is not over 50 miles. Otherwise, a towing dolly should be used. If towed on a dolly from the rear, the ignition key must be in the ACC position, since the steering lock mechanism isn't strong enough to hold the front wheels straight while towing. The tow truck operator will attach a purpose-built steering wheel holder suitable for towing conditions.

Rear jacking position with factory jack - some models have only one mark, like the one shown here; engage the jack with the floorpan ridge directly under the mark. Some models have two marks; on these models, engage the jack with the floorpan ridge directly between the marks

✳✳ **CAUTION:**

Never tow a vehicle with an automatic transaxle from the rear with the front wheels on the ground.

When towing a vehicle equipped with a manual transaxle with all four wheels on the ground, be sure to place the shift lever in neutral and release the parking brake, and turn the ignition key to the first position to unlock the steering wheel. Also, check the transaxle lubricant to make sure it is up to the proper level.

Equipment specifically designed for towing should be used. It should be attached to the main structural members of the vehicle, not the bumpers or brackets.

Safety is a major consideration when towing and all applicable state and local laws must be obeyed. A safety chain system must be used at all times.

#### All-wheel drive models

These models must be transported with all four wheels off the ground, preferably on a flatbed carrier.

Front jacking position with factory jack - remove the plastic cover and engage the jack with the floorpan ridge

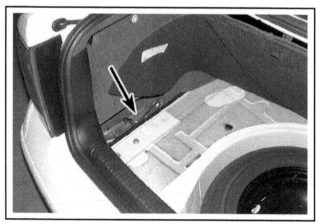

The jack and factory tools are located in the well in the left side of the luggage compartment

## Automotive chemicals and lubricants

A number of automotive chemicals and lubricants are available for use during vehicle maintenance and repair. They include a wide variety of products ranging from cleaning solvents and degreasers to lubricants and protective sprays for rubber, plastic and vinyl.

## CLEANERS

**Carburetor cleaner and choke cleaner** is a strong solvent for gum, varnish and carbon. Most carburetor cleaners leave a dry-type lubricant film which will not harden or gum up. Because of this film it is not recommended for use on electrical components.

**Brake system cleaner** is used to remove brake dust, grease and brake fluid from the brake system, where clean surfaces are absolutely necessary. It leaves no residue and often eliminates brake squeal caused by contaminants.

**Electrical cleaner** removes oxidation, corrosion and carbon deposits from electrical contacts, restoring full current flow. It can also be used to clean spark plugs, carburetor jets, voltage regulators and other parts where an oil-free surface is desired.

**Demoisturants** remove water and moisture from electrical components such as alternators, voltage regulators, electrical connectors and fuse blocks. They are non-conductive and non-corrosive.

**Degreasers** are heavy-duty solvents used to remove grease from the outside of the engine and from chassis components. They can be sprayed or brushed on and, depending on the type, are rinsed off either with water or solvent.

## LUBRICANTS

**Motor oil** is the lubricant formulated for use in engines. It normally contains a wide variety of additives to prevent corrosion and reduce foaming and wear. Motor oil comes in various weights (viscosity ratings) from 0 to 50. The recommended weight of the oil depends on the season, temperature and the demands on the engine. Light oil is used in cold climates and under light load conditions. Heavy oil is used in hot climates and where high loads are encountered. Multi-viscosity oils are designed to have characteristics of both light and heavy oils and are available in a number of weights from 5W-20 to 20W-50.

**Gear oil** is designed to be used in differentials, manual transmissions and other areas where high-temperature lubrication is required.

**Chassis and wheel bearing grease** is a heavy grease used where increased loads and friction are encountered, such as for wheel bearings, ball-joints, tie-rod ends and universal joints.

**High-temperature wheel bearing grease** is designed to withstand the extreme temperatures encountered by wheel bearings in disc brake equipped vehicles. It usually contains molybdenum disulfide (moly), which is a dry-type lubricant.

**White grease** is a heavy grease for metal-to-metal applications where water is a problem. White grease stays soft under both low and high temperatures (usually from -100 to +190-degrees F), and will not wash off or dilute in the presence of water.

**Assembly lube** is a special extreme pressure lubricant, usually containing moly, used to lubricate high-load parts (such as main and rod bearings and cam lobes) for initial start-up of a new engine. The assembly lube lubricates the parts without being squeezed out or washed away until the engine oiling system begins to function.

**Silicone lubricants** are used to protect rubber, plastic, vinyl and nylon parts.

**Graphite lubricants** are used where oils cannot be used due to contamination problems, such as in locks. The dry graphite will lubricate metal parts while remaining uncontaminated by dirt, water, oil or acids. It is electrically conductive and will not foul electrical contacts in locks such as the ignition switch.

**Moly penetrants** loosen and lubricate frozen, rusted and corroded fasteners and prevent future rusting or freezing.

**Heat-sink grease** is a special electrically non-conductive grease that is used for mounting electronic ignition modules where it is essential that heat is transferred away from the module.

## SEALANTS

**RTV sealant** is one of the most widely used gasket compounds. Made from silicone, RTV is air curing, it seals, bonds, waterproofs, fills surface irregularities, remains flexible, doesn't shrink, is relatively easy to remove, and is used as a supplementary sealer with almost all low and medium temperature gaskets.

**Anaerobic sealant** is much like RTV in that it can be used either to seal gaskets or to form gaskets by itself. It remains flexible, is solvent resistant and fills surface imperfections. The difference between an anaerobic sealant and an RTV-type sealant is in the curing. RTV cures when exposed to air, while an anaerobic sealant cures only in the absence of air. This means that an anaerobic sealant cures only after the assembly of parts, sealing them together.

**Thread and pipe sealant** is used for sealing hydraulic and pneumatic fittings and vacuum lines. It is usually made from a Teflon compound, and comes in a spray, a paint-on liquid and as a wrap-around tape.

## CHEMICALS

**Anti-seize compound** prevents seizing, galling, cold welding, rust and corrosion in fasteners. High-temperature anti-seize, usually made with copper and graphite lubricants, is used for exhaust system and exhaust manifold bolts.

**Anaerobic locking compounds** are used to keep fasteners from vibrating or working loose and cure only after installation, in the absence of air. Medium strength locking compound is used for small nuts, bolts and screws that may be removed later. High-strength locking compound is for large nuts, bolts and studs which aren't removed on a regular basis.

**Oil additives** range from viscosity index improvers to chemical treatments that claim to reduce internal engine friction. It should be noted that most oil manufacturers caution against using additives with their oils.

**Gas additives** perform several functions, depending on their chemical makeup. They usually contain solvents that help dissolve gum and varnish that build up on carburetor, fuel injection and intake parts. They also serve to break down carbon deposits that form on the inside surfaces of the combustion chambers. Some additives contain upper cylinder lubricants for valves and piston rings, and others contain chemicals to remove condensation from the gas tank.

## MISCELLANEOUS

**Brake fluid** is specially formulated hydraulic fluid that can withstand the heat and pressure encountered in brake systems. Care must be taken so this fluid does not come in contact with painted surfaces or plastics. An opened container should always be resealed to prevent contamination by water or dirt.

**Weatherstrip adhesive** is used to bond weatherstripping around doors, windows and trunk lids. It is sometimes used to attach trim pieces.

**Undercoating** is a petroleum-based, tar-like substance that is designed to protect metal surfaces on the underside of the vehicle from corrosion. It also acts as a sound-deadening agent by insulating the bottom of the vehicle.

**Waxes and polishes** are used to help protect painted and plated surfaces from the weather. Different types of paint may require the use of different types of wax and polish. Some polishes utilize a chemical or abrasive cleaner to help remove the top layer of oxidized (dull) paint on older vehicles. In recent years many non-wax polishes that contain a wide variety of chemicals such as polymers and silicones have been introduced. These non-wax polishes are usually easier to apply and last longer than conventional waxes and polishes.

## CONVERSION FACTORS

### LENGTH (distance)
| | | | | | |
|---|---|---|---|---|---|
| Inches (in) | X | 25.4 | = Millimeters (mm) | X | 0.0394 | = Inches (in) |
| Feet (ft) | X | 0.305 | = Meters (m) | X | 3.281 | = Feet (ft) |
| Miles | X | 1.609 | = Kilometers (km) | X | 0.621 | = Miles |

### VOLUME (capacity)
| | | | | | |
|---|---|---|---|---|---|
| Cubic inches (cu in; in$^3$) | X | 16.387 | = Cubic centimeters (cc; cm$^3$) | X | 0.061 | = Cubic inches (cu in; in$^3$) |
| Imperial pints (Imp pt) | X | 0.568 | = Liters (l) | X | 1.76 | = Imperial pints (Imp pt) |
| Imperial quarts (Imp qt) | X | 1.137 | = Liters (l) | X | 0.88 | = Imperial quarts (Imp qt) |
| Imperial quarts (Imp qt) | X | 1.201 | = US quarts (US qt) | X | 0.833 | = Imperial quarts (Imp qt) |
| US quarts (US qt) | X | 0.946 | = Liters (l) | X | 1.057 | = US quarts (US qt) |
| Imperial gallons (Imp gal) | X | 4.546 | = Liters (l) | X | 0.22 | = Imperial gallons (Imp gal) |
| Imperial gallons (Imp gal) | X | 1.201 | = US gallons (US gal) | X | 0.833 | = Imperial gallons (Imp gal) |
| US gallons (US gal) | X | 3.785 | = Liters (l) | X | 0.264 | = US gallons (US gal) |

### MASS (weight)
| | | | | | |
|---|---|---|---|---|---|
| Ounces (oz) | X | 28.35 | = Grams (g) | X | 0.035 | = Ounces (oz) |
| Pounds (lb) | X | 0.454 | = Kilograms (kg) | X | 2.205 | = Pounds (lb) |

### FORCE
| | | | | | |
|---|---|---|---|---|---|
| Ounces-force (ozf; oz) | X | 0.278 | = Newtons (N) | X | 3.6 | = Ounces-force (ozf; oz) |
| Pounds-force (lbf; lb) | X | 4.448 | = Newtons (N) | X | 0.225 | = Pounds-force (lbf; lb) |
| Newtons (N) | X | 0.1 | = Kilograms-force (kgf; kg) | X | 9.81 | = Newtons (N) |

### PRESSURE
| | | | | | |
|---|---|---|---|---|---|
| Pounds-force per square inch (psi; lbf/in$^2$; lb/in$^2$) | X | 0.070 | = Kilograms-force per square centimeter (kgf/cm$^2$; kg/cm$^2$) | X | 14.223 | = Pounds-force per square inch (psi; lbf/in$^2$; lb/in$^2$) |
| Pounds-force per square inch (psi; lbf/in$^2$; lb/in$^2$) | X | 0.068 | = Atmospheres (atm) | X | 14.696 | = Pounds-force per square inch (psi; lbf/in$^2$; lb/in$^2$) |
| Pounds-force per square inch (psi; lbf/in$^2$; lb/in$^2$) | X | 0.069 | = Bars | X | 14.5 | = Pounds-force per square inch (psi; lbf/in$^2$; lb/in$^2$) |
| Pounds-force per square inch (psi; lbf/in$^2$; lb/in$^2$) | X | 6.895 | = Kilopascals (kPa) | X | 0.145 | = Pounds-force per square inch (psi; lbf/in$^2$; lb/in$^2$) |
| Kilopascals (kPa) | X | 0.01 | = Kilograms-force per square centimeter (kgf/cm$^2$; kg/cm$^2$) | X | 98.1 | = Kilopascals (kPa) |

### TORQUE (moment of force)
| | | | | | |
|---|---|---|---|---|---|
| Pounds-force inches (lbf in; lb in) | X | 1.152 | = Kilograms-force centimeter (kgf cm; kg cm) | X | 0.868 | = Pounds-force inches (lbf in; lb in) |
| Pounds-force inches (lbf in; lb in) | X | 0.113 | = Newton meters (Nm) | X | 8.85 | = Pounds-force inches (lbf in; lb in) |
| Pounds-force inches (lbf in; lb in) | X | 0.083 | = Pounds-force feet (lbf ft; lb ft) | X | 12 | = Pounds-force inches (lbf in; lb in) |
| Pounds-force feet (lbf ft; lb ft) | X | 0.138 | = Kilograms-force meters (kgf m; kg m) | X | 7.233 | = Pounds-force feet (lbf ft; lb ft) |
| Pounds-force feet (lbf ft; lb ft) | X | 1.356 | = Newton meters (Nm) | X | 0.738 | = Pounds-force feet (lbf ft; lb ft) |
| Newton meters (Nm) | X | 0.102 | = Kilograms-force meters (kgf m; kg m) | X | 9.804 | = Newton meters (Nm) |

### VACUUM
| | | | | | |
|---|---|---|---|---|---|
| Inches mercury (in. Hg) | X | 3.377 | = Kilopascals (kPa) | X | 0.2961 | = Inches mercury |
| Inches mercury (in. Hg) | X | 25.4 | = Millimeters mercury (mm Hg) | X | 0.0394 | = Inches mercury |

### POWER
| | | | | | |
|---|---|---|---|---|---|
| Horsepower (hp) | X | 745.7 | = Watts (W) | X | 0.0013 | = Horsepower (hp) |

### VELOCITY (speed)
| | | | | | |
|---|---|---|---|---|---|
| Miles per hour (miles/hr; mph) | X | 1.609 | = Kilometers per hour (km/hr; kph) | X | 0.621 | = Miles per hour (miles/hr; mph) |

### FUEL CONSUMPTION *
| | | | | | |
|---|---|---|---|---|---|
| Miles per gallon, Imperial (mpg) | X | 0.354 | = Kilometers per liter (km/l) | X | 2.825 | = Miles per gallon, Imperial (mpg) |
| Miles per gallon, US (mpg) | X | 0.425 | = Kilometers per liter (km/l) | X | 2.352 | = Miles per gallon, US (mpg) |

### TEMPERATURE
Degrees Fahrenheit = (°C x 1.8) + 32     Degrees Celsius (Degrees Centigrade; °C) = (°F - 32) x 0.56

*It is common practice to convert from miles per gallon (mpg) to liters/100 kilometers (l/100km), where mpg (Imperial) x l/100 km = 282 and mpg (US) x l/100 km = 235*

## FRACTION/DECIMAL/MILLIMETER EQUIVALENTS

### DECIMALS TO MILLIMETERS

| Decimal | mm | Decimal | mm |
|---|---|---|---|
| 0.001 | 0.0254 | 0.500 | 12.7000 |
| 0.002 | 0.0508 | 0.510 | 12.9540 |
| 0.003 | 0.0762 | 0.520 | 13.2080 |
| 0.004 | 0.1016 | 0.530 | 13.4620 |
| 0.005 | 0.1270 | 0.540 | 13.7160 |
| 0.006 | 0.1524 | 0.550 | 13.9700 |
| 0.007 | 0.1778 | 0.560 | 14.2240 |
| 0.008 | 0.2032 | 0.570 | 14.4780 |
| 0.009 | 0.2286 | 0.580 | 14.7320 |
|  |  | 0.590 | 14.9860 |
| 0.010 | 0.2540 |  |  |
| 0.020 | 0.5080 |  |  |
| 0.030 | 0.7620 |  |  |
| 0.040 | 1.0160 | 0.600 | 15.2400 |
| 0.050 | 1.2700 | 0.610 | 15.4940 |
| 0.060 | 1.5240 | 0.620 | 15.7480 |
| 0.070 | 1.7780 | 0.630 | 16.0020 |
| 0.080 | 2.0320 | 0.640 | 16.2560 |
| 0.090 | 2.2860 | 0.650 | 16.5100 |
|  |  | 0.660 | 16.7640 |
| 0.100 | 2.5400 | 0.670 | 17.0180 |
| 0.110 | 2.7940 | 0.680 | 17.2720 |
| 0.120 | 3.0480 | 0.690 | 17.5260 |
| 0.130 | 3.3020 |  |  |
| 0.140 | 3.5560 |  |  |
| 0.150 | 3.8100 |  |  |
| 0.160 | 4.0640 | 0.700 | 17.7800 |
| 0.170 | 4.3180 | 0.710 | 18.0340 |
| 0.180 | 4.5720 | 0.720 | 18.2880 |
| 0.190 | 4.8260 | 0.730 | 18.5420 |
|  |  | 0.740 | 18.7960 |
| 0.200 | 5.0800 | 0.750 | 19.0500 |
| 0.210 | 5.3340 | 0.760 | 19.3040 |
| 0.220 | 5.5880 | 0.770 | 19.5580 |
| 0.230 | 5.8420 | 0.780 | 19.8120 |
| 0.240 | 6.0960 | 0.790 | 20.0660 |
| 0.250 | 6.3500 |  |  |
| 0.260 | 6.6040 |  |  |
| 0.270 | 6.8580 | 0.800 | 20.3200 |
| 0.280 | 7.1120 | 0.810 | 20.5740 |
| 0.290 | 7.3660 | 0.820 | 21.8280 |
|  |  | 0.830 | 21.0820 |
| 0.300 | 7.6200 | 0.840 | 21.3360 |
| 0.310 | 7.8740 | 0.850 | 21.5900 |
| 0.320 | 8.1280 | 0.860 | 21.8440 |
| 0.330 | 8.3820 | 0.870 | 22.0980 |
| 0.340 | 8.6360 | 0.880 | 22.3520 |
| 0.350 | 8.8900 | 0.890 | 22.6060 |
| 0.360 | 9.1440 |  |  |
| 0.370 | 9.3980 |  |  |
| 0.380 | 9.6520 |  |  |
| 0.390 | 9.9060 | 0.900 | 22.8600 |
| 0.400 | 10.1600 | 0.910 | 23.1140 |
| 0.410 | 10.4140 | 0.920 | 23.3680 |
| 0.420 | 10.6680 | 0.930 | 23.6220 |
| 0.430 | 10.9220 | 0.940 | 23.8760 |
| 0.440 | 11.1760 | 0.950 | 24.1300 |
| 0.450 | 11.4300 | 0.960 | 24.3840 |
| 0.460 | 11.6840 | 0.970 | 24.6380 |
| 0.470 | 11.9380 | 0.980 | 24.8920 |
| 0.480 | 12.1920 | 0.990 | 25.1460 |
| 0.490 | 12.4460 | 1.000 | 25.4000 |

### FRACTIONS TO DECIMALS TO MILLIMETERS

| Fraction | Decimal | mm | Fraction | Decimal | mm |
|---|---|---|---|---|---|
| 1/64 | 0.0156 | 0.3969 | 33/64 | 0.5156 | 13.0969 |
| 1/32 | 0.0312 | 0.7938 | 17/32 | 0.5312 | 13.4938 |
| 3/64 | 0.0469 | 1.1906 | 35/64 | 0.5469 | 13.8906 |
| 1/16 | 0.0625 | 1.5875 | 9/16 | 0.5625 | 14.2875 |
| 5/64 | 0.0781 | 1.9844 | 37/64 | 0.5781 | 14.6844 |
| 3/32 | 0.0938 | 2.3812 | 19/32 | 0.5938 | 15.0812 |
| 7/64 | 0.1094 | 2.7781 | 39/64 | 0.6094 | 15.4781 |
| 1/8 | 0.1250 | 3.1750 | 5/8 | 0.6250 | 15.8750 |
| 9/64 | 0.1406 | 3.5719 | 41/64 | 0.6406 | 16.2719 |
| 5/32 | 0.1562 | 3.9688 | 21/32 | 0.6562 | 16.6688 |
| 11/64 | 0.1719 | 4.3656 | 43/64 | 0.6719 | 17.0656 |
| 3/16 | 0.1875 | 4.7625 | 11/16 | 0.6875 | 17.4625 |
| 13/64 | 0.2031 | 5.1594 | 45/64 | 0.7031 | 17.8594 |
| 7/32 | 0.2188 | 5.5562 | 23/32 | 0.7188 | 18.2562 |
| 15/64 | 0.2344 | 5.9531 | 47/64 | 0.7344 | 18.6531 |
| 1/4 | 0.2500 | 6.3500 | 3/4 | 0.7500 | 19.0500 |
| 17/64 | 0.2656 | 6.7469 | 49/64 | 0.7656 | 19.4469 |
| 9/32 | 0.2812 | 7.1438 | 25/32 | 0.7812 | 19.8438 |
| 19/64 | 0.2969 | 7.5406 | 51/64 | 0.7969 | 20.2406 |
| 5/16 | 0.3125 | 7.9375 | 13/16 | 0.8125 | 20.6375 |
| 21/64 | 0.3281 | 8.3344 | 53/64 | 0.8281 | 21.0344 |
| 11/32 | 0.3438 | 8.7312 | 27/32 | 0.8438 | 21.4312 |
| 23/64 | 0.3594 | 9.1281 | 55/64 | 0.8594 | 21.8281 |
| 3/8 | 0.3750 | 9.5250 | 7/8 | 0.8750 | 22.2250 |
| 25/64 | 0.3906 | 9.9219 | 57/64 | 0.8906 | 22.6219 |
| 13/32 | 0.4062 | 10.3188 | 29/32 | 0.9062 | 23.0188 |
| 27/64 | 0.4219 | 10.7156 | 59/64 | 0.9219 | 23.4156 |
| 7/16 | 0.4375 | 11.1125 | 15/16 | 0.9375 | 23.8125 |
| 29/64 | 0.4531 | 11.5094 | 61/64 | 0.9531 | 24.2094 |
| 15/32 | 0.4688 | 11.9062 | 31/32 | 0.9688 | 24.6062 |
| 31/64 | 0.4844 | 12.3031 | 63/64 | 0.9844 | 25.0031 |
| 1/2 | 0.5000 | 12.7000 | 1 | 1.0000 | 25.4000 |

## Safety first!

Regardless of how enthusiastic you may be about getting on with the job at hand, take the time to ensure that your safety is not jeopardized. A moment's lack of attention can result in an accident, as can failure to observe certain simple safety precautions. The possibility of an accident will always exist, and the following points should not be considered a comprehensive list of all dangers. Rather, they are intended to make you aware of the risks and to encourage a safety conscious approach to all work you carry out on your vehicle.

### ESSENTIAL DOS AND DON'TS

**DON'T** rely on a jack when working under the vehicle. Always use approved jackstands to support the weight of the vehicle and place them under the recommended lift or support points.

**DON'T** attempt to loosen extremely tight fasteners (i.e. wheel lug nuts) while the vehicle is on a jack - it may fall.

**DON'T** start the engine without first making sure that the transmission is in Neutral (or Park where applicable) and the parking brake is set.

**DON'T** remove the radiator cap from a hot cooling system - let it cool or cover it with a cloth and release the pressure gradually.

**DON'T** attempt to drain the engine oil until you are sure it has cooled to the point that it will not burn you.

**DON'T** touch any part of the engine or exhaust system until it has cooled sufficiently to avoid burns.

**DON'T** siphon toxic liquids such as gasoline, antifreeze and brake fluid by mouth, or allow them to remain on your skin.

**DON'T** inhale brake lining dust - it is potentially hazardous (see Asbestos below).

**DON'T** allow spilled oil or grease to remain on the floor - wipe it up before someone slips on it.

**DON'T** use loose fitting wrenches or other tools which may slip and cause injury.

**DON'T** push on wrenches when loosening or tightening nuts or bolts. Always try to pull the wrench toward you. If the situation calls for pushing the wrench away, push with an open hand to avoid scraped knuckles if the wrench should slip.

**DON'T** attempt to lift a heavy component alone - get someone to help you.

**DON'T** rush or take unsafe shortcuts to finish a job.

**DON'T** allow children or animals in or around the vehicle while you are working on it.

**DO** wear eye protection when using power tools such as a drill, sander, bench grinder, etc. and when working under a vehicle.

**DO** keep loose clothing and long hair well out of the way of moving parts.

**DO** make sure that any hoist used has a safe working load rating adequate for the job.

**DO** get someone to check on you periodically when working alone on a vehicle.

**DO** carry out work in a logical sequence and make sure that everything is correctly assembled and tightened.

**DO** keep chemicals and fluids tightly capped and out of the reach of children and pets.

**DO** remember that your vehicle's safety affects that of yourself and others. If in doubt on any point, get professional advice.

### STEERING, SUSPENSION AND BRAKES

These systems are essential to driving safety, so make sure you have a qualified shop or individual check your work. Also, compressed suspension springs can cause injury if released suddenly - be sure to use a spring compressor.

### AIRBAGS

Airbags are explosive devices that can CAUSE injury if they deploy while you're working on the vehicle. Follow the manufacturer's instructions to disable the airbag whenever you're working in the vicinity of airbag components.

### ASBESTOS

Certain friction, insulating, sealing, and other products - such as brake linings, brake bands, clutch linings, torque converters, gaskets, etc. - may contain asbestos or other hazardous friction material. Extreme care must be taken to avoid inhalation of dust from such products, since it is hazardous to health. If in doubt, assume that they do contain asbestos.

### FIRE

Remember at all times that gasoline is highly flammable. Never smoke or have any kind of open flame around when working on a vehicle. But the risk does not end there. A spark caused by an electrical short circuit, by two metal surfaces contacting each other, or even by static electricity built up in your body under certain conditions, can ignite gasoline vapors, which in a confined space are highly explosive. Do not, under any circumstances, use gasoline for cleaning parts. Use an approved safety solvent.

Always disconnect the battery ground (-) cable at the battery before working on any part of the fuel system or electrical system. Never risk spilling fuel on a hot engine or exhaust component. It is strongly recommended that a fire extinguisher suitable for use on fuel and electrical fires be kept handy in the garage or workshop at all times. Never try to extinguish a fuel or electrical fire with water.

### FUMES

Certain fumes are highly toxic and can quickly cause unconsciousness and even death if inhaled to any extent. Gasoline vapor falls into this category, as do the vapors from some cleaning solvents. Any draining or pouring of such volatile fluids should be done in a well ventilated area.

When using cleaning fluids and solvents, read the instructions on the container carefully. Never use materials from unmarked containers.

Never run the engine in an enclosed space, such as a garage. Exhaust fumes contain carbon monoxide, which is extremely poisonous. If you need to run the engine, always do so in the open air, or at least have the rear of the vehicle outside the work area.

### THE BATTERY

Never create a spark or allow a bare light bulb near a battery. They normally give off a certain amount of hydrogen gas, which is highly explosive.

Always disconnect the battery ground (-) cable at the battery before working on the fuel or electrical systems.

If possible, loosen the filler caps or cover when charging the battery from an external source (this does not apply to sealed or maintenance-free batteries). Do not charge at an excessive rate or the battery may burst.

Take care when adding water to a non maintenance-free battery and when carrying a battery. The electrolyte, even when diluted, is very corrosive and should not be allowed to contact clothing or skin.

Always wear eye protection when cleaning the battery to prevent the caustic deposits from entering your eyes.

### HOUSEHOLD CURRENT

When using an electric power tool, inspection light, etc., which operates on household current, always make sure that the tool is correctly connected to its plug and that, where necessary, it is properly grounded. Do not use such items in damp conditions and, again, do not create a spark or apply excessive heat in the vicinity of fuel or fuel vapor.

### SECONDARY IGNITION SYSTEM VOLTAGE

A severe electric shock can result from touching certain parts of the ignition system (such as the spark plug wires) when the engine is running or being cranked, particularly if components are damp or the insulation is defective. In the case of an electronic ignition system, the secondary system voltage is much higher and could prove fatal.

### HYDROFLUORIC ACID

This extremely corrosive acid is formed when certain types of synthetic rubber, found in some O-rings, oil seals, fuel hoses, etc. are exposed to temperatures above 750-degrees F (400-degrees C). The rubber changes into a charred or sticky substance containing the acid. *Once formed, the acid remains dangerous for years. If it gets onto the skin, it may be necessary to amputate the limb concerned.*

When dealing with a vehicle which has suffered a fire, or with components salvaged from such a vehicle, wear protective gloves and discard them after use.

## Troubleshooting

## CONTENTS

This section provides an easy reference guide to the more common problems that may occur during the operation of your vehicle. These problems and their possible causes are grouped under headings denoting various components or systems, such as Engine, Cooling system, etc. They also refer you to the chapter and/or section that deals with the problem.

Remember that successful troubleshooting is not a mysterious art practiced only by professional mechanics. It is simply the result of the right knowledge combined with an intelligent, systematic approach to the problem. Always work by a process of elimination, starting with the simplest solution and working through to the most complex - and never overlook the obvious. Anyone can run the gas tank dry or leave the lights on overnight, so don't assume that you are exempt from such oversights.

Finally, always establish a clear idea of why a problem has occurred and take steps to ensure that it doesn't happen again. If the electrical system fails because of a poor connection, check the other connections in the system to make sure that they don't fail as well. If a particular fuse continues to blow, find out why - don't just replace one fuse after another. Remember, failure of a small component can often be indicative of potential failure or incorrect functioning of a more important component or system.

# ENGINE

### 1 Engine will not rotate when attempting to start

1 Battery terminal connections loose or corroded (Chapter 1).
2 Battery discharged or faulty (Chapter 1).
3 Automatic transaxle not completely engaged in Park (Chapter 7B) or manual transaxle multi-function switch faulty (Chapter 7A).
4 Broken, loose or disconnected wiring in the starting circuit (Chapters 5 and 12).
5 Starter motor pinion jammed in flywheel ring gear (Chapter 5).
6 Starter solenoid faulty (Chapter 5).
7 Starter motor faulty (Chapter 5).
8 Ignition switch faulty (Chapter 12).
9 Starter pinion or flywheel teeth worn or broken (Chapter 5).
10 Transmission Range (TR) switch malfunctioning (Chapter 7B).

### 2 Engine rotates but will not start

1 Fuel tank empty.
2 Battery discharged (engine rotates slowly) (Chapter 5).
3 Battery terminal connections loose or corroded (Chapter 1).
4 Leaking fuel injector(s), faulty fuel pump, pressure regulator, etc. (Chapter 4).
5 Broken or stripped timing belt (Chapter 2).
6 Ignition components damp or damaged (Chapter 5).
7 Worn, faulty or incorrectly gapped spark plugs (Chapter 1).
8 Broken, loose or disconnected wiring in the starting circuit (Chapter 5).
9 Broken, loose or disconnected wires at the ignition coils or faulty coils (Chapter 5).
10 Defective MAF sensor (see Chapter 6).
11 Contaminated fuel.

### 3 Engine hard to start when cold

1 Battery discharged or low (Chapter 1).
2 Malfunctioning fuel system (Chapter 4).
3 Faulty coolant temperature sensor or intake air temperature sensor (Chapter 6).
4 Faulty ignition system (Chapter 5).
5 Defective MAF sensor (see Chapter 6).

### 4 Engine hard to start when hot

1 Air filter clogged (Chapter 1).
2 Fuel not reaching the fuel injection system (Chapter 4).
3 Corroded battery connections, especially ground (Chapter 1).
4 Faulty coolant temperature sensor or intake air temperature sensor (Chapter 6).
5 Low cylinder compression (Chapter 2).

### 5 Starter motor noisy or excessively rough in engagement

1 Pinion or flywheel gear teeth worn or broken (Chapter 5).
2 Starter motor mounting bolts loose or missing (Chapter 5).

### 6 Engine starts but stops immediately

1 Loose or faulty electrical connections at distributor, coil or alternator (Chapter 5).
2 Insufficient fuel reaching the fuel injector(s) (Chapters 1 and 4).
3 Vacuum leak at the gasket between the intake manifold/plenum and throttle body (Chapters 1 and 4).
4 Intake air leaks, broken vacuum lines (see Chapter 4).
5 Contaminated fuel.

### 7 Oil puddle under engine

1 Oil pan gasket and/or oil pan drain bolt washer leaking (Chapter 2).
2 Oil pressure sending unit leaking (Chapter 2).
3 Valve covers leaking (Chapter 2).
4 Engine oil seals leaking (Chapter 2).
5 Oil pump housing leaking (Chapter 2).

### 8 Engine lopes while idling or idles erratically

1 Vacuum leakage (Chapters 2 and 4).
2 Leaking EGR valve (Chapter 6).
3 Air filter clogged (Chapter 1).
4 Fuel pump not delivering sufficient fuel to the fuel injection system (Chapter 4).
5 Leaking head gasket (Chapter 2).
6 Timing belt and/or sprockets worn (Chapter 2).
7 Camshaft lobes worn (Chapter 2).

## 9 Engine misses at idle speed

1 Spark plugs worn or not gapped properly (Chapter 1).
2 Faulty spark plug wires (Chapter 1).
3 Vacuum leaks (Chapters 2 and 4).
4 Uneven or low compression (Chapter 2).
5 Problem with the fuel injection system (Chapter 4).
6 Faulty ignition coils (Chapter 5).

## 10 Engine misses throughout driving speed range

1 Fuel filter clogged and/or impurities in the fuel system (Chapter 1).
2 Low fuel output at the fuel injector(s) (Chapter 4).
3 Faulty or incorrectly gapped spark plugs (Chapter 1).
4 Defective spark plug wires (Chapters 1 or 5).
5 Faulty emission system components (Chapter 6).
6 Low or uneven cylinder compression pressures (Chapter 2).
7 Weak or faulty ignition system (Chapter 5).
8 Vacuum leak in fuel injection system, throttle body, intake manifold or vacuum hoses (Chapter 4).

## 11 Engine stumbles on acceleration

1 Spark plugs fouled (Chapter 1).
2 Problem with fuel injection system (Chapter 4).
3 Fuel filter clogged (Chapters 1 and 4).
4 Intake manifold leak (Chapters 2 and 4).
5 EGR system malfunction (Chapter 6).

## 12 Engine surges while holding accelerator steady

1 Intake air/vacuum leak (Chapter 4).
2 Problem with fuel injection system (Chapter 4).
3 Problem with the emissions control system (Chapter 6).

## 13 Engine stalls

1 Idle speed incorrect (Chapter 4).
2 Fuel filter clogged and/or water and impurities in the fuel system (Chapters 1 and 4).
3 Faulty emissions system components (Chapter 6).
4 Faulty or incorrectly gapped spark plugs (Chapter 1).
5 Faulty spark plug wires (Chapter 1).
6 Vacuum leak in the fuel injection system, intake manifold or vacuum hoses (Chapters 2 and 4).

## 14 Engine lacks power

1 Faulty spark plug wires or coils (Chapters 1 and 5).
2 Faulty or incorrectly gapped spark plugs (Chapter 1).
3 Problem with the fuel injection system (Chapter 4).
4 Plugged air filter (Chapter 1).
5 Brakes binding (Chapter 9).
6 Automatic transaxle fluid level incorrect (Chapter 1).
7 Clutch slipping (Chapter 8).
8 Fuel filter clogged and/or impurities in the fuel system (Chapters 1 and 4).

9 Emission control system not functioning properly (Chapter 6).
10 Low or uneven cylinder compression pressures (Chapter 2).
11 Obstructed exhaust system (Chapters 2 and 4).
12 Defective turbocharger or wastegate (turbo models) (Chapter 4).

## 15 Engine backfires

1 Emission control system not functioning properly (Chapter 6).
2 Faulty plug wires or coils (Chapters 1 and 5).
3 Problem with the fuel injection system (Chapter 4).
4 Vacuum leak at fuel injector(s), intake manifold, air control valve or vacuum hoses (Chapters 2 and 4).
5 Valves sticking (Chapter 2).

## 16 Pinging or knocking engine sounds during acceleration or uphill

1 Incorrect grade of fuel.
2 Ignition timing incorrect (Chapter 5).
3 Fuel injection system faulty (Chapter 4).
4 Improper or damaged spark plugs or wires (Chapter 1).
5 EGR valve not functioning (Chapter 6).
6 Vacuum leak (Chapters 2 and 4).
7 Knock sensor malfunctioning (Chapter 6).

## 17 Engine runs with oil pressure light on

1 Low oil level (Chapter 1).
2 Idle rpm below specification (Chapter 4).
3 Short in wiring circuit (Chapter 12).
4 Faulty oil pressure sender (Chapter 2).
5 Worn engine bearings and/or oil pump (Chapter 2).

## 18 Engine continues to run after switching off

1 Excessive engine operating temperature (Chapter 3).
2 Excessive carbon deposits on valves and pistons (Chapter 2).

# ENGINE ELECTRICAL SYSTEM

## 19 Battery will not hold a charge

1 Alternator drivebelt defective or not adjusted properly (Chapter 1).
2 Battery electrolyte level low (Chapter 1).
3 Battery terminals loose or corroded (Chapter 1).
4 Alternator not charging properly (Chapter 5).
5 Loose, broken or faulty wiring in the charging circuit (Chapter 5).
6 Short in vehicle wiring (Chapter 12).
7 Internally defective battery (Chapters 1 and 5).

## 20 Alternator light fails to go out

1 Faulty alternator or charging circuit (Chapter 5).
2 Alternator drivebelt defective or out of adjustment (Chapter 1).
3 Alternator voltage regulator inoperative (Chapter 5).

**21  Alternator light fails to come on when key is turned on**

1  Warning light bulb defective (Chapter 12).
2  Fault in the printed circuit, dash wiring or bulb holder (Chapter 12).

## FUEL SYSTEM

**22  Excessive fuel consumption**

1  Dirty or clogged air filter element (Chapter 1).
2  Engine management problem (Chapter 6).
3  Emissions system not functioning properly (Chapter 6).
4  Fuel injection system not functioning properly (Chapter 4).
5  Low tire pressure or incorrect tire size (Chapter 1).

**23  Fuel leakage and/or fuel odor**

1  Leaking fuel feed or return line (Chap-ters 1 and 4).
2  Tank overfilled.
3  Problem with the evaporative emissions control system (Chapters 6).
4  Problem with the fuel injection system (Chapter 4).

## COOLING SYSTEM

**24  Overheating**

1  Insufficient coolant in system (Chapter 1).
2  Water pump defective (Chapter 3).
3  Radiator core blocked or grille restricted (Chapter 3).
4  Thermostat faulty (Chapter 3).
5  Drivebelt or tensioner broken (Chapter 1).
6  Electric coolant fan inoperative or blades broken (Chapter 3).
7  Expansion tank cap not maintaining proper pressure (Chapter 3).
8  Blown head gasket (Chapter 2).

**25  Overcooling**

1  Faulty thermostat (Chapter 3).
2  Inaccurate temperature gauge sending unit (Chapter 3)

**26  External coolant leakage**

1  Deteriorated/damaged hoses; loose clamps (Chapters 1 and 3).
2  Water pump defective (Chapter 3).
3  Leakage from radiator core or coolant reservoir tank (Chapter 3).
4  Cylinder head gasket leaking (Chapter 2).

**27  Internal coolant leakage**

1  Leaking cylinder head gasket (Chapter 2).
2  Cracked cylinder bore or cylinder head (Chapter 2).

**28  Coolant loss**

1  Too much coolant in system (Chapter 1).
2  Coolant boiling away because of overheating (Chapter 3).
3  Internal or external leakage (Chapter 3).
4  Faulty expansion tank cap (Chapter 3).

**29  Poor coolant circulation**

1  Defective water pump (Chapter 3).
2  Restriction in cooling system (Chapters 1 and 3).
3  Thermostat sticking (Chapter 3).

## CLUTCH

**30  Pedal travels to floor - no pressure or very little resistance**

1  Hydraulic release system leaking or air in the system (Chapter 8).
2  Broken release bearing or fork (Chapter 8).

**31  Unable to select gears**

1  Faulty transaxle (Chapter 7).
2  Faulty clutch disc or pressure plate (Chapter 8).
3  Faulty release lever or release bearing (Chapter 8).
4  Faulty shift lever assembly or cables (Chapter 8).
5  Faulty clutch release system.

**32  Clutch slips (engine speed increases with no increase in vehicle speed)**

1  Clutch plate worn (Chapter 8).
2  Clutch plate is oil soaked by leaking rear main seal (Chapter 2 and 8).
3  Clutch plate not seated (Chapter 8).
4  Warped pressure plate or flywheel (Chapters 2 and 8).
5  Weak diaphragm spring in pressure plate (Chapter 8).
6  Clutch plate overheated. Allow to cool.
7  Piston stuck in bore of clutch release cylinder, preventing clutch from fully engaging (Chapter 8).

**33  Grabbing (chattering) as clutch is engaged**

1  Oil on clutch plate lining, burned or glazed facings (Chapter 8).
2  Worn or loose engine or transaxle mounts (Chapters 2 and 7).
3  Worn splines on clutch plate hub (Chapter 8).
4  Warped pressure plate or flywheel (Chapter 8).
5  Burned or smeared resin on flywheel or pressure plate (Chapter 8).

**34  Transaxle rattling (clicking)**

1  Release fork loose (Chapter 8).
2  Low engine idle speed (Chapter 1).

## 35  Noise in clutch area

Faulty bearing (Chapter 8).

## 36  Clutch pedal stays on floor

1  Broken release bearing or fork (Chapter 8).
2  Hydraulic release system leaking or air in the system (Chapter 8).
3  Over-center spring in clutch pedal assembly broken (Chapter 8).

## 37  High pedal effort

1  Piston binding in bore of release cylinder (Chapter 8).
2  Pressure plate faulty (Chapter 8).

# MANUAL TRANSAXLE

## 38  Knocking noise at low speeds

Worn driveaxle constant velocity (CV) joints (Chapter 8).

## 39  Noise most pronounced when turning

Differential gear noise (Chapter 7A).*

## 40  Clunk on acceleration or deceleration

1  Loose engine or transaxle mounts (Chapters 2 and 7A).
2  Worn differential pinion shaft in case.*
3  Worn or damaged driveaxle inboard CV joints (Chapter 8).

## 41  Clicking noise in turns

Worn or damaged outboard CV joint (Chapter 8).

## 42  Vibration

1  Rough wheel bearing (Chapter 10).
2  Damaged driveaxle (Chapter 8).
3  Out-of-round tires (Chapter 1).
4  Tire out of balance (Chapters 1 and 10).
5  Worn CV joint (Chapter 8).

## 43  Noisy in neutral with engine running

1  Damaged input gear bearing (Chapter 7A).*
2  Damaged clutch release bearing (Chapter 8).

## 44  Noisy in one particular gear

1  Damaged or worn constant mesh gears (Chapter 7A).*
2  Damaged or worn synchronizers (Chapter 7A).*
3  Bent reverse fork (Chapter 7A).*
4  Damaged fourth speed gear or output gear (Chapter 7A).*
5  Worn or damaged reverse idler gear or idler bushing (Chapter 7A).*

## 45  Noisy in all gears

1  Insufficient lubricant (Chapter 7A).
2  Damaged or worn bearings (Chapter 7A).*
3  Worn or damaged input gear shaft and/or output gear shaft (Chapter 7A).*

## 46  Slips out of gear

1  Worn or improperly adjusted linkage (Chapter 7A).
2  Transaxle loose on engine (Chapter 7A).
3  Shift linkage does not work freely, binds (Chapter 7A).
4  Input gear bearing retainer broken or loose (Chapter 7A).*
5  Dirt between clutch cover and engine housing (Chapter 7A).
6  Worn shift fork (Chapter 7A).*

## 47  Leaks lubricant

1  Side gear shaft seals worn (Chapter 7A).
2  Excessive amount of lubricant in transaxle (Chapters 1 and 7A).
3  Loose or broken input gear shaft bearing retainer (Chapter 7A).*
4  Vehicle speed sensor O-ring leaking (Chapter 6).

## 48  Hard to shift

Shift linkage out of adjustment (Chapter 7A).
*Although the corrective action necessary to remedy the symptoms described is beyond the scope of this manual, the above information should be helpful in isolating the cause of the condition so that the owner can communicate clearly with a professional mechanic.*

# AUTOMATIC TRANSAXLE

➡ **Note: Due to the complexity of the automatic transaxle, it is difficult for the home mechanic to properly diagnose and service this component. For problems other than the following, the vehicle should be taken to a dealer or transaxle shop.**

## 49  Fluid leakage

1  Automatic transaxle fluid in these models is transparent yellow in color. Fluid leaks should not be confused with engine oil, which can easily be blown onto the transaxle by air flow.

2  To pinpoint a leak, first remove all built-up dirt and grime from the transaxle housing with degreasing agents and/or steam cleaning. Then drive the vehicle at low speeds so air flow will not blow the leak far from its source. Raise the vehicle and determine where the leak is coming from. Common areas of leakage are:

    *a) Pan (Chapters 1 and 7B).*
    *b) Transaxle oil lines (Chapter 7B).*
    *c) Vehicle speed sensor (Chapter 6).*
    *e) Driveaxle oil seals (Chapter 7A).*

## 50  Transaxle fluid brown or has a burned smell

Transaxle fluid overheated (Chapter 1).

## 51 General shift mechanism problems

1  Chapter 7, Part B, deals with checking and adjusting the shift cable on automatic transaxles. Common problems that may be attributed to poorly adjusted cable are:

a) *Engine starting in gears other than Park or Neutral.*
b) *Indicator on shifter pointing to a gear other than the one actually being used.*
c) *Vehicle moves when in Park.*

2  Refer to Chapter 7B for the shift cable adjustment procedure.

## 52 Transaxle will not downshift with accelerator pedal pressed to the floor

The transaxle is electronically controlled. This type of problem - which is caused by a malfunction in the control unit, a sensor or solenoid, or the circuit itself - is beyond the scope of this book. Take the vehicle to a dealer service department or a competent automatic transmission shop.

## 53 Engine will start in gears other than Park or Neutral

Transmission Range (TR) switch malfunctioning (Chapter 7B).

## 54 Transaxle slips, shifts roughly, is noisy or has no drive in forward or reverse gears

There are many probable causes for the above problems, but the home mechanic should be concerned with only one possibility - fluid level. Before taking the vehicle to a repair shop, check the level and condition of the fluid as described in Chapter 1. Correct the fluid level as necessary or change the fluid and filter if needed. If the problem persists, have a professional diagnose the cause.

# DRIVEAXLES

## 55 Clicking noise in turns

Worn or damaged outboard CV joint (Chapter 8).

## 56 Shudder or vibration during acceleration

1  Excessive toe-in (Chapter 10).
2  Incorrect spring heights (Chapter 10).
3  Worn or damaged inboard or outboard CV joints (Chapter 8).
4  Sticking inboard CV joint assembly (Chapter 8).

## 57 Vibration at highway speeds

1  Out-of-balance front wheels and/or tires (Chapters 1 and 10).
2  Out-of-round front tires (Chapters 1 and 10).
3  Worn CV joint(s) (Chapter 8).

# BRAKES

**➡ Note: Before assuming that a brake problem exists, make sure that:**

a) *The tires are in good condition and properly inflated (Chapter 1).*
b) *The front end alignment is correct (Chapter 10).*
c) *The vehicle is not loaded with weight in an unequal manner.*

## 58 Vehicle pulls to one side during braking

1  Incorrect tire pressures (Chapter 1).
2  Front end out of alignment (have the front end aligned).
3  Front, or rear, tire sizes not matched to one another.
4  Restricted brake lines or hoses (Chap-ter 9).
5  Malfunctioning caliper assembly (Chapter 9).
6  Loose suspension parts (Chapter 10).
7  Loose calipers (Chapter 9).
8  Excessive wear of brake pad material or disc on one side.

## 59 Noise (high-pitched squeal when the brakes are applied)

Front and/or rear disc brake pads worn out (Chapter 9).

## 60 Brake roughness or chatter (pedal pulsates)

1  Excessive lateral runout (Chapter 9).
2  Uneven pad wear (Chapter 9).
3  Defective disc (Chapter 9).

## 61 Excessive brake pedal effort required to stop vehicle

1  Malfunctioning power brake booster (Chapter 9).
2  Partial system failure (Chapter 9).
3  Excessively worn pads (Chapter 9).
4  Piston in caliper stuck or sluggish (Chapter 9).
5  Brake pads contaminated with oil or grease (Chapter 9).
6  Brake disc grooved and/or glazed (Chapter 1).
7  New pads installed and not yet seated. It will take a while for the new material to seat against the disc.

## 62 Excessive brake pedal travel

1  Partial brake system failure (Chapter 9).
2  Insufficient fluid in master cylinder (Chapters 1 and 9).
3  Air trapped in system (Chapters 1 and 9).

## 63 Dragging brakes

1  Master cylinder pistons not returning correctly (Chapter 9).
2  Restricted brakes lines or hoses (Chapters 1 and 9).
3  Incorrect parking brake adjustment (Chapter 9).

## 64 Grabbing or uneven braking action

1 Malfunction of proportioning valve (Chapter 9).
2 Malfunction of power brake booster unit (Chapter 9).
3 Binding brake pedal mechanism (Chapter 9).

## 65 Brake pedal feels spongy when depressed

1 Air in hydraulic lines (Chapter 9).
2 Master cylinder mounting bolts loose (Chapter 9).
3 Master cylinder defective (Chapter 9).

## 66 Brake pedal travels to the floor with little resistance

1 Little or no fluid in the master cylinder reservoir caused by leaking caliper piston(s) (Chapter 9).
2 Malfunctioning master cylinder (Chapter 9).
3 Loose, damaged or disconnected brake lines (Chapter 9).

## 67 Parking brake does not hold

Parking brake cables improperly adjusted (Chapter 9).

# SUSPENSION AND STEERING SYSTEMS

➡ **Note: Before attempting to diagnose the suspension and steering systems, perform the following preliminary checks:**

a) *Tires for wrong pressure and uneven wear.*
b) *Steering universal joints from the column to the rack-and-pinion for loose connectors or wear.*
c) *Front and rear suspension and the rack-and-pinion assembly for loose or damaged parts.*
d) *Out-of-round or out-of-balance tires, bent rims and loose and/or rough wheel bearings.*

## 68 Vehicle pulls to one side

1 Mismatched or uneven tires (Chapter 10).
2 Broken or sagging springs (Chapter 10).
3 Wheel alignment out of specifications (Chapter 10).
4 Front brake dragging (Chapter 9).

## 69 Abnormal or excessive tire wear

1 Wheel alignment out of specifications (Chapter 10).
2 Sagging or broken springs (Chapter 10).
3 Tire out-of-balance (Chapter 10).
4 Worn shock absorber (Chapter 10).
5 Overloaded vehicle.
6 Tires not rotated regularly.

## 70 Wheel makes a thumping noise

1 Blister or bump on tire (Chapter 10).
2 Improper shock absorber action (Chapter 10).

## 71 Shimmy, shake or vibration

1 Tire or wheel out-of-balance or out-of-round (Chapter 10).
2 Loose or worn wheel bearings (Chap-ters 1, 8 and 10).
3 Worn tie-rod ends (Chapter 10).
4 Worn balljoints (Chapters 1 and 10).
5 Excessive wheel runout (Chapter 10).
6 Blister or bump on tire (Chapter 10).

## 72 Hard steering

1 Defective balljoints, tie-rod ends or rack-and-pinion assembly (Chapter 10).
2 Front wheel alignment out of specifications (Chapter 10).
3 Low tire pressure(s) (Chapters 1 and 10).

## 73 Poor returnability of steering to center

1 Defective balljoints or tie-rod ends (Chapter 10).
2 Binding in steering gear or column (Chapter 10).
3 Lack of lubricant in steering system (Chapter 1).
4 Front wheel alignment out of specifications (Chapter 10).

## 74 Abnormal noise at the front end

1 Defective balljoints or tie-rod ends (Chapters 1 and 10).
2 Damaged shock absorber mounting (Chapter 10).
3 Worn control arm bushings or tie-rod ends (Chapter 10).
4 Loose stabilizer bar (Chapter 10).
5 Loose wheel bolts (Chapter 1 Specifications).
6 Loose suspension bolts (Chapter 10).

## 75 Wander or poor steering stability

1 Mismatched or uneven tires (Chapter 10).
2 Defective balljoints or tie-rod ends (Chapters 1 and 10).
3 Worn shock absorber assemblies (Chapter 10).
4 Loose stabilizer bar (Chapter 10).
5 Broken or sagging springs (Chapter 10).
6 Wheels out of alignment (Chapter 10).

## 76 Erratic steering when braking

1 Wheel bearings worn (Chapter 10).
2 Broken or sagging springs (Chapter 10).
3 Defective caliper (Chapter 9).
4 Warped discs (Chapter 9).
5 Front end alignment incorrect.

## 77 Excessive pitching and/or rolling around corners or during braking

1 Loose stabilizer bar (Chapter 10).
2 Worn shock absorbers or mountings (Chapter 10).
3 Broken or sagging springs (Chapter 10).
4 Overloaded vehicle.
5 Front end alignment incorrect.

### 78 Suspension bottoms

1 Overloaded vehicle.
2 Worn shock absorbers (Chapter 10).
3 Incorrect, broken or sagging springs (Chapter 10).

### 79 Cupped tires

1 Worn shock absorbers (Chapter 10).
2 Wheel bearings worn (Chapter 10).
3 Excessive tire or wheel runout (Chapter 10).
4 Worn balljoints (Chapter 10).

### 80 Excessive tire wear on outside edge

1 Inflation pressures incorrect (Chapter 1).
2 Excessive speed in turns.
3 Front end alignment incorrect (excessive toe-in or camber). Have professionally aligned.
4 Suspension arm bent or twisted (Chapter 10).

### 81 Excessive tire wear on inside edge

1 Inflation pressures incorrect (Chapter 1).

2 Front end alignment incorrect (toe-out or negative camber). Have professionally aligned.
3 Loose or damaged steering components (Chapter 10).

### 82 Tire tread worn in one place

1 Tires out-of-balance.
2 Damaged wheel. Inspect and replace if necessary.
3 Defective tire (Chapter 1).

### 83 Excessive play or looseness in steering system

1 Wheel bearing(s) worn (Chapter 10).
2 Tie-rod end loose or worn (Chapter 10).
3 Steering gear loose or worn (Chapter 10).
4 Worn or loose steering intermediate shaft (Chapter 10).

### 84 Rattling or clicking noise in steering gear

1 Steering gear loose or worn (Chapter 10).
2 Steering gear defective.

**Section**

**Reference to other Chapters**

CHECK ENGINE light on - See Chapter 6

# 1

# TUNE-UP AND ROUTINE MAINTENANCE

## 1   Maintenance schedule

The following maintenance intervals are based on the assumption that the vehicle owner will be doing the maintenance or service work, as opposed to having a dealer service department do the work. These are the minimum maintenance intervals recommended by the factory for vehicles that are driven daily. If you wish to keep your vehicle in peak condition at all times, you may wish to perform some of these procedures even more often. Because frequent maintenance enhances the efficiency, performance and resale value of your car, we encourage you to do so. If you drive in dusty areas, tow a trailer, idle or drive at low speeds for extended periods or drive for short distances (less than four miles) in below freezing temperatures, shorter intervals are also recommended.

When the vehicle is new, follow the maintenance schedule to the letter, record the maintenance performed in your owner's manual and keep all receipts to protect the new vehicle warranty. In many cases the initial maintenance check is done at no cost to the owner (check with your dealer service department for more information).

### ✳✳ CAUTION 1:

**These models are equipped with an anti-theft radio. Before performing a procedure that requires disconnecting the battery, make sure you have the proper activation code.**

### ✳✳ CAUTION 2:

**Disconnecting the battery can cause driveability problems that require a scan tool to rectify. Additionally, disconnecting the battery may cause one or more warning lights on the instrument panel to illuminate, which will also require the use of a scan tool to turn off. Most scan tools available to the public do not have the capability to perform either of these tasks, which will necessitate taking the vehicle to a dealer service department or other properly equipped repair facility after service work has been performed. See Chapter 5, Section 1 for the use of an auxiliary voltage input device ("memory saver") before disconnecting the battery and for other precautions related to battery disconnection.**

### EVERY 250 MILES (400 KM) OR WEEKLY, WHICHEVER COMES FIRST

Check the engine oil level (Section 4)
Check the coolant level (Section 4)
Check the windshield washer fluid level (Section 4)
Check the brake and clutch fluid levels (Section 4)
Check the tires and tire pressures (Section 5)

### EVERY 3000 MILES (4800 KM) OR 3 MONTHS, WHICHEVER COMES FIRST

*All items listed above, plus . . .*
Check the power steering fluid level (Section 6)
Change the engine oil and filter (Section 7)

### EVERY 6000 MILES (9600 KM) OR 6 MONTHS, WHICHEVER COMES FIRST

*All items listed above, plus . . .*
Check the seat belts (Section 8)
Inspect the windshield wiper blades (Section 9)
Check and service the battery (Section 10)
Check the engine drivebelt(s) (Section 11)
Inspect the underhood hoses (Section 12)
Check the cooling system (Section 13)
Rotate the tires (Section 14)

### EVERY 15,000 MILES (24,000 KM) OR 12 MONTHS, WHICHEVER COMES FIRST

*All items listed above, plus . . .*
Check the fuel system (Section 15)
Check the brake system (Section 16)*
Check the exhaust system (Section 17)
Check the steering, suspension and driveaxle boots (Section 18)
Check the manual transmission lubricant level (Section 19)
Check the automatic transaxle fluid level (Section 20)
Check the front differential lubricant level (automatic transaxle models) (Section 21)
Check the center differential lubricant level (all-wheel drive models with automatic transaxles) (Section 22)
Check the rear differential lubricant level (all-wheel drive models) (Section 23)
Replace the interior ventilation filter (Section 24)*

### EVERY 30,000 MILES (48,000 KM) OR 24 MONTHS, WHICHEVER COMES FIRST

*All items listed above, plus . . .*
Change the automatic transaxle fluid (Section 25)
Replace the air filter (Section 26)*
Replace the spark plugs (Section 27)
Inspect/replace the spark plug wires (V6 engines) (Section 28)
Service the cooling system (drain, flush and refill) (Section 29)

## EVERY TWO YEARS, REGARDLESS OF MILEAGE

Change the brake fluid (Section 30)

## EVERY 60,000 MILES (96,000 KM) OR 48 MONTHS, WHICHEVER COMES FIRST

Replace the fuel filter (Section 31)
Replace the timing belt and tensioner (Chapter 2)
Change the manual transaxle lubricant (Section 32)
Change the center differential lubricant level (all-wheel drive models with automatic transaxles) (Section 33)**
Change the rear differential lubricant level (all-wheel drive models) (Section 34)
Replace the oxygen sensor(s) (Chapter 6)
Service the cooling system (drain, flush and refill) and replace the cooling system hoses (Section 29)

\* This item is affected by "severe" operating conditions, as described below. If the vehicle is operated under severe conditions, perform all maintenance indicated with an asterisk (*) at half the indicated intervals. Severe conditions exist if you mainly operate the vehicle . . .

in dusty areas
towing a trailer
idling for extended periods
driving at low speeds when outside temperatures remain below freezing and most trips are less than four miles long.

\*\* Perform this procedure at half the recommended interval if operated under one or more of the following conditions:

in heavy city traffic where the outside temperature regularly reaches 90-degrees F or higher
in hilly or mountainous terrain
frequent trailer towing
if the vehicle has been driven through deep water

## 2  Introduction

This Chapter is designed to help the home mechanic maintain the Audi A4/Volkswagen Passat with the goals of maximum performance, economy, safety and reliability in mind.

Included is a master maintenance schedule, followed by procedures dealing specifically with each item on the schedule. Visual checks, adjustments, component replacement and other helpful items are included. Refer to the accompanying illustrations of the engine compartment and the underside of the vehicle for the locations of various components.

Servicing your vehicle in accordance with the mileage/time maintenance schedule and the step-by-step procedures will result in a planned maintenance program that should produce a long and reliable service life. Keep in mind that it's a comprehensive plan, so maintaining some items but not others at the specified intervals will not produce the same results.

As you service your vehicle, you will discover that many of the procedures can - and should - be grouped together because of the nature of the particular procedure you're performing or because of the close proximity of two otherwise unrelated components to one another.

For example, if the vehicle is raised for chassis lubrication, you should inspect the exhaust, suspension, steering and fuel systems while you're under the vehicle. When you're rotating the tires, it makes good sense to check the brakes since the wheels are already removed. Finally, let's suppose you have to borrow or rent a torque wrench. Even if you only need it to tighten the spark plugs, you might as well check the torque of as many critical fasteners as time allows.

The first step in this maintenance program is to prepare yourself before the actual work begins. Read through all the procedures you're planning to do, then gather up all the parts and tools needed. If it looks like you might run into problems during a particular job, seek advice from a mechanic or an experienced do-it-yourselfer.

## OWNER'S MANUAL AND VECI LABEL INFORMATION

Your vehicle Owner's Manual was written for your year and model and contains very specific information on component locations, specifications, fuse ratings, part numbers, etc. The Owner's Manual is an important resource for the do-it-yourselfer to have; if one was not supplied with your vehicle, it can generally be ordered from a dealer parts department.

Among other important information, the Vehicle Emissions Control Information (VECI) label contains specifications and procedures for tune-up adjustments (if applicable) and, in some instances, spark plugs (see Chapter 6 for more information on the VECI label). The information on this label is the exact maintenance data recommended by the manufacturer. This data often varies by intended operating altitude, local emissions regulations, month of manufacture, etc.

This Chapter contains procedural details, safety information and more ambitious maintenance intervals than you might find in manufacturer's literature. However, you may also find procedures or specifications in your Owner's Manual or VECI label that differ with what's printed here. In these cases, the Owner's Manual or VECI label can be considered correct, since it is specific to your particular vehicle.

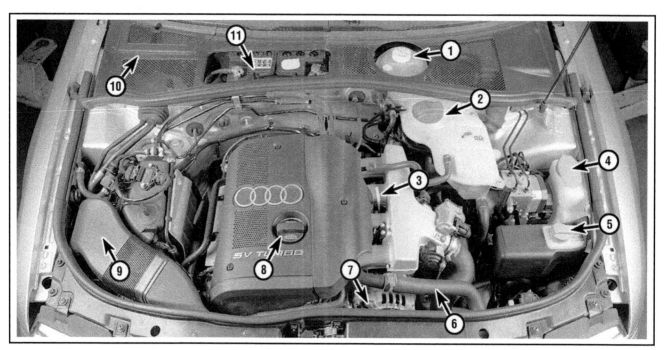

**Engine compartment components - four-cylinder model**

| | | | | | |
|---|---|---|---|---|---|
| 1 | Brake fluid reservoir | 5 | Power steering fluid reservoir | 9 | Air filter housing |
| 2 | Coolant expansion tank | 6 | Radiator hose | 10 | Interior ventilation filter housing |
| 3 | Engine oil dipstick | 7 | Drivebelt (not visible in photo) | 11 | Battery |
| 4 | Windshield wiper fluid reservoir | 8 | Engine oil filler cap | | |

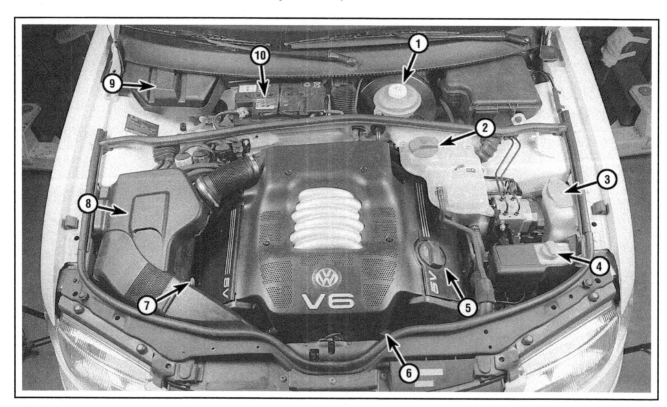

**Engine compartment components - DOHC V6 model**

| | | | | | |
|---|---|---|---|---|---|
| 1 | Brake fluid reservoir | 5 | Engine oil filler cap | 8 | Air filter housing |
| 2 | Coolant expansion tank | 6 | Drivebelt (not visible in photo) | 9 | Interior ventilation filter housing |
| 3 | Windshield wiper fluid reservoir | 7 | Engine oil dipstick | 10 | Battery |
| 4 | Power steering fluid reservoir | | | | |

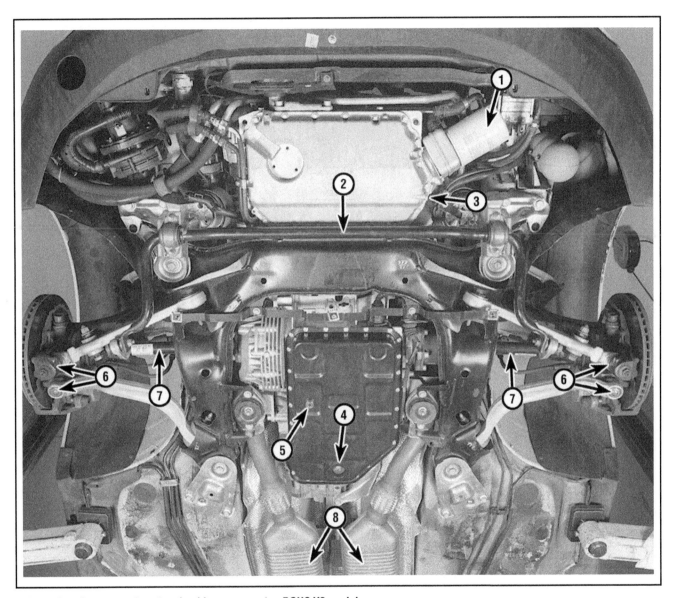

**Typical engine compartment underside components - DOHC V6 model**

| | | | | | |
|---|---|---|---|---|---|
| 1 | Engine oil filter | 4 | Automatic transaxle check plug | 7 | Driveaxles |
| 2 | Stabilizer bar | 5 | Automatic transaxle fluid drain plug | 8 | Catalytic converters |
| 3 | Engine oil drain plug | 6 | Lower control arm balljoints | | |

**Rear underside components (2WD model)**

| | | | | | |
|---|---|---|---|---|---|
| 1 | Muffler | 4 | Shock absorber | 6 | Rear axle assembly |
| 2 | Fuel tank | 5 | Coil spring | 7 | Fuel filter |
| 3 | Brake caliper | | | | |

**Rear underside components - all-wheel drive Audi A4**

| | | | | | |
|---|---|---|---|---|---|
| 1 | Lower control arm | 4 | Track rod | 7 | Stabilizer bar |
| 2 | Rear knuckle and hub assembly | 5 | Upper control arm | 8 | Differential drain plug |
| 3 | Shock absorber/coil spring assembly | 6 | Stabilizer bar clamp | 9 | Differential lubricant check/fill plug (not visible in photo) |

**Rear underside components - all-wheel drive VW Passat Wagon**

| | | | | | |
|---|---|---|---|---|---|
| 1 | Lower control arm | 4 | Track rod | 7 | Stabilizer bar |
| 2 | Rear knuckle and hub assembly | 5 | Upper control arm | 8 | Differential drain plug |
| 3 | Shock absorber/coil spring assembly | 6 | Stabilizer bar clamp | 9 | Differential lubricant check/fill plug |

## 3 Tune-up general information

The term tune-up is used in this manual to represent a combination of individual operations rather than one specific procedure that will maintain a gasoline engine in proper tune.

If, from the time the vehicle is new, the routine maintenance schedule is followed closely and frequent checks are made of fluid levels and high wear items, as suggested throughout this manual, the engine will be kept in relatively good running condition and the need for additional work will be minimized.

More likely than not, however, there may be times when the engine is running poorly due to lack of regular maintenance. This is even more likely if a used vehicle, which has not received regular and frequent maintenance checks, is purchased. In such cases, an engine tune-up will be needed outside of the regular routine maintenance intervals.

The first step in any tune-up or diagnostic procedure to help correct a poor-running engine is a cylinder compression check. A compression check (see Chapter 2C) will help determine the condition of internal engine components and should be used as a guide for tune-up and repair procedures. If, for instance, the compression check indicates serious internal engine wear, a conventional tune-up won't improve the performance of the engine and would be a waste of time and money. Because of its importance, the compression check should be done by someone with the right equipment and the knowledge to use it properly.

The following procedures are those most often needed to bring a generally poor-running engine back into a proper state of tune.

### MINOR TUNE-UP

*Check all engine related fluids (Section 4)*
*Clean, inspect and test the battery Section 10)*
*Check the engine drivebelt(s) (Section 11)*
*Check all underhood hoses (Section 12)*
*Check the cooling system (Section 13)*
*Check the air filter (Section 26)*
*Replace the spark plugs (Section 27)*
*Inspect the spark plug wires (Section 28)*

### MAJOR TUNE-UP

All items listed under Minor tune-up, plus . . .
*Replace the air filter (Section 26)*
*Replace the spark plug wires (Section 28)*
*Check the ignition system (Chapter 5)*
*Check the charging system (Chapter 5)*

## 4 Fluid level checks (every 250 miles [400 km] or weekly)

➡**Note: The following are fluid level checks to be done on a 250 mile or weekly basis. Additional fluid level checks can be found in specific maintenance procedures that follow. Regardless of intervals, be alert to fluid leaks under the vehicle, which would indicate a fault to be corrected immediately.**

1   Fluids are an essential part of the lubrication, cooling, brake, clutch and windshield washer systems. Because the fluids gradually become depleted and/or contaminated during normal operation of the vehicle, they must be periodically replenished. See *Recommended lubricants and fluids* at the end of this Chapter before adding fluid to any of the following components.

➡**Note: The vehicle must be on level ground when fluid levels are checked.**

### ENGINE OIL

▶ **Refer to illustrations 4.2a, 4.2b, 4.4 and 4.6**

2   The engine oil level is checked with a dipstick that extends through a tube and into the oil pan at the bottom of the engine (see illustrations).

3   The oil level should be checked before the vehicle has been driven, or about 5 minutes after the engine has been shut off. If the oil is checked immediately after driving the vehicle, some of the oil will remain in the upper engine components, resulting in an inaccurate reading on the dipstick.

4   Pull the dipstick out of the tube and wipe all the oil from the end

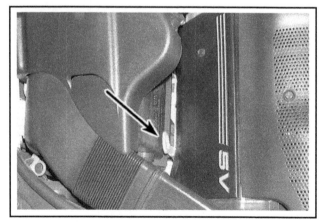

**4.2a  Engine oil dipstick location - V6 engine**

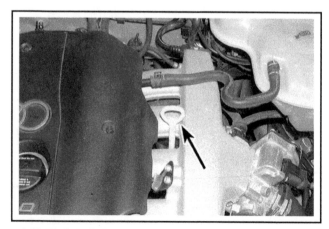

**4.2b  Engine oil dipstick location - four-cylinder engine**

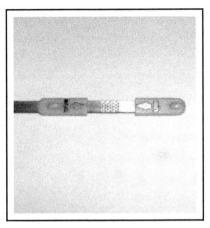

**4.4 The oil level must be maintained between the marks at all times - it takes one quart of oil to raise the level from the MIN to MAX mark**

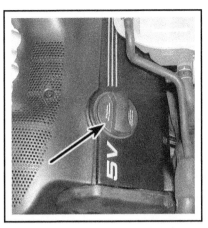

**4.6 Oil is added to the engine after unscrewing the oil filler cap (DOHC V6 engine shown) - always make sure the area around the opening is clean before removing the cap to prevent dirt from contaminating the engine**

**4.9 The coolant expansion tank is located on the left (driver's) side - keep the level near the MAX mark or MIN mark on the side of the reservoir, depending on engine temperature**

with a clean rag or paper towel. Insert the clean dipstick all the way back into the tube, then pull it out again. Note the oil at the end of the dipstick. Add oil as necessary to keep the level in the shaded area on the dipstick (see illustration).

5   Do not overfill the engine by adding too much oil since this may result in oil-fouled spark plugs, catalytic converter failure, oil leaks or oil seal failures.

6   Oil is added to the engine after unscrewing a cap from the valve cover (see illustration). A funnel may help to reduce spills.

7   Checking the oil level is an important preventive maintenance step. A consistently low oil level indicates oil leakage through damaged seals, defective gaskets or past worn rings or valve guides. If the oil looks milky or has water droplets in it, the cylinder head gasket may be blown or the head or block may be cracked. The engine should be checked immediately. The condition of the oil should also be checked. Whenever you check the oil level, slide your thumb and index finger up the dipstick before wiping off the oil. If you see small dirt or metal particles clinging to the dipstick, the oil should be changed (see Section 8).

## ENGINE COOLANT

▶ Refer to illustration 4.9

### ✳✳ WARNING:

**Do not allow antifreeze to come in contact with your skin or painted surfaces of the vehicle. Rinse off spills immediately with plenty of water. Antifreeze is highly toxic if ingested. Never leave antifreeze lying around in an open container or in puddles on the floor; children and pets are attracted by its sweet smell and may drink it. Check with local authorities on disposing of used anti-freeze. Many communities have collection centers that will see that antifreeze is disposed of safely.**

➡**Note: Non-toxic antifreeze is now manufactured and available at local auto parts stores, but even this type should be disposed of properly.**

### ✳✳ CAUTION:

**Never mix green-colored ethylene glycol anti-freeze and red-colored silicate-free coolant because doing so will destroy the efficiency of the red coolant, which could result in engine damage.**

8   All vehicles covered by this manual are equipped with a coolant expansion tank, located at the left side of the engine compartment, and connected by hoses to the radiator and cooling system.

9   The coolant level in the expansion tank should be checked regularly.

### ✳✳ WARNING:

**Do not remove the pressure cap to check the coolant level when the engine is warm.**

The level of coolant in the expansion tank varies with the temperature of the engine. When the engine is cold, the coolant level should be at or slightly above the MIN mark on the tank (see illustration). Once the engine has warmed up, the level should be at or near the MAX mark. If it isn't, add coolant to the expansion tank. To add coolant simply twist open the cap and add a 50/50 mixture of the specified coolant (see **Caution** above).

10   Drive the vehicle and recheck the coolant level. If only a small amount of coolant is required to bring the system up to the proper level, water can be used. However, repeated additions of water will dilute the antifreeze and water solution. In order to maintain the proper ratio of antifreeze and water, always top up the coolant level with the correct mixture. An empty plastic milk jug or bleach bottle makes an excellent container for mixing coolant. Do not use rust inhibitors or additives.

11   If the coolant level drops consistently, there may be a leak in the system. Inspect the radiator, hoses, filler cap, drain plugs and water pump (see Section 13). If no leaks are noted, have the pressure cap tested by a service station.

12   If you have to remove the pressure cap, wait until the engine has cooled completely, then wrap a thick cloth around the cap and unscrew

**4.14 Flip open the cap (arrow) to check the fluid level in the windshield washer tank**

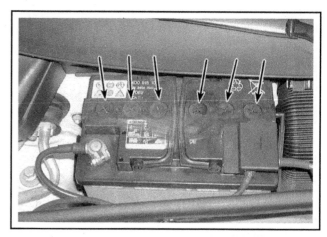

**4.17 Remove the screw-in caps (arrows) to check the battery electrolyte level**

16 To help prevent icing in cold weather, warm the windshield with the defroster before using the washer.

## BATTERY ELECTROLYTE

▶ **Refer to illustration 4.17**

17 These vehicles are equipped with a battery which has a translucent plastic case. You should be able to read the electrolyte level by looking at the side of the case and comparing the level to the Minimum and Maximum markings on the case, but due to the location of the battery these marks may be difficult to see. Instead, read the fluid level by removing the screw-in plastic caps on top of the battery (see illustration). This check is most critical during the warm summer months. Add only distilled water to any battery, and fill only to the ledge below the filler caps.

➡**Note: The filler caps are equipped with O-rings. Make sure the O-rings are in place when reinstalling the caps.**

## BRAKE AND CLUTCH FLUID

▶ **Refer to illustration 4.19**

18 The brake master cylinder is mounted on the brake booster in the cowl area. The clutch master cylinder on manual transaxle vehicles is mounted at the clutch pedal assembly, but receives its hydraulic fluid directly from the brake master cylinder reservoir, via a hose. Checking the fluid at the brake master cylinder reservoir effectively checks the level of brake and clutch fluid.

19 The translucent plastic reservoir allows the fluid inside to be checked without removing the cap (see illustration). Be sure to wipe the top of the reservoir with a clean rag to prevent contamination of the brake and/or clutch system before removing the cap.

20 When adding fluid, pour it carefully into the reservoir to avoid spilling it on surrounding painted surfaces. Be sure the specified fluid is used, since mixing different types of brake fluid can cause damage to the system. See *Recommended lubricants and fluids* at the end of this Chapter or your owner's manual.

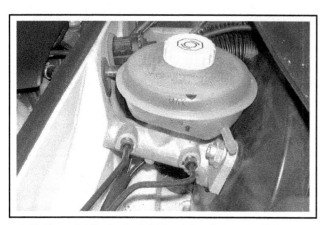

**4.19 Never let the brake fluid level drop below the MIN mark**

it slowly. If coolant or steam escapes, let the engine cool down longer, then remove the cap. ·

13 Check the condition of the coolant as well. It should be relatively clear. If it is brown or rust colored, the system should be drained, flushed and refilled. Even if the coolant appears to be normal, the corrosion inhibitors wear out, so it must be replaced at the specified intervals.

## WINDSHIELD WASHER FLUID

▶ **Refer to illustration 4.14**

14 Fluid for the windshield washer system is located in a plastic reservoir in the left side of the engine compartment (see illustration).

15 In milder climates, plain water can be used in the reservoir, but it should be kept no more than 2/3 full to allow for expansion if the water freezes. In colder climates, use windshield washer system antifreeze, available at any auto parts store, to lower the freezing point of the fluid. Mix the antifreeze with water in accordance with the manufacturer's directions on the container.

### ✳✳ CAUTION:

**Don't use cooling system antifreeze - it will damage the vehicle's paint.**

## ✳✳ WARNING:

**Brake fluid can harm your eyes and damage painted surfaces, so use extreme caution when handling or pouring it. Do not use brake fluid that has been standing open or is more than one year old. Brake fluid absorbs moisture from the air. Moisture in the system can cause a dangerous loss of brake performance.**

21  At this time, the fluid and master cylinder can be inspected for contamination. The system should be drained and refilled if deposits, dirt particles or water droplets are seen in the fluid.

22  After filling the reservoir to the proper level, make sure the cap is on tight to prevent fluid leakage.

23  The brake fluid level in the master cylinder will drop slightly as the pads at the front wheels wear down during normal operation. If the master cylinder requires repeated additions to keep it at the proper level, it's an indication of leakage in the brake system or clutch release system, which should be corrected immediately. Check all brake lines and connections (see Section 16 for more information). Inspect the clutch release system, too (see Chapter 8).

24  If, upon checking the master cylinder fluid level, you discover the reservoir empty or nearly empty, the brake system should be bled and thoroughly inspected (see Chapter 9). Inspect the clutch release system, too (see Chapter 8).

## 5   Tire and tire pressure checks (every 250 miles [400 km] or weekly)

▶ **Refer to illustrations 5.2, 5.3, 5.4a, 5.4b, 5.8a and 5.8b**

1  Periodic inspection of the tires may spare you the inconvenience of being stranded with a flat tire. It can also provide you with vital information regarding possible problems in the steering and suspension systems before major damage occurs.

2  The original tires on this vehicle are equipped with 1/2-inch wide wear bands that will appear when tread depth reaches 1/16-inch, at which point the tires can be considered worn out. Tread wear can be monitored with a simple, inexpensive device known as a tread depth indicator (see illustration).

3  Note any abnormal tread wear (see illustration). Tread pattern irregularities such as cupping, flat spots and more wear on one side than the other are indications of front end alignment and/or balance problems. If any of these conditions are noted, take the vehicle to a tire shop or service station to correct the problem.

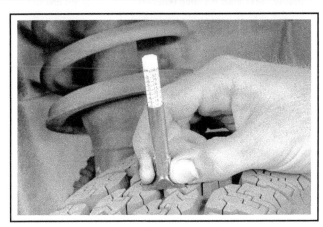

**5.2  Use a tire tread depth indicator to monitor tire wear - they are available at auto parts stores and service stations and cost very little**

**UNDERINFLATION**

**INCORRECT TOE-IN OR EXTREME CAMBER**

**CUPPING**

Cupping may be caused by:
- Underinflation and/or mechanical irregularities such as out-of-balance condition of wheel and/or tire, and bent or damaged wheel.
- Loose or worn steering tie-rod or steering idler arm.
- Loose, damaged or worn front suspension parts.

**OVERINFLATION**

**FEATHERING DUE TO MISALIGNMENT**

**5.3  This chart will help you determine the condition of the tires and the probable cause(s) of abnormal wear**

**5.4a If a tire loses air on a steady basis, check the valve stem core first to make sure it's snug (special inexpensive wrenches are commonly available at auto parts stores)**

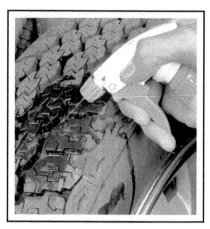

**5.4b If the valve stem core is tight, raise the corner of the vehicle with the low tire and spray a soapy water solution onto the tread as the tire is turned slowly - leaks will cause small bubbles to appear**

**5.8a To extend the life of the tires, check the air pressure at least once a week with an accurate gauge (don't forget the spare!)**

4   Look closely for cuts, punctures and embedded nails or tacks. Sometimes a tire will hold air pressure for a short time or leak down very slowly after a nail has embedded itself in the tread. If a slow leak persists, check the valve stem core to make sure it's tight (see illustration). Examine the tread for an object that may have embedded itself in the tire or for a "plug" that may have begun to leak (radial tire punctures are repaired with a plug that's installed in a puncture). If a puncture is suspected, it can be easily verified by spraying a solution of soapy water onto the puncture area (see illustration). The soapy solution will bubble if there's a leak. Unless the puncture is unusually large, a tire shop or service station can usually repair the tire.

5   Carefully inspect the inner sidewall of each tire for evidence of brake fluid leakage. If you see any, inspect the brakes immediately.

6   Correct air pressure adds miles to the lifespan of the tires, improves mileage and enhances overall ride quality. Tire pressure cannot be accurately estimated by looking at a tire, especially if it's a radial. A tire pressure gauge is essential. Keep an accurate gauge in the vehicle. The pressure gauges attached to the nozzles of air hoses at gas stations are often inaccurate.

7   Always check tire pressure when the tires are cold. Cold, in this case, means the vehicle has not been driven over a mile in the three hours preceding a tire pressure check. A pressure rise of four to eight pounds is not uncommon once the tires are warm.

8   Unscrew the valve cap protruding from the wheel or hubcap and push the gauge firmly onto the valve stem (see illustration). Note

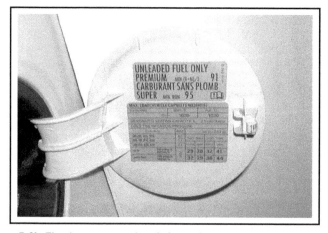

**5.8b The tire pressure chart is located on the fuel filler lid**

the reading on the gauge and compare the figure to the recommended tire pressure shown on the placard on the fuel filler door (see illustration). Be sure to reinstall the valve cap to keep dirt and moisture out of the valve stem mechanism. Check all four tires and, if necessary, add enough air to bring them up to the recommended pressure.

9   Don't forget to keep the spare tire inflated to the specified pressure (refer to your owner's manual or the tire sidewall).

## 6   Power steering fluid level check (every 3000 miles [4800 km] or 3 months)

♦ **Refer to illustrations 6.2 and 6.6**

1   Unlike manual steering, the power steering system relies on fluid which may, over a period of time, require replenishing.

2   The fluid reservoir for the power steering pump is located at the left front corner of the engine compartment (see illustration).

3   For the check, the front wheels should be pointed straight ahead and the engine should be off.

4   Use a clean rag to wipe off the reservoir cap and the area around the cap. This will help prevent any foreign matter from entering the reservoir during the check.

5   Twist off the cap and check the temperature of the fluid at the end of the dipstick with your finger.

6   Wipe off the fluid with a clean rag, reinsert the dipstick, then withdraw it and read the fluid level. The fluid should be at the proper level, depending on whether it was checked hot or cold (see illustration). Never allow the fluid level to drop below the lower mark on the dipstick.

7   If additional fluid is required, pour the specified type directly into the reservoir, using a funnel to prevent spills.

8   If the reservoir requires frequent fluid additions, all power steer-ing hoses, hose connections, steering gear and the power steering pump should be carefully checked for leaks.

**6.2  The power steering fluid reservoir is located in the left front corner of the engine compartment - turn the cap counterclockwise to remove it**

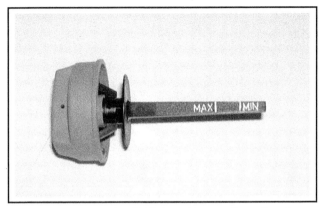

**6.6  The power steering fluid dipstick has marks on it so the fluid can be checked hot (MAX) or cold (MIN)**

## 7   Engine oil and filter change (every 3000 miles [4800 km] or 3 months)

▶ **Refer to illustrations 7.3, 7.7, 7.9, 7.14 and 7.18**

1   Frequent oil changes are the most important preventive mainte-nance procedures that can be done by the home mechanic. As engine oil ages, it becomes diluted and contaminated, which leads to prema-ture engine wear.

2   Although some sources recommend oil filter changes every other oil change, we feel that the minimal cost of an oil filter and the relative ease with which it is installed dictate that a new filter be installed every time the oil is changed.

3   Gather together all necessary tools and materials before begin-ning this procedure (see illustration).

4   You should have plenty of clean rags and newspapers handy to mop up any spills. Access to the underside of the vehicle may be improved if the vehicle can be lifted on a hoist, driven onto ramps or supported by jackstands.

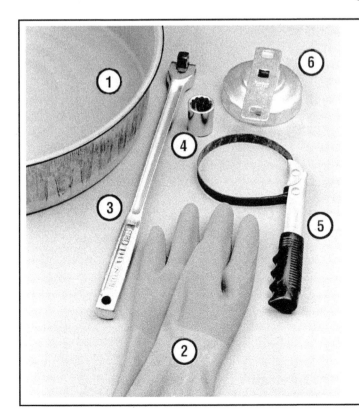

**7.3  These tools are required when changing the engine oil and filter**

*1   **Drain pan** - It should be fairly shallow in depth, but wide to prevent spills*

*2   **Rubber gloves** - When removing the drain plug and filter, you will get oil on your hands (the gloves will prevent burns)*

*3   **Breaker bar** - Sometimes the oil drain plug is tight, and a long breaker bar is needed to loosen it*

*4   **Socket** – To be used with the breaker bar or a ratchet (must be the correct size to fit the drain plug - six-point preferred)*

*5   **Filter wrench** - This is a metal band-type wrench, which requires clearance around the filter to be effective*

*6   **Filter wrench** - This type fits on the bottom of the filter and can be turned with a ratchet or breaker bar (different-size wrenches are available for different types of filters)*

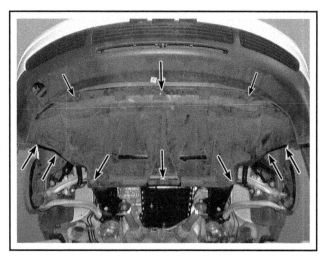

**7.7 Remove the fasteners (arrows) and lower the engine splash shield to access the oil filter**

## ❊❊ WARNING:

**Do not work under a vehicle which is supported only by a bumper, hydraulic or scissors-type jack.**

5   Warm the engine to normal operating temperature. If the new oil or any tools are needed, use this warm-up time to gather everything necessary for the job. The correct type of oil for your application can be found in *Recommended lubricants and fluids* at the end of this Chapter.

6   With the engine oil warm (warm engine oil will drain better and more built-up sludge will be removed with it), raise and support the vehicle. Make sure it's safely supported!

7   Remove the splash shield under the engine (see illustration).

8   Move all necessary tools, rags and newspapers under the vehicle. Set the drain pan under the drain plug. Keep in mind that the oil will initially flow from the pan with some force; position the pan accordingly.

9   Being careful not to touch any of the hot exhaust components, use a wrench to remove the drain plug near the bottom of the oil pan (see illustration). Depending on how hot the oil is, you may want to

wear gloves while unscrewing the plug the final few turns.

10   Allow the oil to drain into the pan. It may be necessary to move the pan as the oil flow slows to a trickle.

11   After all the oil has drained, wipe off the drain plug with a clean rag. Small metal particles may cling to the plug and would immediately contaminate the new oil.

12   Clean the area around the drain plug opening and reinstall the plug. Tighten the plug securely with the wrench. If a torque wrench is available, use it to tighten the plug to the torque listed in this Chapter's Specifications.

13   Move the drain pan into position under the oil filter.

14   Use the oil filter wrench to loosen the oil filter (see illustrations).

15   Completely unscrew the old filter. Be careful; it's full of oil. Empty the oil inside the filter into the drain pan, then lower the filter.

16   Compare the old filter with the new one to make sure they're the same type.

17   Use a clean rag to remove all oil, dirt and sludge from the area where the oil filter mounts to the engine. Check the old filter to make sure the rubber gasket isn't stuck to the engine. If the gasket is stuck to the engine (use a flashlight if necessary), remove it.

18   Apply a light coat of clean oil to the rubber gasket on the new oil filter (see illustration).

19   Attach the new filter to the engine, following the tightening directions printed on the filter canister or packing box. Most filter manufacturers recommend against using a filter wrench due to the possibility of overtightening and damage to the seal.

20   Remove all tools, rags, etc. from under the vehicle, being careful not to spill the oil in the drain pan, reinstall the splash shield, then lower the vehicle.

21   Move to the engine compartment and locate the oil filler cap.

22   Pour the fresh oil through the filler opening. A funnel may be helpful.

23   Refer to the engine oil capacity in this Chapter's Specifications and add the proper amount of fresh oil into the engine. Wait a few minutes to allow the oil to drain into the pan, then check the level on the oil dipstick (see Section 4 if necessary). If the oil level is above the hatched area, start the engine and allow the new oil to circulate.

24   Run the engine for only about a minute and then shut it off. Immediately look under the vehicle and check for leaks at the oil pan drain plug and around the oil filter.

25   With the new oil circulated and the filter now completely full,

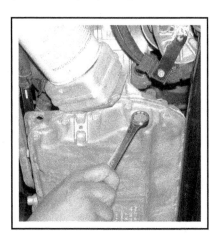

**7.9 Use the proper size box-end wrench or socket to remove the oil drain plug to avoid rounding it off**

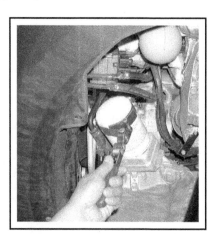

**7.14 Since the oil filter (arrow) is on very tight, you'll need a special wrench for removal - DO NOT use the wrench to tighten the new filter**

**7.18 Lubricate the oil filter gasket with clean engine oil before installing the filter on the engine**

recheck the level on the dipstick and add more oil as necessary.

26  During the first few trips after an oil change, make it a point to check frequently for leaks and proper oil level.

27  The old oil drained from the engine cannot be reused in its present state and should be disposed of. Check with your local auto parts store, disposal facility or environmental agency to see if they will accept the oil for recycling. After the oil has cooled it can be drained into a container (capped plastic jugs, topped bottles, milk cartons, etc.) for transport to one of these disposal sites. Don't dispose of the oil by pouring it on the ground or down a drain!

## 8  Seat belt check (every 6000 miles [9600 km]

### ✳✳ WARNING:

**The lap and shoulder belt adjusters incorporate a small explosive device that is triggered by the airbag system. The adjusters automatically tighten up the seat belts in case of a collision sufficient to set off the airbags. Do not use a hammer or impact driver of any kind near the seat belt or airbag module unless the airbag system has been disabled (see Chapter 12).**

1  Check the seat belts, buckles, latch plates and guide loops for obvious damage and signs of wear.

2  Where the seat belt receptacle bolts to the floor of the vehicle, check that the bolts are secure.

3  See if the seat belt reminder light comes on when the key is turned to the Run or Start position. A chime should also sound.

## 9  Wiper blade inspection and replacement (every 6000 miles [9600 km] or 6 months)

▶ **Refer to illustration 9.3**

1  The windshield wiper blade elements should be checked periodically for cracks and deterioration.

2  Lift the wiper blade assembly away from the glass.

3  Press the release lever and slide the blade assembly out of the hook in the end of the wiper arm (see illustration).

4  Squeeze the two rubber prongs at the end of the blade element, then slide the element out of the frame.

➡**Note: These elements can be replaced by hand, without pliers.**

5  Compare the new element with the old for length, design, etc. Some replacement elements come in a three-piece design (two metal strips, one on either side of the rubber) that is held together by several small plastic sleeves. Keep the sleeves in place on this design until you start sliding the element into the frame. Remove each of the plastic sleeves as needed when they reach the frame.

6  Slide the new element into the frame, notched end last and secure the clips into the notches of the frame.

7  Reinstall the blade assembly on the arm, wet the windshield and test for proper operation.

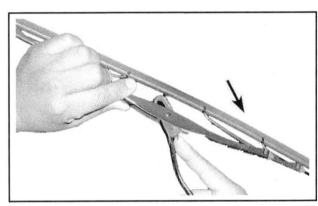

**9.3  Depress the release lever (finger is on it here) and slide the wiper assembly down the wiper arm and out of the hook in the end of the arm**

## 10  Battery check, maintenance and charging (every 6000 miles [9600 km] or 6 months)

▶ **Refer to illustrations 10.1, 10.5, 10.7a and 10.7b**

### ✳✳ WARNING:

**Certain precautions must be followed when checking and servicing the battery. Hydrogen gas, which is highly flammable, is always present in the battery cells, so keep lighted tobacco and all other open flames and sparks away from the battery. The electrolyte inside the battery is actually dilute sulfuric acid, which will cause injury if splashed on your skin or in your eyes. It will also ruin clothes and painted surfaces. When removing the battery cables, always detach the negative cable first and hook it up last!**

### ✳✳ CAUTION:

**Disconnecting the battery can cause driveability problems that require a scan tool to rectify. Additionally, disconnecting the battery may cause one or more warning lights on the instrument panel to illuminate, which will also require the use of a scan tool to turn off. Most scan tools available to the public do not have the capability to perform either of these tasks, which will necessitate taking the vehicle to a dealer service department or other properly equipped repair facility after service work has been performed. See Chapter 5, Section 1 for the use of an auxiliary voltage input device ("memory saver") before disconnecting the battery and for other precautions related to battery disconnection.**

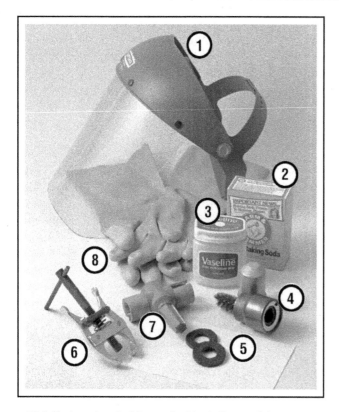

**10.1  Tools and materials required for battery maintenance**

*1*   **Face shield/safety goggles** - *When removing corrosion with a brush, the acidic particles can easily fly up into your eyes*
*2*   **Baking soda** - *A solution of baking soda and water can be used to neutralize corrosion*
*3*   **Petroleum jelly** - *A layer of this on the battery posts will help prevent corrosion*
*4*   **Battery post/cable cleaner** - *This wire brush cleaning tool will remove all traces of corrosion from the battery posts and cable clamps*
*5*   **Treated felt washers** - *Placing one of these on each post, directly under the cable clamps, will help prevent corrosion*
*6*   **Puller** - *Sometimes the cable clamps are very difficult to pull off the posts, even after the nut/bolt has been completely loosened. This tool pulls the clamp straight up and off the post without damage*
*7*   **Battery post/cable cleaner** - *Here is another cleaning tool that is a slightly different version of Number 4 above, but it does the same thing*
*8*   **Rubber gloves** - *Another safety item to consider when servicing the battery; remember that's acid inside the battery!*

1   A routine preventive maintenance program for the battery in your vehicle is the only way to ensure quick and reliable starts. But before performing any battery maintenance, make sure that you have the proper equipment necessary to work safely around the battery (see illustration).

2   There are also several precautions that should be taken whenever battery maintenance is performed. Before servicing the battery, always turn the engine and all accessories off and disconnect the cable from the negative terminal of the battery.

3   The battery produces hydrogen gas, which is both flammable and explosive. Never create a spark, smoke or light a match around the battery. Always charge the battery in a ventilated area.

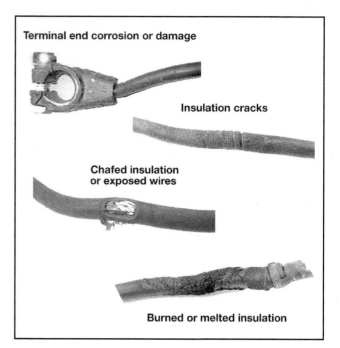

**10.5  Typical battery cable problems**

4   Electrolyte contains poisonous and corrosive sulfuric acid. Do not allow it to get in your eyes, on your skin or on your clothes. Never ingest it. Wear protective safety glasses when working near the battery. Keep children away from the battery.

5   Note the external condition of the battery. If the positive terminal and cable clamp on your vehicle's battery is equipped with a rubber protector, make sure that it's not torn or damaged. It should completely cover the terminal. Look for any corroded or loose connections, cracks in the case or cover or loose hold-down clamps. Also check the entire length of each cable for cracks and frayed conductors (see illustration).

6   If corrosion, which looks like white, fluffy deposits is evident, particularly around the terminals, the battery should be removed for cleaning. Loosen the cable bolts with a wrench, being careful to remove the ground cable first, and slide them off the terminals. Refer to Chapter 5 for specific battery removal procedures.

➡**Note: Read the Cautions in Section 1 before disconnecting the battery cables.**

7   Clean the cable ends thoroughly with a battery brush or a terminal cleaner and a solution of warm water and baking soda. Wash the terminals and the battery case with the same solution but make sure that the solution doesn't get into the battery. When cleaning the cables, terminals and battery case, wear safety goggles and rubber gloves to prevent any solution from coming in contact with your eyes or hands. Wear old clothes too - even diluted, sulfuric acid splashed onto clothes will burn holes in them. If the terminals have been corroded, clean them up with a terminal cleaner (see illustrations). Thoroughly wash all cleaned areas with plain water.

8   Make sure that the battery tray is in good condition and the hold-down clamp bolt is tight. If the battery is removed from the tray, make sure no parts remain in the bottom of the tray when the battery is reinstalled. When reinstalling the hold-down clamp bolt, do not overtighten it.

9   Any metal parts of the vehicle damaged by corrosion should be covered with a zinc-based primer, then painted.

**10.7a  A tool like this one (available at auto parts stores) is used to clean the top-terminal type battery posts**

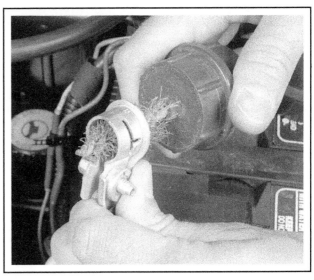

**10.7b  Use the brush side of the tool to clean inside the cable end**

10 Information on removing and installing the battery can be found in Chapter 5. Information on jump starting can be found at the front of this manual.

## CHARGING

### ❋❋ WARNING:

**When batteries are being charged, hydrogen gas, which is very explosive and flammable, is produced. Do not smoke or allow open flames near a charging or a recently charged battery. Wear eye protection when near the battery during charging. Also, make sure the charger is unplugged before connecting or disconnecting the battery from the charger.**

➡**Note: The manufacturer recommends the battery be removed from the vehicle for charging because the gas that escapes during this procedure can damage the paint. Fast charging with the battery cables connected can result in damage to the electrical system.**

11 Slow-rate charging is the best way to restore a battery that's discharged to the point where it will not start the engine. It's also a good way to maintain the battery charge in a vehicle that's only driven a few miles between starts. Maintaining the battery charge is particularly important in the winter when the battery must work harder to start the engine and electrical accessories that drain the battery are in greater use.

12 It's best to use a one or two-amp battery charger (sometimes called a "trickle" charger). They are the safest and put the least strain on the battery. They are also the least expensive. For a faster charge, you can use a higher amperage charger, but don't use one rated more than 1/10th the amp/hour rating of the battery. Rapid boost charges that

claim to restore the power of the battery in one to two hours are hardest on the battery and can damage batteries not in good condition. This type of charging should only be used in emergency situations.

13 The average time necessary to charge a battery should be listed in the instructions that come with the charger. As a general rule, a trickle charger will charge a battery in 12 to 16 hours.

14 Remove all the cell caps (if equipped) and cover the holes with a clean cloth to prevent spattering electrolyte. Disconnect the negative battery cable and hook the battery charger cable clamps up to the battery posts (positive to positive, negative to negative), then plug in the charger. Make sure it is set at 12-volts if it has a selector switch.

15 If you're using a charger with a rate higher than two amps, check the battery regularly during charging to make sure it doesn't overheat. If you're using a trickle charger, you can safely let the battery charge overnight after you've checked it regularly for the first couple of hours.

16 If the battery has removable cell caps, measure the specific gravity with a hydrometer every hour during the last few hours of the charging cycle. Hydrometers are available inexpensively from auto parts stores - follow the instructions that come with the hydrometer. Consider the battery charged when there's no change in the specific gravity reading for two hours and the electrolyte in the cells is gassing (bubbling) freely. The specific gravity reading from each cell should be very close to the others. If not, the battery probably has a bad cell(s).

17 Some batteries with sealed tops have built-in hydrometers on the top that indicate the state of charge by the color displayed in the hydrometer window. Normally, a bright-colored hydrometer indicates a full charge and a dark hydrometer indicates the battery still needs charging.

18 If the battery has a sealed top and no built-in hydrometer, you can hook up a digital voltmeter across the battery terminals to check the charge. A fully charged battery should read 12.5 volts or higher.

19 Further information on the battery and jump-starting can be found in Chapter 5 and at the front of this manual.

## 11 Drivebelt check and replacement (every 6000 miles [9600 km] or 6 months)

1   A serpentine, ribbed drivebelt is located at the front of the engine and plays an important role in the overall operation of the engine and its components. Due to its function and material make up, the belt is prone to wear and should be periodically inspected. On V6 engines the drivebelt drives the cooling fan, alternator, power steering pump, water pump and air conditioning compressor. On 1.8L four-cylinder engines, one ribbed belt drives the cooling fan, alternator and power steering pump, while the other ribbed belt drives the air conditioning compressor; a V-belt turned by the power steering pump pulley drives the water pump.

## CHECK

♦ **Refer to illustrations 11.2a and 11.2b**

2   With the engine off, open the hood and use your fingers (and a flashlight, if necessary), to move along the belt checking for cracks and separation of the belt plies. Also check for fraying and glazing, which gives the belt a shiny appearance (see illustrations). Both sides of the belt should be inspected, which means you will have to twist the belt to check the underside.

3   Check the ribs on the underside of the belt. They should all be the same depth, with none of the surface uneven.

4   On V6 engines the tension of the ribbed drivebelt is maintained by a spring-loaded tensioner assembly and isn't adjustable. On four-cylinder engines, the ribbed belt that drives the cooling fan, alternator and power steering pump is also tensioned by a spring loaded adjuster. The ribbed belt that drives the air conditioning compressor must be adjusted occasionally, by a moveable tensioner pulley. The V-belt that drives the water pump isn't adjustable; if it appears loose, it's time to change it.

## REPLACEMENT

### Ribbed belt

➡**Note: If the belt is being removed for access to other components and will be re-used, mark the direction of rotation on the belt so it can be reinstalled the same way.**

### All except air conditioning drivebelt on four-cylinder engines

♦ **Refer to illustrations 11.6a and 11.6b**

5   If you're working on a V6 engine, remove the engine cover.

6   Rotate the tensioner to release belt tension (see illustrations).

7   Remove the belt from the tensioner and auxiliary components and slowly release the tensioner.

8   Route the new belt over the various pulleys, again rotating the tensioner to allow the belt to be installed, then release the belt tensioner. Install the engine cover.

### Air conditioning drivebelt (four-cylinder models)

♦ **Refer to illustration 11.10**

9   Raise the front of the vehicle and support it securely on jackstands. Remove the under-vehicle splash shield.

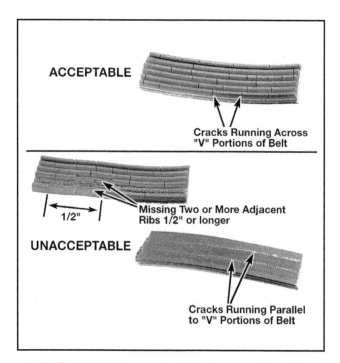

**11.2a  Check ribbed (serpentine) belts for signs of wear like these - if it looks worn, replace it**

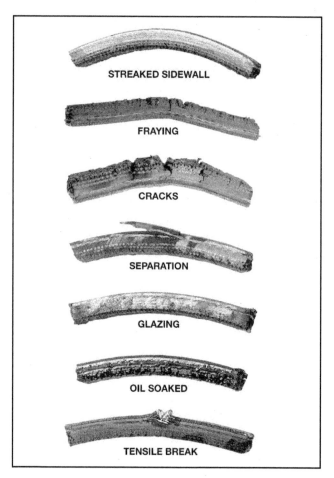

**11.2b  Look for these signs of wear or damage on V-belt drivebelts**

**11.6a  Use a socket and breaker bar to rotate the tensioner - front of vehicle removed for clarity (V6 model shown)**

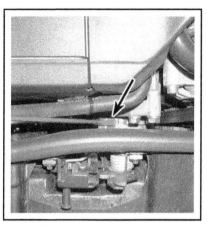

**11.6b  Rotating the drivebelt tensioner (arrow) on a four-cylinder model**

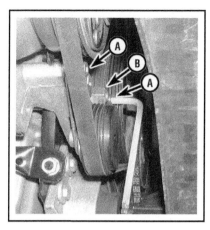

**11.10  Details of the tensioner for the air conditioning compressor drivebelt**

*A    Allen bolt*          *B    Boss for wrench*

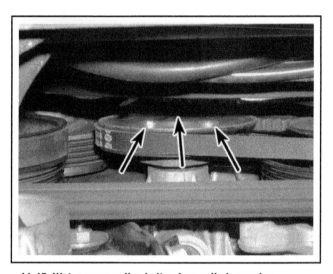

**11.15  Water pump pulley bolts - four-cylinder engine**

**11.17  To remove the tensioner on a V6 engine, remove this bolt; when installing, make sure the lug on the back of the tensioner engages its corresponding hole**

10  Loosen the two Allen bolts securing the tensioner, then rotate the tensioner up and remove the belt from the pulleys (see illustration).

11  Installation is the reverse of removal, making sure the belt seats properly on the pulleys. To tension the belt, rotate it downwards, then lock it in place by tightening the Allen bolts securely. The belt is properly tensioned if it only deflects approximately 3/16-inch (4.7 mm) when a force of 25 pounds is applied between the air conditioning compressor pulley and the tensioner pulley.

### V-belt (four-cylinder models)

▶ **Refer to illustration 11.15**

12  Place the radiator support panel in the service position (see Chapter 11).

13  Remove the serpentine drivebelt as described previously.

14  Remove the engine cooling fan and fan clutch assembly (see Chapter 3).

15  Unscrew the water pump pulley bolts (see illustration). To prevent

the pulley from turning, insert a drift into the power steering pump pulley from the back side. Remove the front half of the water pump pulley, followed by the V-belt.

16  To install the belt, reverse the removal procedure. You may have to rotate the power steering pump and the water pump as the pulley bolts are tightened, to allow the belt to seat properly.

### TENSIONER REPLACEMENT

▶ **Refer to illustration 11.17**

17  To replace a tensioner that doesn't fall into the proper tension range, even with a new belt, or that exhibits binding, remove the two mounting bolts (four-cylinder engine) or the single mounting bolt (V6 engines) (see illustration).

18  Installation is the reverse of the removal procedure. Tighten the mounting bolt(s) to the torque listed in this Chapter's Specifications.

## 12 Underhood hose check and replacement (every 6000 miles [9600 km] or 6 months)

### GENERAL

> ❋❋ **WARNING:**
>
> Replacement of air conditioning hoses must be left to a dealer service department or air conditioning shop that has the equipment to depressurize the system safely and recover the refrigerant. Never remove air conditioning components or hoses until the system has been depressurized.

1  High temperatures in the engine compartment can cause the deterioration of the rubber and plastic hoses used for engine, accessory and emission systems operation. Periodic inspection should be made for cracks, loose clamps, material hardening and leaks. Information specific to the cooling system hoses can be found in Section 13.

2  Some, but not all, hoses are secured to their fittings with clamps. Where clamps are used, check to be sure they haven't lost their tension, allowing the hose to leak. If clamps aren't used, make sure the hose has not expanded and/or hardened where it slips over the fitting, allowing it to leak.

### VACUUM HOSES

3  It's quite common for vacuum hoses, especially those in the emissions system, to be color-coded or identified by colored stripes molded into them. Various systems require hoses with different wall thickness, collapse resistance and temperature resistance. When replacing hoses, be sure the new ones are made of the same material.

4  Often the only effective way to check a hose is to remove it completely from the vehicle. If more than one hose is removed, be sure to label the hoses and fittings to ensure correct installation.

5  When checking vacuum hoses, be sure to include any plastic T-fittings in the check. Inspect the fittings for cracks and the hose where it fits over the fitting for distortion, which could cause leakage.

6  A small piece of vacuum hose (1/4-inch inside diameter) can be used as a stethoscope to detect vacuum leaks. Hold one end of the hose to your ear and probe around vacuum hoses and fittings with the other end, listening for the "hissing" sound characteristic of a vacuum leak.

> ❋❋ **WARNING:**
>
> When probing with the vacuum hose stethoscope, be very careful not to come into contact with moving engine components such as the drivebelt, cooling fan, etc.

### FUEL HOSE

> ❋❋ **WARNING:**
>
> Gasoline is extremely flammable, so take extra precautions when you work on any part of the fuel system. Don't smoke or allow open flames or bare light bulbs near the work area, and don't work in a garage where a gas-type appliance (such as a water heater or clothes dryer) is present. Since gasoline is carcinogenic, wear fuel-resistant gloves when there's a possibility of being exposed to fuel, and, if you spill any fuel on your skin, rinse it off immediately with soap and water. Mop up any spills immediately and do not store fuel-soaked rags where they could ignite. When you perform any kind of work on the fuel system, wear safety glasses and have a Class B type fire extinguisher on hand. The fuel system is under pressure, so if any lines must be disconnected, the pressure in the system must be relieved first (see Chapter 4 for more information).

7  Check all fuel lines for deterioration and chafing. Check especially for cracks in areas where the hose bends and just before fittings, such as where a hose attaches to the fuel filter and fuel rail.

8  High quality fuel line, specifically designed for high-pressure fuel injection applications, must be used for fuel line replacement. Never, under any circumstances, use regular fuel line, unreinforced vacuum line, clear plastic tubing or water hose for fuel lines.

9  Spring-type (pinch) clamps are commonly used on fuel lines. These clamps often lose their tension over a period of time, and can be "sprung" during removal. Replace all spring-type clamps with screw clamps whenever a hose is replaced.

### METAL LINES

10  Sections of metal line may be routed along the frame, between the fuel tank and the engine. Check carefully to be sure the line has not been bent or crimped and no cracks have started in the line.

11  If a section of metal fuel line must be replaced, only seamless steel tubing should be used, since copper and aluminum tubing don't have the strength necessary to withstand normal engine vibration.

12  Check the metal brake lines where they enter the master cylinder and brake proportioning unit for cracks in the lines or loose fittings. Any sign of brake fluid leakage calls for an immediate and thorough inspection of the brake system.

## 13 Cooling system check (every 6000 miles [9600 km] or 6 months)

▶ **Refer to illustration 13.4**

> ❋❋ **CAUTION:**
>
> Never mix green-colored ethylene glycol anti-freeze and red-colored silicate-free coolant because doing so will destroy the efficiency of the red coolant, which could cause engine damage.

1  Many major engine failures can be attributed to a faulty cooling system. If the vehicle is equipped with an automatic transmission, the cooling system also cools the transmission fluid and thus plays an important role in prolonging transmission life.

2  The cooling system should be checked with the engine cold. Do this before the vehicle is driven for the day or after it has been shut off for at least three hours.

3   Remove the pressure cap on the expansion tank by slowly turning it counterclockwise. If you hear any hissing sounds (indicating there is still pressure in the system), wait until it stops. Thoroughly clean the cap, inside and out, with clean water. Also clean the filler neck on the tank. All traces of corrosion should be removed. The coolant inside the tank should be relatively transparent. If it is rust colored, the system should be drained, flushed and refilled (see Section 29). If the coolant level is not up to the top, add additional antifreeze/coolant mixture (see Section 4).

4   Carefully check the large radiator hoses along with any smaller diameter heater hoses that run from the engine to the firewall. Inspect each hose along its entire length, replacing any hose that is cracked, swollen or shows signs of deterioration. Cracks may become more apparent if the hose is squeezed (see illustration).

5   Make sure all hose connections are tight. A leak in the cooling system will usually show up as white or rust-colored deposits on the areas adjoining the leak. If wire-type clamps are used at the ends of the hoses, it may be wise to replace them with more secure, screw-type clamps.

6   Use compressed air or a soft brush to remove bugs, leaves, etc. from the front of the radiator or air conditioning condenser. Be careful not to damage the delicate cooling fins or cut yourself on them.

7   Every other inspection, or at the first indication of cooling system problems, have the cap and system pressure tested. If you don't have a pressure tester, most gas stations and repair shops will do this for a minimal charge.

**Check for a chafed area that could fail prematurely.**

**Check for a soft area indicating the hose has deteriorated inside.**

**13.4  Hoses, like drivebelts, have a habit of failing at the worst possible time - to prevent the inconvenience of a blown radiator or heater hose, inspect them carefully as shown here**

**Overtightening the clamp on a hardened hose will damage the hose and cause a leak.**

**Check each hose for swelling and oil-soaked ends. Cracks and breaks can be located by squeezing the hose.**

## 14   Tire rotation (every 6000 miles [9600 km] or 6 months)

▶ **Refer to illustrations 14.2a and 14.2b**

1   The tires should be rotated at the specified intervals and whenever uneven wear is noticed.

2   Radial tires must be rotated in the recommended pattern (see illustrations).

➡ **Note: Most vehicles are sold with non-directional radial tires, but some replacement performance tires are available that are directional, and have a different rotation pattern. Directional tires have an arrow on the sidewall indicating the direction they must turn when mounted on the vehicle.**

3   Refer to the information in *Jacking and towing* at the front of this manual for the proper procedures to follow when raising the vehicle and changing a tire. If the brakes are to be checked, don't apply the parking brake as stated. Make sure the tires are blocked to prevent the vehicle from rolling as it's raised.

4   Preferably, the entire vehicle should be raised at the same time. This can be done on a hoist or by jacking up each corner and then lowering the vehicle onto jackstands placed under the frame rails. Always use four jackstands and make sure the vehicle is safely supported.

5   After rotation, check and adjust the tire pressures as necessary. Tighten the wheel bolts to the torque listed in this Chapter's Specifications.

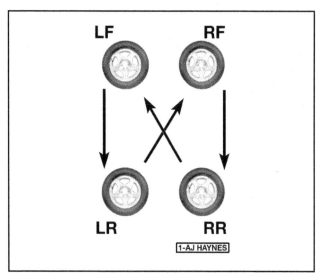

**14.2a The recommended four-tire rotation pattern for non-directional radial tires**

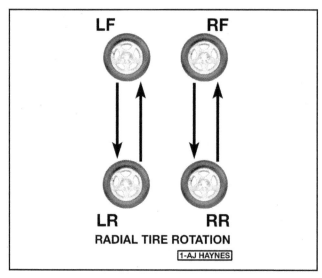

**14.2b The recommended four-tire rotation pattern for directional radial tires**

## 15 Fuel system check (every 15,000 miles [24,000 km] or 12 months)

▶ Refer to illustrations 15.9a and 15.9b

❊❊ **WARNING:**

**Gasoline is extremely flammable, so take extra precautions when you work on any part of the fuel system. Don't smoke or allow open flames or bare light bulbs near the work area, and don't work in a garage where a gas-type appliance (such as a water heater or clothes dryer) is present. Since gasoline is carcinogenic, wear fuel-resistant gloves when there's a possibility of being exposed to fuel, and, if you spill any fuel on your skin, rinse it off immediately with soap and water. Mop up any spills immediately and do not store fuel-soaked rags where they could ignite. When you perform any kind of work on the fuel system, wear safety glasses and have a Class B type fire extinguisher on hand. The fuel system is under constant pressure, so, before any lines are disconnected, the fuel system pressure must be relieved (see Chapter 4).**

1   If you smell fuel while driving or after the vehicle has been sitting in the sun, inspect the fuel system immediately.

2   Remove the fuel filler cap and inspect it
for damage and corrosion. The gasket should have an unbroken sealing imprint. If the gasket is damaged or corroded, install a new cap.

3   Inspect the fuel feed and return lines for cracks. Make sure that the connections between the fuel lines and the fuel injection system are tight.

❊❊ **WARNING:**

**Your vehicle is fuel injected, so you must relieve the fuel system pressure before servicing fuel system components. The fuel system pressure relief procedure is outlined in Chapter 4.**

4   If the fuel injectors are visible, look for signs of fuel leakage (wet

spots) around any of the injectors; they may need new O-rings (see Chapter 4).

5   Since some components of the fuel system - the fuel tank and part of the fuel feed and return lines, for example - are underneath the vehicle, they can be inspected more easily with the vehicle raised on a hoist. If that's not possible, raise the vehicle and support it securely on jackstands.

6   With the vehicle raised and safely supported, inspect the fuel tank and filler neck for punctures, cracks and other damage. The connection between the filler neck and the tank is particularly critical. Inspect all fuel tank mounting brackets and straps to be sure that the tank is securely attached to the vehicle.

❊❊ **WARNING:**

**Do not, under any circumstances, try to repair a fuel tank (except rubber components).**

7   Carefully check all rubber hoses and nylon lines leading away from the fuel tank. Check for loose connections, deteriorated hoses, crimped lines and other damage. Repair or replace damaged sections as necessary (see Chapter 4).

8   The evaporative emissions control system can also be a source of fuel odors. The function of the system is to store fuel vapors from the fuel tank in a charcoal canister until they can be routed to the intake manifold where they mix with incoming air before being burned in the combustion chambers.

9   The most common symptom of a faulty evaporative emissions system is a strong odor of fuel. If a fuel odor has been detected, and you have already checked the areas described above, check the charcoal canister, located under the vehicle below the spare tire well. A cover must be removed for access (see illustrations).

**15.9a  Remove this cover . . .**

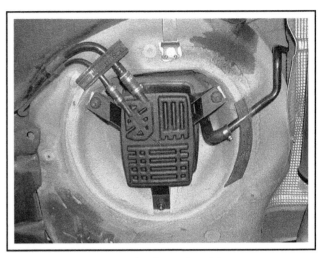

**15.9b  . . . for access to the charcoal canister**

## 16  Brake system check (every 15,000 miles [24,000 km] or 12 months)

▶ **Refer to illustrations 16.7a, 16.7b, 16.9 and 16.11**

### ❊❊ WARNING :

**The dust created by the brake system is harmful to your health. Never blow it out with compressed air and don't inhale any of it. An approved filtering mask should be worn when working on the brakes. Do not, under any circumstances, use petroleum-based solvents to clean brake parts. Use brake system cleaner only! Try to use non-asbestos replacement parts whenever possible.**

➡**Note: For detailed photographs of the brake system, refer to Chapter 9.**

1   In addition to the specified intervals, the brakes should be inspected every time the wheels are removed or whenever a defect is suspected.

2   Any of the following symptoms could indicate a potential brake system defect: The vehicle pulls to one side when the brake pedal is depressed; the brakes make squealing or dragging noises when applied; brake pedal travel is excessive; the pedal pulsates; or brake fluid leaks, usually onto the inside of the tire or wheel.

3   Loosen the wheel bolts.

4   Raise the vehicle and place it securely on jackstands.

5   Remove the wheels (see *Jacking and towing* at the front of this book, or your owner's manual, if necessary).

6   There are two pads (an outer and an inner) in each caliper. The pads are visible with the wheels removed. The vehicles covered by this manual have disc brakes front and rear.

7   Check the pad thickness by looking at each end of the caliper and through the inspection window in the caliper body (see illustrations). If the lining material is less than the thickness listed in this Chapter's Specifications, replace the pads.

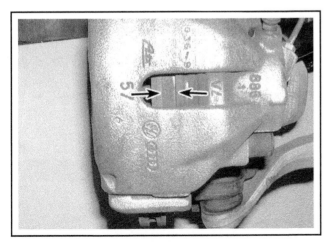

**16.7a  With the wheel off, check the thickness of the inner pad (arrows) through the inspection hole (front caliper shown, rear caliper similar)**

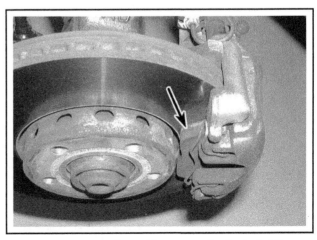

**16.7b  The outer pad (arrow) is more easily checked at the edge of the caliper**

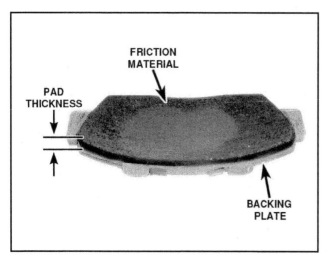

**16.9 If a more precise measurement of pad thickness is necessary, remove the pads and measure the remaining friction material**

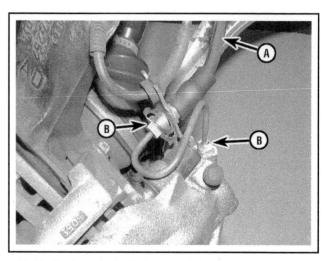

**16.11 Check along the brake hoses (A) and at each fitting (B, caliper end) for deterioration, cracks and leakage**

→Note: Keep in mind that the lining material is riveted or bonded to a metal backing plate and the metal portion is not included in this measurement.

8   If it is difficult to determine the exact thickness of the remaining pad material by the above method, or if you are at all concerned about the condition of the pads, remove the caliper(s), then remove the pads from the calipers for further inspection (refer to Chapter 9).

9   Once the pads are removed from the calipers, clean them with brake cleaner and re-measure them with a ruler or a vernier caliper (see illustration).

10  Measure the disc thickness with a micrometer to make sure that it still has service life remaining. If any disc is thinner than the specified minimum thickness, replace it (refer to Chapter 9). Even if the disc has service life remaining, check its condition. Look for scoring, gouging and burned spots. If these conditions exist, remove the disc and have it resurfaced (see Chapter 9).

11  Before installing the wheels, check all brake lines and hoses for damage, wear, deformation, cracks, corrosion, leakage, bends and twists, particularly in the vicinity of the rubber hoses at the calipers (see illustration). Check the clamps for tightness and the connections for leakage. Make sure that all hoses and lines are clear of sharp edges, moving parts and the exhaust system. If any of the above conditions are noted, repair, reroute or replace the lines and/or fittings as necessary (see Chapter 9).

## BRAKE BOOSTER CHECK

12  Sit in the driver's seat and perform the following sequence of tests.

13  With the brake fully depressed, start the engine - the pedal should move down a little when the engine starts.

14  With the engine running, depress the brake pedal several times - the travel distance should not change.

15  Depress the brake, stop the engine and hold the pedal in for about 30 seconds - the pedal should neither sink nor rise.

16  Restart the engine, run it for about a minute and turn it off. Then firmly depress the brake several times - the pedal travel should decrease with each application.

17  If your brakes do not operate as described, the brake booster has failed. Refer to Chapter 9 for the replacement procedure.

## PARKING BRAKE

18  The parking brake cables operate the rear calipers mechanically, and there is no adjustment needed at the rear brakes unless cables or other parts are being replaced. One method of checking the parking brake is to park the vehicle on a steep hill with the parking brake set and the transmission in Neutral (stay in the vehicle for this check!). If the parking brake cannot prevent the vehicle from rolling, it's in need of adjustment (see Chapter 9).

## 17  Exhaust system check (every 15,000 miles [24,000 km] or 12 months)

♦ Refer to illustration 17.2

1   With the engine cold (at least three hours after the vehicle has been driven), check the complete exhaust system from the manifold to the end of the tailpipe. Be careful around the catalytic converter, which may be hot even after three hours or so. The inspection should be done

with the vehicle on a hoist to permit unrestricted access. If a hoist isn't available, raise the vehicle and support it securely on jackstands.

2   Check the exhaust pipes and connections for signs of leakage and/or corrosion indicating a potential failure. Make sure that all brackets and hangers are in good condition and tight (see illustration).

3   Inspect the underside of the body for holes, corrosion, open seams, etc. which may allow exhaust gasses to enter the passenger compartment. Seal all body openings with silicone sealant or body putty.

4   Rattles and other noises can often be traced to the exhaust system, especially the hangers, mounts and heat shields. Try to move the pipes, mufflers and catalytic converter. If the components can come in contact with the body or suspension parts, secure the exhaust system with new brackets and hangers.

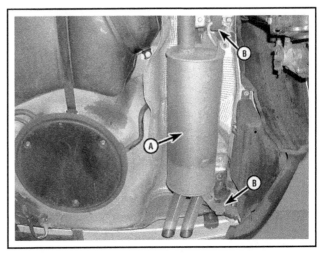

**17.2  Inspect the muffler (A) for signs of deterioration, and all hangers (B)**

## 18  Suspension, steering and driveaxle boot check (every 15,000 miles [24,000 km] or 12 months)

➡**Note: The steering linkage and suspension components should be checked periodically. Worn or damaged suspension and steering components can result in excessive and abnormal tire wear, poor ride quality and vehicle handling and reduced fuel economy. For detailed illustrations of the steering and suspension components, refer to Chapter 10.**

### SHOCK ABSORBER CHECK

▸ **Refer to illustration 18.6**

1   Park the vehicle on level ground, turn the engine off and set the parking brake. Check the tire pressures.

2   Push down at one corner of the vehicle, then release it while noting the movement of the body. It should stop moving and come to rest in a level position within one or two bounces.

3   If the vehicle continues to move up-and-down or if it fails to return to its original position, a worn or weak shock absorber is probably the reason.

4   Repeat the above check at each of the three remaining corners of the vehicle.

5   Raise the vehicle and support it securely on jackstands.

6   Check the shock absorbers for evidence of fluid leakage (see illustration). A light film of fluid is no cause for concern. Make sure that any fluid noted is from the shocks and not from some other source. If leakage is noted, replace the shock absorbers as a set.

7   Check the shocks to be sure that they are securely mounted and undamaged. Check the upper mounts for damage and wear. If damage or wear is noted, replace the shocks as a set (front or rear).

8   If the shocks must be replaced, refer to Chapter 10 for the procedure.

### STEERING AND SUSPENSION CHECK

▸ **Refer to illustrations 18.9a, 18.9b, 18.9c and 18.11**

9   Visually inspect the steering and suspension components (front and rear) for damage and distortion. Look for damaged seals, boots and

**18.6  Check for signs of fluid leakage at this point on the shock absorbers (front shock shown)**

bushings and leaks of any kind. Examine the bushings where the control arms meet the chassis (see illustrations).

10  Clean the lower end of the steering knuckle. Have an assistant grasp the lower edge of the tire and move the wheel in-and-out while you look for movement at each steering knuckle-to-control arm balljoint. If there is any movement the suspension balljoint(s) must be replaced.

11  Grasp each front tire at the front and rear edges, push in at the front, pull out at the rear and feel for play in the steering system components. If any freeplay is noted, check the steering gear mounts and the tie-rod ends for looseness (see illustration).

12  Additional steering and suspension system information and illustrations can be found in Chapter 10.

## DRIVEAXLE BOOT CHECK

▶ **Refer to illustration 18.14**

13  The driveaxle boots are very important because they prevent dirt, water and foreign material from entering and damaging the constant velocity (CV) joints. Oil and grease can cause the boot material to deteriorate prematurely, so it's a good idea to wash the boots with soap and water. Because it constantly pivots back and forth following the steering action of the front hub, the outer CV boot wears out sooner and should be inspected regularly.

.14  Inspect the boots for tears and cracks as well as loose clamps (see illustration). If there is any evidence of cracks or leaking lubricant, they must be replaced as described in Chapter 8.

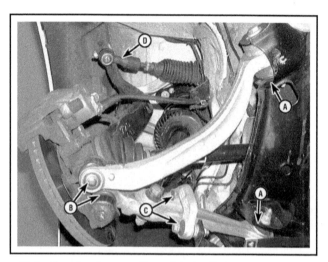

**18.9a  Examine the mounting points for the lower control arms (A) on the front suspension subframe, the lower balljoints (B), the stabilizer link bushings (C) and the tie-rod end (D)**

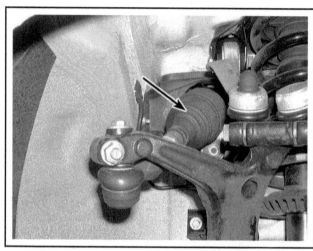

**18.9b  Inspect the steering gear boots (arrow) for signs of cracking or lubricant leakage (which would indicate a seal failure)**

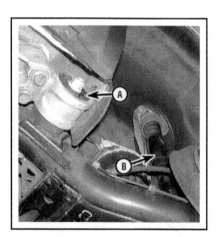

**18.9c  At the rear suspension on 2WD models, check the rear axle pivot bushings (A) and the shock absorbers (B)**

**18.11  With the steering wheel in the locked position and the vehicle raised, grasp the front tire as shown and try to move it back-and-forth - if any play is noted, check the steering gear mounts and tie-rod ends for looseness**

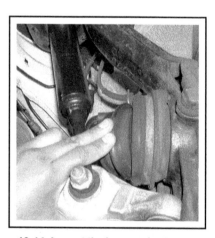

**18.14  Inspect the inner and outer driveaxle boots for loose clamps, cracks or signs of leaking lubricant (inner boot shown)**

## 19 Manual transaxle lubricant level check (every 15,000 miles [24,000 km] or 12 months)

▶ **Refer to illustration 19.2**

1  The manual transaxle has a filler plug which must be removed to check the lubricant level. If the vehicle is raised to gain access to the plug, be sure to support it safely on jackstands - DO NOT crawl under a vehicle that is supported only by a jack! Be sure the vehicle is level or the check may be inaccurate.

2  Using the appropriate wrench, unscrew the check/fill plug from the transaxle (see illustration).

3  Use your little finger to reach inside the housing to feel the lubricant level. The level should be at or near the bottom of the plug hole. If it isn't, add the recommended lubricant through the plug hole with a syringe or squeeze bottle.

4  Install and tighten the plug. Check for leaks after the first few miles of driving.

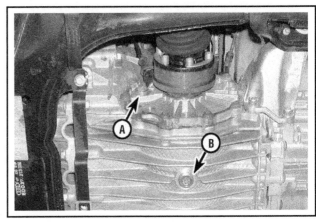

19.2  Location of the manual transaxle check/fill plug (A) - (B) is the transaxle drain plug

## 20 Automatic transaxle fluid level check (every 15,000 miles [24,000 km] or 24 months)

### ✳✳ WARNING:

This procedure is potentially dangerous and is best left to a professional shop with a safe lifting apparatus. The vehicle must be kept level while being safely raised high enough for access to the check plug on the transaxle.

➡**Note 1:** When the work is done at a dealership, the factory scan tool is used to read the transmission fluid temperature. To perform the job at home, you may need a cooking or darkroom thermometer that has a long probe, which you can insert in the level check hole to read the fluid temperature.

➡**Note 2:** The engine must be running when checking the fluid level and when adding fluid.

### CHECK

▶ **Refer to illustration 20.3**

1  Correct automatic transaxle fluid level is extremely important for proper transaxle operation. Low fluid level can lead to slipping or loss of drive, while overfilling can cause foaming and loss of fluid.

➡**Note:** Checking the fluid on these vehicles isn't easy. The transaxle is considered by the manufacturer to be a "sealed" unit, to which fluid doesn't need to be added unless a leak is evident. There is no conventional dipstick in the engine compartment, but rather a level inspection plug accessible only from below the transaxle. Make sure you have a new seal for the plug before you begin. If fluid does have to be added to an 01N transaxle, a new filler cap seal will be required, too.

2  Transaxle fluid expands as it warms up, and the fluid check should only be performed at the specified temperature range of 95 to 113-degrees F (35 to 45-degrees C).

➡**Note:** The manufacturer states that a scan tool must be used to monitor the temperature of the transaxle fluid. While this is certainly the best way, you can get a fairly accurate reading

after driving the vehicle a short distance (starting with the drivetrain cold) - approximately one trip around the block or so. The transaxle pan should feel warm to the touch, but not hot enough to cause pain.

3  Raise the vehicle; if you don't have access to a vehicle hoist, support the vehicle securely on four jackstands (the fluid level must be checked with the vehicle level). Start the engine, then move the shift lever through all the gear ranges, ending in Park. When the transaxle temperature reaches 95 to 113-degrees F (35 to 45-degrees C), remove the level check plug (see illustration). If fluid just comes out of the hole as drips, the fluid level is OK. If more fluid comes out, the transmission may have been overfilled. If no fluid comes out at 113-degrees F, fluid will have to be added.

### ADDING FLUID

#### Models with an 01N transaxle

4  On models with an 01N transaxle, pry off the cap that secures the

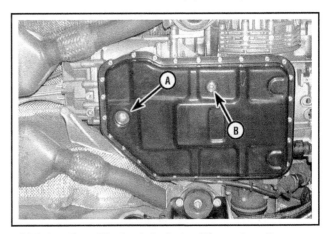

20.3  Location of the level check plug (A) on automatic transaxles; (B) is the drain plug

fill plug in its tube, then remove the plug (the short filler tube protrudes from the left front corner of the transaxle fluid pan). Add fluid slowly until it begins to drip from the check hole, then install the check plug and fill plug. Secure the fill plug with a new cap. Tighten the check plug to the torque listed in this Chapter's Specifications.

### Models with an 01V transaxle

♦ **Refer to illustration 20.5**

5   On these transaxles, the fluid is added through the check hole. A special tool is available for this purpose, but a suction gun equipped with an angled nozzle can be used to pump the fluid up into the pan. The nozzle must pass through the opening in the deflector cap (see illustration). Be careful not to knock the cap out of position with the nozzle.

6   Slowly add fluid until it begins to drip from the check hole, then install a new seal on the plug. Install the plug and tighten it to the torque listed in this Chapter's Specifications.

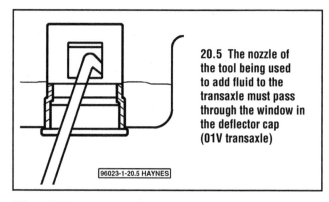

20.5  The nozzle of the tool being used to add fluid to the transaxle must pass through the window in the deflector cap (01V transaxle)

### All models

7   The condition of the fluid should also be checked along with the level. If the fluid is a dark reddish-brown color, or if it smells burned, it should be changed. If you are in doubt about the condition of the fluid, purchase some new fluid and compare the two for color and smell.

## 21   Front differential lubricant level check (automatic transaxles) (every 15,000 miles [24,000 km] or 12 months)

♦ **Refer to illustration 21.2**

➡**Note: On manual transaxle models, the differential lubricant is not separated from the transmission lubricant; filling the transmission fills the differential, too (see Section 19). The differential has a separate filler, and different lubricant, on automatic transaxle models.**

1   Loosen the right front wheel lug bolts. Raise the vehicle and support it securely on jackstands. Remove the right front wheel.

2   Using the appropriate wrench, unscrew the plug from the transaxle (see illustration).

3   Use your little finger to reach inside the housing to feel the lubricant level. The level should be at or near the bottom of the plug hole. If it isn't, add the recommended lubricant through the plug hole with a syringe or squeeze bottle.

4   Install and tighten the plug. Check for leaks after the first few miles of driving.

21.2  On models with an automatic transaxle, the check/fill plug for the automatic transaxle differential is located on right side of the transaxle, just in front of the driveaxle

## 22   Center differential lubricant level check (all-wheel drive models with automatic transaxles) (every 15,000 miles [24,000 km] or 12 months)

♦ **Refer to illustration 22.2**

➡**Note: On manual transaxle models, the center differential lubricant is not separated from the transmission lubricant; filling the transmission fills the center differential, too (see Section 19).**

1   Raise the vehicle and support it securely on jackstands.

2   Using the appropriate wrench, unscrew the plug from the center differential (see illustration).

3   Use your little finger to reach inside the housing to feel the lubricant level. The level should be at or near the bottom of the plug hole. If it isn't, add the recommended lubricant through the plug hole with a syringe or squeeze bottle.

4   Install and tighten the plug. Check for leaks after the first few miles of driving.

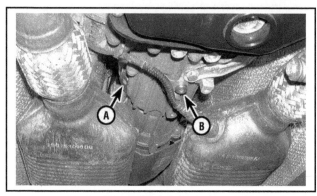

22.2  Location of the check/fill plug for the center differential (A) - B is the drain plug (all-wheel drive models with automatic transaxles)

## 23  Rear differential lubricant level check (all-wheel drive models) (every 15,000 miles [24,000 km] or 12 months)

▶ **Refer to illustration 23.2**

1  Raise the vehicle and support it securely on jackstands.

2  Using the appropriate wrench, unscrew the plug from the rear differential (see illustration).

3  Use your little finger to reach inside the housing to feel the lubricant level. The level should be at or near the bottom of the plug hole. If it isn't, add the recommended lubricant through the plug hole with a syringe or squeeze bottle.

4  Install and tighten the plug. Check for leaks after the first few miles of driving.

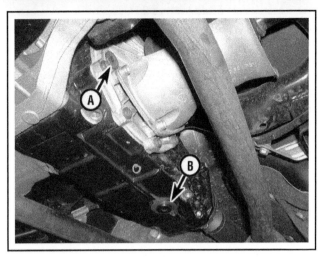

**23.2  Location of the rear differential check/fill plug (A) - B is the drain plug (all-wheel drive models)**

## 24  Interior ventilation filter replacement (every 15,000 miles [24,000 km] or 12 months)

▶ **Refer to illustrations 24.2a, 24.2b and 24.2c**

1  The models covered by this manual are equipped with an air filter in a housing in the cowl area that cleans the air before it enters the passenger compartment.

2  Remove the screw, lift off the cover and pull the filter out (see illustrations).

3  Installation is the reverse of the removal procedure. Be sure to install the filter with the airflow arrows on the side of the filter pointing down.

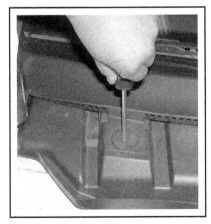

**24.2a  Remove the screw . . .**

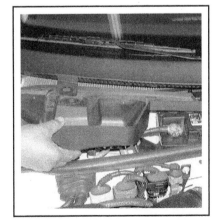

**24.2b  . . . lift off the cover . . .**

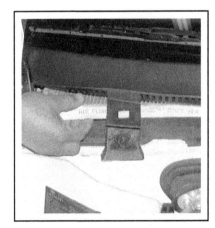

**24.2c  . . . and take out the interior ventilation filter**

## 25  Automatic transaxle fluid change (every 30,000 miles [48,000 km] or 24 months)

### ✻✻ WARNING:

♦ **Refer to the Warning and Notes in Section 20 before performing this procedure.**

1   At the specified intervals, the transmission fluid should be drained and replaced. Since the fluid will remain hot long after driving, perform this procedure only after the engine has cooled down completely.

2   Before beginning work, purchase the specified transmission fluid (see *Recommended lubricants and fluids* at the end of this Chapter).

3   Raise the vehicle and support it securely on jackstands.

4   Place the drain pan underneath the transmission pan.

5   Unscrew the drain plug and allow the fluid to drain (see illustration 20.3).

6   Install a new seal on the drain plug, then install the plug and tighten it to the torque listed in this Chapter's Specifications.

7   If you're working on a model with an 01N transaxle, add approximately 3 quarts of the specified type of automatic transmission fluid through the filler tube (see Section 20). If you're working on a model with an 01V transaxle, add fluid through the check/fill hole until it runs out (see Section 20).

8   Start the engine and allow it to run for a few minutes.

9   With the transmission in Park and the parking brake set, run the engine at a fast idle, but don't race it.

10  Move the gear selector through each range and back to Park, then let the engine idle for a few minutes. Check the fluid level. It will probably be low. Add enough fluid to bring the level up until it just drips out of the level check hole.

➡**Note: The fluid level must be checked at a certain temperature (see Section 20).**

11  Check under the vehicle for leaks during the first few trips.

## 26  Air filter replacement (every 30,000 miles [48,000 km] or 24 months)

♦ **Refer to illustrations 26.3a, 26.3b, 26.4a and 26.4b**

1   At the specified intervals, the air filter element should be replaced with a new one.

2   On all models, the air filter is housed in a black plastic box mounted on the inner fenderwell on the right side of the engine compartment.

3   Remove the trim cover from above the air filter housing, then detach the air intake duct from the housing (see illustrations).

4   Disengage the clips and pull the air filter housing cover up, then lift the air filter element out of the housing (see illustrations). Wipe out the inside of the air filter housing with a clean rag.

➡**Note: Some models are equipped with a strainer at the bottom of the filter housing. If your vehicle has this strainer, remove the screw, lift out the strainer and clean it with compressed air.**

5   While the cover is off, be careful not to drop anything down into the air filter housing.

6   Place the new filter element in the air filter housing. Make sure it seats properly in the groove of the housing.

7   Installation is the reverse of removal.

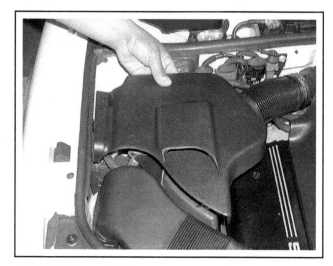

**26.3a  Pull up on the trim cover to remove it . . .**

**26.3b  . . . then detach the intake duct from the air filter housing**

**26.4a  Disengage the four clips . . .**

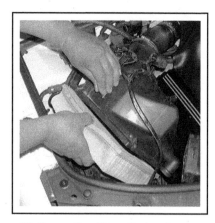

**26.4b  . . . then lift up the housing and remove the filter element**

## 27 Spark plug replacement (every 30,000 miles [48,000 km] or 24 months)

♦ **Refer to illustrations 27.1, 27.4a, 27.4b, 27.7, 27.8a, 27.8b and 27.9**

1   In most cases, the tools necessary for spark plug replacement include a spark plug socket which fits onto a ratchet (spark plug sockets are padded inside to prevent damage to the porcelain insulators on the new plugs), various extensions and a gap gauge to check and adjust the gaps on the new plugs (see illustration). A special plug boot removal tool is available for separating the wire boots from the spark plugs. A torque wrench should be used to tighten the new plugs.

2   The best approach when replacing the spark plugs is to purchase the new ones in advance, adjust them to the proper gap and replace them one at a time. When buying the new spark plugs, be sure to obtain the correct plug type for your particular engine. This information can be found in the Specifications at the end of this Chapter.

3   Allow the engine to cool completely before attempting to remove any of the plugs. Remove the plastic cover at the top of the engine for access. While you're waiting for the engine to cool, check the new plugs for defects and adjust the gaps.

4   The gap is checked by inserting the proper thickness gauge between the electrodes at the tip of the plug (see illustration). The gap between the electrodes should be the same as the one specified in this Chapter's Specifications. The gauge should just slide between the electrodes. If the gap is incorrect, use the adjuster on the gauge body to bend the curved side electrode slightly until the proper gap is obtained (see illustration). If the side electrode is not exactly over the center electrode, bend it with the adjuster until it is. Check for cracks in the porcelain insulator (if any are found, the plug should not be used).

➡**Note: Manufacturers recommend using a tapered thickness gauge when checking platinum-type spark plugs. Other types of gauges may scrape the thin platinum coating from the electrodes, thus dramatically shortening the life of the plugs.**

5   With the engine cool, remove the spark plug wire from one spark plug (V6 engines) or the ignition coil from over the spark plug (four-cylinder engines - see Chapter 5). When removing spark plug wires, pull only on the boot at the end of the wire - do not pull on the wire.

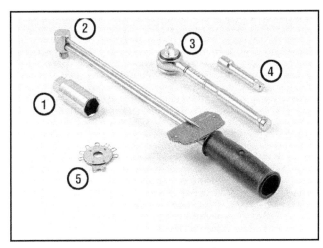

**27.1  Tools required for changing spark plugs**

*1*   ***Spark plug socket*** *- This will have special padding inside to protect the spark plug's porcelain insulator*
*2*   ***Torque wrench*** *- Although not mandatory, using this tool is the best way to ensure the plugs are tightened properly*
*3*   ***Ratchet*** *- Standard hand tool to fit the spark plug socket*
*4*   ***Extension*** *- Depending on model and accessories, you may need special extensions and universal joints to reach one or more of the plugs*
*5*   ***Spark plug gap gauge*** *- This gauge for checking the gap comes in a variety of styles. Make sure the gap for your engine is included*

6   If compressed air is available, use it to blow any dirt or foreign material away from the spark plug hole. The idea here is to eliminate the possibility of debris falling into the cylinder as the spark plug is removed.

7   Place the spark plug socket over the plug and remove it from the engine by turning it in a counterclockwise direction (see illustration).

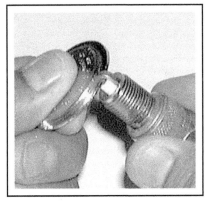

**27.4a  Spark plug manufacturers recommend using a tapered thickness gauge when checking the gap - slide the thin side into the gap and turn until the gauge just fills the gap, then read the thickness on the gauge - do not force the tool into the gap or use the tapered portion to widen a gap**

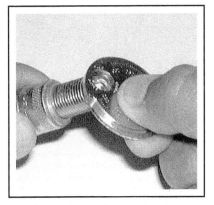

**27.4b  To change the gap, bend the side electrode only, using the adjuster hole in the tool, and be very careful not to crack or chip the porcelain insulator surrounding the center electrode**

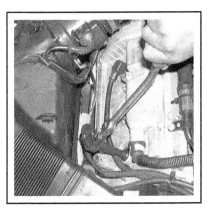

**27.7  Use a socket and extension to unscrew the spark plugs**

A normally worn spark plug should have light tan or gray deposits on the firing tip.

A carbon fouled plug, identified by soft, sooty, black deposits, may indicate an improperly tuned vehicle. Check the air cleaner, ignition components and engine control system.

An oil fouled spark plug indicates an engine with worn piston rings and/or bad valve seals allowing excessive oil to enter the chamber.

This spark plug has been **left in the engine too long,** as evidenced by the extreme gap. Plugs with such an extreme gap can cause misfiring and stumbling accompanied by a noticeable lack of power.

A physically damaged spark plug may be evidence of severe detonation in that cylinder. Watch that cylinder carefully between services, as a continued detonation will not only damage the plug, but could also damage the engine.

A bridged or almost bridged spark plug, identified by a build-up between the electrodes caused by excessive carbon or oil build-up on the plug.

**27.8a Inspect the spark plug to determine engine running conditions**

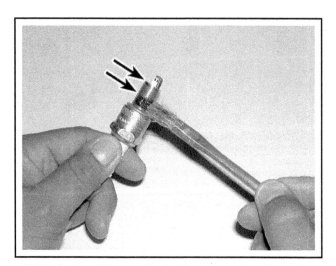

**27.8b Apply a thin coat of anti-seize compound to the spark plug threads, being careful not to get any near the lower threads (arrows)**

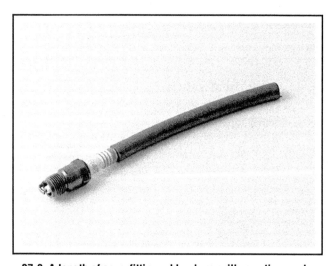

**27.9 A length of snug-fitting rubber hose will save time and prevent damaged threads when installing the spark plugs**

➡**Note: On V6 engines, you'll have to remove the upper half of the air filter housing (see Section 26) to remove the plug from cylinder number 3, and the coolant reservoir (see Section 5 in Chapter 3) to remove the plugs for cylinder numbers 5 and 6.**

8   Compare the spark plug with the chart (see illustration) to get an indication of the general running condition of the engine. Before installing the new plugs, it is a good idea to apply a thin coat of anti-seize compound to the threads (see illustration). Don't get any on the bottom threads or it could run (during hot engine operation) down onto the porcelain or electrodes and potentially ruin the plug.

9   Thread one of the new plugs into the hole until you can no longer

turn it with your fingers, then tighten it with a torque wrench (if available) or the ratchet. Its a good idea to slip a short length of rubber hose over the end of the plug to use as a tool to thread it into place (see illustration). The hose will grip the plug well enough to turn it, but will start to slip if the plug begins to cross-thread in the hole - this will prevent damaged threads and the accompanying repair costs.

10 Attach the plug wire or ignition coil to the new spark plug, again using a twisting motion on the boot until it's seated on the spark plug.

11 Repeat the procedure for the remaining spark plugs, replacing them one at a time to prevent mixing up the spark plug wires.

## 28 Spark plug wire check and replacement (V6 engines) (every 30,000 miles [48,000 km] or 24 months)

♦ **Refer to illustration 28.6**

1   The spark plug wires should be checked at the recommended intervals and whenever new spark plugs are installed in the engine.

2   Remove the engine cover (see Chapter 2B). Begin this procedure by making a visual check of the spark plug wires while the engine is running. In a darkened garage (make sure there is adequate ventilation) start the engine and observe each plug wire. Be careful not to come into contact with any moving engine parts. If there is a break in the wire, you will see arcing or a small spark at the damaged area.

3   Disconnect the plug wire from one spark plug (with the engine Off). To do this, grab the rubber boot, twist slightly and pull the wire free. Do not pull on the wire itself, only on the rubber boot.

4   Check inside the boot for corrosion, which will look like a white crusty powder. Push the wire and boot back onto the end of the spark plug. It should be a tight fit on the plug. If it isn't, remove the wire and use a pair of pliers to carefully crimp the metal connector inside the boot until it fits securely on the end of the spark plug.

5   Using a clean rag, wipe the entire length of the wire to remove any built-up dirt and grease. Once the wire is clean, check for holes, burned areas, cracks and other damage. Don't bend the wire excessively or the conductor inside might break.

6   Disconnect the wire from the coil pack. Pull the wire straight off the coil - a boot pulling tool is helpful (see illustration). Pull only on the rubber boot during removal. Check for corrosion and a tight fit in the same manner as the spark plug end. Reattach the wire to the coil.

7   Check the remaining spark plug wires one at a time, making sure they are securely fastened at both ends when the check is complete.

8   If new spark plug wires are required, purchase a new set for your specific engine model. Wire sets are available pre-cut, with the rubber boots already installed. Remove and replace the wires one at a time to avoid mix-ups in the firing order. The wire routing is extremely important, so be sure to note exactly how each wire is situated before removing it. Compare your old ones to the new ones to insure obtaining the correct replacements. Consult the cylinder and coil numbering illustration at the end of this Chapter for the correct routing of the wires.

**28.6  When detaching the spark plug wires from the ignition coil pack, pull only on the boots**

## 29 Cooling system servicing (draining, flushing and refilling) (every 30,000 miles [48,000 km] or 24 months)

### ✳✳ WARNING:

**Do not allow antifreeze to come in contact with your skin or painted surfaces of the vehicle. Rinse off spills immediately with plenty of water. Antifreeze is highly toxic if ingested. Never leave antifreeze lying around in an open container or in puddles on the floor; children and pets are attracted by its sweet smell and may drink it. Check with local authorities on disposing of used anti-freeze. Many communities have collection centers that will see that antifreeze is disposed of safely.**

➡**Note: Non-toxic antifreeze is now manufactured and available at local auto parts stores, but even this type should be disposed of properly.**

### ✳✳ CAUTION:

**Never mix green-colored ethylene glycol anti-freeze and red-colored silicate-free coolant because doing so will destroy the efficiency of the red coolant, which could cause engine damage.**

### DRAINING

♦ **Refer to illustrations 29.3, 29.4a and 29.4b**

1   Periodically, the cooling system should be drained, flushed and refilled to replenish the antifreeze mixture and prevent formation of rust and corrosion, which can impair the performance of the cooling system and cause engine damage. When the cooling system is serviced, all hoses and the expansion tank cap should be checked and replaced if necessary.

2   Apply the parking brake and block the wheels. Raise the front of the vehicle and support it securely on jackstands, then remove the under-vehicle splash shield (see illustration 7.7).

### ✳✳ WARNING:

**If the vehicle has just been driven, wait several hours to allow the engine to cool down before beginning this procedure.**

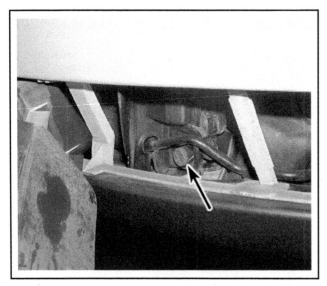

**29.3 The radiator drain valve (arrow) can be accessed by removing the grille panel from the lower left side of the bumper cover**

**29.4a Location of the water pump drain plug (four-cylinder engine)**

3  Move a large container under the radiator drain to catch the coolant. The coolant can be drained either by detaching the lower radiator hose from the radiator or by removing the grille panel from the lower left (driver's) side of the bumper cover, attaching a hose to the radiator drain valve, then turning the knob on the valve (see illustration). Remove the cap from the coolant expansion tank and allow the coolant to drain.

4  After coolant stops flowing out of the radiator, move the container under the engine. If you're working on a model with a four-cylinder engine, remove the drain plug from the water pump (see illustration). If you're working on a model with a V6 engine, remove the drain plug from the bottom of the engine block, to the rear of the subframe crossmember (see illustration). When the coolant flow ceases, install a new O-ring on the drain plug, reinstall the drain plug and tighten it to the torque listed in this Chapter's Specifications.

5  While the coolant is draining, check the condition of the radiator hoses, heater hoses and clamps (refer to Section 13 if necessary).

6  Replace any damaged clamps or hoses.

## FLUSHING

7  Fill the cooling system with clean water, following the *Refilling* procedure (see Step 13).

8  Start the engine and allow it to reach normal operating temperature, then rev up the engine a few times.

9  Turn the engine off and allow it to cool completely, then drain the system as described earlier.

10  Repeat Steps 7 through 9 until the water being drained is free of contaminants.

11  In severe cases of contamination or clogging of the radiator, remove the radiator (see Chapter 3) and have a radiator repair facility clean and repair it if necessary.

12  Many deposits can be removed by the chemical action of a cleaner available at auto parts stores. Follow the procedure outlined in the manufacturer's instructions.

**29.4b Location of the engine block drain plug (V6 engines)**

➡Note: When the coolant is regularly drained and the system refilled with the correct antifreeze/water mixture, there should be no need to use chemical cleaners or descalers.

## REFILLING

♦ Refer to illustrations 29.15, 29.17a and 29.17b

13  Reinstall the under-vehicle splash shield, then lower the vehicle. Place the heater temperature control in the maximum heat position. If you're working on a V6 model, remove the engine cover.

14  Remove the screws and raise the expansion tank approximately 4-inches (100 mm). Support it in this position.

15  Remove the cover from the heater hoses at the firewall. Loosen

**29.15 Loosen the clamp and slide the right-side heater hose off the heater core pipe until this small hole is exposed**

**29.17a The bleeder screw in the rear coolant pipe is located at the rear of the left cylinder head, just inboard of the expansion tank (V6 engines)**

17 If you're working on a V6 engine, loosen the bleeder screw on the coolant pipe at the rear of the left (driver's side) cylinder head (see illustration). Add coolant to the expansion tank until coolant flows from this bleeder, then tighten the bleeder screw to the torque listed in this Chapter's Specifications. Now loosen the bleeder screw at the front coolant pipe and add coolant until it flows from this bleeder (see illustration), then tighten the bleeder screw to the torque listed in this Chapter's Specifications.

18 Fill the expansion tank with the recommended coolant mixture up to the MAX mark, then install the cap.

19 Start the engine and allow it to run for ten minutes at idle.

20 Raise the engine speed to 2000 rpm for five minutes, then return it to idle. Check the lower radiator hose to make sure it's hot, indicating that the engine is at normal operating temperature.

21 Check the coolant level in the expansion tank. When the engine is at normal operating temperature the level must be at the MAX mark. If it isn't, allow the engine to cool, then slowly unscrew the expansion tank cap and add coolant.

**29.17b Location of the front coolant pipe bleeder screw**

the hose clamp and slide the right-side heater hose (at the heater core) off the heater core pipe just until the small hole in the hose is exposed (see illustration).

16 Slowly fill the expansion tank with the recommended mixture of antifreeze and water until coolant flows from the small hole, then slide the hose back into place and secure it with the clamp. Be sure to use the proper coolant listed in this Chapter's Specifications.

### ✻✻ WARNING:

**If you hear a hissing sound when unscrewing the expansion tank cap, stop unscrewing the cap and let the pressure in the system dissipate.**

➡**Note: When the engine is cold, the level of coolant in the expansion tank must be between the MIN and MAX marks.**

## 30 Brake fluid change (every two years, regardless of mileage)

### ✻✻ WARNING:

**Brake fluid can harm your eyes and damage painted surfaces, so use extreme caution when handling or pouring it. Do not use brake fluid that has been standing open or is more than one year old. Brake fluid absorbs moisture from the air. Excess moisture can cause a dangerous loss of braking effectiveness.**

1 At the specified intervals, the brake fluid should be drained and replaced. Since the brake fluid may drip or splash when pouring it, place plenty of rags around the master cylinder to protect any surrounding painted surfaces.

2 Before beginning work, purchase the specified brake fluid (see *Recommended lubricants and fluids* at the end of this Chapter).

3 Remove the cap from the master cylinder reservoir.

4   Using a hand suction pump or similar device, withdraw the fluid from the master cylinder reservoir.

5   Add new fluid to the master cylinder until it rises to the line indicated on the reservoir.

6   Bleed the brake system as described in Chapter 9 at all four brakes until new and uncontaminated fluid is expelled from each bleeder screw. Be sure to maintain the fluid level in the master cylinder as you perform the bleeding process. If you allow the master cylinder to run dry, air will enter the system.

7   Refill the master cylinder with fluid and check the operation of the brakes. The pedal should feel solid when depressed, with no sponginess.

### ✳✳ WARNING:

**Do not operate the vehicle if you are in doubt about the effectiveness of the brake system.**

### 31   Fuel filter replacement (every 60,000 miles [96,000 km] or 48 months)

♦ Refer to illustrations 31.1a, 31.1b and 31.1c

### ✳✳ WARNING:

Gasoline is extremely flammable, so take extra precautions when you work on any part of the fuel system. Don't smoke or allow open flames or bare light bulbs near the work area, and don't work in a garage where a gas-type appliance (such as a water heater or clothes dryer) is present. Since gasoline is carcinogenic, wear fuel-resistant gloves when there's a possibility of being exposed to fuel, and, if you spill any fuel on your skin, rinse it off immediately with soap and water. Mop up any spills immediately and do not store fuel-soaked rags where they could ignite. When you perform any kind of work on the fuel system, wear safety glasses and have a Class B type fire extinguisher on hand. The fuel system is under pressure, so if any lines must be disconnected, the pressure in the system must be relieved first (see Chapter 4 for more information).

### ✳✳ CAUTION 1:

These models are equipped with an anti-theft radio. Before performing a procedure that requires disconnecting the battery, make sure you have the proper activation code.

### ✳✳ CAUTION 2:

Disconnecting the battery can cause driveability problems that require a scan tool to rectify. Additionally, disconnecting the battery may cause one or more warning lights on the instrument panel to illuminate, which will also require the use of a scan tool to turn off. Most scan tools available to the public do not have the capability to perform either of these tasks, which will necessitate taking the vehicle to a dealer service department or other properly equipped repair facility after service work has been performed. See Chapter 5, Section 1 for the use of an auxiliary voltage input device ("memory saver") before disconnecting the battery and for other precautions related to battery disconnection.

1   The fuel filter is located under the rear of the vehicle, ahead of the right-rear wheel or adjacent to the fuel tank (see illustrations).

2   The manufacturer does not give a replacement interval, but our suggested interval is based on experience with many other vehicles; replacing it is like inexpensive insurance and may prevent an untimely breakdown.

3   Depressurize the fuel system (see Chapter 4), then disconnect the cable from the negative terminal of the battery. Raise the vehicle and support it securely on jackstands.

4   Use compressed air or brake system cleaner to clean any dirt surrounding the fuel inlet and outlet line fittings.

➡ Note: On some models it is necessary to remove a plastic splash shield for access.

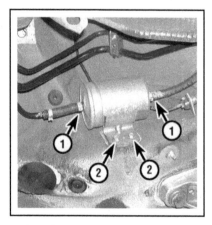

**31.1a  Fuel filter mounting details - 2WD Passat**

*1   Fuel hose clamps*
*2   Filter mounting bolts*

**31.1b  Fuel filter mounting details - all-wheel drive Passat wagon**

*1   Banjo bolt*     *2   Clamp bolt*

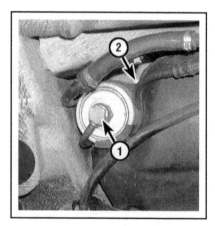

**31.1c  Fuel filter mounting details - all-wheel drive A4**

*1   Banjo bolt*     *2   Clamp bolt*

5   On some models, the fuel lines are secured to the fuel filter by crimp-type hose clamps; cut off the clamps and discard them, then unbolt the mounting bracket and remove the filter.

6   On other models, banjo fittings are used. On Passat wagons both fittings are accessible with the filter in place. On A4 models, unscrew the lower banjo bolt, remove the clamp bolt, slide the filter down out of its bracket then remove the upper banjo bolt.

➡ **Note: Have spare rags or a small container to catch or wipe up extra gasoline that will spill from the filter.**

7   Installation is the reverse of removal. On models with banjo fittings, use new sealing washers and tighten the banjo bolts to the torque values listed in this Chapter's Specifications. Make sure the arrow on the side of the new filter is pointing toward the engine side of the fuel system, and check for leaks after running the vehicle.

## 32  Manual transaxle lubricant change (every 60,000 miles [96,000 km] or 48 months)

1   This procedure should be performed after the vehicle has been driven so the lubricant will be warm and therefore will flow out of the transmission more easily. Raise the vehicle and support it securely on jackstands.

2   Move a drain pan, rags, newspapers and wrenches under the transaxle.

3   Remove the transaxle drain plug and allow the lubricant to drain into the pan (see illustration 19.2).

4   After the lubricant has drained completely, reinstall the plug and tighten it securely.

5   Remove the fill plug from the side of the transmission case (see Section 19). Using a hand pump, syringe or squeeze bottle, fill the transmission with the specified lubricant until it just reaches the bottom edge of the hole. Reinstall the fill plug and tighten it securely.

6   Lower the vehicle.

7   Drive the vehicle for a short distance, then check the drain and fill plugs for leakage.

## 33  Center differential lubricant change, (all-wheel drive models with automatic transaxles) (every 60,000 miles [96,000 km] or 48 months)

1   This procedure should be performed after the vehicle has been driven, so the lubricant will be warm and therefore will flow out of the differential more easily.

2   Raise the vehicle and support it securely on jackstands. Move a drain pan, rags, newspapers and wrenches under the vehicle.

3   Remove the drain plug and allow the lubricant to drain into the pan (see illustration 22.2).

4   Install the drain plug and tighten it to the torque listed in this Chapter's Specifications.

5   Remove the check/fill plug. Use a hand pump, syringe or squeeze bottle to fill the center differential housing with the specified lubricant, through the fill plug hole (see illustration 22.2).

6   When the level is correct, reinstall the check/fill plug and tighten it to the torque listed in this Chapter's Specifications.

## 34  Rear differential lubricant change (all-wheel drive models) (every 60,000 miles [96,000 km] or 48 months)

1   This procedure should be performed after the vehicle has been driven, so the lubricant will be warm and therefore will flow out of the differential more easily.

2   Raise the vehicle and support it securely on jackstands. Move a drain pan, rags, newspapers and wrenches under the vehicle.

3   Remove the drain plug and allow the lubricant to drain into the pan (see illustration 23.2).

4   Install the drain plug and tighten it to the torque listed in this Chapter's Specifications.

5   Remove the check/fill plug. Use a hand pump, syringe or squeeze bottle to fill the rear differential housing with the specified lubricant, through the fill plug hole (see illustration 23.2).

6   When the level is correct, reinstall the check/fill plug and tighten it to the torque listed in this Chapter's Specifications.

## 35  Service reminder indicator resetting (A4 models)

The service reminder indicator illuminates at specified maintenance intervals to notify the driver that it is time for certain maintenance procedures to be performed. These intervals are the ones set by the manufacturer, and don't necessary correlate to the maintenance intervals found at the beginning of this Chapter.

Unfortunately, there is no way to reset this indicator without a very expensive scan tool. This tool is generally only available to dealer service departments and other qualified repair shops. Take the vehicle to a dealer service department or other repair facility equipped with the necessary tool.

## Specifications

### Recommended lubricants and fluids

➡Note: Listed here are manufacturer recommendations at the time this manual was written. Manufacturers occasionally upgrade their fluid and lubricant specifications, so check with your local auto parts store for current recommendations.

| | |
|---|---|
| Engine oil | API "certified for gasoline engines" |
| Viscosity | See accompanying chart |
| Fuel | Unleaded gasoline, (91) octane minimum |
| Automatic transaxle fluid | VW automatic transmission fluid or ESSO EGL 71141 only |
| Manual transaxle lubricant | SAE 75W90 synthetic gear oil |
| Differential, front (automatic transaxles) | SAE 75W90 synthetic gear oil |
| Differential, rear (all-wheel drive models) | API GL5, SAE 90 synthetic gear oil |
| Differential, center (all-wheel drive models with automatic transaxle) | API GL5, SAE 75W90 synthetic gear oil |
| Power steering fluid | Hydraulic mineral fluid |
| Brake fluid | DOT 4 brake fluid |
| Engine coolant | |
|     Vehicle manufacture date of 6/96 or earlier | 50/50 mixture of G 011 (green) antifreeze and demineralized water |
|     Vehicle manufacture date of 7/96 or later | 50/50 mixture of G 012 (red) antifreeze and demineralized water |
| Hood and trunk hinge lubricant | Lubriplate, lubricant aerosol spray |
| Door hinge and check spring grease | NLGI no. 2 multi-purpose grease |
| Key lock cylinder lubricant | Graphite spray |
| Hood latch assembly lubricant | NLGI no. 2 multi-purpose grease |
| Door latch lubricant | NLGI no. 2 multi-purpose grease or equivalent |

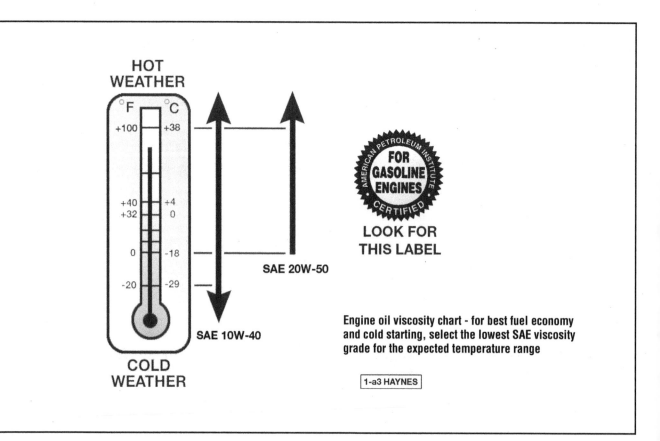

**Engine oil viscosity chart - for best fuel economy and cold starting, select the lowest SAE viscosity grade for the expected temperature range**

1-a3 HAYNES

## Capacities*

Engine oil (including filter)

    Four-cylinder engine                          4.1 quarts (3.9 liters)

    V6 engine

        A4                                5.5 quarts (5.2 liters)

        Passat

            1998                      6.0 quarts (5.7 liters)

            1999                      6.3 quarts (6.0 liters)

            2000 and later          6.5 quarts (6.2 liters)

Manual transaxle

    2WD                               2.38 quarts (2.25 liters)

    All-wheel drive                 2.91 quarts (2.75 liters)

Automatic transaxle

    01V transaxle

        Drain and refill            2.7 quarts (2.6 liters)

        From dry                  9.5 quarts (9.0 liters)

        Front differential        0.8 quart (0.75 liter)

    01N transaxle

        Drain and refill            3.7 quarts (3.5 liters)

        From dry                  5.7 quarts (5.5 liters)

        Front differential        0.8 quart (0.75 liter)

Differential, center (all-wheel drive A4

    with manual transaxle)          0.79 quart (0.75 liter)

Differential, rear (all-wheel drive models)

    A4                                2.1 quarts (1.9 liters)

    Passat                          1.6 quarts (1.5 liters)

Cooling system

    Four-cylinder engine                     7.4 quarts (7.0 liters)

    V6 engines

        SOHC                       9.0 quarts (8.5 liters)

        DOHC                     6.4 quarts (6.02 liters)

*All capacities approximate. Add as necessary to bring to appropriate level.*

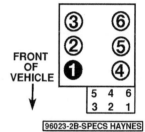

72031-1-SPECS HAYNES

**Engine cylinder location
(four-cylinder engine)**

**Engine cylinder location and
coil terminal numbering**

96023-2B-SPECS HAYNES

---

## Brakes

Disc brake pad wear limit (lining only)

    Front                               1/8 inch (3 mm)

    Rear                               5/64 inch (2 mm)

---

## Ignition system

Spark plug type

    Four-cylinder engine                     NGK BKUR6ET

    V6 engines

        SOHC                       BKUR6ET-10

        DOHC                     BKR6EKUB

Spark plug gap

    Four-cylinder engine                     0.032 inch (0.8 mm)

    V6 engines

        SOHC                       0.040 inch (1.02 mm)

        DOHC                     0.063 inch (1.6 mm)

Firing order

    Four-cylinder engine                     1-3-4-2

    V6 engines                            1-4-3-6-2-5

| Torque specifications | Ft-lbs | Nm |
|---|---|---|
| Engine oil drain plug | | |
|     Four-cylinder engine | 37 | 50 |
|     V6 engines | | |
|         SOHC | 30 | 40 |
|         DOHC | 22 | 30 |
| Automatic transaxle | | |
|     Check/fill plug | 59 | 80 |
|     Drain plug | 30 | 40 |
|     Front differential check/fill plug | 18 | 25 |
|     Center differential (all-wheel drive models) | | |
|         Check/fill plug | 26 | 35 |
|         Drain plug | 15 | 20 |
| Manual transaxle | | |
|     Check/fill plug | 18 | 25 |
|     Drain plug | 26 | 35 |
| Rear differential | | |
|     Check/fill plug | 26 | 35 |
|     Drain plug | 26 | 35 |
| Spark plugs | 22 | 30 |
| Drivebelt tensioner mounting bolt(s) | 18 | 25 |
| Water pump pulley bolts (four-cylinder engine) | 18 | 25 |
| Water pump drain plug (four-cylinder engine) | 22 | 30 |
| Engine block drain plug (V6 engines) | 15 | 20 |
| Coolant pipe bleeder screws (V6 engines) | 15 | 20 |
| Fuel filter banjo bolts | | |
|     M12 bolts | 15 | 20 |
|     M14 bolts | 22 | 30 |
| Wheel bolts | 89 | 120 |

## Section

## Reference to other Chapters

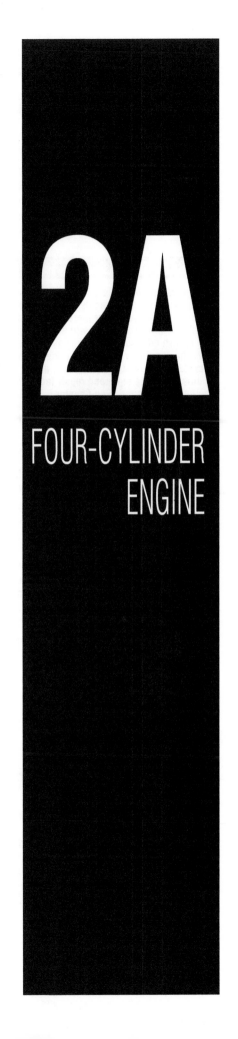

# 2A

## FOUR-CYLINDER ENGINE

## 1   General information

◆ Refer to illustration 1.1

### ❊❊ CAUTION 1:

These models are equipped with an anti-theft radio. Before performing a procedure that requires disconnecting the battery, make sure you have the proper activation code.

### ❊❊ CAUTION 2:

Disconnecting the battery can cause driveability problems that require a scan tool to rectify. Additionally, disconnecting the battery may cause one or more warning lights on the instrument panel to illuminate, which will also require the use of a scan tool to turn off. Most scan tools available to the public do not have the capability to perform either of these tasks, which will necessitate taking the vehicle to a dealer service department or other properly equipped repair facility after service work has been performed. See Chapter 5, Section 1 for the use of an auxiliary voltage input device ("memory saver") before disconnecting the battery and for other precautions related to battery disconnection.

➡Note: The engine cover must removed before performing many of the procedures in this Chapter (see illustration).

This Part of Chapter 2 is devoted to in-vehicle repair procedures for the 1.8L four-cylinder gasoline engine. This engine utilizes a cast-iron engine block with an aluminum cylinder head. The 1.8L engine is turbocharged and utilizes dual overhead camshafts. Hydraulic lifters are used to actuate the valves. The aluminum cylinder head is equipped with pressed-in valve guides and hardened valve seats. The oil pump on early (AEB and ATW) engines is mounted at the rear of the engine and is driven by an idler shaft from the timing belt. On later (AUG and AWM) engines, the manufacturer used a modified oil pump that is mounted below the front of the engine and is chain driven from

1.1  Engine cover mounting screws

the crankshaft. The three-letter engine identification is found on the underhood vehicle identification sticker. This can be verified by removing the engine cover and checking the stamp on top of the front engine lift point.

All information concerning engine removal and installation and engine block and cylinder head overhaul can be found in Part C of this Chapter.

The following repair procedures are based on the assumption that the engine is installed in the vehicle. If the engine has been removed from the vehicle and mounted on a stand, many of the steps outlined in this Part of Chapter 2 will not apply.

The Specifications included in this Part of Chapter 2 apply only to the procedures contained in this Part. Part C of Chapter 2 contains the Specifications necessary for cylinder head and engine block rebuilding.

## 2   Repair operations possible with the engine in the vehicle

Many major repair operations can be accomplished without removing the engine from the vehicle.

Clean the engine compartment and the exterior of the engine with some type of degreaser before any work is done. It will make the job easier and help keep dirt out of the internal areas of the engine.

Depending on the components involved, it may be helpful to remove the hood to improve access to the engine as repairs are performed (refer to Chapter 11 if necessary). Cover the fenders to prevent damage to the

paint. Special pads are available, but an old bedspread or blanket will also work.

If vacuum, exhaust, oil or coolant leaks develop, indicating a need for gasket or seal replacement, the repairs can generally be made with the engine in the vehicle. The intake and exhaust manifold gaskets, oil pan gasket, crankshaft oil seals and cylinder head gasket are all accessible with the engine in place.

Exterior engine components, such as the intake and exhaust manifolds, the oil pan, the oil pump, the water pump, the starter motor, the alternator and the fuel system components can be removed for repair with the engine in place.

Since the cylinder head can be removed without pulling the engine, camshaft and valve component servicing can also be accomplished with the engine in the vehicle. Replacement of the timing belt and pulleys is also possible with the engine in the vehicle.

In extreme cases caused by a lack of necessary equipment, repair or replacement of piston rings, pistons, connecting rods and rod bearings is possible with the engine in the vehicle. However, this practice is not recommended because of the cleaning and preparation work that must be done to the components involved.

## 3   Top Dead Center (TDC) for number one piston - locating

♦ **Refer to illustration 3.6**

1   Top Dead Center (TDC) is the highest point in the cylinder that each piston reaches as it travels up-and-down when the crankshaft turns. Each piston reaches TDC on the compression stroke and again on the exhaust stroke, but TDC generally refers to piston position on the compression stroke. The timing marks on the vibration damper/crankshaft pulley installed on the front of the crankshaft are referenced to the number one piston at TDC.

2   Positioning the piston(s) at TDC is an essential part of procedures such as timing belt and sprocket replacement.

3   In order to bring any piston to TDC, the crankshaft must be turned using one of the methods outlined below. When looking at the timing belt end of the engine, normal crankshaft rotation is clockwise.

### ❋❋ WARNING:

**Before beginning this procedure, be sure to place the transmission in Park or Neutral, set the parking brake and remove the ignition key.**

**3.6 Align the notch on the crankshaft drivebelt pulley with the arrow on the timing belt cover**

a) *The preferred method is to turn the crankshaft with a large socket and breaker bar attached to the large bolt threaded into the center of the crankshaft pulley.*

b) *A remote starter switch, which may save some time, can also be used. Attach the switch leads to the S (switch) and B (battery) terminals on the starter motor. Once the piston is close to TDC, use a socket and breaker bar as described in the previous paragraph.*

c) *If an assistant is available to turn the ignition switch to the Start position in short bursts, you can get the piston close to TDC without a remote starter switch. Use a socket and breaker bar as described in Paragraph a) to complete the procedure.*

4   Disable the ignition and fuel systems by disconnecting the primary electrical connectors at the ignition coil modules (see Chapter 5) and removing the fuel pump fuse.

5   Remove the spark plugs (see Chapter 1) and install a compression gauge in the number one cylinder. Turn the crankshaft clockwise with a socket and breaker bar as described above.

6   When the piston approaches TDC, compression will be noted on the compression gauge. Continue turning the crankshaft until the notch in the crankshaft damper is aligned with the TDC mark on the front cover (see illustration). At this point number one cylinder is at TDC on the compression stroke. If the marks aligned but there was no compression, the piston was on the exhaust stroke; continue rotating the crankshaft 360-degrees (1-turn) and line-up the marks.

➡**Note: If a compression gauge is not available, TDC for the No.1 piston can be obtained by simultaneously aligning the marks on the camshaft (timing belt) sprocket with the marks on the rear timing cover and the marks on the crankshaft damper with the TDC mark on the front cover (see illustration 5.10).**

7   After the number one piston has been positioned at TDC on the compression stroke, TDC for any of the remaining cylinders can be located by turning the crankshaft 180 degrees and following the firing order (refer to the Specifications). Rotating the engine 180 degrees past TDC #1 will put the engine at TDC compression for cylinder #3.

## 4    Valve cover - removal and installation

### REMOVAL

▸ **Refer to illustrations 4.4 and 4.7**

1    Remove the engine cover (see illustration 1.1). Remove the piping for the PCV system and the combination valve, if so equipped.
2    If equipped, loosen the clip securing the crankcase breather hose to the breather pipe, then loosen the bolts securing the pipe to the engine and remove the breather pipe.
3    Remove the ignition coils (see Chapter 5).
4    Remove the heat shield from the valve cover (see illustration).
5    Remove the upper timing belt cover from the engine (see illustrations 5.9a and 5.9b).
6    Remove the ground strap from the valve cover.
7    Remove the retaining nuts and detach the valve cover from the cylinder head (see illustration).

**4.4  Unscrew the heat shield nut and bolts**

**4.7  Removing the valve cover**

8    If the cover is stuck to the head, bump the end with a block of wood and a hammer to jar it loose. If that doesn't work, try to slip a flexible putty knife between the head and cover to break the seal.

### ✳✳ CAUTION:

**Don't pry at the cover-to-head joint or damage to the sealing surfaces may occur, leading to oil leaks after the cover is reinstalled.**

### INSTALLATION

▸ **Refer to illustrations 4.10, 4.11 and 4.12**

9    The mating surfaces of the housing or cylinder head and cover must be clean when the cover is installed. Use a gasket scraper to remove all traces of sealant and old gasket material including the spark plug tube seal gasket, then clean the mating surfaces with lacquer thinner or acetone. If there's residue or oil on the mating surfaces when the cover is installed, oil leaks may develop. Also inspect the rubber end plug at the rear of the cylinder head on for cracks and damage. Now would be a good time to replace it, if damage has occurred.
10 Apply RTV sealant to the corners of the camshaft front bearing cap and to the camshaft drive chain tensioner where they meet the cylinder head (see illustration).
11 Position a new valve cover gasket over the studs on the cylinder head. Install the spark plug tube grommet gasket over the studs on the cylinder head with the index marks facing the timing belt end of the engine (see illustration).
12 Install the valve cover (see illustration) and any brackets removed, then tighten the retaining nuts to the torque listed in this Chapter's Specifications in several steps.
13 Reinstall the remaining parts, run the engine and check for oil leaks.

**4.10  Apply sealant to the corners of the camshaft drive chain tensioner and to the two points at the front camshaft bearing cap where they meet the cylinder head**

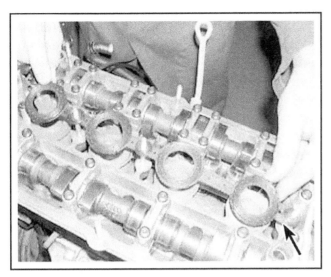

**4.11 Install the spark plug tube seal gasket with the tab (arrow) facing the timing belt end of the engine**

**4.12 Installing the valve cover**

## 5  Timing belt and sprockets - removal, inspection and installation

### ❋❋ WARNING:

Wait until the engine is completely cool before beginning this procedure.

### ❋❋ CAUTION:

Do not rotate the crankshaft or the camshaft separately during this procedure with the timing belt removed as damage to valves may occur. Only rotate the camshaft a few degrees as necessary to align the camshaft sprocket marks with the marks on the rear timing cover.

### REMOVAL

### ❋❋ CAUTION ❋❋

The timing system is complex. Severe engine damage will occur if you make any mistakes. Do not attempt this procedure unless you are highly experienced with this type of repair. If you are at all unsure of your abilities, consult an expert. Double-check all your work and be sure everything is correct before you attempt to start the engine.

⭢ Refer to illustrations 5.8a, 5.8b, 5.9a, 5.9b, 5.10, 5.12a, 5.12b, 5.13, 5.14a, 5.14b, 5.14c, 5.14d, 5.16a, 5.16b, 5.17a and 5.17b

1   Disconnect the cable from the negative terminal of the battery.

### ❋❋ CAUTION 1:

These models are equipped with an anti-theft radio. Before performing a procedure that requires disconnecting the battery, make sure you have the proper activation code.

### ❋❋ CAUTION 2:

Disconnecting the battery can cause driveability problems that require a scan tool to rectify. Additionally, disconnecting the battery may cause one or more warning lights on the instrument panel to illuminate, which will also require the use of a scan tool to turn off. Most scan tools available to the public do not have the capability to perform either of these tasks, which will necessitate taking the vehicle to a dealer service department or other properly equipped repair facility after service work has been performed. See Chapter 5, Section 1 for the use of an auxiliary voltage input device ("memory saver") before disconnecting the battery and for other precautions related to battery disconnection.

2   Raise the front of the vehicle and support it securely on jackstands.

3   Working under the vehicle, remove the lower splash shield below the engine.

4   Place the radiator support panel in the service position (see Chapter 11).

⭢Note: Access to the front of the engine is limited, therefore it will be necessary to remove the front bumper and position the radiator support panel forward enough to allow removal of the timing belt and the surrounding components.

5   Working from above in the engine compartment, remove the engine cover (see illustration 1.1).

6  Remove the cooling fan (see Chapter 3).

7  Remove the spark plugs and the drivebelts (see Chapter 1).

8  Remove the drivebelt tensioner from the front of the engine (see illustrations).

9  Unclip and remove the upper timing belt cover from the engine (see illustrations).

10  Rotate the engine in the normal direction of rotation (clockwise) until the No.1 cylinder is located at TDC (see Section 3). Verify that the camshaft sprocket mark is aligned with the mark on the rear timing belt cover (see illustration).

11  Use a strap wrench to hold the crankshaft pulley from rotating. Loosen the crankshaft drive sprocket retaining bolt and the crankshaft pulley bolts, then remove the crankshaft pulley (see Section 11). After the bolts are loosened, verify that the crankshaft has not moved from TDC.

➡**Note: Loosening the drive sprocket bolt is only required if the crankshaft drive sprocket is expected to be removed. It is not typically necessary to remove the drive sprocket when you're simply replacing a timing belt, but it will need to be removed if you are replacing the crankshaft front oil seal or housing. If you do remove the drive sprocket, obtain a new bolt (the manufacturer doesn't recommend re-using it).**

12  Detach the retaining bolts from the lower timing belt cover and remove the cover (see illustrations).

13  If you plan to re-use the timing belt, apply match marks on the sprocket and belt and an arrow indicating direction of travel on the belt (see illustration).

14  The tension from the timing belt tensioner must be released. There are three types of belt tensioners used:

  1) *For type 1 early models, use a Torx wrench to loosen the lock bolt and then rotate the tensioner clockwise (see illustration).*

  2) *Type 2 is a non-adjustable tensioner. To release it, screw a length of M5 threaded rod (approximately 55 mm, or 2-3/16 inches), install a washer and nut on the rod, then tighten the nut to align the small holes in the piston and the tensioner housing (see illustrations). When the holes are lined up, put a piece of wire through the hole to lock the piston in the retracted position.*

  3) *Later models use a type 3 adjusting tensioner. For this type, put an Allen wrench into one of the tensioner holes and very slowly and steadily rotate it counter-clockwise far enough to insert a lock plate (Volkswagen part # T10008 or a similar fabricated piece) (see illustrations). Loosen the lock nut on the tensioner and use a pin wrench (Volkswagen part #3387 or similar) or right-angle snap-ring pliers to rotate the tensioner clockwise and release the belt tension.*

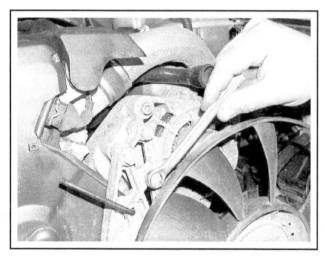

**5.8a  Unscrew the bolts . . .**

**5.8b  . . . and remove the drivebelt tensioner**

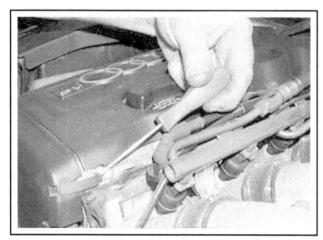

**5.9a  Detach the clips . . .**

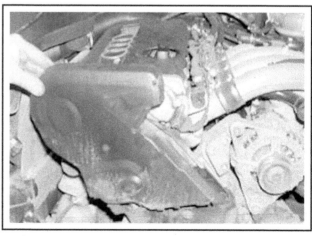

**5.9b  . . . and pull straight up to remove the upper timing belt cover**

5.10  When the engine is positioned at TDC for the No.1 cylinder on the compression stroke, the camshaft sprocket mark will be aligned with the mark on the rear timing belt cover

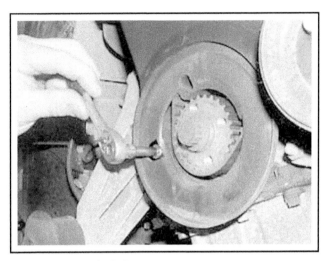

5.12a  Remove the bolts . . .

5.12b  . . . and detach the lower timing belt cover from the engine

5.13  If you intend to re-use the timing belt, apply directional marks on the belt and the rear timing belt cover

5.14a  Use a Torx bit to loosen the timing belt tensioner adjustment bolt

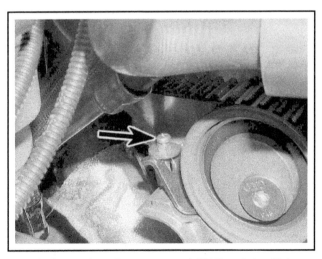

5.14b  On type 2 tensioners, screw an M5 threaded rod into the tensioner, install a washer and nut on the rod . . .

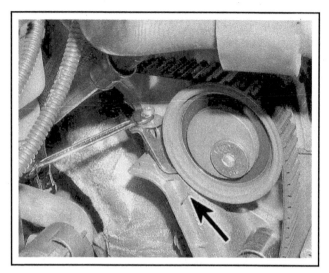

5.14c . . . then turn the nut on the threaded rod until the piston can be locked in position using a metal pin or drill bit inserted through the hole in the housing

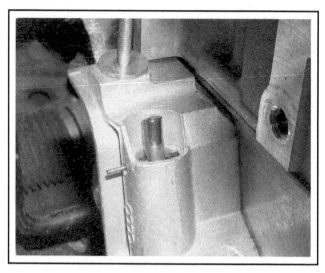

5.14d View showing drill bit inserted through the hole in the tensioner piston and housing (tensioner removed for clarity)

5.16a With the crankshaft drive sprocket retaining bolt removed . . .

15 Remove the timing belt from the engine, taking care to avoid twisting or kinking it excessively.

➡ **Note: If you're removing the upper part of the belt only, for camshaft seal replacement or cylinder head removal, it isn't necessary to detach the belt from the crankshaft sprocket.**

16 If the crankshaft sprocket is worn or damaged, or if you need to replace the crankshaft front oil seal, remove the drive sprocket retaining bolt which was loosened in Step 11 and detach the crankshaft sprocket from the crankshaft (see illustrations).

17 If the camshaft sprocket is damaged or needs to be removed for other procedures such as cylinder head removal, use a spanner wrench or similar tool to hold the sprocket in place as the sprocket retaining bolt is loosened, then remove the camshaft sprocket from the end of the camshaft (see illustration).

➡ **Note: The intermediate shaft sprocket may also removed in the same manner as the camshaft sprocket if necessary (see illustration).**

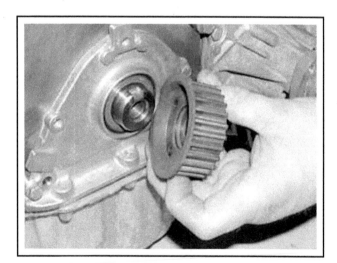

5.16b . . . the crankshaft sprocket is easily detached from the engine

5.17a If necessary, the camshaft sprocket bolt and the intermediate shaft sprocket bolt can be loosened while holding the sprocket in place with a spanner wrench

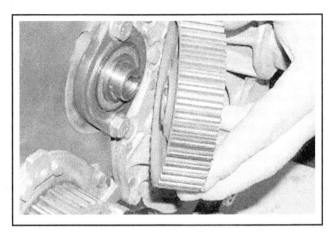

**5.17b  Removing the intermediate shaft sprocket**

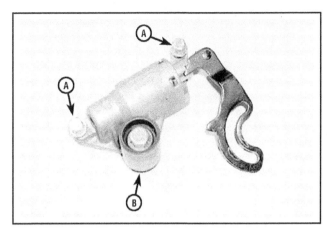

**5.18  Type 1 timing belt tensioner-to-engine block mounting bolts (A) - make sure the idler wheel (B) and the tensioner pulley on the engine block spin freely**

## INSPECTION

♦ **Refer to illustrations 5.18 and 5.19**

### ✳✳ CAUTION:

**Do not bend, twist or turn the timing belt inside out. Do not allow it to come in contact with oil, coolant or fuel. Do not turn the crankshaft or camshaft more than a few degrees (if necessary for tooth alignment) while the timing belt is removed.**

18  Spin the timing belt tensioner pulley (which is the large pulley bolted to the engine block) and the idler wheel (which is the small roller mounted on the tensioner body) and check the bearings for smooth operation and excessive play. Also inspect the remaining timing belt sprockets for any obvious damage. Replace all worn parts as necessary (see illustration).

19  Examine the belt for evidence of contamination by coolant or lubricant. If this is the case, find the source of the contamination before progressing any further. Check the belt for signs of wear or damage, particularly around the leading edges of the belt teeth (see illustration).

### ✳✳ CAUTION:

**If the belt appears to be in good condition and can be re-used, it is essential that it is reinstalled the same way around, otherwise accelerated wear will result, leading to premature failure.**

20  Replace the belt if its condition is in doubt; the cost of belt replacement is negligible compared with potential cost of the engine repairs, should the belt fail in service. Similarly, if the belt is known to have covered more than 60,000 miles, it is prudent to replace it regardless of condition, as a precautionary measure.

## INSTALLATION

### ✳✳ CAUTION ✳✳

**Before starting the engine, carefully rotate the crankshaft by hand through at least two full revolutions (use a socket and breaker bar on the crankshaft pulley centerbolt). If you feel any resistance, STOP! There is something wrong - most likely, valves are contacting the pistons. You must find the problem before proceeding. Check your work and see if any updated repair information is available.**

♦ **Refer to illustrations 5.26a and 5.26b**

21  Ensure that the crankshaft is still set to TDC on No. 1 cylinder, as described in Section 3. If any of the timing sprockets or the tensioner pulley were removed for inspection or needed replacement, install them back onto the engine now. If the timing belt tensioner was removed, reinstall it now and torque the bolts to the Specifications listed at the end of this Chapter, then install the tensioner pulley adjustment bolt loosely.

22  Align the camshaft sprocket mark with the mark on the rear timing belt cover (see illustration 5.10).

23  Loop the timing belt loosely under the crankshaft sprocket.

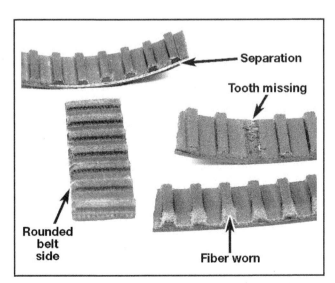

Separation

Tooth missing

Rounded belt side

Fiber worn

**5.19  Check the timing belt for cracked and missing teeth - wear on one side of the belt indicates sprocket misalignment problems**

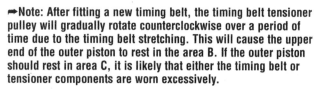

## ✳✳ CAUTION:

**Observe the direction of rotation markings on the belt.**

24 Engage the timing belt teeth with the crankshaft sprocket, then maneuver it into position over the intermediate shaft sprocket and the camshaft sprocket. Ensure the belt teeth seat correctly on the sprockets, then install the belt around the timing belt tensioner pulley.

**➡Note: Slight adjustments to the position of the camshaft sprocket may be necessary to achieve this.**

25 Ensure that the 'front run' of the belt is taut and all the slack is in the section of the belt that passes over the tensioner pulley.

26 On early-model type 1 tensioners, tighten the timing belt by inserting a sturdy of pair right-angled snap-ring pliers against the tensioner pulley guide pin and the raised lug on the tensioner arm. Rotate the tensioner arm clockwise until the spring-tensioned inner piston is fully extended and the outer piston lifts approximately 0.039 inch (1.0 mm), then tighten the adjusting bolt. Check that the notched area A coincides with the upper end of the outer piston (see illustrations). If necessary, loosen the adjusting bolt and re-adjust the tensioner. The distance from the top of the outer piston to the top of the inner piston eye must be between 0.984 to 1.142 inch (25.0 to 29.0 mm).

**➡Note: After fitting a new timing belt, the timing belt tensioner pulley will gradually rotate counterclockwise over a period of time due to the timing belt stretching. This will cause the upper end of the outer piston to rest in the area B. If the outer piston should rest in area C, it is likely that either the timing belt or tensioner components are worn excessively.**

27 On type 2 tensioners, simply pull out the wire and remove the tightening bolt you installed in Step 14. This tensioner is not adjustable.

28 On type 3 tensioners, turn the pin wrench or snap-ring pliers counterclockwise until you can remove the lock plate. After removing the lock plate, turn the wrench clockwise until a 0.31 inch (8 mm) drill can be used to set the distance between the lever and the tensioner housing. Tighten the lock bolt.

29 At this point install the lower timing belt cover and the crankshaft pulley. Double check to make sure that the crankshaft is still set to TDC on No. 1 cylinder (see Section 3) and the camshaft sprocket mark is aligned with the mark on the rear timing belt cover.

30 Rotate the crankshaft through two complete revolutions. Reset the engine to TDC on No. 1 cylinder, with reference to Section 3 and check the alignment marks again. Also re-check the timing belt tension and adjust it, if necessary.

31 Install the upper timing belt cover, the drivebelt tensioner, the cooling fan and the accessory drivebelts.

32 The remainder of the installation is the reverse of removal.

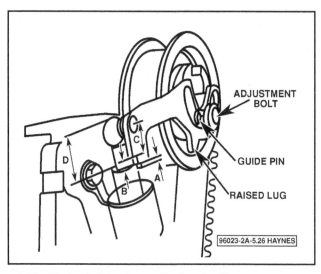

**5.26a  Timing belt tensioner adjustment settings**

A   Correct adjustment zone
B   Wear zone
C   Out of adjustment zone
D   Distance from the top of the outer piston to the top of the inner piston

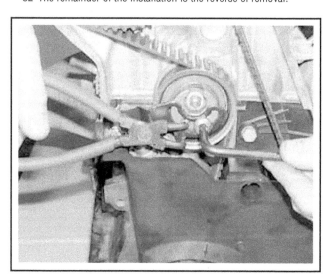

**5.26b  Setting the timing belt tension**

## 6   Camshafts and lifters - removal and installation

## ✳✳ CAUTION:

**Performing this procedure may cause a trouble code to be set, which will require taking the vehicle to a dealer service department (or other repair shop equipped with the necessary scan tool) to have the cam sensor synchronized and the trouble code cleared.**

**➡Note: The camshafts and lifters should always be thoroughly inspected before installation and camshaft endplay should always be checked prior to camshaft removal. Although the hydraulic lifters are self adjusting and require no periodic service, there is an in-vehicle procedure for checking excessively noisy hydraulic lifters. Refer to Chapter 2C for the camshaft and lifter inspection procedures.**

## REMOVAL

**▶ Refer to illustrations 6.2, 6.4, 6.5, 6.6a, 6.6b, 6.7, 6.9a and 6.9b**

1   Remove the engine cover (see illustration 1.1).

2   Remove the valve cover (see Section 4). Also remove the oil deflector(s) to expose the camshaft(s) (see illustration).

3   Remove the timing belt and camshaft sprocket (see Section 5).

4   Remove the camshaft position sensor (see Chapter 6). Remove the camshaft sensor reluctor ring from the end of the intake camshaft (see illustration).

5   Mark the position of the camshaft drive chain in relationship to the sprockets and the marks on the rear bearing cap (see illustration). This will ensure that the drive chain is installed in exactly the same direction and position from which it was removed.

**6.2  The plastic oil deflector is easily removed by simply lifting it off the cylinder head**

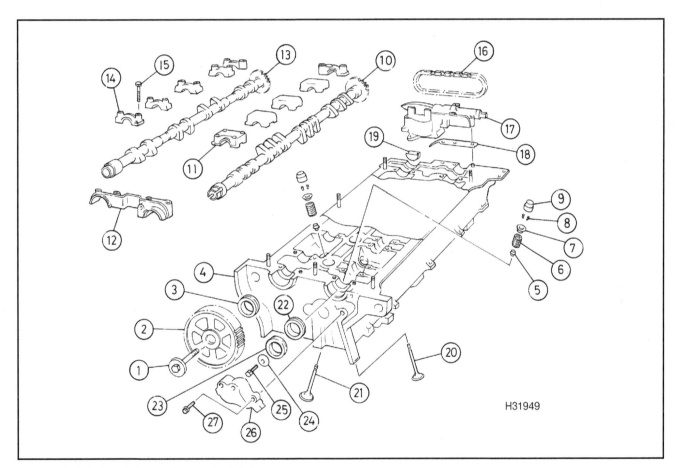

H31949

**6.4  Exploded view of the camshafts and related components**

| | | | | | |
|---|---|---|---|---|---|
| 1 | Camshaft sprocket bolt | 10 | Intake camshaft | 19 | End plug |
| 2 | Camshaft sprocket | 11 | Intake camshaft bearing cap | 20 | Exhaust valve |
| 3 | Oil seal | 12 | No.1 bearing cap | 21 | Intake valve |
| 4 | Cylinder head | 13 | Exhaust camshaft | 22 | Oil seal |
| 5 | Valve stem seal | 14 | Exhaust camshaft bearing cap | 23 | Reluctor ring |
| 6 | Valve spring | 15 | Bolt | 24 | Washer |
| 7 | Valve spring retainer | 16 | Camshaft drive chain | 25 | Bolt |
| 8 | Valve keepers | 17 | Camshaft drive chain tensioner | 26 | Camshaft position sensor |
| 9 | Hydraulic lifter | 18 | Gasket | 27 | Bolt |

6   Compress the camshaft drive chain tensioner. This can be accomplished by purchasing a special tool from the dealer service department which is specifically made for this purpose or by fabricating a home made tool using a threaded rod, several nuts, a plastic cable tie and a small metal plate (see illustrations).

7   Mark the location of the camshaft bearing caps from 1 to 6 (and I for intake, E for exhaust), starting with the double bearing cap at the front (timing belt) end. Also mark arrows indicating the front of the engine (see illustration). Loosen the bearing cap nuts alternately in the following order:

1)  *Loosen and remove the No. 2 and 4 bearing caps from the intake and exhaust camshafts*
2)  *Loosen and remove the No. 1 bearing cap*
3)  *Loosen and remove the No. 6 bearing caps from the intake and exhaust camshafts*
4)  *Loosen and remove the drive chain tensioner mounting bolts*
5)  *Loosen and remove the No. 3 and 5 bearing caps from the exhaust camshafts*
6)  *Loosen and remove the No. 3 and 5 bearing caps from the intake camshafts*

8   Remove the camshafts and the drive chain tensioner as an assembly from the cylinder head.

### ✳✳ CAUTION:

**Keep the caps in order. They must go back in the same location they were removed from (and face in the same direction). Separate the tensioner and the drive chain from the camshafts on a workbench.**

9   Remove the lifters from the cylinder head, keeping the them in order with their respective valve and cylinder (see illustrations).

### ✳✳ CAUTION:

**Keep the lifters in order. They must go back in the same location they were removed from.**

10  Inspect the camshaft and lifters as described in Chapter 2C.

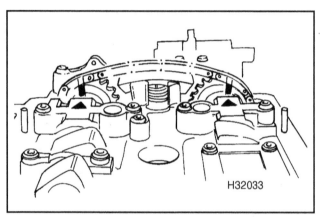

**6.5  With the notches in the rear drive chain sprockets aligned with the arrows on the rear bearing caps, apply match marks on the chain with a permanent marker - be sure to wipe the oil from the chain and sprockets first, so the marker will adhere to the components - the number of rollers from each mark should be exactly 16**

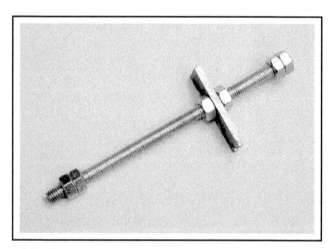

**6.6a  A home made tool can be fabricated to compress the camshaft drive chain tensioner**

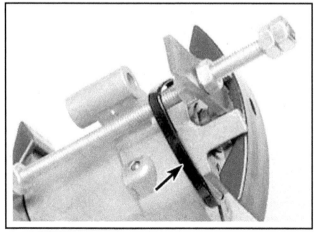

**6.6b  The plastic cable tie (arrow) is used to help secure the home made tool in position as the tensioner is compressed**

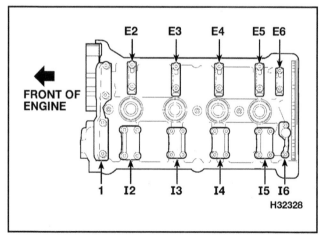

**6.7  The camshaft bearing caps should be marked as shown with a number and letter stamp or a marker to ensure correct reinstallation**

## INSTALLATION

▶ **Refer to illustrations 6.11, 6.12a, 6.12b, 6.16, 6.17 and 6.19**

11  Apply clean engine oil onto the sides and underside of the hydraulic lifters, and install them into position in their bores in the cylinder head (see illustration). Push them down until they contact the valves, then lubricate the camshaft lobe contact surfaces.

12  Clean the mating surfaces of the drive chain tensioner and the cylinder head, then install the half-round end plugs at the rear of the cylinder head and a new drive chain tensioner gasket (see illustrations).

13  Align the marks on the drive chain (made previously) with the notches on the camshaft drive gears and install the chain over the drive gears. If you're installing a new drive chain, install the drive chain with exactly 16 rollers between the notches on the drive gears. Note that the exhaust camshaft notch is slightly off center. In either case verify that there are 16 rollers between the notches of the intake and exhaust camshaft.

14  Compress the camshaft drive chain tensioner with the special tool and insert it between the drive chain and the camshafts.

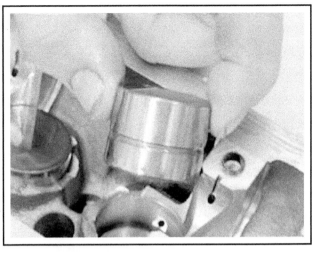

**6.9a  The lifters can be removed from the cylinder head by hand or with a magnet . . .**

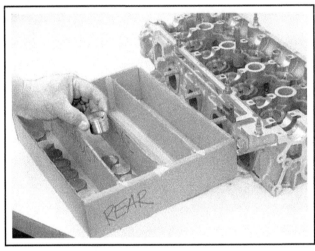

**6.9b  . . . and stored in individually marked plastic bags or a divided box as shown - be sure to keep them in order with their respective valves**

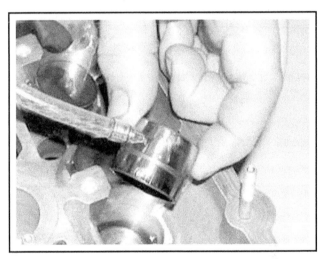

**6.11  Lubricate the lifters with clean engine oil before installing them into the cylinder head**

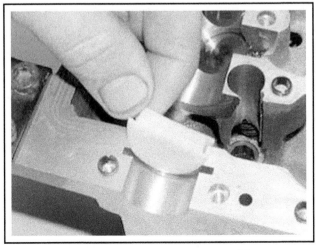

**6.12a  Install the rubber end plugs at the rear of the cylinder head**

**6.12b  When installing the drive chain tensioner gasket apply RTV sealant at the shaded area**

15 Lubricate the camshaft and cylinder head bearing journals with clean engine oil. Then carefully lower the camshafts, drive chain and tensioner as an assembly into position on the cylinder head with the No.1 camshaft lobes facing up. Support the ends of the shaft as it is inserted, to avoid damaging the lobes and journals.

16 Install the drive chain tensioner over the dowels on the cylinder head and tighten the bolts to the torque listed in this Chapter's Specifications (see illustration).

17 Install the camshaft bearing caps in the reverse order of removal (see Step 7). Be sure to apply a small amount of RTV sealant to the mating surface of the front bearing cap before installing it (see illustration). After the bearing caps have been tightened, remove the drive chain tensioning tool from the tensioner. Reconfirm that there are 16 rollers between the notches of the intake and exhaust camshaft and that the notches align with the arrows on the caps and that the No.1 camshaft lobes face up.

18 Clean the oil seal housing bores and lubricate the lip of a new camshaft oil seals with clean engine oil and locate them over the end of the camshafts. Slide the seals along the camshaft until they locate squarely in the housing bores.

19 Using a socket with an outside diameter slightly smaller than the outside diameter of the seal, carefully drive the new seals into place with a hammer (see illustration). Make sure they're installed squarely and driven in to the same depth as the original. If a socket isn't available, a short section of pipe will also work.

➡Note: Be sure to install both camshaft oil seals, one for the intake camshaft and one for the exhaust camshaft.

20 Install the camshaft sprocket on the exhaust camshaft and the camshaft position sensor reluctor ring, conical washer and retaining bolt on the intake camshaft. Tighten the bolts to the torque listed in this Chapter's Specifications.

21 Install the timing belt (see Section 5). When installing the timing belt, make sure the crankshaft is at TDC for the No. 1 cylinder and the camshaft sprocket mark is aligned with the rear timing cover.

22 The remainder of installation is the reverse of removal.

**✳✳ CAUTION:**

**If new lifters were used, wait at least 30 minutes before starting the vehicle to allow the lifters to bleed down. Failure to do so will result in serious engine damage.**

**6.16  Align the drive chain tensioner over the dowels on the cylinder head**

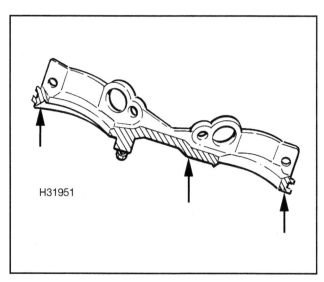

**6.17  Apply a small amount of RTV sealant to the No.1 bearing cap at the shaded areas**

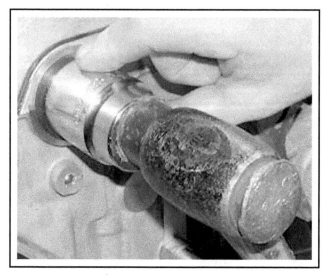

**6.19  Gently drive the new camshaft oil seals into place with the spring side facing the engine**

## 7   Valve springs, retainers and seals - replacement

▶ **Refer to illustrations 7.4, 7.8, 7.10, 7.15a, 7.15b, 7.15c, 7.16 and 7.17**

➡**Note: Broken valve springs and defective valve stem seals can be replaced without removing the cylinder heads. Two special tools and a compressed air source are normally required to perform this operation, so read through this Section carefully and rent or buy the tools before beginning the job.**

1   Remove the valve cover referring to Section 4. Then refer to Section 6 and remove the camshaft and lifters.

2   Remove the spark plug from the cylinder which has the defective component. If all of the valve stem seals are being replaced, all of the spark plugs should be removed.

3   Turn the crankshaft until the piston in the affected cylinder is at top dead center (TDC) on the compression stroke (see Section 3 for instructions). If you're replacing all of the valve stem seals, begin with cylinder number one and work on the valves for one cylinder at a time. Move from cylinder-to-cylinder following the firing order sequence (see the Specifications listed at the end of this Chapter).

4   Thread an adapter into the spark plug hole (see illustration) and connect an air hose from a compressed air source to it. Most auto parts stores can supply the air hose adapter.

➡**Note: Many cylinder compression gauges utilize a screw-in fitting that may work with your air hose quick-disconnect fitting.**

5   Apply compressed air to the cylinder. The valves should be held in place by the air pressure.

### ❊❊ WARNING:

**If the cylinder isn't exactly at TDC, air pressure may force the piston down, causing the engine to quickly rotate. DO NOT leave a wrench on the crankshaft drive sprocket bolt or you may be injured by the tool.**

6   Stuff shop rags into the cylinder head holes around the valves to prevent parts and tools from falling into the engine.

7   Using a socket and a hammer gently tap on the top of the each valve spring retainer several times. This will break the bond between the valve keeper and the spring retainer and allow the keeper to separate from the valve spring retainer as the valve spring is compressed.

8   Use a valve spring compressor to compress the spring. Remove the keepers with small needle-nose pliers or a magnet (see illustration).

➡**Note: Several different types of tools are available for compressing the valve springs with the head in place. Be sure to purchase or rent the "Import type" that bolts to the top of the cylinder head. This type uses a support bar across the cylinder head for leverage as the valve spring is compressed. The lack of clearance surrounding the valve springs on these engines prohibits other types of valve spring compressors from being used.**

9   Remove the valve spring and retainer.

➡**Note: If air pressure fails to retain the valve in the closed position during this operation, the valve face or seat may be damaged. If so, the cylinder head will have to be removed for repair.**

10   Remove the old valve stem seals, noting differences between the intake and exhaust seals (see illustration).

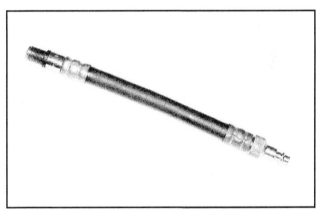

**7.4   You'll need an air hose adapter this long to reach down into the spark plug wells on the cylinder head - they're commonly available at auto parts stores**

**7.8   While the valve spring tool is compressing the spring, remove the keepers with a small magnet or pliers**

**7.10   The old valve stem seals can be removed with a pair of needle-nose pliers**

11 Wrap a rubber band or tape around the top of the valve stem so the valve won't fall into the combustion chamber, then release the air pressure.

12 Inspect the valve stem for damage. Rotate the valve in the guide and check the end for eccentric movement, which would indicate that the valve is bent.

13 Move the valve up-and-down in the guide and make sure it doesn't bind. If the valve stem binds, either the valve is bent or the guide is damaged. In either case, the head will have to be removed for repair.

14 Reapply air pressure to the cylinder to retain the valve in the closed position, then remove the tape or rubber band from the valve stem.

15 Lubricate the valve stem with engine oil and install a new seal on the valve guide (see illustrations).

16 Install the valve spring and the spring retainer in position over the valve (see illustration).

17 Compress the valve spring and carefully position the keepers in the groove. Apply a small dab of grease to the inside of each keeper to hold it in place (see illustration).

18 Remove the pressure from the spring tool and make sure the keepers are seated.

19 Disconnect the air hose and remove the adapter from the spark plug hole.

20 Install the camshaft, lifters, timing belt and the valve cover by referring to the appropriate Sections.

21 Install the spark plug(s) and ignition coil(s).

22 Start and run the engine, then check for oil leaks and unusual sounds coming from the valve cover area.

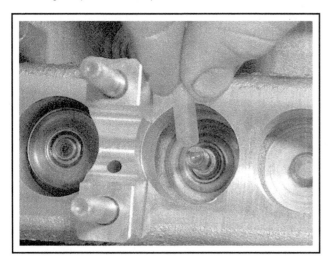

**7.15a  Install the protective plastic sleeve over the valve end face to avoid damage to the valve seal as the seal is installed**

**7.15b  Push a new valve stem seal over the valve and down to the top of the guide, then remove the plastic installation tool**

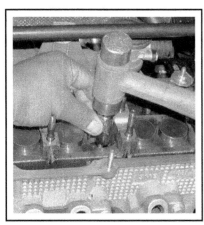

**7.15c  Gently tap the new seal in place on the guide with a socket**

**7.16  Install the valve spring and the retainer over the valve**

**7.17  Apply a small dab of grease to each keeper as shown here before installation - it'll hold them in place on the valve stem as the spring is released**

## 8 Intake manifold - removal and installation

▶ **Refer to illustrations 8.8 and 8.9**

**✳✳ WARNING:**

**Wait until the engine is completely cool before beginning this procedure.**

1   Remove the engine cover (see illustration 1.1).
2   Partially drain the engine coolant and remove the coolant expansion tank (see Chapter 3).
3   On early models, disconnect the throttle and cruise control cables.

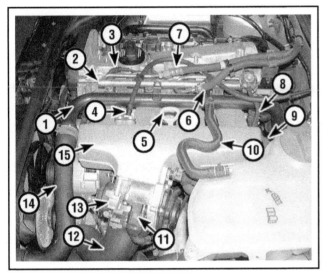

**8.8 Overview of the intake manifold and related components**

*1   Upper coolant pipe*
*2   Fuel rail, fuel injectors and fuel pressure regulator*
*3   Fuel injector electrical connectors*
*4   Accelerator cable and retaining bracket*
*5   Oil dipstick*
*6   Fuel return line*
*7   Fuel feed line*
*8   Vacuum hose to fuel pressure regulator*
*9   Vacuum hose to brake booster*
*10   Coolant hose to expansion tank*
*11   Throttle body electrical connectors*
*12   Air intake duct*
*13   Throttle body/throttle valve control module*
*14   Upper radiator hose*
*15   Intake manifold*

4   Label and detach the electrical connectors from the throttle valve control module and the intake air temperature sensor.
5   Remove the upper intercooler hose from the throttle body.
6   Relieve the fuel system pressure and remove the fuel rail and injectors (see Chapter 4).
7   Remove the upper coolant pipe from the engine (see Chapter 3) and the coolant hoses from the throttle body (see Chapter 4). Also remove the oil dipstick.
8   Label and detach any remaining vacuum lines (such as the brake booster vacuum hose) or electrical wiring that would interfere with removal of the intake manifold (see illustration).
9   Remove the intake manifold lower support brace (see illustration).
10   Remove the mounting nuts/bolts then detach the manifold and the throttle body as an assembly from the engine.
11   Use a scraper to remove all traces of old gasket material and sealant from the manifold and cylinder head, then clean the mating surfaces with lacquer thinner or acetone. If the gasket was leaking, have the manifold checked for warpage at an automotive machine shop and resurfaced if necessary.
12   Install a new gasket, then position the manifold on the head and install the nuts/bolts.
13   Tighten the nuts/bolts in three or four equal steps to the torque listed in this Chapter's Specifications. Work from the center out towards the ends to avoid warping the manifold.
14   Install the remaining parts in the reverse order of removal.
15   Check the coolant and add some, if necessary, to bring it to the appropriate level. Run the engine and check for coolant and vacuum leaks.
16   Road test the vehicle and check for proper operation of all accessories, including the cruise control system.

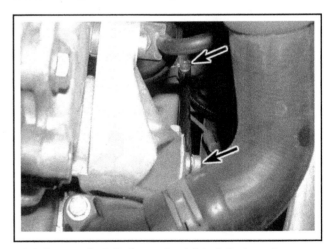

**8.9 Remove the bolts (arrows) and detach the lower brace from the intake manifold**

## 9  Exhaust manifold - removal and installation

♦ Refer to illustration 9.8

### ✷✷ WARNING:

The engine must be completely cool before beginning this procedure.

### REMOVAL

1  Remove the engine cover (see illustration 1.1).
2  Remove the air intake duct and the air cleaner housing (see Chapter 4).
3  Raise the front of the vehicle and support it securely on jackstands.
4  Remove the splash guard from below the engine compartment.
5  Refer to Chapter 4 and loosen the turbocharger support bracket bolts several turns, then detach the turbocharger oil supply line from the exhaust manifold heat shield if equipped.
6  Remove the exhaust manifold heat shield.
7  Remove the three bolts securing the turbocharger to the exhaust manifold and lower the turbocharger slightly. Remove the turbocharger to exhaust manifold gasket and cover the opening of the turbocharger with a rag to prevent dirt particles and foreign objects from entering the turbocharger.
8  Remove the nuts/bolts and detach the exhaust manifold and gasket (see illustration). If necessary, apply penetrating oil to the manifold mounting nuts/bolts to help facilitate removal.

### INSTALLATION

9  Use a scraper to remove all traces of old gasket material and carbon deposits from the manifold and cylinder head mating surfaces. If the gasket was leaking, have the manifold checked for warpage at an automotive machine shop and resurfaced if necessary.

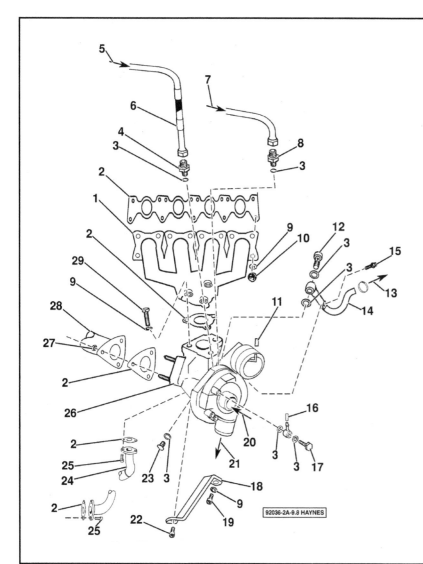

**9.8 Exploded view of the exhaust manifold and related components**

1   Nut
2   Gasket
3   Oil seal
4   Union
5   From oil filter adapter
6   Oil supply line
7   From coolant line on intake manifold
8   Union
9   Washer
10  Nut
11  Vacuum hose to wastegate bypass actuator
12  Banjo bolt
13  Coolant line from cylinder block
14  Coolant return line
15  Bolt
16  Hose to wastegate bypass regulator
17  Banjo bolt
18  Turbocharger support bracket
19  Bolt
20  From air cleaner
21  To intercooler
22  Bolt
23  Plug
24  Oil return line
25  Bolt
26  Turbocharger
27  Nut
28  Front exhaust pipe
29  Bolt

92036-2A-9.8 HAYNES

10 Position a new gasket over the cylinder head studs.

11 Install the manifold and thread the mounting nuts/bolts into place. Make sure to use hi-temp anti-seize compound on the exhaust manifold fasteners.

12 Working from the center out, tighten the nuts/bolts to the torque listed in this Chapter's Specifications in three or four equal steps.

13 Reinstall the remaining parts in the reverse order of removal.

14 Run the engine and check for exhaust leaks.

## 10  Cylinder head - removal and installation

### ✳✳ WARNING:

The engine must be completely cool before beginning this procedure.

### ✳✳ CAUTION:

Performing this procedure may cause a trouble code to be set, which will require taking the vehicle to a dealer service department (or other repair shop equipped with the necessary scan tool) to have the cam sensor synchronized and the trouble code cleared.

➡Note: The cylinder head can be removed with the exhaust manifold attached.

## REMOVAL

♦ **Refer to illustrations 10.8, 10.9a and 10.9b**

1  Disconnect the cable from the negative terminal of the battery.

### ✳✳ CAUTION 1:

These models are equipped with an anti-theft radio. Before performing a procedure that requires disconnecting the battery, make sure you have the proper activation code.

### ✳✳ CAUTION 2:

Disconnecting the battery can cause driveability problems that require a scan tool to rectify. Additionally, disconnecting the battery may cause one or more warning lights on the instrument panel to illuminate, which will also require the use of a scan tool to turn off. Most scan tools available to the public do not have the capability to perform either of these tasks, which will necessitate taking the vehicle to a dealer service department or other properly equipped repair facility after service work has been performed. See Chapter 5, Section 1 for the use of an auxiliary voltage input device ("memory saver") before disconnecting the battery and for other precautions related to battery disconnection.

2  Drain the engine coolant (see Chapter 1). Refer to Section 3 and set the engine to TDC. Rotate the crankshaft an additional 90-degrees to ensure that no pistons are at Top Dead Center.

3  Remove the intake manifold (see Section 8).

4  Refer to Section 5 and remove the timing belt. Follow Steps 1 through 15.

5  Refer to Section 9 and detach the turbocharger from the exhaust manifold. Follow Steps 1 through 7.

6  Remove the valve cover (see Section 4).

7  Unplug all electrical connectors and the heater hose from the elbow at the rear of the cylinder head, labeling each wiring connector or hose to aid the installation process.

8  Working in the reverse of the sequence shown in illustration 10.23a, progressively loosen the cylinder head bolts, by half a turn at a time, until all bolts can be unscrewed by hand. Discard the bolts - new ones must be installed on reassembly.

➡Note: Several types of cylinder head bolts have been employed over the production years of this engine (see illustration). Therefore two different types of socket head drivers may be necessary to remove the old bolts and install the new ones.

9  Check that nothing remains connected to the cylinder head, then lift the head away from the cylinder block; seek assistance if possible, as it is very heavy, especially when being removed with the exhaust manifold (see illustration). If resistance is felt, carefully pry the cylinder

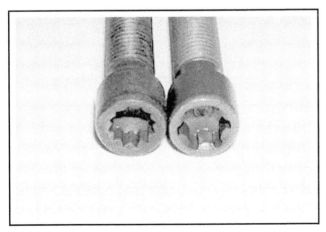

**10.8  Two types of cylinder head bolts**

**10.9a  Lift the cylinder head off the engine with the exhaust manifold attached**

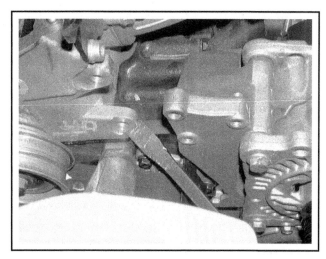

**10.9b If the cylinder head is stuck, it may be necessary to pry upward on the casting protrusion to dislodge the head from the block**

head upward, beyond the gasket surface, at a casting protrusion (see illustration).

10 Remove the gasket from the top of the block. Do not discard the gasket - it will be needed for identification purposes.

11 If the cylinder head is to be disassembled for service, separate the exhaust manifold as described in Section 9. Disregard the steps that do not apply since the cylinder head is already removed from the vehicle, then proceed to Chapter 2C for overhaul procedures. Be sure to reinstall the manifold back onto the cylinder head before installing the cylinder head on the vehicle.

## INSTALLATION

▶ **Refer to illustrations 10.19a, 10.19b, 10.23a and 10.23b**

12 The mating faces of the cylinder head and cylinder block must be perfectly clean before installing the head. Use a hard plastic or wood scraper to remove all traces of gasket and carbon; also clean the piston crowns. Take particular care during the cleaning operations, as aluminum alloy is easily damaged. Also, make sure that the carbon is not allowed to enter the oil and water passages - this is particularly important for the lubrication system, as carbon could block the oil supply to the engine's components. Using adhesive tape and paper, seal the water, oil and bolt holes in the cylinder block.

13 Check the mating surfaces of the cylinder block and the cylinder head for nicks, deep scratches and other damage. If slight, they may be removed carefully with abrasive paper.

14 If warpage of the cylinder head gasket surface is suspected, use a straight-edge to check it for distortion, but note that head machining will not be possible - refer to Chapter 2C.

15 Clean out the cylinder head bolt holes using a suitable tap. Be sure they're clean and dry before installation of the head bolts.

16 It is possible for the piston crowns to strike and damage the valve heads if the camshaft is rotated with the timing belt removed and the crankshaft set to TDC. For this reason, the crankshaft must be set to a position other than TDC on No. 1 cylinder before the cylinder head is reinstalled. Use a wrench and socket on the crankshaft pulley center bolt to turn the crankshaft in the opposite direction of rotation (counterclockwise), until all four pistons are positioned halfway down their bores - approximately 90-degrees before TDC.

17 If the cylinder head has been resurfaced, make sure the valve seats have been reworked by the same amount to allow the correct piston to valve clearance before installing the cylinder head. See Chapter 2C for further information.

18 Cut off the heads from two of the old cylinder head bolts to use as alignment dowels during cylinder head installation. Also cut a slot in the end of these two bolts, big enough for a screwdriver blade, so that the alignment dowels can be removed after the cylinder head is installed. A simple hand-held hacksaw can be used to fabricate the alignment dowels.

19 Install the alignment dowels in the outer rear holes of the cylinder block and position the new head gasket on the cylinder block, engaging it with the locating dowels. Ensure that the manufacturer's "TOP" and part number markings are face up (see illustrations).

20 With the help of an assistant, place the cylinder head centrally on the cylinder block, ensuring that the locating dowels engage with the recesses in the cylinder head. Check that the head gasket is correctly

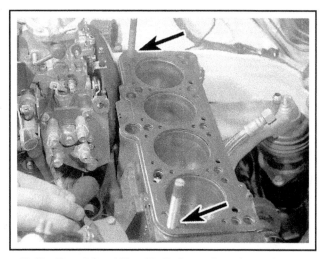

**10.19a Two of the old head bolts (arrows) can be used as cylinder head alignment dowels**

**10.19b Be sure the "TOP" mark faces upward**

seated before allowing the full weight of the cylinder head to rest upon it.

➡**Note: If the cylinder head had been disassembled for repair, be sure the camshaft(s) are reinstalled on the cylinder head with the No.1 cylinder camshaft lobes pointing upward.**

21  Install each cylinder head bolt into its relevant hole and screw them in hand tight. Be sure to use NEW cylinder head bolts, as the old bolts are stretch-type fasteners that will not provide the correct torque readings if reused.

22  Unscrew the homemade alignment dowels, using a flat-bladed screwdriver and install the remaining two bolts hand tight.

23  Working in the sequence shown (see illustration), tighten the cylinder head bolts in four steps to the torque and angle of rotation listed in this Chapter's Specifications.

➡**Note: It is recommended that an angle-measuring gauge be used during the final stages of the tightening, to ensure accu-**
racy (see illustration). If a gauge is not available, use white paint to make alignment marks between the bolt head and cylinder head prior to tightening; the marks can then be used to check that the bolt has been rotated through the correct angle during tightening.

24  Rotate the crankshaft counterclockwise 90 degrees to set the engine back at TDC. Be sure the alignment mark on the crankshaft pulley aligns with the mark on the front cover. Refer to Section 3, if necessary.

25  Install the timing belt tensioner and the camshaft sprocket on the engine if removed.

26  Install and adjust the timing belt as described in Section 5.

27  The remainder of the installation is the reverse of removal.

28  Change the engine oil and coolant (see Chapter 1). Run the engine and check for leaks.

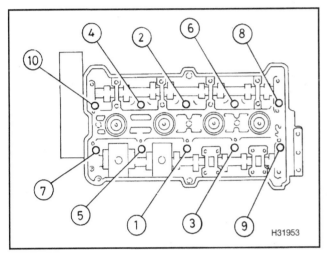

**10.23a  Cylinder head bolt TIGHTENING sequence**

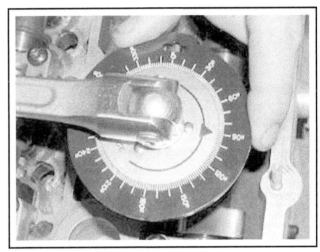

**10.23b  Using an angle measurement gauge during the final stages of tightening**

## 11  Crankshaft pulley - removal and installation

▶ **Refer to illustrations 11.5, 11.6, and 11.7**

1  Raise the front of the vehicle and support it securely on jackstands.

2  Working under the vehicle, remove the lower splash shield below the engine.

3  Place the radiator support panel in the service position (see Chapter 11).

➡**Note: Access to the front of the engine is limited, therefore it will be necessary to remove the front bumper and position the radiator support panel forward enough to allow removal of the surrounding components.**

4  Remove the serpentine drivebelts (see Chapter 1).

5  Remove the bolts from the front of the crankshaft pulley and detach it from the engine (see illustration).

**11.5  Crankshaft pulley retaining bolts (arrows)**

6   If you're removing the crankshaft pulley for other procedures in this manual, such as crankshaft front oil seal removal, loosen the drive sprocket retaining bolt first, before removing the crankshaft pulley (see illustration).

➡**Note: If you remove this bolt, obtain a new one (the manufacturer doesn't recommend re-using it). Upon installation, be sure to tighten the crankshaft drive sprocket bolt to the torque and angle of rotation listed in this Chapter's Specifications.**

**11.6  If you're removing the crankshaft pulley to access the front oil seal, hold the crankshaft pulley with a chain or strap wrench and loosen the crankshaft drive sprocket retaining bolt first (wrap the pulley with a piece of old drivebelt)**

7   Position the crankshaft pulley/balancer on the crankshaft drive sprocket and align the mounting holes. Note that the pulley can only go on one way (see illustration).

8   Install the pulley mounting bolts and tighten them to the torque listed in this Chapter's Specifications.

9   The remaining installation steps are the reverse of removal.

**11.7  Align the crankshaft pulley mounting holes and install the bolts**

## 12  Crankshaft front oil seal and housing - replacement

▶ **Refer to illustrations 12.2 and 12.4**

1   Remove the timing belt and crankshaft sprocket (see Section 5).

2   Note how far the seal is recessed in the bore, then carefully pry it out of the front cover with a screwdriver or seal removal tool. Don't scratch the housing bore or damage the crankshaft in the process (if the crankshaft is damaged, the new seal will end up leaking).

➡**Note: If a seal removal tool is unavailable, you can thread two self tapping screws (180 degrees apart from one another) into the front seal to pry the seal out (see illustration).**

3   Clean the bore in the housing and coat the outer edge of the new seal with engine oil or multi-purpose grease. Apply multi-purpose grease to the seal lip.

4   Using a socket with an outside diameter slightly smaller than the outside diameter of the seal, carefully drive the new seal into place with a hammer (see illustration). Make sure it's installed squarely and driven in to the same depth as the original. If a socket isn't available, a short section of large diameter pipe will also work. Check the seal after installation to make sure the spring didn't pop out of place.

5   If the front oil seal housing needs to be removed for access to other components, simply loosen the housing mounting bolts and remove the housing from the engine while noting the installed position of the fasteners. The front oil seal housing can be removed with or without the front oil seal. In some instances the front oil seal removal and installation is easier with the front housing removed, since the seal can be placed on a workbench and driven straight in and out of the bore with no special tools or adapters.

6   If the front of the oil pan gasket was damaged while removing the housing, use a razor blade or utility knife to cut the pan gasket off flush with front edge of the cylinder block. This part of the oil pan gasket will be replaced with RTV sealant upon installation of the cover.

7   Before installing the front cover, make sure the mating surfaces of the cover, the cylinder block and the oil pan rail are perfectly clean. Use a hard plastic or wood scraper to remove all traces of gasket material. Take particular care when cleaning the front cover, as aluminum alloy is easily damaged.

8   Apply a 3/16-inch (5 mm) bead of RTV sealant to the oil pan flange.

9   Locate the oil seal housing gasket over the dowels on the engine block and install the oil seal housing.

➡**Note: Be sure to lubricate the oil seal lip before installing the front cover onto the engine. This will aid the installation process and prevent dry start ups, which may damage the seal and lead to future oil leaks.**

10  Tighten the front oil seal housing bolts in several steps to the torque listed in this Chapter's Specifications.

11  Reinstall the crankshaft sprocket and timing belt (see Section 5).

12  Run the engine and check for oil leaks at the front seal.

**12.2  If a seal removal tool is unavailable, the front seal can also be removed with self tapping screws (as shown) to pry the seal out**

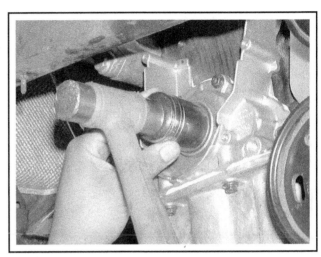

**12.4  Lubricate the seal lip and drive the new crankshaft seal into place with a seal driver or a large socket and a hammer**

# 13  Oil pan - removal and installation

▶ **Refer to illustrations 13.19a, 13.19b, 13.20a and 13.20b**

## REMOVAL

1   Set the parking brake and block the rear wheels.

2   Raise the front of the vehicle and support it securely on jackstands.

3   Remove the splash shield under the engine, if equipped.

4   Place the radiator support panel in the service position (see Chapter 11).

➡**Note: Access to the front of the engine is limited, therefore it will be necessary to remove the front bumper and position the radiator support panel forward enough to allow removal of the surrounding components.**

5   Drain the engine oil (see Chapter 1). Remove the oil dipstick.

6   Remove the serpentine drivebelts (see Chapter 1).

7   Remove the engine cooling fan (see Chapter 3).

8   On air-conditioned models, unbolt the air conditioning compressor drivebelt tensioner roller.

9   Remove the torque rod support bracket side braces (see Section 18).

10  Unbolt the torque rod support bracket. On models equipped with air conditioning, this bracket is attached to the front of the engine block just below the crankshaft pulley. On models without air conditioning, this bracket is attached to the bottom of the radiator support panel.

11  Detach the starter motor cables from under the engine mount by cutting the plastic cable ties.

12  Disconnect the electrical connector from the oil temperature sending unit at the bottom of the oil pan if equipped.

13  Loosen the rear bolt on the right-hand transmission mount a few turns, then unscrew and remove the front bolt. On automatic transmission models, repeat this procedure on the left-hand transmission mount as well.

14  On manual transmission models, unscrew the left-hand transmission mount nut until it is flush with the end of the bolt (approximately four turns).

15  Connect a suitable hoist to the engine, then raise it as far as possible without damaging or stretching the coolant hoses and wiring.

16  Remove the left engine mount and the bracket from the engine block (see Section 18).

17  Unscrew the flange bolts and disconnect the turbocharger oil return line from the side of the oil pan (see Chapter 4).

18 Position the oil return line aside and remove the gasket. Be sure to replace the gasket with a new one upon installation.

19 Unscrew and remove the oil pan bolts. Note that on manual transmission models, the two rear oil pan bolts are accessed through a cut-out in the flywheel - turn the flywheel as necessary to align the cut-out (see illustrations).

20 Remove the oil pan and the gasket if equipped. If it is stuck, tap it gently with a mallet to free it (see illustrations).

## INSTALLATION

♦ **Refer to illustrations 13.25, 13.26 and 13.28**

21 Use a scraper to remove all traces of old sealant from the block and oil pan. Clean the mating surfaces with lacquer thinner or acetone.

➡**Note: Some models use RTV sealant to seal the oil pan-to-cylinder block mating surface while others use a solid gasket mate-**

rial to seal the oil pan-to-cylinder block mating surface. Always reseal the oil pan-to-cylinder block mating surface with the same type of sealing material that it was originally equipped.

22 Make sure the threaded bolt holes in the block are clean.

23 Check the oil pan flange for distortion, particularly around the bolt holes. Remove any nicks or burs as necessary.

24 Inspect the oil pump pick-up tube assembly for cracks and a blocked strainer. If the pick-up was removed, clean it thoroughly and install it now, using a new O-ring or gasket. Tighten the nuts/bolts to the torque listed in this Chapter's Specifications.

25 If the vehicle originally used RTV sealant, apply a 3/16-inch (5 mm) bead of RTV sealant to the oil pan flange (see illustration).

➡**Note: The oil pan must be installed within 5 minutes once the sealant has been applied.**

26 If the vehicle originally used a solid gasket, apply a 3/16 inch (5 mm) bead of RTV sealant to the corners where the front and rear oil seal housings meet the engine block (see illustration). Then place the gasket

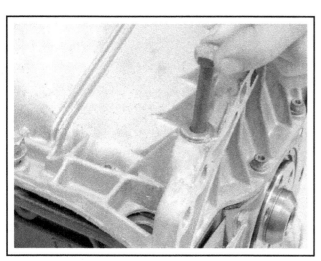

**13.19a  Removing the oil pan bolts**

**13.19b  On models with a manual transmission and an aluminum oil pan, it will be necessary to align the cut-outs in the flywheel with the cut-outs on the oil pan to access the rear bolts**

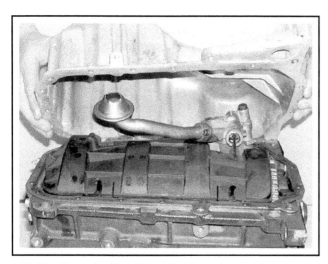

**13.20a  Removing the oil pan . . .**

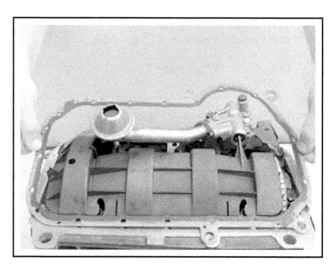

**13.20b  . . . and gasket**

on the oil pan and install the four corner bolts to align the bolt holes in the gasket with the bolt holes in the oil pan.

27  Carefully position the oil pan on the engine block and install the oil pan-to-engine block bolts loosely.

28  On models equipped with an aluminum oil pan, install the oil pan to bellhousing bolts and tighten them just a little more than finger tight. This should draw the oil pan flush with the bellhousing. If the transmission is not installed in the vehicle use a straightedge to align the rear surface of the oil pan with the rear face of the intermediate plate (see illustration).

29  Working from the center out, tighten the oil pan-to-engine block bolts to the torque listed in this Chapter's Specifications in three or four steps.

30  On models equipped with an aluminum oil pan, final tighten the oil pan-to-bellhousing bolts to the torque listed in this Chapter's Specifications.

31  The remainder of installation is the reverse of removal.

➡ **Note: Be sure to follow the sealant manufacturer's recommendations on curing times and allow the sealant to properly cure before adding oil.**

32  Run the engine and check for oil pressure and leaks.

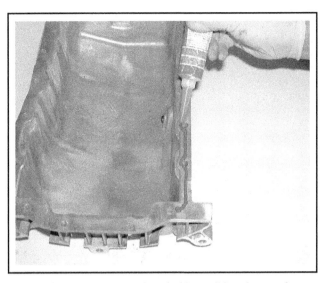

**13.25  On vehicles not equipped with a solid gasket, apply a 3/16-inch (5 mm) bead of RTV sealant as shown to the oil pan sealing flange**

**13.26  On vehicles equipped with a solid gasket, apply RTV sealant to the corners where the front and rear oil seal housings meet the engine block**

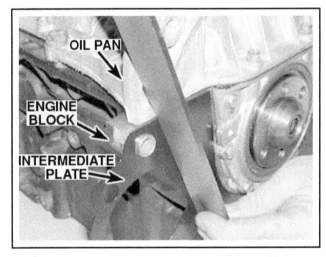

**13.28  Using a straight edge to align the rear of the oil pan with the intermediate plate**

---

## 14  Oil pump - removal, inspection and installation

♦ **Refer to illustrations 14.2 and 14.12**

### AEB AND ATW ENGINES

#### Removal

1  Remove the oil pan (see Section 13).

2  Release and remove the baffle plate from the bottom of the crankcase (see illustration).

3  Unscrew the bolts and remove the suction pipe from the oil pump. Remove the O-ring seal.

4  Unscrew and remove the large oil pump mounting bolts, then withdraw the pump from the block. If necessary, pull the subframe downward to provide sufficient room to remove the oil pump.

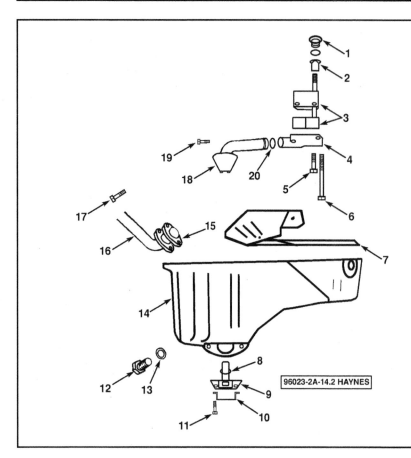

**14.2 Oil pump and related components, AEB and ATW engines**

1 Plug
2 Drive gear
3 Oil pump housing and gears
4 Oil pump cover
5 Oil pump cover bolts
6 Oil pump mounting bolt
7 Baffle plate
8 O-ring
9 Oil level/temperature sender
10 Cover plate
11 Bolt
12 Drain plug
13 Gasket
14 Oil pan
15 Gasket
16 Oil return line from turbocharger
17 Bolt
18 Oil pick-up
19 Bolt
20 O-ring

96023-2A-14.2 HAYNES

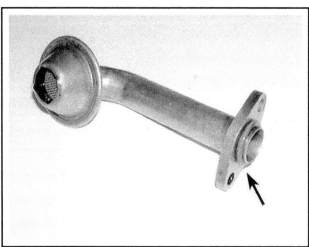

**14.12 Always replace the pick-up tube O-ring (arrow)**

## Inspection

5 Unscrew the two bolts and lift off the cover. Note that the cover incorporates the pressure relief valve.

6 Clean all components with solvent, then inspect them for wear and damage.

7 Using a feeler blade, check the backlash between the gears, and compare that with the Specifications listed at the end of this Chapter. Similarly check the endplay of the gears, using a straight edge across the end face of the pump.

8 If damage or wear is noted, replacement of the entire oil pump assembly is recommended.

## Installation

9 Clean the contact faces, then install the cover to the oil pump and tighten the bolts to the specified torque.

10 Prime the pump with oil by pouring oil into the pick-up tube aperture while turning the driveshaft.

11 Clean the oil pump and block, then install the oil pump, insert the mounting bolts, and tighten them to the specified torque.

12 Locate a new O-ring seal on the end of the pick-up tube (see illustration). Insert the tube into the oil pump, then install the bolts and tighten them to the specified torque.

13 Reinstall the remaining parts in the reverse order of removal.

14 Add oil, start the engine and check for oil pressure and leaks.

15 Recheck the engine oil level.

## AUG AND AWM ENGINES

### Removal

▶ **Refer to illustrations 14.20, 14.21, 14.24 and 14.25**

16 Remove the timing belt and the crankshaft drive sprocket (see Section 5).

17 After the timing belt has been removed, reinstall the passenger side engine mount to support the engine during the removal and installation of the oil pan and pump.

18 Remove the oil pan (see Section 13).

**14.20 Remove the oil pump drive chain tensioner (AUG and AWM engines)**

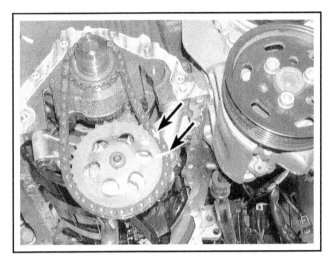

**14.21 Mark the face of the oil pump drive chain and the driven sprocket so the chain can be installed in the same position and direction of travel (AUG and AWM engines)**

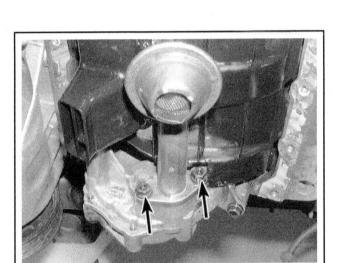

**14.24 Oil pump pick-up tube mounting bolts (AUG and AWM engines)**

**14.25 Oil pump mounting bolts (AUG and AWM engines)**

19 Remove the crankshaft front oil seal housing (see Section 12).

20 Remove the oil pump drive chain tensioner (see illustration).

21 Mark the face of the oil pump driven sprocket and chain so they can be installed the same way (see illustration).

**➡Note: Installing the oil pump drive chain in a different direction from which it was originally installed will accelerate wear and cause premature failure.**

22 Wedge a screwdriver in one of the holes of the oil pump driven sprocket to hold the sprocket from turning as the driven sprocket retaining bolt is loosened, then remove the oil pump driven sprocket retaining bolt.

23 Remove the oil pump drive chain and driven sprocket from the engine.

24 Detach the two retaining bolts and separate the oil pump pick-up tube from the oil pump body. Lift out the pick-up tube and the O-ring (see illustration).

25 Remove the mounting bolts and separate the oil pump and the oil

pan baffle from the engine (see illustration).

## Inspection

26 Clean all components with solvent, then inspect them for wear and damage.

27 If damage or wear is noted, replacement of the entire oil pump assembly is recommended.

## Installation

**♦ Refer to illustration 14.35**

28 Use a scraper to remove all traces of sealant and old gasket material from the pump case and engine block, then clean the mating surfaces with lacquer thinner or acetone.

29 Place the oil pump on the engine block over the dowel pins and install the two bolts facing the front of the engine block.

30 Position the oil pan baffle plate in place and install the remaining mounting bolt.

31 Tighten the bolts to the torque listed in this Chapter's Specifications in several steps. Follow a criss-cross pattern to avoid warping the case.

32 Using a new O-ring, install the oil pick-up tube assembly (see illustration 14.12). Tighten the fasteners to the torque listed in this Chapter's Specifications.

33 Reinstall the oil pump drive chain and the driven sprocket in the original direction from which it was removed. Be sure to tighten the driven sprocket bolt to the correct torque Specifications.

34 Install the drive chain tensioner and bolt loosely on the engine block.

35 Engage the tab on the tensioner spring on the inside lip of the engine block (see illustration). Allow the tensioner to apply spring tension to the drive chain and tighten the tensioner retaining bolt.

36 Reinstall the remaining parts in the reverse order of removal. Be sure to install the crankshaft front oil seal housing to the engine block before installing the oil pan.

37 Add oil, start the engine and check for oil pressure and leaks.

38 Recheck the engine oil level.

**14.35 Engage the tab on the drive chain tensioner spring on the inside lip of the engine block (AUG and AWM engines)**

## 15 Flywheel/driveplate - removal and installation

▸ Refer to illustrations 15.3, 15.5 and 15.10

### ✳✳ CAUTION:

**The manufacturer recommends replacing the flywheel/driveplate bolts with new ones whenever they are removed.**

## REMOVAL

1 Raise the vehicle and support it securely on jackstands, then refer to Chapter 7 and remove the transaxle. If it's leaking, now would be a very good time to replace the front pump seal/O-ring (automatic transaxle) or input shaft seal (manual transaxle).

2 On manual transaxle equipped vehicles, remove the pressure plate and clutch disc (see Chapter 8). Now is a good time to check/replace the clutch components.

3 Use a center punch or paint to make alignment marks on the flywheel/driveplate and crankshaft to ensure correct alignment during reinstallation (see illustration).

4 Remove the bolts that secure the flywheel/driveplate to the crankshaft. If the crankshaft turns, wedge a screwdriver in the ring gear teeth to jam the flywheel.

5 Remove the flywheel/driveplate from the crankshaft. Since the flywheel is fairly heavy, be sure to support it while removing the last bolt. Automatic transaxle equipped vehicles have spacers on both sides of the driveplate (see illustration). Keep them with the driveplate.

**15.3 Mark the flywheel/driveplate and the crankshaft so they can be reassembled in the same relative positions**

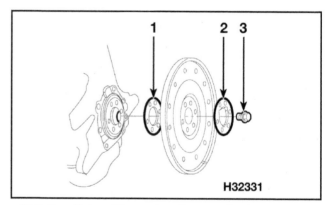

H32331

**15.5 On vehicles with an automatic transaxle, there is a spacer plate on each side of the driveplate - mark each plate as it is removed so they can be installed back in the same position**

| | |
|---|---|
| 1 Backing plate | 3 Bolt |
| 2 Shim | |

## INSTALLATION

6   Clean the flywheel to remove grease and oil. Inspect the surface for cracks, rivet grooves, burned areas and score marks. Light scoring can be removed with emery cloth. Check for cracked and broken ring gear teeth. Lay the flywheel on a flat surface and use a straightedge to check for warpage.

7   Clean and inspect the mating surfaces of the flywheel/driveplate and the crankshaft. If the crankshaft rear seal is leaking, replace it before reinstalling the flywheel/driveplate.

8   Position the flywheel/driveplate and spacer (if used) against the crankshaft. Be sure to align the marks made during removal. Note that some engines have an alignment dowel or staggered bolt holes to ensure correct installation. Before installing the bolts, apply thread locking compound to the threads.

9   Wedge a screwdriver in the ring gear teeth to keep the flywheel/driveplate from turning and tighten the bolts to the torque listed in this Chapter's Specifications. Follow a criss-cross pattern and work up to the final torque in three or four steps.

10   On vehicles equipped with automatic transaxles, measure the installed height at three equal places around the driveplate and compare the average measurement to this Chapter's Specifications (see illustra-

tion). If the measurement is incorrect, the driveplate must be removed and shimmed to the proper height.

11   The remainder of installation is the reverse of the removal procedure.

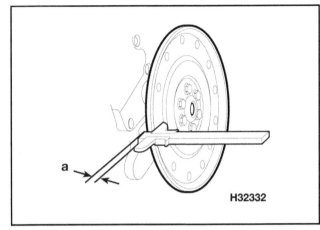

**15.10 The driveplate installed height can be measured with a machinist's ruler**

## 16  Rear main oil seal - replacement

▶ **Refer to illustrations 16.3, 16.5, 16.7, 16.8a and 16.8b**

1   The transaxle and the flywheel/driveplate must be removed from the vehicle for this procedure (see Chapter 7 and Section 15).

2   The rear oil seal and housing are an integral part which must be removed and replaced together as a unit, however the rear seal and housing can be replaced without removing the oil pan. On vehicles equipped with a solid oil pan gasket, it will be necessary to remove the oil pan and fit a new gasket if the oil pan gasket is damaged while removing the housing. Be sure to reinstall the oil pan after installing the housing.

➡**Note: Some models use RTV sealant to seal the oil pan-to-cylinder block mating surface while others use a solid gasket material to seal the oil pan-to-cylinder block mating surface. Always reseal the oil pan to cylinder block mating surface with the same type of sealing material that it was originally equipped.**

3   Remove the rear oil seal housing mounting bolts and remove the housing from the engine (see illustration).

4   Before installing a new rear seal and housing, make sure the mating surfaces of the cover, the cylinder block and the oil pan rail are perfectly clean. Use a hard plastic or wood scraper to remove all traces

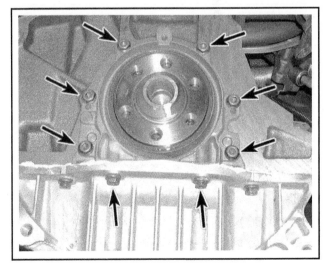

**16.3  Crankshaft rear oil seal housing mounting bolts (arrows)**

of gasket material. Take particular care when cleaning the rear housing, as aluminum alloy is easily damaged.

5    Install a new gasket on the cylinder block (see illustration).

6    If the vehicle was originally equipped with a solid oil pan gasket material and it was not damaged during removal, apply a 3/16-inch (5 mm) bead of RTV sealant to the corners where the oil pan gasket meets the engine block.

7    If the vehicle was originally equipped with RTV sealant as the oil pan gasket material, apply a 3/16-inch (5 mm) bead of RTV sealant to the oil pan sealing surface on the rear oil seal housing flange (see illustration).

➡**Note: Do not apply sealant to the oil pan as this will cause the sealant to be pushed inward into the oil pan as the rear oil seal housing is positioned on the block.**

8    Install the rear oil seal housing over the dowels onto the engine.

➡**Note 1: Be sure to lubricate the oil seal lip before installing the housing onto the engine. This will aid the installation process and prevent dry start ups, which may damage the seal and to lead to future oil leaks.**

➡**Note 2: Some oil seals come equipped with a installation tool to prevent damage to the oil seal as it is being installed. First locate the tool on the end of the crankshaft and slide the oil seal and housing into place on the block (see illustrations).**

9    Tighten the rear oil seal housing bolts evenly, in several steps to the torque listed in this Chapter's Specifications. Be sure to follow the sealant manufacturers recommendations on curing times and allow the sealant to properly cure before adding oil.

10   The remaining steps are the reverse of removal.

16.5  Installing a new gasket on the cylinder block

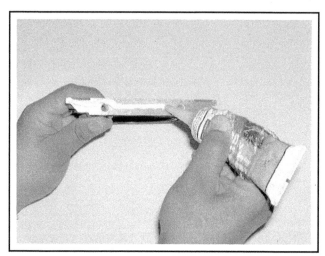

16.7  If the vehicle originally used RTV sealant as the oil pan gasket material, apply a 3/16-inch (5 mm) bead of RTV sealant to the oil pan sealing surface on real oil seal housing

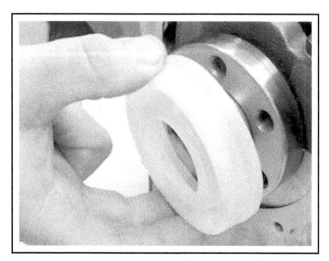

16.8a  Install the seal installation tool over the end of the crankshaft . . .

16.8b  . . . and slide the oil seal and housing over the tool

## 17 Intermediate shaft oil seal - replacement

1   Remove the intermediate shaft sprocket as described in Section 5.

2   Drill two small holes into the existing oil seal, diagonally opposite each other. Thread two self-tapping screws into the holes, and using two pairs of pliers, pull on the heads of the screws to extract the oil seal (see illustration 12.2). Take great care to avoid drilling through into the seal flange. An alternative method is to unbolt the flange, remove the inner O-ring from the inner groove, and press out the seal.

3   Clean out the seal flange and sealing surface of the camshaft by wiping it with a lint-free cloth. Remove any chips or burrs that may cause the seal to leak.

4   Lubricate the lip and outer edge of the new oil seal with clean engine oil, and start it in its housing by hand initially making sure that the closed end of the seal is facing outwards.

5   Using a hammer and a socket of suitable diameter, drive the seal squarely into its housing (see illustration 12.4).

➡**Note: Select a socket that bears only on the hard outer surface of the seal, not the inner lip which can easily be damaged.**

6   If the flange has been removed, replace the O-ring, then reinstall the flange and tighten the bolts to the torque listed in this Chapter's Specifications.

7   Reinstall the intermediate shaft sprocket (see Section 5).

## 18 Engine mounts - check and replacement

1   Engine mounts seldom require attention, but broken or deteriorated mounts should be replaced immediately or the added strain placed on the driveline components may cause damage or wear.

## CHECK

2   During the check, the engine must be raised slightly to remove the weight from the mounts.

3   Raise the vehicle and support it securely on jackstands, then position a jack under the engine oil pan. Place a large block of wood between the jack head and the oil pan, then carefully raise the engine just enough to take the weight off the mounts. Do not position the wood block under the drain plug.

### ✳✳ WARNING:

**DO NOT place any part of your body under the engine when it's supported only by a jack!**

4   Remove the splash shield under the engine, if equipped. Check the mounts to see if the rubber is cracked, hardened or separated from the metal plates. Sometimes the rubber will split right down the center.

5   Check for relative movement between the mount plates and the engine or frame (use a large screwdriver or pry bar to attempt to move the mounts). If movement is noted, lower the engine and tighten the mount fasteners.

6   Rubber preservative should be applied to the mounts to slow deterioration.

## REPLACEMENT

7   Raise the vehicle and support it securely on jackstands (if not already done). Support the engine as described in Step 3.

8   Remove the splash shield under the engine, if equipped.

### Front torque rod (models without air conditioning)

▶ **Refer to illustration 18.11**

9   Unbolt the bracket from the bottom of the radiator support.

10   Pull the rubber stopper off the torque rod and replace it with a new one if it's damaged.

11   If necessary, the torque rod can be removed or replaced by unbolting it from the front of the cylinder block (see illustration).

12   Installation is the reverse of the removal procedure.

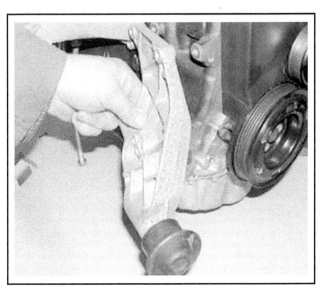

**18.11 Removing the front torque rod from the engine (non-air-conditioned models)**

### Front torque rod (models with air conditioning)

♦ **Refer to illustration 18.13**

13 Unscrew the bolts and detach the aluminum stop from the torque rod bracket on the front of the engine (see illustration). Slide the aluminum stop over the rubber, then remove the rubber from the torque rod and remove the stop.

14 If necessary, the torque rod bracket can be unbolted from the front of the engine, and the side supports also removed. Note that the air conditioning drivebelt tensioner must removed first to access some of the bracket bolts.

15 Install the new rubber and bracket using the reverse of the removal procedure.

### Driver and passenger side engine mounts

♦ **Refer to illustrations 18.21, 18.22 and 18.23**

---

### ❊❊ WARNING:

**The weight of the entire engine will be supported by the transaxle mounts and the front torque rod during this procedure. Never place any part of your body directly under the engine when performing this procedure.**

---

16 Set the parking brake and block the rear wheels.

17 Raise the front of the vehicle and support it securely on jackstands.

18 Remove the splash shield under the engine, if equipped.

19 Detach the starter motor cables from the lower engine mount brackets by cutting the plastic cable ties and maneuvering the wires out of the plastic retainers.

20 Mark the relationship of the engine mount locating dowel to the bottom of the aluminum engine mount bracket for installation purposes, then unscrew and remove the nuts from the bottom of left and right engine mounts.

21 Support the front stabilizer bar with a floor jack, then unscrew and remove the engine mount bracket bolts. The front two bolts must be unscrewed first, then the rear bolts (see illustration). Lower the aluminum engine mount brackets together with the stabilizer bar to the ground.

➡**Note: Lowering the stabilizer bar and the lower engine mount brackets will allow the front half of the subframe to be lowered slightly. Never loosen the rear subframe bolts as front wheel alignment will be changed.**

22 Remove the engine mount from the bracket on the engine block (see illustration).

23 If necessary, unbolt the mounting bracket from the side of the cylinder block (see illustration).

24 Installation is the reverse removal procedure.

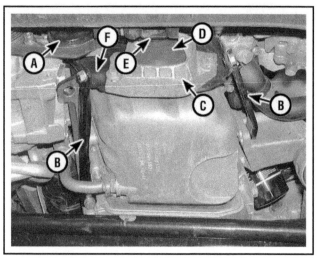

**18.13 Front torque rod and related components (air-conditioned models)**

A    Air conditioning compressor drivebelt tensioner
B    Side braces
C    Aluminum stop
D    Rubber stop
E    Torque rod
F    Torque rod bracket

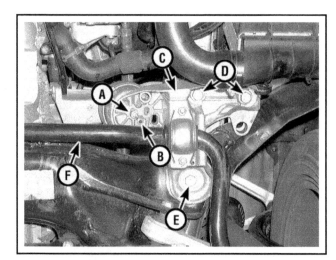

**18.21 Lower engine mount bracket and related components**

A    Engine mount retaining nut (lower)
B    Locating dowel
C    Lower engine mount bracket
D    Lower engine mount bracket bolts (front)
E    Lower engine mount bracket bolt (rear)
F    Front stabilizer bar

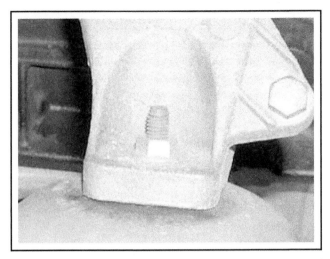

**18.22  Engine mount upper retaining nut**

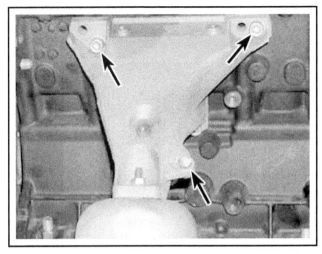

**18.23  Upper engine mount bracket bolts**

## Specifications

### General

| | |
|---|---|
| Engine designation | AEB, ATW, AUG, AWM |
| Displacement | 109 cubic inches (1.8 liters) |
| Bore and stroke | 3.189 x 3.401 inches (81 x 86 mm) |
| Cylinder numbers (drivebelt end-to-transaxle end) | 1-2-3-4 |
| Firing order | 1-3-4-2 |

### Camshaft

| | |
|---|---|
| Endplay | 0.008 inch (0.2 mm) |
| Journal oil clearance | 0.004 inch (0.1 mm) |
| Runout | 0.0004 inch (0.01 mm) |

### Driveplate

| | |
|---|---|
| Driveplate installed height | 1.024 to 1.102 inches (26 to 28 mm) |

FRONT
OF
VEHICLE

④
③
②
❶

72031-1-SPECS HAYNES

**Cylinder numbering**

| Torque specifications | Ft-lbs (unless otherwise indicated) | Nm |
|---|---|---|

➡**Note: One foot-pound (ft-lb) of torque is equivalent to 12 inch-pounds (in-lbs) of torque. Torque values below approximately 15 ft-lbs are expressed in inch-pounds, since most foot-pound torque wrenches are not accurate at these smaller values.**

| | | |
|---|---|---|
| Camshaft sprocket bolt | 48 | 65 |
| Camshaft bearing cap bolts/nuts | 84 in-lbs | 10 |
| Camshaft drive chain tensioner | 84 in-lbs | 10 |
| Crankshaft drive sprocket bolt* | | |
| Step one | 66 | 90 |
| Step two | Tighten an additional 90 degrees | |
| Crankshaft pulley bolts | 18 | 25 |
| Crankshaft oil seal housing bolts (front and rear) | | |
| M6 | 84 in-lbs | 10 |
| M8 | 18 | 25 |
| Cylinder head bolts* (in sequence - see illustration 10.23a) | | |
| Step one | 30 | 40 |
| Step two | 44 | 60 |
| Step three | Tighten an additional 90 degrees | |
| Step four | Tighten an additional 90 degrees | |
| Flywheel/driveplate bolts* | | |
| Step one | 44 | 60 |
| Step two | Tighten an additional 90 degrees | |
| Intake manifold bolts | 84 in-lbs | 10 |
| Exhaust manifold nuts | 22 | 30 |
| Intermediate shaft sprocket bolt | 48 | 65 |
| Oil pan-to-engine block bolts | | |
| M6 | 132 in-lbs | 15 |
| M10 | 33 | 45 |
| Oil pan-to-bellhousing bolts | 33 | 45 |
| Oil pump cover bolts | 84 in-lbs | 10 |
| Oil pump pick-up tube | 84 in-lbs | 10 |
| Oil pump mounting bolts | | |
| AEB and ATW engines | 18 | 25 |
| AUG and AWM engines | 132 in-lbs | 15 |
| Oil pump sprocket bolt | 15 | 21 |
| Timing belt cover-to-block bolts | 84 in-lbs | 10 |
| Timing belt tensioner | | |
| Damper-to-engine block bolts | 84 in-lbs | 10 |
| Pulley adjustment screw | 18 | 25 |
| Pulley retaining nut | 84 in-lbs | 10 |
| Idler wheel-to-damper body | 18 | 25 |
| Valve cover-to-cylinder head nuts | 84 in-lbs | 10 |
| Water pump housing-to-engine block bolts | 22 | 30 |
| Engine mount bolts/nuts* | | |
| M6 | 84 in-lbs | 10 |
| M8 | 18 | 25 |
| M10 | 33 | 45 |
| M12 | 44 | 60 |
| Engine mount brackets-to-engine block bolts | 22 | 30 |

*Replace with new fastener(s)*

## Section

## Reference to other Chapters

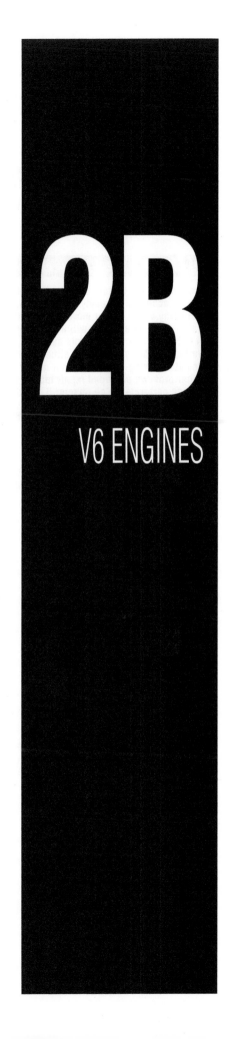

# 2B

## V6 ENGINES

## 1   General information

♦ **Refer to illustrations 1.1 and 1.2**

### ※※ CAUTION 1:

These models are equipped with an anti-theft radio. Before performing a procedure that requires disconnecting the battery, make sure you have the proper activation code.

### ※※ CAUTION 2:

Disconnecting the battery can cause driveability problems that require a scan tool to rectify. Additionally, disconnecting the battery may cause one or more warning lights on the instrument panel to illuminate, which will also require the use of a scan tool to turn off. Most scan tools available to the public do not have the capability to perform either of these tasks, which will necessitate taking the vehicle to a dealer service department or other properly equipped repair facility after service work has been performed. See Chapter 5, Section 1 for the use of an auxiliary voltage input device ("memory saver") before disconnecting the battery and for other precautions related to battery disconnection.

➡Note: The engine cover must removed before performing many of the procedures in this Chapter (see illustration).

This Part of Chapter 2 is devoted to in-vehicle repair procedures for the 2.8L Single Overhead Camshaft (SOHC) and Dual Overhead Camshaft (DOHC) V6 engines. Both engines utilize a cast-iron engine block with aluminum cylinder heads. SOHC engines use two valves per cylinder, while DOHC engines use five valves per cylinder (three intake and two exhaust). Hydraulic lifters that ride directly below the camshafts are used to actuate the valves on both engines. The aluminum cylinder heads are equipped with pressed-in valve guides and hardened valve seats. The oil pump on SOHC engines and early DOHC engines is driven off the front of the crankshaft and mounted to the front of the engine block. On later model DOHC engines, the manufacturer used a modified oil pump, which is mounted below the front of the engine and is chain driven from the crankshaft. To positively identify SOHC engines, locate the engine code letters stamped on the right hand side of the cylinder block between the cylinder head and the power steering pump. To positively identify DOHC engines, locate the engine code on the front of the right cylinder head. To positively identify an early model DOHC engine with a gear driven oil pump from a later model DOHC engine with a chain driven oil pump visually inspect the valve covers (see illustration).

All information concerning engine removal and installation and engine block and cylinder head overhaul can be found in Part C of this Chapter.

The following repair procedures are based on the assumption that the engine is installed in the vehicle. If the engine has been removed from the vehicle and mounted on a stand, many of the steps outlined in this Part of Chapter 2 will not apply.

The Specifications included in this Part of Chapter 2 apply only to the procedures contained in this Part. Part C of Chapter 2 contains the Specifications necessary for cylinder head and engine block rebuilding.

**1.1  The engine cover is fastened to the engine with four retaining screws (A) and the side covers are fastened with two screws (B) - note that it will be necessary to remove the oil filler cap before removing the left side cover on this model (DOHC engine shown, SOHC engine similar)**

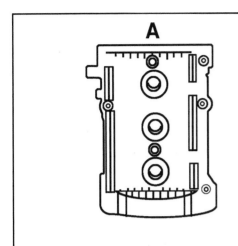 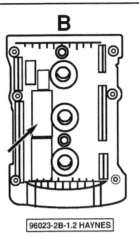

**1.2** Later model DOHC engines equipped with a chain driven oil pump can be identified by the raised area on the valve cover (B) - the earlier version DOHC engine (A) has a flat smooth appearing valve cover

96023-2B-1.2 HAYNES

## 2 Repair operations possible with the engine in the vehicle

Many major repair operations can be accomplished without removing the engine from the vehicle.

Clean the engine compartment and the exterior of the engine with some type of degreaser before any work is done. It will make the job easier and help keep dirt out of the internal areas of the engine.

Depending on the components involved, it may be helpful to remove the hood to improve access to the engine as repairs are performed (refer to Chapter 11 if necessary). Cover the fenders to prevent damage to the paint. Special pads are available, but an old bedspread or blanket will also work.

If vacuum, exhaust, oil or coolant leaks develop, indicating a need for gasket or seal replacement, the repairs can generally be made with the engine in the vehicle. The intake and exhaust manifold gaskets, oil pan gasket, crankshaft oil seals and cylinder head gasket are all accessible with the engine in place.

Exterior engine components, such as the intake and exhaust manifolds, the oil pan, the oil pump, the water pump, the starter motor, the alternator and the fuel system components can be removed for repair with the engine in place.

Since the cylinder head can be removed without pulling the engine, camshaft and valve component servicing can also be accomplished with the engine in the vehicle. Replacement of the timing belt and pulleys is also possible with the engine in the vehicle.

In extreme cases caused by a lack of necessary equipment, repair or replacement of piston rings, pistons, connecting rods and rod bearings is possible with the engine in the vehicle. However, this practice is not recommended because of the cleaning and preparation work that must be done to the components involved.

## 3 Top Dead Center (TDC) - locating

**♦ Refer to illustration 3.6**

1   Top Dead Center (TDC) is the highest point in the cylinder that each piston reaches as it travels up-and-down when the crankshaft turns. Each piston reaches TDC on the compression stroke and again on the exhaust stroke, but TDC generally refers to piston position on the compression stroke. The timing marks on the vibration damper/crankshaft pulley at the front of the engine are typically referenced to the number one piston at TDC, but on the engines covered by this manual the SOHC V6 uses the number one piston to reference TDC and the DOHC engine uses the number three cylinder to reference TDC.

2   Positioning the piston(s) at TDC is an essential part of procedures such as timing belt and sprocket replacement.

3   In order to bring any piston to TDC, the crankshaft must be turned using one of the methods outlined below. When looking at the timing belt end of the engine, normal crankshaft rotation is clockwise.

### ✳✳ WARNING:

**Before beginning this procedure, be sure to place the transmission in Park or Neutral, set the parking brake and remove the ignition key.**

a)  *The preferred method is to turn the crankshaft with a large socket and breaker bar attached to the large bolt threaded into the center of the crankshaft pulley.*

b)  *A remote starter switch, which may save some time, can also be used. Attach the switch leads to the S (switch) and B (battery) terminals on the starter motor. Once the piston is close to TDC, use a socket and breaker bar as described in the previous paragraph.*

c)  *If an assistant is available to turn the ignition switch to the Start position in short bursts, you can get the piston close to TDC without a remote starter switch. Use a socket and breaker bar as described in Paragraph a) to complete the procedure.*

4   Disable the ignition and fuel systems by disconnecting the primary electrical connectors at the ignition coil modules (see Chapter 5) and removing the fuel pump fuse.

5   Remove the spark plugs (see Chapter 1) and install a compression gauge in the number one cylinder on SOHC engines or the number three cylinder on DOHC engines. Turn the crankshaft clockwise with a socket and breaker bar as described above.

6   When the piston approaches TDC, compression will be noted on the compression gauge. Continue turning the crankshaft until the notch in the crankshaft damper is aligned with the TDC mark on the front

cover (see illustration). At this point the number one cylinder should be at TDC on the compression stroke on SOHC engines or the number three cylinder should be at TDC on the compression stroke on DOHC engines. If the marks aligned but there was no compression, the piston was on the exhaust stroke; continue rotating the crankshaft 360-degrees (1-turn) and line-up the marks.

➡**Note: If a compression gauge is not available, TDC can be obtained by simultaneously aligning the marks on the camshaft retainer plates and the marks on the crankshaft damper with the TDC mark on the front cover (see illustrations 5.11a and 5.11b).**

7    After the engine is set at TDC on the compression stroke, TDC for any of the remaining cylinders can be located by turning the crankshaft 120 degrees and following the firing order (refer to the Specifications). Rotating the engine 120 degrees past TDC #1 on SOHC engines will put the engine at TDC compression for cylinder #4. Rotating the engine 120 degrees past TDC #3 on DOHC engines will put the engine at TDC compression for cylinder #6.

**3.6  Align the notch on the crankshaft drivebelt pulley with the arrow on the timing belt cover**

## 4    Valve cover - removal and installation

### REMOVAL

▶ **Refer to illustrations 4.4 and 4.5**

1    Remove the engine cover and side covers if equipped (see illustration 1.1).

2    Remove the spark plug wires from the spark plugs.

3    If equipped, loosen the clip securing the crankcase breather hose to the valve cover, then position the breather hose aside.

4    Remove any electrical wiring, ground straps or hoses attached to the valve cover (see illustration).

5    Remove the retaining nuts and detach the valve cover from the cylinder head (see illustration).

6    If the cover is stuck to the head, bump the end with a block of wood and a hammer to jar it loose. If that doesn't work, try to slip a flexible putty knife between the head and cover to break the seal.

### ❋❋ CAUTION:

**Don't pry at the cover-to-head joint or damage to the sealing surfaces may occur, leading to oil leaks after the cover is reinstalled.**

### INSTALLATION

▶ **Refer to illustrations 4.9 and 4.11**

7    The mating surfaces of the housing or cylinder head and cover must be clean when the cover is installed. Use a gasket scraper to remove all traces of sealant and old gasket material including the spark plug tube seal gasket, then clean the mating surfaces with lacquer thinner or acetone. If there's residue or oil on the mating surfaces when the cover is installed, oil leaks may develop. On DOHC engines, inspect the rubber end plug at the rear of the cylinder head on for cracks and damage. Now would be a good time to replace it, if damage has occurred.

8    On SOHC engines, apply RTV sealant to the corners of the front and rear camshaft bearing caps where they meet the cylinder head.

9    On DOHC engines, apply RTV sealant to the corners of the camshaft front bearing cap and to the camshaft drive chain tensioner where

they meet the cylinder head (see illustration).

10  Position a new valve cover gasket over the studs on the cylinder head.

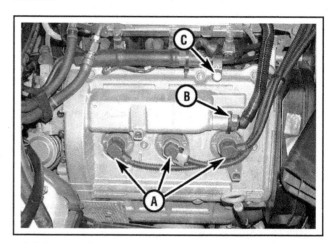

**4.4  Remove the spark plug wires (A), the crankcase breather hose (B) and the power steering hose retaining tab (C) (right valve cover on a DOHC engine shown)**

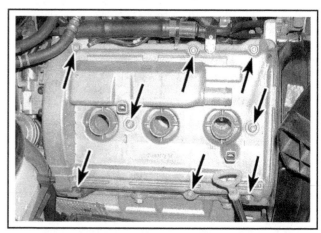

**4.5  Valve cover mounting bolts (DOHC engine)**

**4.9 Apply sealant to the corners of the camshaft drive chain tensioner and to the two points at the front camshaft bearing cap where they meet the cylinder head**

**4.11 Install the spark plug tube seal gasket with the tab (arrow) facing the timing belt end of the engine**

11  On DOHC engines, install the spark plug tube grommet gasket over the studs on the cylinder head with the index marks facing the timing belt end of the engine (see illustration).

12  Install the valve cover and tighten the retaining nuts to the torque listed in this Chapter's Specifications in several steps.

13  Reinstall the remaining parts, run the engine and check for oil leaks.

## 5   Timing belt and sprockets - removal, inspection and installation

### ✳ WARNING:

Wait until the engine is completely cool before beginning this procedure.

### ✳ CAUTION:

Do not rotate the crankshaft or the camshaft separately during this procedure with the timing belt removed as damage to valves may occur. Only rotate the camshaft a few degrees as necessary to align the camshaft alignment tool with the holes in the camshaft retainer plate.

➡Note 1: Two special tools are required to complete this procedure, so read through the entire Section and obtain the special tools before beginning work.

➡Note 2: This procedure is illustrated using a DOHC engine. The procedure for SOHC engines is technically the exact same except where noted in the text, although the physical appearance of the parts and the exact location of clips and fasteners on SOHC engines may vary from what is depicted.

## REMOVAL

### ✳ CAUTION ✳
The timing system is complex. Severe engine damage will occur if you make any mistakes. Do not attempt this procedure unless you are highly experienced with this type of repair. If you are at all unsure of your abilities, consult an expert. Double-check all your work and be sure everything is correct before you attempt to start the engine.

▶ Refer to illustrations 5.8, 5.9a, 5.9b, 5.9c, 5.10, 5.11a, 5.11b, 5.12a, 5.12b, 5.14, 5.15, 5.16, 5.18a, 5.18b, 5.19a and 5.19b

1   Disconnect the cable from the negative terminal of the battery.

### ✳ CAUTION 1:

These models are equipped with an anti-theft radio. Before performing a procedure that requires disconnecting the battery, make sure you have the proper activation code.

### ✳ CAUTION 2:

Disconnecting the battery can cause driveability problems that require a scan tool to rectify. Additionally, disconnecting the battery may cause one or more warning lights on the instrument panel to illuminate, which will also require the use of a scan tool to turn off. Most scan tools available to the public do not have the capability to perform either of these tasks, which will necessitate taking the vehicle to a dealer service department or other properly equipped repair facility after service work has been performed. See Chapter 5, Section 1 for the use of an auxiliary voltage input device ("memory saver") before disconnecting the battery and for other precautions related to battery disconnection.

2   Raise the front of the vehicle and support it securely on jackstands.

3   Working under the vehicle, remove the lower splash shield below the engine.

4   Place the radiator support panel in the service position (see Chapter 11).

**5.8  Drivebelt tensioner retaining bolt (arrow)**

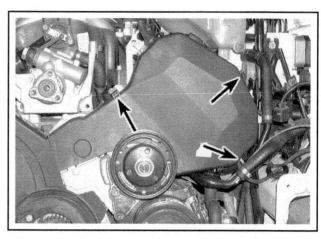

**5.9a  Left timing belt cover retaining clips (arrows)**

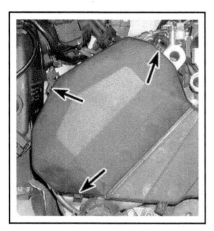

**5.9b  Right timing belt cover retaining clips (arrows)**

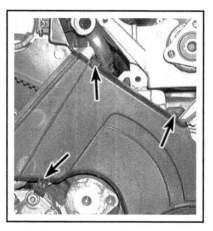

**5.9c  Center timing belt cover retaining clips (arrows) (DOHC engine only)**

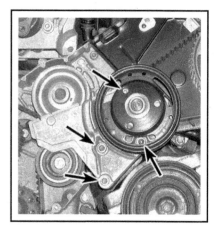

**5.10  Mechanical cooling fan pulley bracket bolts**

➡Note: Access to the front of the engine is limited, therefore it will be necessary to remove the front bumper and position the radiator support panel forward enough to allow removal of the timing belt and the surrounding components. In some instances from the lack of sufficient working room and from a procedural standpoint it may be easier to remove the entire support panel to gain full access to the components at the front of the engine. The extra time spent removing the panel will expedite some procedures significantly.

5   Working from above in the engine compartment, remove the engine cover (see illustration 1.1) and the spark plugs (see Chapter 1 if necessary).

6   On DOHC engines, remove the mechanical cooling fan (see Chapter 3).

7   Loosen the power steering pump pulley bolts (see Chapter 10) then remove the accessory drivebelt (see Chapter 1).

8   Remove the power steering pump pulley and the drivebelt tensioner from the front of the engine (see illustration).

9   Unclip and remove the upper timing belt covers from the engine (see illustrations).

➡Note: SOHC engines have two upper timing belt covers (a left and a right), while DOHC engines use three upper timing belt covers (a left, a right and a center cover).

10   On DOHC engines, remove the cooling fan drivebelt pulley

bracket (see illustration).

11   Rotate the engine in the normal direction of rotation (clockwise) until the engine is located at TDC (see Section 3). Verify that the larger holes in the camshaft sprocket plates are aligned inward facing each other (see illustrations). If the larger holes are facing outward, rotate the

**5.11a  When the engine is positioned at TDC, the larger holes on the camshaft sprocket plate (arrow) . . .**

5.11b . . . should face inward toward each other

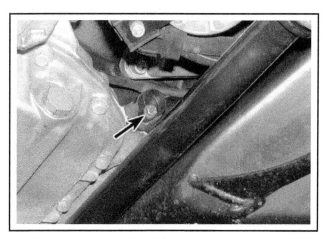

5.12a  Remove the sealing plug . . .

5.12b . . . and screw the TDC locking
tool into place to keep the crankshaft
from turning

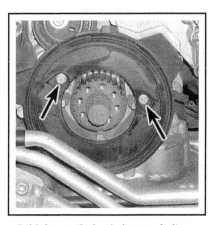

5.14  Lower timing belt cover bolts

5.15  If you intend to re-use the
timing belt, apply directional marks
on the belt and the rear timing
belt cover

crankshaft one complete revolution further (360 degrees) until the mark
on the crankshaft pulley is aligned again with the mark on the lower
cover.

12 Remove the crankshaft position sensor (SOHC) or the lower seal-

5.16  On DOHC engines, rotate the timing belt tensioner
clockwise until the holes in the tensioner housing and the
tensioner arm align, then insert a small diameter drill bit to
lock the tensioner in place

ing plug (DOHC) and install the TDC locking tool (see illustrations).

➥**Note: The drilled hole in the crankshaft must be aligned with
the sight window in the oil pan before installing the TDC locking
tool.**

13 Loosen the crankshaft drive sprocket retaining bolt and the crankshaft pulley bolts, then remove the crankshaft pulley (see Section 11).

➥**Note: Loosening the drive sprocket bolt is only required if the
crankshaft drive sprocket is expected to be removed. It is not
typically necessary to remove the drive sprocket when you're
simply replacing a timing belt, but it will need to be removed if
you are replacing the crankshaft front oil seal or housing. If you
do remove the drive sprocket, obtain a new bolt (the manufacturer doesn't recommend re-using it).**

14 Detach the retaining bolts from the lower timing belt cover and
remove the cover (see illustration).

15 If you plan to re-use the timing belt, apply match marks on the
sprocket and belt and an arrow indicating direction of travel on the belt
(see illustration).

16 Release the timing belt tensioner (see illustration).

➥**Note: On SOHC engines the tensioner pulley is released by
simply loosening the center bolt.**

17 Remove the timing belt from the engine, taking care to avoid
twisting or kinking it excessively.

5.18a Using the camshaft alignment tool to hold the camshaft sprockets as the retaining bolt is loosened

5.18b Using a two jaw puller to separate the camshaft sprocket from the taper on the camshaft

5.19a With the crankshaft drive sprocket retaining bolt removed . . .

5.19b . . . the crankshaft sprocket is easily detached from the engine

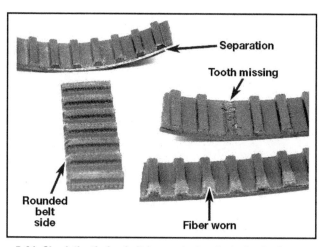

5.21 Check the timing belt for cracked and missing teeth - wear on one side of the belt indicates sprocket misalignment problems

➡Note: If you're removing the upper part of the belt only, for camshaft seal replacement or cylinder head removal, it isn't necessary to detach the belt from the crankshaft sprocket.

18 The camshaft sprockets must be loosened and separated from the taper on the camshafts so they rotate freely during the timing belt tensioning process. Using the camshaft alignment tool to hold the sprockets in place, loosen the camshaft sprocket retaining bolts. Remove the crankshaft locking tool and rotate the crankshaft counterclockwise slightly so the engine is not at TDC, then remove the camshaft alignment tool. Use a two jaw puller to separate the sprockets from the taper on the camshafts and reinstall the sprocket bolts finger tight, making sure that the sprockets rotate freely on the taper of the camshafts (see illustrations). If the camshaft sprockets are damaged or need to be removed for other procedures such as cylinder head or camshaft removal, simply remove the camshaft sprocket and the sprocket retaining plate from the end of the camshaft.

19 If the crankshaft sprocket is worn or damaged, or if you need to replace the crankshaft front oil seal, remove the drive sprocket retaining bolt which was loosened in Step 13 and detach the crankshaft sprocket from the crankshaft (see illustrations).

## INSPECTION

▶ Refer to illustration 5.21

### ✳✳ CAUTION:

Do not bend, twist or turn the timing belt inside out. Do not allow it to come in contact with oil, coolant or fuel. Do not turn the crankshaft or camshaft more than a few degrees (if necessary for tooth alignment) while the timing belt is removed.

20 Spin the timing belt tensioner pulley (which is the pulley mounted to the left as your facing the front of the engine) and the idler pulley (which is the pulley mounted to the right) and check the bearings for smooth operation and excessive play. Also inspect the remaining timing belt sprockets, the tensioner and the tensioner lever (if equipped) for any obvious damage. Replace all worn parts as necessary.

21 Examine the belt for evidence of contamination by coolant or lubricant. If this is the case, find the source of the contamination before progressing any further. Check the belt for signs of wear or damage, particularly around the leading edges of the belt teeth (see illustration).

**5.25  Timing belt routing**

**5.27  Camshaft alignment tool properly installed - be sure the camshaft sprockets can rotate freely from the taper on the camshaft**

### ✳✳ CAUTION:

**If the belt appears to be in good condition and can be re-used, it is essential that it is reinstalled the same way around, otherwise accelerated wear will result, leading to premature failure.**

22  Replace the belt if its condition is in doubt; the cost of belt replacement is negligible compared with potential cost of the engine repairs, should the belt fail in service. Similarly, if the belt is known to have covered more than 60,000 miles, it is prudent to replace it regardless of condition, as a precautionary measure.

## INSTALLATION

### ✳✳ CAUTION ✳✳

**Before starting the engine, carefully rotate the crankshaft by hand through at least two full revolutions (use a socket and breaker bar on the crankshaft pulley centerbolt). If you feel any resistance, STOP! There is something wrong - most likely, valves are contacting the pistons. You must find the problem before proceeding. Check your work and see if any updated repair information is available.**

▶ **Refer to illustrations 5.25, 5.27 and 5.29**

23  If any of the timing sprockets where removed for inspection or needed replacement, install them back onto the engine now making sure the camshaft sprockets rotate freely on the taper and the sprocket plate is engage with the flat(s) on the nose of the camshaft. Install the camshaft alignment tool onto the camshafts, then rotate the crankshaft clockwise slightly to realign TDC and install the crankshaft locking tool again. If the timing belt tensioner, the tensioner lever (DOHC) or the tensioner pulley were removed, reinstall them now and tighten the bolts to the torque listed at the end of this Chapter.

24  Loop the timing belt loosely under the crankshaft sprocket.

### ✳✳ CAUTION:

**Observe the direction of rotation markings on the belt.**

25  Engage the timing belt teeth with the crankshaft sprocket, then

**5.29  On DOHC engines, final timing belt tension is achieved by pretensioning the pulley with torque wrench in the counterclockwise direction to 132 in-lbs (15 Nm)**

maneuver it into position over the idler pulley and around the camshaft sprocket. Ensure the belt teeth seat correctly on the sprocket, then install the belt around the water pump pulley and over the remaining camshaft sprocket and then around the timing belt tensioner pulley (see illustration).

26  Ensure that the 'front run' of the belt is taut and all the slack is in the section of the belt that passes over the tensioner pulley.

27  Install the camshaft alignment tool into the holes on the camshaft sprocket plates. This synchronizes the camshafts in the proper position for timing belt installation (see illustration).

28  On SOHC engines, tension the timing belt by rotating the tensioner pulley clockwise with an Allen wrench until the belt is tight, then tighten the center bolt. The belt is correctly tensioned when you can rotate the belt 90 degrees using your thumb and index finger half-way between the water pump and one of the camshaft sprockets. If the belt won't rotate 90 degrees the belt is too tight. If it rotates more than 90 degrees it's too loose.

29  On DOHC engines, tension the timing belt by rotating the tensioner pulley clockwise with an Allen wrench until you can release the drill bit and unlock the tensioner. Take a minute or two to let the tensioner find it's extended position, then using a torque wrench rotate the tensioner pulley counterclockwise until a torque value of 132 in-lbs (15 Nm) is reached. This pretensions the belt and allows the tensioner to extend to its final destination (see illustration).

30 Tighten the camshaft sprocket bolts to the torque listed in this Chapter's Specifications.

31 Remove the camshaft alignment tool and the TDC locking tool.

32 Rotate the crankshaft through two complete revolutions. Reset the engine to TDC with reference to Section 3. Reinstall the TDC locking tool and the camshaft alignment tool and check the alignment of the camshaft sprockets with the crankshaft locked at TDC. The dowels on the camshaft alignment tool should fit into the holes on the sprocket plates easily. Also re-check the timing belt tension and adjust it, if necessary.

33 Installation of the remaining components is the reverse of removal.

## 6  Camshafts and lifters - removal and installation

### ✳✳ CAUTION:

**Performing this procedure may cause a trouble code to be set, which will require taking the vehicle to a dealer service department (or other repair shop equipped with the necessary scan tool) to have the cam sensor(s) synchronized and the trouble code cleared.**

➡Note: The camshafts and lifters should always be thoroughly inspected before installation and camshaft endplay should always be checked prior to camshaft removal. Although the hydraulic lifters are self adjusting and require no periodic service, there is an in-vehicle procedure for checking excessively noisy hydraulic lifters. Refer to Chapter 2C for the camshaft and lifter inspection procedures.

### REMOVAL

▶ Refer to illustrations 6.4, 6.6, 6.8, 6.9, 6.10, 6.11, 6.12, 6.13a and 6.13b

1  Remove the engine cover (see illustration 1.1).

2  Remove the valve cover (see Section 4). Also remove the oil deflector(s) to expose the camshaft(s) if equipped.

3  Remove the timing belt and the camshaft sprockets making sure the engine is rotated counterclockwise slightly off TDC when removing the camshaft sprockets (see Section 5).

4  Remove the rear timing belt cover on the cylinder head from which the camshaft(s) are to be removed (see illustration).

### SOHC engine

5  On SOHC engines, remove the cam-shaft position sensor from the left cylinder head and the camshaft plug from the end of the right cylinder head. If only one camshaft is to be removed, only remove the

**6.4  Rear timing belt cover mounting bolts (DOHC engine)**

affected part from the camshaft you intend to remove.

6  Mark the location of the camshaft bearing caps from 1 to 4, starting at the front (timing belt) end. Also mark arrows indicating the front of the engine (see illustration).

7  Loosen the center two bearing caps (No.2 and No.3), then alternately loosen the No.1 and No.4 bearing caps. Detach the caps and remove the camshaft(s) from the cylinder head.

### ✳✳ CAUTION:

**Keep the caps in order. They must go back in the same location and direction they were removed from.**

### DOHC engine

8  Remove the camshaft position sensor (see Chapter 6) and the camshaft sensor reluctor ring from the end of the intake camshaft (see illustration) on the cylinder head which the camshafts are to be

**6.6  On SOHC engines, simply mark the bering caps from 1 to 4 starting with the front (timing belt end of the engine)**

**6.8  On DOHC engines remove the camshaft position reluctor ring (arrow)**

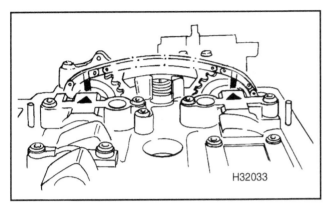

**6.9 With the notches in the rear drive chain sprockets aligned with the arrows on the rear bearing caps, apply match marks on the chain with a permanent marker - be sure to wipe the oil from the chain and sprockets first, so the marker will adhere to the components - the number of rollers between the marks should be exactly 16**

**6.10 Using a special tool to compress the camshaft drive chain tensioner**

removed. If all four camshafts are to be removed, remove both camshaft position sensors and reluctor rings.

9    Mark the position of the camshaft drive chain in relationship to the sprockets and the marks on the rear bearing cap (see illustration). This will ensure that the drive chain is installed in exactly the same direction and position from which it was removed.

10  Compress the camshaft drive chain tensioner. This can be accomplished by purchasing a special tool from the dealer service department which is specifically made for this purpose (see illustration) or by fabricating a home made tool using a threaded rod, several nuts, a plastic cable tie and a small metal plate (see Chapter 2A for illustrations of the homemade tool).

11  Mark the location of the camshaft bearing caps from 1 to 7. On the right cylinder head start with the double bearing cap at the front (timing belt) end. Also mark arrows indicating the front of the engine. On the left cylinder head start with the caps just behind the drive chain (see illustration) also marking arrows indicating the front of the engine. The camshaft adjuster and the bearing cap adjacent to the adjuster will be numbered No.6 and No.7 on both cylinder heads. Loosen the bearing caps nuts alternately in the following order:

1)  *Loosen and remove the No. 6 camshaft adjuster*
2)  *Loosen and remove the No. 1, 3, 5 and 7 bearing caps from the intake and exhaust camshafts (when loosening the intake camshaft bearing caps, loosen them diagonally)*
3)  *Loosen and remove the No. 2 and 4 bearing caps from the intake and exhaust camshafts*

12  Remove the camshafts and the drive chain tensioner as an assembly from the cylinder head (see illustration).

### ❋❋ CAUTION:

**Keep the caps in order. They must go back in the same location and direction they were removed from. Separate the tensioner and the drive chain from the camshafts on a workbench.**

➡**Note: Do not mix the camshafts for the left and right cylinder heads together - they are specific to each cylinder head and will not interchange.**

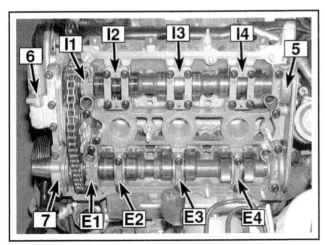

**6.11 The camshaft bearing caps should be marked as shown with a number and letter stamp or a marker to ensure correct reinstallation**

**6.12 Removing the camshafts and the drive chain tensioner as an assembly from the cylinder head**

6.13a The lifters can be removed from the cylinder head by hand or with a magnet . . .

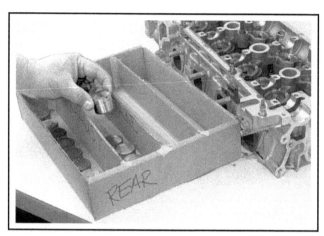

6.13b . . . and stored in individually marked plastic bags or a divided box as shown - be sure to keep them in order with their respective valves

6.18 On SOHC engines apply a small amount of RTV sealant to the mating surfaces of the No.1 and No.4 bearing caps

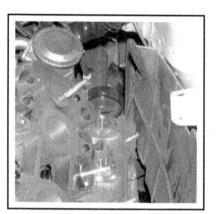

6.22a A metal end plug is installed at the rear of each cylinder head in the exhaust camshaft bore (DOHC engines)

6.22b A half-round rubber seal is used to seal oil below the camshaft adjuster on each cylinder head

## All engines

13 Remove the lifters from the cylinder head, keeping the them in order with their respective valve and cylinder (see illustrations).

### ❋❋ CAUTION:

**Keep the lifters in order. They must go back in the same location they were removed from.**

14 Inspect the camshafts and lifters as described in Chapter 2C.

## INSTALLATION

▶ **Refer to illustrations 6.18, 6.22a, 6.22b, 6.22c, 6.23, 6.27a, 6.27b and 6.29**

15 Apply clean engine oil onto the sides of the hydraulic lifters, and install them into position in their bores in the cylinder head. Push them down until they contact the valves, then lubricate the camshaft lobe contact surfaces.

## SOHC engine

16 Lubricate the camshaft and cylinder head bearing journals with

clean engine oil. Then carefully lower the camshaft into position on the cylinder head so that the camshaft plate is aligned horizontally in the cylinder head with the big hole facing inward (see illustration 5.11a). Support the ends of the shaft as it is inserted, to avoid damaging the lobes and journals.

17 If you're working on the camshaft in the left cylinder head, place the camshaft position sensor oil seal over the rear of the camshaft and install it in position before the bearing caps.

18 Oil the upper surfaces of the camshaft bearing journals, then apply a small amount of RTV sealant to the mating surface of No.1 and No.4 bearing caps and install them over the camshaft (see illustration). Tighten the retaining nuts alternately and diagonally to 100 in-lbs.

19 Install the No.2 and No.3 bearing caps and tighten the retaining nuts alternately and diagonally to 100 in-lbs.

20 Tighten the nuts to the final specified torque starting with the center caps and working towards the end caps.

21 If removed, install a new O-ring around the end plug and install the end plug in the rear of the right cylinder head.

## DOHC engine

22 Clean the mating surfaces of the drive chain tensioner and the cylinder head, then install the end plugs at the front and rear of the cylinder head and a new drive chain tensioner gasket (see illustrations).

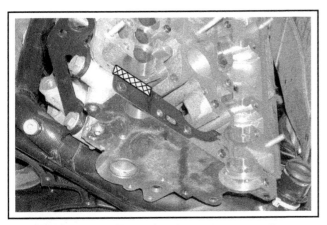

**6.22c Install a new drive chain tensioner/camshaft adjuster gasket and apply a small amount of RTV sealant at the shaded area**

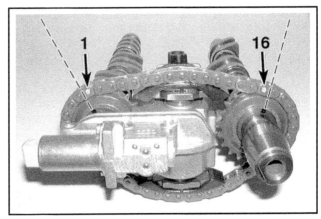

**6.23 With the drive chain tensioner installed, the drive chain should have 16 rollers between the notches on the camshafts**

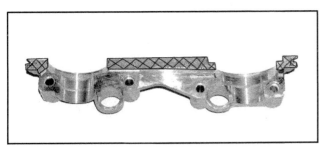

**6.27a Apply RTV sealant to the double bearing cap at the shaded areas . . .**

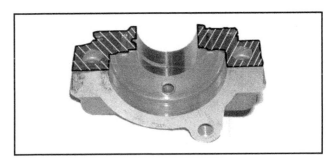

**6.27b . . . and to the bearing cap adjacent to the camshaft adjuster**

Apply a small amount of RTV sealant to the designated area on the drive chain tensioner gasket (see illustration).

23 Align the marks on the drive chain (made previously) with the notches on the camshaft drive gears and install the chain over the drive gears. If you're installing a new drive chain, install the drive chain with exactly 16 rollers between the notches on the drive gears. Note that the exhaust camshaft notch is slightly off center. In either case verify that there are 16 rollers between the notches of the intake and exhaust camshafts (see illustration).

24 Compress the camshaft drive chain tensioner with the special tool and insert it between the drive chain and the camshafts.

25 Lubricate the camshaft and cylinder head bearing journals with clean engine oil. If you're working on the camshafts in the left cylinder head, place the camshaft position sensor oil seal over the end of the intake camshaft and install it with the camshafts. Carefully lower the camshafts, drive chain and tensioner and the camshaft position sensor oil seal as an assembly into position on the cylinder head. Support the ends of the shaft as it is inserted, to avoid damaging the lobes and journals.

26 Install the drive chain tensioner over the dowels on the cylinder head and tighten the bolts to the torque listed in this Chapter's Specifications.

27 Install the camshaft bearing caps in the reverse order of removal (see Step 11). Be sure to apply a thin film of RTV sealant to the mating surfaces of the bearing caps that face the outside of the cylinder head before installing them (see illustrations). After the bearing caps have been tightened, remove the drive chain tensioning tool from the tensioner. Reconfirm that there are 16 rollers between the notches of the intake and exhaust camshaft and that the notches align with the arrows on the caps.

**6.29 Gently drive the new camshaft oil seals into place with the spring side facing the engine**

## All engines

28 Clean the oil seal housing bores and lubricate the lips of the new camshaft oil seals with clean engine oil and locate them over the ends of the camshafts. Slide the seals along the camshaft until they locate squarely in the housing bores.

29 Using a socket with an outside diameter slightly smaller than the outside diameter of the seal, carefully drive the new seals into place with a hammer (see illustration). Make sure they're installed squarely and driven in to the same depth as the original. If a socket isn't available, a short section of pipe will also work.

➡️**Note 1: On DOHC engines this will include installing two camshaft oil seals at the front of the right cylinder head and one on the left cylinder head.**

→**Note 2: On SOHC engines this will include installing two camshaft oil seals, one at the front of the right cylinder head and one on the front of the left cylinder head.**

30 On DOHC engines, install the camshaft position sensors (see Chapter 6).

31 Install the camshaft sprockets and tighten the bolts finger tight. Be sure the camshaft sprockets rotate freely on the camshaft taper.

32 Install the timing belt (see Section 5). When installing the timing belt, install the camshaft alignment tool on the camshafts first, then rotate the crankshaft clockwise slightly, to TDC and install crankshaft locking tool.

33 The remainder of installation is the reverse of removal.

### ✳✳ CAUTION:

**If new lifters were used, wait at least 30 minutes before starting the vehicle to allow the lifters to bleed down. Failure to do so will result in serious engine damage.**

## 7    Valve springs, retainers and seals - replacement

▶ **Refer to illustrations 7.4, 7.8, 7.10, 7.15a, 7.15b, 7.15c and 7.17**

→**Note: Broken valve springs and defective valve stem seals can be replaced without removing the cylinder heads. Two special tools and a compressed air source are normally required to perform this operation, so read through this Section carefully and rent or buy the tools before beginning the job.**

1 Remove the valve cover referring to Section 4. Then refer to Section 6 and remove the camshaft and lifters.

2 Remove the spark plug from the cylinder which has the defective component. If all of the valve stem seals are being replaced, all of the spark plugs should be removed.

3 Turn the crankshaft until the piston in the affected cylinder is at TDC. If you're replacing all of the valve stem seals, begin with cylinder number one and work on the valves for one cylinder at a time. Move from cylinder-to-cylinder following the firing order sequence (see the Specifications listed at the end of this Chapter).

→**Note: On DOHC engines start with the number three cylinder and follow the firing order back around to the number four cylinder. This is because DOHC engines synchronize TDC for the timing belt removal procedure off the number three cylinder (see Sections 3 and 5).**

4 Thread an adapter into the spark plug hole (see illustration) and connect an air hose from a compressed air source to it. Most auto parts stores can supply the air hose adapter.

→**Note: Many cylinder compression gauges utilize a screw-in fitting that may work with your air hose quick-disconnect fitting.**

5 Apply compressed air to the cylinder. The valves should be held in place by the air pressure.

### ✳✳ WARNING:

**If the cylinder isn't exactly at TDC, air pressure may force the piston down, causing the engine to quickly rotate. DO NOT leave a wrench on the crankshaft drive sprocket bolt or you may be injured by the tool.**

6 Stuff shop rags into the cylinder head holes around the valves to prevent parts and tools from falling into the engine.

7 Using a socket and a hammer gently tap on the top of the each valve spring retainer several times. This will break the bond between the valve keeper and the spring retainer and allow the keeper to separate from the valve spring retainer as the valve spring is compressed.

8 Use a valve spring compressor to compress the spring. Remove the keepers with small needle-nose pliers or a magnet (see illustration).

→**Note: Several different types of tools are available for compressing the valve springs with the head in place. Be sure to purchase or rent the "Import type" that bolts to the top of the**

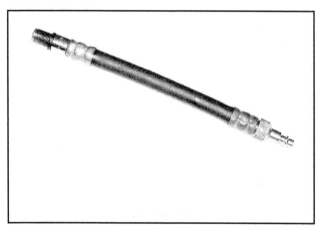

**7.4  You'll need an air hose adapter this long to reach down into the spark plug wells on the cylinder head - they're commonly available at auto parts stores**

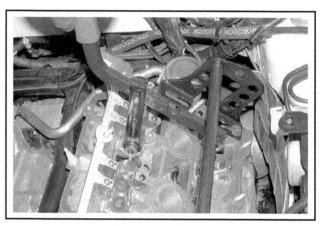

**7.8  While the valve spring tool is compressing the spring, remove the keepers with a small magnet or pliers**

7.10 The old valve stem seals can be removed with a pair of needle-nose pliers

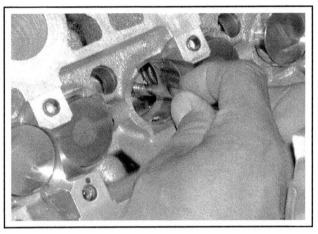

7.15a Install the protective plastic sleeve over the valve end face to avoid damage to the valve seal as the seal is installed

cylinder head. This type uses a support bar across the cylinder head for leverage as the valve spring is compressed. The lack of clearance surrounding the valve springs on these engines prohibits other types of valve spring compressors from being used.

9   Remove the valve spring and retainer.

➡**Note: If air pressure fails to retain the valve in the closed position during this operation, the valve face or seat may be damaged. If so, the cylinder head will have to be removed for repair.**

10 Remove the old valve stem seals, noting differences between the intake and exhaust seals (see illustration).

11 Wrap a rubber band or tape around the top of the valve stem so the valve won't fall into the combustion chamber, then release the air pressure.

12 Inspect the valve stem for damage. Rotate the valve in the guide and check the end for eccentric movement, which would indicate that the valve is bent.

13 Move the valve up-and-down in the guide and make sure it doesn't bind. If the valve stem binds, either the valve is bent or the guide is damaged. In either case, the head will have to be removed

for repair.

14 Reapply air pressure to the cylinder to retain the valve in the closed position, then remove the tape or rubber band from the valve stem.

15 Lubricate the valve stem with engine oil and install a new seal on the valve guide (see illustrations).

16 Install the valve spring and the spring retainer in position over the valve.

17 Compress the valve spring and carefully position the keepers in the groove. Apply a small dab of grease to the inside of each keeper to hold it in place (see illustration).

18 Remove the pressure from the spring tool and make sure the keepers are seated.

19 Disconnect the air hose and remove the adapter from the spark plug hole.

20 Install the camshaft, lifters, timing belt and the valve cover by referring to the appropriate Sections.

21 Install the spark plug(s) and hook up the wire(s).

22 Start and run the engine, then check for oil leaks and unusual sounds coming from the valve cover area.

7.15b Push a new valve stem seal over the valve and down to the top of the guide, then remove the plastic installation tool

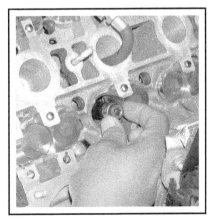

7.15c Gently tap the new seal in place on the guide with a socket

7.17 Apply a small dab of grease to each keeper as shown here before installation - it'll hold them in place on the valve stem as the spring is released

## 8  Intake manifold - removal and installation

▶ **Refer to illustrations 8.4, 8.9 and 8.11**

### ❋❋ WARNING:

**Wait until the engine is completely cool before beginning this procedure.**

1  Remove the engine cover (see illustration 1.1).
2  Partially drain the engine coolant (see Chapter 3).
3  Refer to Chapter 4 and remove the throttle cable, the cruise control actuator rod (if equipped) and the throttle body
4  Detach the crankcase breather tube from the valve covers and the rear of the intake manifold (see illustration).
5  Remove the ignition coil pack (see Chapter 5).
6  Relieve the fuel system pressure and remove the fuel rail and injectors (see Chapter 4).
7  On SOHC engines remove the EGR valve from the side of the manifold (see Chapter 6).
8  Label and detach any remaining vacuum lines or electrical wiring that would interfere with removal of the intake manifold.
9  Remove the mounting nuts/bolts then detach the manifold from the engine (see illustration). On SOHC engines it will be necessary to remove the upper plenum first, since several of the bolts protrude through the lower manifold and into the cylinder head.
10  On SOHC engines, use a scraper to remove all traces of gasket material from the manifold and the cylinder head. On DOHC engines, peel the O-ring gaskets from the receiver grooves on the intake manifold. Clean the mating surfaces on the cylinder head and the manifold with lacquer thinner or acetone. If the gaskets were leaking, check the manifold for warpage with a straight edge. If the manifold is warped, replacement is the only alternative.
11  On DOHC engines, install the new gaskets in the receiver grooves on the manifold (see illustration), then position the manifold on the engine and install the bolts. On SOHC engines, apply some contact cement on the manifold side of the intake gaskets and affix the gaskets to the manifold. This will hold the gaskets to the manifold as the manifold is lowered on the engine. Lower the manifold onto the engine and

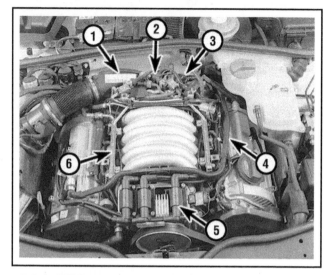

**8.4  On DOHC engines, remove the following components to allow removal of the intake manifold**

| | | | |
|---|---|---|---|
| *1* | *Air intake duct* | *4* | *Crankcase breather hoses* |
| *2* | *Solenoid valve bracket* | *5* | *Ignition coil pack* |
| *3* | *Throttle body* | *6* | *Fuel rail and injectors* |

place the upper plenum gasket in place, then position the upper plenum on top of the lower manifold and install the bolts.
12  Tighten the nuts/bolts in three or four equal steps to the torque listed in this Chapter's Specifications. Work from the center out towards the ends to avoid warping the manifold.
13  Install the remaining parts in the reverse order of removal.
14  Before starting the engine, check the throttle linkage for smooth operation.
15  Refill and bleed the cooling system (see Chapter 1). Run the engine and check for coolant and vacuum leaks.
16  Road test the vehicle and check for proper operation of all accessories, including the cruise control system.

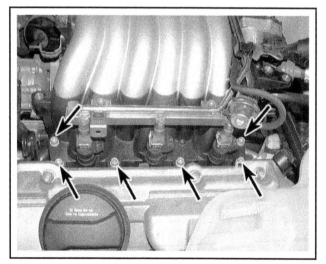

**8.9  Intake manifold bolts on a DOHC engine (left side shown, right side similar)**

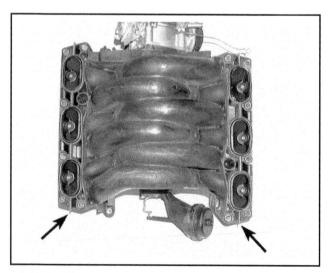

**8.11  On DOHC engines, press the O-ring type intake manifold gaskets into the receiver groove on the manifold**

## 9 Exhaust manifold - removal and installation

▸ **Refer to illustrations 9.7a, 9.7b and 9.8**

### ✳✳ WARNING 1:

The engine must be completely cool before beginning this procedure.

### ✳✳ WARNING 2:

The air conditioning system is under high pressure. DO NOT loosen any fittings or remove any components until after the system has been discharged. Air conditioning refrigerant should be properly discharged into an EPA-approved container at a dealership service department or an automotive air conditioning facility. Always wear eye protection when disconnecting air conditioning system fittings.

## REMOVAL

1   Remove the engine cover (see illustration 1.1).
2   Remove the air intake duct and the air cleaner housing (see Chapter 4).
3   Raise the front of the vehicle and support it securely on jackstands.
4   Remove the splash guard from below the engine compartment.

5   Refer to Chapter 4 and detach the front exhaust pipe from the exhaust manifold.
6   Have the air conditioning refrigerant discharged and recycled by an air conditioning technician (see **Warning** above), then remove the radiator support panel (see Chapter 11).
7   Remove the exhaust manifold heat shields (see illustrations). If you're removing the right manifold detach oil level dipstick tube.
8   Detach the nuts and remove the exhaust manifold and gasket through the front of the engine compartment (see illustration). If necessary, apply penetrating oil to the manifold mounting nuts/bolts to help facilitate removal.

## INSTALLATION

9   Use a scraper to remove all traces of old gasket material and carbon deposits from the manifold and cylinder head mating surfaces. If the gasket was leaking, have the manifold checked for warpage at an automotive machine shop and resurfaced if necessary.
10   Position a new gasket over the cylinder head studs.
11   Install the manifold and thread the mounting nuts into place. Make sure to use hi-temp anti-seize compound on the exhaust manifold fasteners.
12   Working from the center out, tighten the nuts/bolts to the torque listed in this Chapter's Specifications in three or four equal steps.
13   Reinstall the remaining parts in the reverse order of removal.
14   Run the engine and check for exhaust leaks.

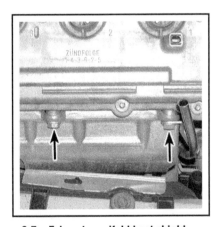

9.7a  Exhaust manifold heat shield upper (arrows) . . .

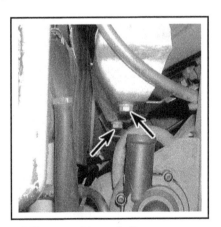

9.7b  . . . and lower bolts

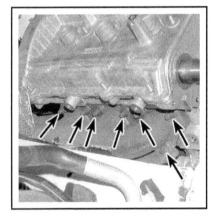

9.8  Exhaust manifold mounting nuts (lower nuts not visible in this photo)

## 10 Cylinder head - removal and installation

### ✳✳ CAUTION 1:

Performing this procedure may cause a trouble code to be set, which will require taking the vehicle to a dealer service department (or other repair shop equipped with the necessary scan tool) to have the cam sensor(s) synchronized and the trouble code cleared.

### ✳✳ CAUTION 2:

The engine must be completely cool before beginning this procedure.

➥Note: The cylinder head can be removed with the exhaust manifold attached.

## REMOVAL

♦ **Refer to illustrations 10.5, 10.7 and 10.10**

1   Disconnect the cable from the negative terminal of the battery.

### ※ CAUTION 1:

**These models are equipped with an anti-theft radio. Before performing a procedure that requires disconnecting the battery, make sure you have the proper activation code.**

### ※ CAUTION 2:

**Disconnecting the battery can cause driveability problems that require a scan tool to rectify. Additionally, disconnecting the battery may cause one or more warning lights on the instrument panel to illuminate, which will also require the use of a scan tool to turn off. Most scan tools available to the public do not have the capability to perform either of these tasks, which will necessitate taking the vehicle to a dealer service department or other properly equipped repair facility after service work has been performed. See Chapter 5, Section 1 for the use of an auxiliary voltage input device ("memory saver") before disconnecting the battery and for other precautions related to battery disconnection.**

2   Drain the engine coolant (see Chapter 1).

3   Refer to Chapter 4 and detach the front exhaust pipe from the exhaust manifold.

4   Remove the valve cover(s) (see Section 4).

5   Remove the intake manifold (see Section 8). Also remove the power steering hose if you're removing the right cylinder head (see illustration).

6   Refer to Section 5 and remove the timing belt and the rear timing belt covers.

7   On SOHC engines, remove the coolant pipe from the rear of cylinder heads. On DOHC engines, remove the secondary air injection pipes from the check valves at the rear of the cylinder head. Also remove the front and rear coolant pipes (see illustration).

**10.5 Detach the power steering hose at three places (arrows) (DOHC engine shown, SOHC engine similar)**

8   Unplug all electrical connectors and coolant hoses labeling each wiring connector or hose to aid the installation process.

9   Working in the reverse of the sequence shown in illustration 10.22a, progressively loosen the cylinder head bolts, by half a turn at a time, until all bolts can be unscrewed by hand. Discard the bolts - new ones must be installed on reassembly.

10  Check that nothing remains connected to the cylinder head, then lift the head away from the cylinder block; seek assistance if possible, as it is very heavy, especially when being removed with the exhaust manifold. If the head is stuck to the block, carefully pry the cylinder head upward, at a casting protrusion beyond the gasket surface (see illustration).

11  Remove the gasket from the top of the block. Do not discard the gasket - it will be needed for identification purposes.

12  If the cylinder head is to be disassembled for service, remove the exhaust manifold, then proceed to Chapter 2C for overhaul procedures. Be sure to reinstall the manifold back onto the cylinder head before installing the cylinder head on the vehicle.

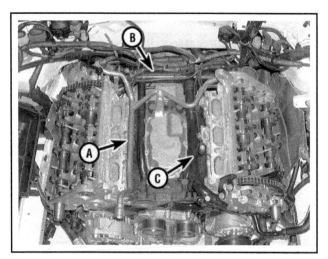

**10.7 On DOHC engines it will be necessary to remove the secondary air injection pipe (A), the rear coolant pipe (B) and the front coolant pipe (C)**

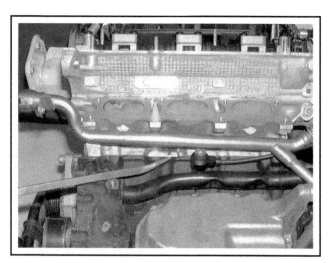

**10.10 If the cylinder head is stuck, it may be necessary to pry upward on the casting protrusion to dislodge the head from the block**

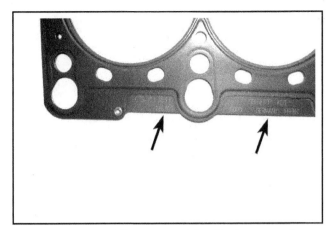

**10.19** Be sure the "TOP" mark and Part No. face upward

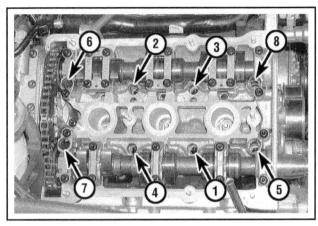

**10.22a** Cylinder head bolt TIGHTENING sequence (DOHC engine shown, SOHC engine uses the same sequence)

## INSTALLATION

♦ **Refer to illustrations 10.19, 10.22a and 10.22b**

13 The mating faces of the cylinder head and cylinder block must be perfectly clean before installing the head. Use a hard plastic or wood scraper to remove all traces of gasket and carbon; also clean the piston crowns. Take particular care during the cleaning operations, as aluminum alloy is easily damaged. Also, make sure that the carbon is not allowed to enter the oil and water passages - this is particularly important for the lubrication system, as carbon could block the oil supply to the engine's components. Using adhesive tape and paper, seal the water, oil and bolt holes in the cylinder block.

14 Check the mating surfaces of the cylinder block and the cylinder head for nicks, deep scratches and other damage. If slight, they may be removed carefully with abrasive paper.

15 If warpage of the cylinder head gasket surface is suspected, use a straight-edge to check it for distortion. If it's distorted beyond the amount listed in this Chapter's Specifications it must be resurfaced. The specific resurfacing dimension is listed in Chapter 2C Specifications.

16 Clean out the cylinder head bolt holes using a suitable tap. Be sure they're clean and dry before installation of the head bolts.

17 It is possible for the piston crowns to strike and damage the valve heads if the camshaft is rotated with the timing belt removed and the crankshaft set to TDC. For this reason, the crankshaft must be set to a position other than TDC before the cylinder head is reinstalled. Use a wrench and socket on the crankshaft pulley center bolt to turn the crankshaft in the opposite direction of rotation (counterclockwise), until all six pistons are positioned down in their bores - approximately 90-degrees before TDC.

18 If the cylinder head has been resurfaced, make sure the valve seats have been reworked by the same amount to allow the correct piston to valve clearance before installing the cylinder head. See Chapter 2C for further information.

19 Position the new head gasket on the cylinder block, engaging it with the locating dowels. Ensure that the manufacturer's "TOP" and part number markings are face up (see illustration).

20 With the help of an assistant, place the cylinder head on the cylinder block, ensuring that the locating dowels engage with the recesses in the cylinder head. Check that the head gasket is correctly seated before allowing the full weight of the cylinder head to rest upon it.

➡**Note: If the cylinder head had been disassembled for repair, be sure the camshaft(s) are installed in the cylinder head as**

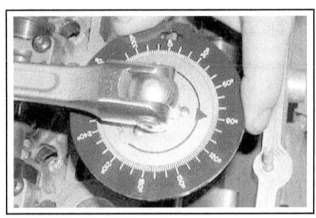

**10.22b** Using an angle measurement gauge during the final stages of tightening

described in Section 6.

21 Install each cylinder head bolt into its relevant hole and screw them in hand tight. Be sure to use NEW cylinder head bolts, as the old bolts are stretch-type fasteners that will not provide the correct torque readings if reused.

22 Working progressively and in the sequence shown (see illustration), tighten the cylinder head bolts in three steps to the torque and angle of rotation listed in this Chapter's Specifications.

➡**Note: It is recommended that an angle-measuring gauge be used during the final stages of the tightening, to ensure accuracy (see illustration). If a gauge is not available, use white paint to make alignment marks between the bolt head and cylinder head prior to tightening; the marks can then be used to check the bolt has been rotated through the correct angle during tightening.**

23 Rotate the crankshaft in the normal direction of rotation (clockwise) 90 degrees to TDC. Be sure the alignment mark on the crankshaft pulley aligns with the mark on the front cover. Refer to Section 3, if necessary.

24 Install the timing belt tensioner and the camshaft sprockets on the engine if removed.

25 Install and adjust the timing belt as described in Section 5.

26 The remainder of the installation is the reverse of removal.

27 Change the engine oil and coolant (see Chapter 1). Run the engine and check for leaks.

## 11  Crankshaft pulley - removal and installation

♦ **Refer to illustration 11.1**

The procedure for removal and installation of the crankshaft pulley on V6 engines is essentially the same as it is on the four cylinder engine, except for the fact that the V6 pulley uses eight retaining bolts instead of four retaining bolts (see illustration). Refer to Chapter 2A for this procedure. When tightening the bolts, use the torque specification listed at the end of this Chapter.

**11.1  Crankshaft pulley retaining bolts (arrows)**

## 12  Crankshaft front oil seal and housing - replacement

♦ **Refer to illustrations 12.2, 12.4, 12.6a, 12.6b, 12.6c and 12.8**

1   Remove the timing belt and crankshaft sprocket (see Section 5).
2   Note how far the seal is recessed in the bore, then carefully pry it out of the front cover with a screwdriver or seal removal tool (see illustration). Don't scratch the housing bore or damage the crankshaft in the process (if the crankshaft is damaged, the new seal will end up leaking).

➡**Note: If a seal removal tool is unavailable, you can thread two self tapping screws (180 degrees apart from one another) into the front seal to pry the seal out.**

3   Clean the bore in the housing and coat the outer edge of the new seal with engine oil or multi-purpose grease. Apply multi-purpose grease to the seal lip.

4   Using a socket with an outside diameter slightly smaller than the outside diameter of the seal, carefully drive the new seal into place with a hammer (see illustration). Make sure it's installed squarely and driven in to the same depth as the original. If a socket isn't available, a short section of large diameter pipe will also work. Check the seal after installation to make sure the spring didn't pop out of place.

5   If the front oil seal/oil pump housing on SOHC and early DOHC engines needs to be removed for access to other components, refer to oil pump removal (see Section 14), disregarding the steps that do not apply since the timing belt is already removed.

6   If the front oil seal housing on a later DOHC engines needs to

**12.2  If a seal removal tool is unavailable, the front seal can also be pried out with screwdriver**

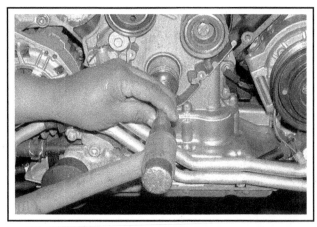

**12.4  Lubricate the seal lip and drive the new crankshaft seal into place with a seal driver or a large socket and a hammer**

**12.6a  The timing belt tensioner (A), the tensioner lever (B), the tensioner pulley (C) and the idler pulley (D) must be removed to access the front oil seal housing bolts**

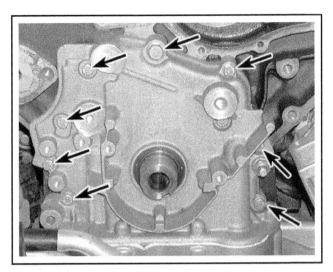

**12.6b  Front oil seal housing upper retaining bolts (arrows)**

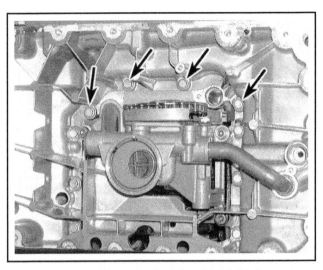

**12.6c  Front oil seal housing lower retaining bolts (arrows) - note that the lower bolts are accessed from the bottom after the lower oil pan is removed (later DOHC engine)**

be removed for access to other components, detach the timing belt tensioner, the tensioner pulley, the tensioner lever and the idler pulley from the housing then remove the lower oil pan (see Section 13). Loosen the housing mounting bolts and remove the housing from the engine while noting the installed position of the fasteners (see illustrations). The front oil seal housing can be removed with or without the front oil seal. In some instances the front oil seal removal and installation is easier with the front housing removed, since the seal can be placed on a workbench and driven straight in and out of the bore with no special tools or adapters.

7    Before installing the front cover, make sure the mating surfaces of the cover, the cylinder block and the oil pan rail are perfectly clean. Use a hard plastic or wood scraper to remove all traces of gasket material. Take particular care when cleaning the front cover, as aluminum alloy is easily damaged.

8    Apply a 3/16-inch (5 mm) bead of RTV sealant to the oil pan flange on the front oil seal housing (see illustration) and to the front of the block where the oil pan and the engine block meet.

9    Locate the oil seal housing gasket over the dowels on the engine block and install the oil seal housing.

➡**Note: Be sure to lubricate the oil seal lip before installing the front cover onto the engine. This will aid the installation process and prevent dry start ups, which may damage the seal and lead to future oil leaks.**

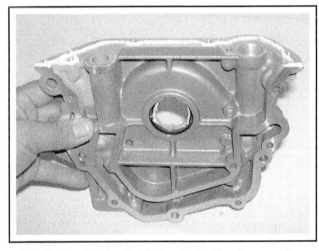

**12.8  Apply a 3/16 bead of RTV sealant to the oil pan mating surface on the front cover**

10  Tighten the front oil seal housing bolts in several steps to the torque listed in this Chapter's Specifications.

11  Reinstall the crankshaft sprocket and timing belt (see Section 5).

12  Run the engine and check for oil leaks at the front seal.

## 13  Oil pan - removal and installation

➡**Note: V6 engines are equipped with a two-piece oil pan. The upper half of the oil pan can only be removed with the engine out of the vehicle. Therefore the following procedure pertains only to the removal and installation of the lower oil pan in the vehicle. Refer to Chapter 2C for the removal and installation of the upper oil pan.**

### REMOVAL

⬥ **Refer to illustration 13.6**

1    Set the parking brake and block the rear wheels.

2    Raise the front of the vehicle and support it securely on jackstands.

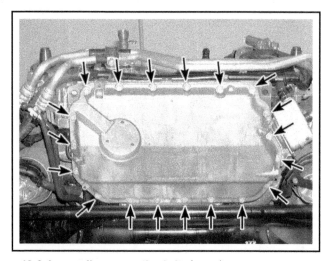

**13.6 Lower oil pan mounting bolts (arrow)**

**13.8 Use a scraper to remove all traces of old sealant from the upper and lower oil pan**

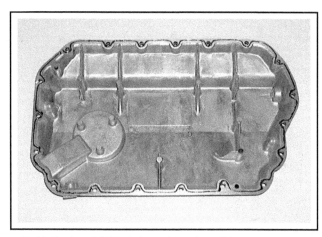

**13.12 Apply a 3/16-inch (5 mm) bead of RTV sealant as shown to the oil pan sealing flange**

3  Remove the splash shield under the engine, if equipped.

4  Drain the engine oil (see Chapter 1). Remove the oil dipstick.

5  Remove the stabilizer bar bushing bolts and lower the stabilizer bar to the ground (see Chapter 10).

6  Unscrew and remove the oil pan bolts (see illustration).

7  Remove the oil pan and the gasket if equipped. If it is stuck, tap it gently with a mallet to free it.

➡**Note: On SOHC engines it will be necessary to remove the oil pump pick-up cover before removing the gasket.**

## INSTALLATION

▶ **Refer to illustrations 13.8 and 13.12**

8  Use a scraper to remove all traces of old sealant or gasket material from the upper and lower oil pan (see illustration). Clean the mating surfaces with lacquer thinner or acetone.

9  Make sure the threaded bolt holes in the upper oil pan are clean.

10  Check the oil pan flange for distortion, particularly around the bolt holes. Remove any nicks or burs as necessary.

11  Inspect the oil pick-up screen for blockage.

12  On SOHC and early DOHC engines affix the lower oil pan gasket to the upper oil pan then install the pick-up cover. On later DOHC engines with a chain driven oil pump, apply a 3/16-inch (5 mm) bead of RTV sealant to the oil pan flange (see illustration).

➡**Note: The oil pan must be installed within 5 minutes once the sealant has been applied.**

13  Carefully position the oil pan on the upper oil pan and install the oil pan bolts loosely. Tighten the bolts to the torque listed in this Chapter's Specifications.

14  The remainder of installation is the reverse of removal.

➡**Note: Be sure to follow the sealant manufacturers recommendations on curing times and allow the sealant to properly cure before adding oil.**

15  Run the engine and check for oil pressure and leaks.

## 14  Oil pump - removal and installation

➡**Note: On later DOHC engines the oil pump can be removed from the engine without removing the timing belt and the front oil seal housing, by simply removing the lower oil pan, then disconnecting the large oil supply pipe and removing the oil pump. The oil pump drive chain, however, cannot be removed and inspected unless the timing belt and the front oil seal housing are removed first.**

## REMOVAL

1  Remove the timing belt and the crankshaft sprocket (see Section 5).

2  Remove the lower oil pan (see Section 13).

3  Remove the timing belt tensioner, the tensioner pulley, the tensioner lever and the idler pulley (see illustration 12.6a).

**14.7 Oil pump drive chain cover mounting bolt (arrow) (later DOHC engine)**

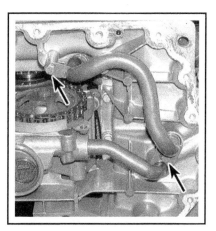

**14.8 Large oil supply tube mounting bolts (arrows)**

**14.9 Oil pump driven sprocket retaining bolt (arrow)**

**14.10 Oil pump mounting bolts (arrows) (later DOHC engine)**

**14.11 The oil pump drive chain can now be removed by lifting it off the upper sprocket and moving it over the end of the crankshaft**

### SOHC and early DOHC engines

4   Remove the oil pump mounting bolts. Note that the lower four bolts are accessed through the lower oil pan opening just like the front oil seal housing bolts on later DOHC engines (see illustration 12.6c).

5   Remove the oil pump housing from the front of the engine.

### Later DOHC engine (with chain-driven oil pump)

▶ **Refer to illustrations 14.7, 14.8, 14.9, 14.10 and 14.11**

6   Remove the front oil seal housing as described in Section 12. Disregard the steps that do not apply since the timing belt and crankshaft sprocket are already removed.

7   Remove the oil pump drive chain cover (see illustration).

8   Remove the long oil supply tube making sure the O-rings are removed with the tube (see illustration).

9   Mark the face of the drive chain and the lower sprocket so the chain can be installed in the same position and direction of travel. Loosen and remove the oil pump sprocket bolt (see illustration). Separate the lower drive sprocket from the shaft on the oil pump and disengage it from the chain.

10  Remove the oil pump mounting bolts and detach the oil pump with the short supply tube (see illustration). Be sure the O-ring is

removed with the short supply tube. If not, remove the supply tube O-rings from the upper oil pan.

11  Disengage the drive chain from the upper sprocket and remove it from the engine (see illustration). Remove the upper drive chain sprocket with a two or three jaw puller.

## INSPECTION

12  Clean all components with solvent, then inspect them for wear and damage.

13  If damage or wear is noted, replacement of the entire oil pump assembly (and the drive chain and sprockets), if equipped, is recommended.

## INSTALLATION

### SOHC and early DOHC engines (with gear driven oil pump)

14  Before installing the oil pump housing, make sure the mating surfaces of the housing, the cylinder block and the upper oil pan rail are perfectly clean. Use a hard plastic or wood scraper to remove all traces

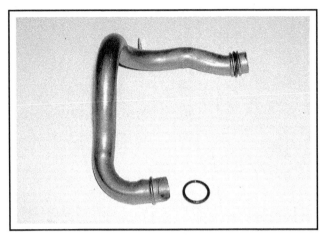

**14.25 Always replace the oil supply tube O-rings (later DOHC engine)**

of gasket material. Take particular care when cleaning the oil pump housing and the section of the upper oil pan, as aluminum alloy is easily damaged.

15 Apply a 3/16-inch (5 mm) bead of RTV sealant to the oil pan flange on the oil pump housing and to the front of the block where the oil pan and the engine block meet.

16 Locate the oil pump housing gasket over the dowels on the engine block and install the oil pump onto the engine while aligning the flats on the crankshaft with the flats on the inner rotor of the oil pump.

➡**Note: Be sure to lubricate the oil seal lip before installing the oil pump onto the engine. This will aid the installation process and prevent dry start ups, which may damage the seal and lead to future oil leaks.**

17 Tighten the oil pump housing bolts in several steps to the torque listed in this Chapter's Specifications.

18 Reinstall the remaining parts in the reverse order of removal.

19 Add oil (and coolant, if the hoses were disconnected), start the engine and check for oil pressure and leaks (see Chapter 1).

20 Recheck the engine oil level.

### Later DOHC engine (with chain-driven oil pump)

▶ **Refer to illustration 14.25**

21 Install new O-rings on the short supply tube and install the short supply tube into the oil pump, then install the oil pump onto the engine and tighten the bolts to the torque listed in this Chapter's Specifications.

22 If removed, install the upper drive chain sprocket on the crankshaft. Engage the drive chain around the teeth of the upper sprocket and between the oil pump drive chain tensioner (see illustration 14.11). Be sure the marks made earlier on the drive chain and the sprocket are facing the front of the engine.

23 Engage the teeth of the oil pump sprocket in the rollers of the drive chain. Pull the drive chain and the sprocket downward and insert the sprocket onto the oil pump while aligning the flat on the sprocket with the flat on the oil pump shaft.

24 Install the oil pump driven sprocket retaining bolt and tighten it to the torque listed in this Chapter's Specifications.

25 Install new O-rings on the long supply tube and install the long supply tube into the holes in the upper oil pan and tighten the bolts to the torque listed in this Chapter's Specifications (see illustration).

26 Reinstall the remaining parts in the reverse order of removal.

27 Add oil (and coolant, if hoses were disconnected), start the engine and check for oil pressure and leaks (see Chapter 1).

28 Recheck the engine oil level.

## 15  Flywheel/driveplate - removal and installation

The procedure for removal and installation of the flywheel/driveplate on V6 engines is essentially the same as it is on the four cylinder engine. Refer to Chapter 2A for this procedure (but use the torque specification listed at the end of this Chapter).

## 16  Rear main oil seal - replacement

The procedure for removal and installation of the rear main oil seal on V6 engines is essentially the same as it is on the four cylinder engine. Refer to Chapter 2A for this procedure (but use the torque specification listed at the end of this Chapter).

## 17  Engine mounts - check and replacement

1 Engine mounts seldom require attention, but broken or deteriorated mounts should be replaced immediately or the added strain placed on the driveline components may cause damage or wear.

### CHECK

2 During the check, the engine must be raised slightly to remove the weight from the mounts.

3 Raise the vehicle and support it securely on jackstands, then position a jack under the engine oil pan. Place a large block of wood between the jack head and the oil pan, then carefully raise the engine

just enough to take the weight off the mounts. Do not position the wood block under the drain plug.

### ✳✳ WARNING:

**DO NOT place any part of your body under the engine when it's supported only by a jack!**

4 Check the mounts to see if the rubber is cracked, hardened or separated from the metal plates. Sometimes the rubber will split right down the center.

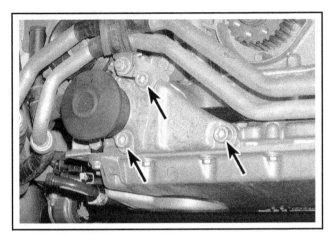

**17.11 Front torque rod mounting bolts (arrows)**

5 Check for relative movement between the mount plates and the engine or frame (use a large screwdriver or pry bar to attempt to move the mounts). If movement is noted, lower the engine and tighten the mount fasteners.

6 Rubber preservative should be applied to the mounts to slow deterioration.

## REPLACEMENT

7 Raise the vehicle and support it securely on jackstands (if not already done). Support the engine as described in Step 3.

8 Remove the splash shield under the engine, if equipped.

## FRONT TORQUE ROD

♦ **Refer to illustration 17.11**

9 Place the radiator support panel in the service position (see Chapter 11).

10 Pull the rubber stopper off the torque rod and replace it with a new one if it's damaged.

11 If necessary, the torque rod can be removed or replaced by unbolting it from the front of the engine (see illustration).

12 Installation is the reverse of the removal procedure.

## DRIVER AND PASSENGER SIDE ENGINE MOUNTS

♦ **Refer to illustrations 17.18 and 17.19**

### ✴✴ WARNING:

**The weight of the entire engine will be supported by the transaxle mounts and the front torque rod during this procedure. Never place any part of your body directly under the engine when performing this procedure.**

13 Set the parking brake and block the rear wheels.

14 Raise the front of the vehicle and support it securely on jackstands.

15 Remove the splash shield under the engine, if equipped.

16 Detach the starter motor cables from the lower engine mount

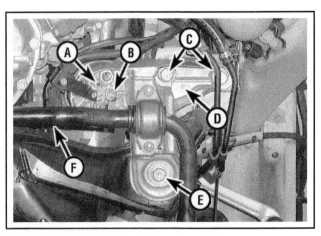

**17.18 Lower engine mount bracket and related components**

| | | | |
|---|---|---|---|
| A | Locating dowel | D | Lower engine mount bracket |
| B | Engine mount retaining nut (lower) | E | Lower engine mount bracket bolt (rear) |
| C | Lower engine mount bracket bolts (front) | F | Front stabilizer bar |

**17.19 Engine mount upper retaining**

brackets by cutting the plastic cable ties and maneuvering the wires out of the plastic retainers.

17 Mark the relationship of the engine mount locating dowel to the bottom of the aluminum engine mount bracket for installation purposes, then unscrew and remove the nuts from the bottom of left and right engine mounts.

18 Support the front stabilizer bar with a floor jack, then unscrew and remove the engine mount bracket bolts. The front two bolts must be unscrewed first, then the rear bolts (see illustration). Lower the aluminum engine mount brackets together with the stabilizer bar to the ground.

➡**Note: Lowering the stabilizer bar and the lower engine mount brackets will allow the front half of the subframe to be lowered slightly. Never loosen the rear subframe bolts as front wheel alignment will be changed.**

19 Remove the engine mount from the bracket on the engine block (see illustration).

20 If necessary, unbolt the mounting bracket from the side of the cylinder block.

21 Installation is the reverse removal procedure.

## Specifications

### General

Engine designation
    2.8L SOHC                  AFC
    2.8L DOHC                AHA, ACK, ALG, APR, ATQ and AQD
Displacement                   171 cubic inches (2.8 liters)
Bore and stroke               3.248 x 3.401 inches (82.5 x 86.4 mm)
Cylinder numbers (front to rear)
    Left bank                 4-5-6
    Right bank              1-2-3
Firing order                    1-4-3-6-2-5

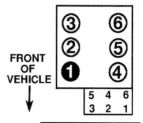

**Cylinder and coil
terminal numbering**

### Camshaft

Endplay
    SOHC
        New              0.0016 to 0.0059 inch (0.04 to 0.15 mm)
        Wear limit        0.0138 inch (0.35 mm)
    DOHC                 0.0080 inch (0.20 mm)
Journal diameter          N/A
Journal oil clearance (in cylinder head)    0.004 inch (0.10 mm)
Lobe lift
    Intake                  N/A
    Exhaust               N/A
Runout                      0.0004 inch (0.01 mm)

### Cylinder head

Resurfacing dimension (minimum)
    SOHC                 5.226 inches (132.75 mm)
    DOHC                 5.480 inches (139.20 mm)
Warpage limit
    SOHC                 0.002 inch (0.05 mm)
    DOHC                 0.004 inch (0.10 mm)

### Driveplate

Driveplate installed height
    SOHC                 0.713 to 0.716 inch (18.1 to 19.7 mm)
    DOHC                 0.484 inch (12.3 mm)

| Torque specifications | Ft-lbs (unless otherwise indicated) | Nm |
| --- | --- | --- |

➡ **Note: One foot-pound (ft-lb) of torque is equivalent to 12 inch-pounds (in-lbs) of torque. Torque values below approximately 15 ft-lbs are expressed in inch-pounds, since most foot-pound torque wrenches are not accurate at these smaller values.**

| | | |
| --- | --- | --- |
| Camshaft sprocket bolt | | |
|   SOHC | 52 | 70 |
|   DOHC | 41 | 55 |
| Camshaft bearing cap bolts/nuts | | |
|   SOHC | 156 in-lbs | 17 |
|   DOHC | 84 in-lbs | 10 |
| Camshaft drive chain tensioner (DOHC) | 84 in-lbs | |
| Crankshaft drive sprocket bolt (with bolt oiled)* | | |
|   Step one | 148 | 200 |
|   Step two | Tighten an additional 180 degrees | |

| Torque specifications (continued) | Ft-lbs (unless otherwise indicated) | Nm |
|---|---|---|
| Crankshaft pulley bolts | 18 | 25 |
| Crankshaft rear oil seal housing bolts | 84 in-lbs | 10 |
| Cylinder head bolts* (in sequence - see illustration 10.22a) | | |
|     Step one | 44 | 60 |
|     Step two | Tighten an additional 90 degrees | |
|     Step three | Tighten an additional 90 degrees | |
| Driveplate bolts* | | |
|     Step one | 44 | 60 |
|     Step two | Tighten an additional 90 degrees | |
| Flywheel bolts* | | |
|     Step one | 44 | 60 |
|     Step two | Tighten an additional 180 degrees | |
| Exhaust manifold nuts | 18 | 25 |
| Intake manifold bolts | | |
|     SOHC | | |
|         Upper plenum | | |
|             M6 | 84 in-lbs | 10 |
|             M8 | 15 | 20 |
|         Lower manifold | 15 | 20 |
|     DOHC | 84 in-lbs | 10 |
| Oil pan bolts | | |
|     M6 | 84 in-lbs | 10 |
|     M8 | 18 | 25 |
| Oil pan-to-bellhousing bolts | 33 | 45 |
| Oil pump | | |
|     SOHC and early DOHC | | |
|         Oil pick-up cover bolts | 84 in-lbs | 10 |
|         Oil pump bolts | 84 in-lbs | 10 |
|         Spray jet valve | | |
|             SOHC | 22 | 30 |
|             Early DOHC | 30 | 40 |
|     Later DOHC | | |
|         Supply tube bracket bolts | 84 in-lbs | 10 |
|         Oil pump bolts | 15 | 20 |
|         Chain sprocket bolt | 18 | 25 |
| Timing belt | | |
|     SOHC | | |
|         Cover-to-block bolts | 84 in-lbs | 10 |
|         Tensioner bolts | 84 in-lbs | 10 |
|         Tensioner pulley bolt | 33 | 45 |
|         Idler pulley | 18 | 25 |
|     DOHC | | |
|         Tensioner bolts | 84 in-lbs | 10 |
|         Tensioner pulley bolt | 15 | 20 |
|         Idler pulley | 33 | 45 |
| Valve cover-to-cylinder head nuts | 84 in-lbs | 10 |
| Engine mount bolts/nuts* | | |
|     M6 | 84 in-lbs | 10 |
|     M8 | 18 | 25 |
|     M10 | 33 | 45 |
|     M12 | 44 | 60 |
| Engine mount brackets-to engine block bolts | 22 | 30 |

*Always replace with new bolt or nut*

**Notes**

**Section**

**Reference to other Chapters**

CHECK ENGINE light - See Chapter 6

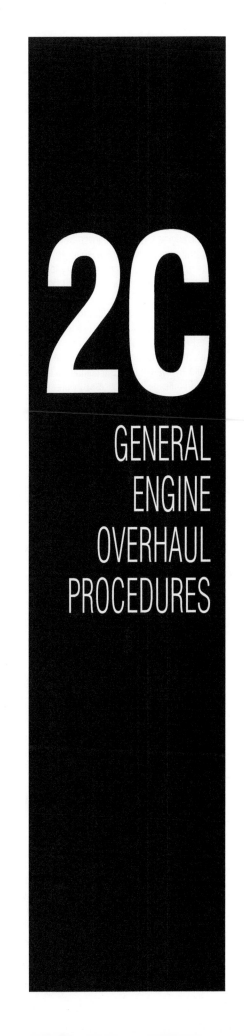

# 2C

## GENERAL ENGINE OVERHAUL PROCEDURES

## 1  General information - engine overhaul

Included in this portion of Chapter 2 are the general overhaul procedures for the cylinder head and internal engine components.

The information ranges from advice concerning preparation for an overhaul and the purchase of replacement parts to detailed, step-by-step procedures covering removal and installation of internal engine components and the inspection of parts.

The following Sections have been written based on the assumption that the engine has been removed from the vehicle. For information concerning in-vehicle engine repair, as well as removal and installation of the external components necessary for the overhaul, see Chapter 2A (four cylinder engines) or Chapter 2B (V6 engines), and Section 8 of this Chapter.

The Specifications included in this Part are only those necessary for the inspection and overhaul procedures which follow. Refer to Chapter 2, Part A or Part B for additional Specifications.

It's not always easy to determine when, or if, an engine should be completely overhauled, as a number of factors must be considered.

High mileage is not necessarily an indication that an overhaul is needed, while low mileage doesn't preclude the need for an overhaul. Frequency of servicing is probably the most important consideration. An engine that's had regular and frequent oil and filter changes, as well as other required maintenance, will most likely give many thousands of miles of reliable service. Conversely, a neglected engine may require an overhaul very early in its life.

Excessive oil consumption is an indication that piston rings, valve seals and/or valve guides are in need of attention. Make sure that oil leaks aren't responsible before deciding that the rings and/or guides are bad. Perform a cylinder compression check to determine the extent of the work required (see Section 3). Also check the vacuum readings under various conditions (see Section 4).

Loss of power, rough running, knocking or metallic engine noises, excessive valve train noise and high fuel consumption rates may also point to the need for an overhaul, especially if they're all present at the same time. If a complete tune-up doesn't remedy the situation, major mechanical work is the only solution.

An engine overhaul involves restoring the internal parts to the specifications of a new engine. During an overhaul, the piston rings are replaced and the cylinder walls are reconditioned (re-bored and/or honed). If a re-bore is done by an automotive machine shop, new over-size pistons will also be installed. The main bearings and connecting rod bearings are generally replaced with new ones and, if necessary, the crankshaft may be reground to restore the journals. Generally, the valves are serviced as well, since they're usually in less-than-perfect condition at this point. While the engine is being overhauled, other components, such as the starter and alternator, can be rebuilt as well. The end result should be a like new engine that will give many trouble free miles.

➡**Note: Critical cooling system components such as the hoses, drivebelts, thermostat and water pump should be replaced with new parts when an engine is overhauled. The radiator should be checked carefully to ensure that it isn't clogged or leaking (see Chapter 3). If you purchase a rebuilt engine or short block, some rebuilders will not warranty their engines unless the radiator has been professionally flushed. Also, we don't recommend overhauling the oil pump - always install a new one when an engine is rebuilt.**

Before beginning the engine overhaul, read through the entire procedure to familiarize yourself with the scope and requirements of the job. Overhauling an engine isn't difficult, but it is time-consuming. Plan on the vehicle being tied up for a minimum of two weeks, especially if parts must be taken to an automotive machine shop for repair or reconditioning. Check on availability of parts and make sure that any necessary special tools and equipment are obtained in advance. Most work can be done with typical hand tools, although a number of precision measuring tools are required for inspecting parts to determine if they must be replaced. Often an automotive machine shop will handle the inspection of parts and offer advice concerning reconditioning and replacement.

➡**Note: Always wait until the engine has been completely disassembled and all components, especially the engine block, have been inspected before deciding what service and repair operations must be performed by an automotive machine shop.**

Since the block's condition will be the major factor to consider when determining whether to overhaul the original engine or buy a rebuilt one, never purchase parts or have machine work done on other components until the block has been thoroughly inspected. As a general rule, time is the primary cost of an overhaul, so it doesn't pay to install worn or substandard parts.

As a final note, to ensure maximum life and minimum trouble from a rebuilt engine, everything must be assembled with care in a spotlessly-clean environment.

## 2  Oil pressure check

♦ **Refer to illustrations 2.2a and 2.2b**

1   Low engine oil pressure can be a sign of an engine in need of rebuilding. A "low oil pressure" indicator (often called an "idiot light") is not a test of the oiling system. Such indicators only come on when the oil pressure is dangerously low. Even a factory oil pressure gauge in the instrument panel is only a relative indication, although much better for driver information than a warning light. A better test is with a mechanical (not electrical) oil pressure gauge. When used in conjunction with an accurate tachometer, an engine's oil pressure performance can be compared to the manufacturers Specifications.

2   Find the oil pressure indicator sending unit (see illustrations).

3   Remove the oil pressure sending unit and install a fitting which will allow you to directly connect your hand-held, mechanical oil pressure gauge. Use Teflon tape or sealant on the threads of the adapter and the fitting on the end of your gauge's hose.

4   Connect an accurate tachometer to the engine, according to the tachometer manufacturer's instructions.

5   Check the oil pressure with the engine running (full operating temperature) at the specified engine speed, and compare it to this Chapter's Specifications. If it's extremely low, the bearings and/or oil pump are probably worn out.

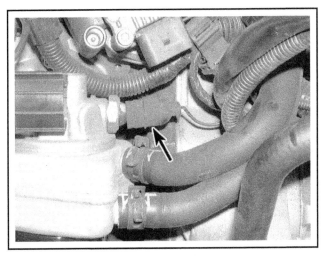

**2.2a  The oil pressure sending unit (arrow) on 1.8L engines is mounted on the oil filter adapter housing which is located on the side of the engine block**

**2.2b  The oil pressure sending unit (arrow) on V6 engines is mounted at the front of the engine on the front oil seal/oil pump housing**

## 3  Cylinder compression check

▶ **Refer to illustration 3.6**

### ✳✳ CAUTION:

**The following test may set a trouble code and turn on the CHECK ENGINE light. If this happens, the vehicle will have to be taken to a dealership service department or other qualified repair shop to have the trouble code cleared and the light turned off.**

1  A compression check will tell you what mechanical condition the upper end (pistons, rings, valves, head gaskets) of the engine is in. Specifically, it can tell you if the compression is down due to leakage caused by worn piston rings, defective valves and seats or a blown head gasket.

➡**Note: The engine must be at normal operating temperature and the battery must be fully charged for this check.**

2  Begin by cleaning the area around the spark plugs before you remove them. Compressed air should be used, if available, otherwise a small brush will work. The idea is to prevent dirt from getting into the cylinders as the compression check is being done.

3  Remove all of the spark plugs (see Chapter 1) from the engine.

4  Block the throttle wide open.

5  Disable the fuel and ignition systems by disconnecting the primary electrical connectors at the ignition coil pack/modules (see Chapter 5) and the electrical connectors at the fuel injectors (see Chapter 4).

6  Install the compression gauge in the number one spark plug hole (see illustration).

7  Crank the engine over at least seven compression strokes and watch the gauge. The compression should build up quickly in a healthy engine. Low compression on the first stroke, followed by gradually increasing pressure on successive strokes, indicates worn piston rings.

A low compression reading on the first stroke, which doesn't build up during successive strokes, indicates leaking valves or a blown head gasket (a cracked head could also be the cause). Deposits on the undersides of the valve heads can also cause low compression. Record the highest gauge reading obtained.

8  Repeat the procedure for the remaining cylinders, turning the engine over for the same length of time for each cylinder, and compare the results to this Chapter's Specifications.

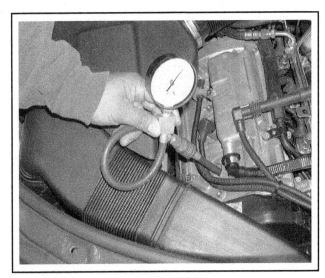

**3.6  A compression gauge with a threaded fitting for the spark plug hole is preferred over the type that requires hand pressure to maintain the seal - be sure to open the throttle valve as far as possible during the compression check**

9  If the readings are below normal, add some engine oil (about three squirts from a plunger-type oil can) to each cylinder, through the spark plug hole, and repeat the test.

10  If the compression increases after the oil is added, the piston rings are definitely worn. If the compression doesn't increase significantly, the leakage is occurring at the valves or head gasket. Leakage past the valves may be caused by burned valve seats and/or faces or warped, cracked or bent valves.

11  If two adjacent cylinders have equally low compression, there's a strong possibility the head gasket between them is blown. The appearance of coolant in the combustion chambers or the crankcase would verify this condition.

12  If one cylinder is about 20-percent lower than the others, and the engine has a slightly rough idle, a worn exhaust lobe on the camshaft could be the cause.

13  If the compression is unusually high, the combustion chambers are probably coated with carbon deposits. If that's the case, the cylinder heads should be removed and decarbonized.

14  If compression is way down or varies greatly between cylinders, it would be a good idea to have a leak-down test performed by an automotive repair shop. This test will pinpoint exactly where the leakage is occurring and how severe it is.

15  Installation of the remaining components is the reverse of removal.

## 4   Vacuum gauge diagnostic checks

▶ **Refer to illustrations 4.4 and 4.6**

1   A vacuum gauge provides valuable information about what is going on in the engine at a low cost. You can check for worn rings or cylinder walls, leaking head or intake manifold gaskets, incorrect carburetor adjustments, restricted exhaust, stuck or burned valves, weak valve springs, improper ignition or valve timing and ignition problems.

2   Unfortunately, vacuum gauge readings are easy to misinterpret, so they should be used in conjunction with other tests to confirm the diagnosis.

3   Both the gauge readings and the rate of needle movement are important for accurate interpretation. Most gauges measure vacuum in inches of mercury (in-Hg). As vacuum increases (or atmospheric pressure decreases), the reading will increase. Also, for every 1,000-foot increase in elevation above sea level, the gauge readings will decrease about one inch of mercury.

4   Connect the vacuum gauge directly to intake manifold vacuum, not to ported vacuum (see illustration). Be sure no hoses are left disconnected during the test or false readings will result.

5   Before you begin the test, allow the engine to warm up completely. Block the wheels and set the parking brake. With the transaxle in Park, start the engine and allow it to run at normal idle speed.

6   Read the vacuum gauge; an average, healthy engine should normally produce about 17 to 22 inches of vacuum with a fairly steady needle. Refer to the following vacuum gauge readings and what they indicate about the engine's condition (see illustration).

7   A low, steady reading usually indicates a leaking gasket between the intake manifold and carburetor or throttle body, a leaky vacuum hose, late ignition timing or incorrect camshaft timing. Eliminate all other possible causes, utilizing the tests provided in this Chapter before you remove the timing belt cover to check the timing marks.

8   If the reading is three to eight inches below normal and it fluctuates at that low reading, suspect an intake manifold gasket leak at an intake port.

9   If the needle has regular drops of about two to four inches at a steady rate, the valves are probably leaking. Perform a compression or leak-down test to confirm this.

10  An irregular drop or down-flick of the needle can be caused by a sticking valve or an ignition misfire. Perform a compression or leak-down test and read the spark plugs.

11  A rapid vibration of about four inches-Hg vibration at idle combined with exhaust smoke indicates worn valve guides. Perform a leak-down test to confirm this. If the rapid vibration occurs with an increase in engine speed, check for a leaking intake manifold gasket or head gasket, weak valve springs, burned valves or ignition misfire.

12  A slight fluctuation, say one inch up and down, may mean ignition problems. Check all the usual tune-up items and, if necessary, run the engine on an ignition analyzer.

13  If there is a large fluctuation, perform a compression or leak-down test to look for a weak or dead cylinder or a blown head gasket.

14  If the needle moves slowly through a wide range, check for a clogged PCV system, throttle body or intake manifold gasket leaks.

15  Check for a slow return after revving the engine by quickly snapping the throttle open until the engine reaches about 2,500 rpm and let it shut. Normally the reading should drop to near zero, rise above normal idle reading (about 5 in-Hg over) and then return to the previous idle reading. If the vacuum returns slowly and doesn't peak when the throttle is snapped shut, the rings may be worn. If there is a long delay, look for a restricted exhaust system (often the muffler or catalytic converter). An easy way to check this is to temporarily disconnect the exhaust ahead of the suspected part and re-test.

**4.4 A simple vacuum gauge can be very handy in diagnosing engine condition and performance**

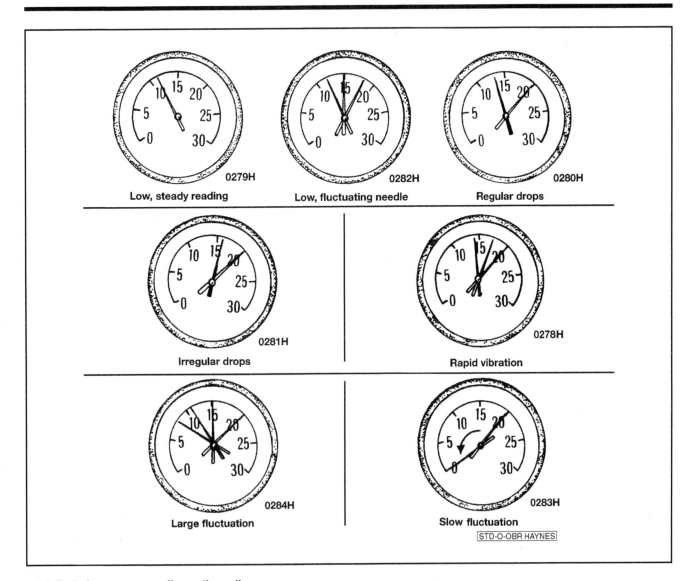

**4.6  Typical vacuum gauge diagnostic readings**

## 5    Engine rebuilding alternatives

The home mechanic is faced with a number of options when performing an engine overhaul. The decision to replace the engine block, piston/connecting rod assemblies and crankshaft depends on a number of factors, with the number one consideration being the condition of the block. Other considerations are cost, access to machine shop facilities, parts availability, time required to complete the project and the extent of prior mechanical experience.

Some of the rebuilding alternatives include:

**Individual parts** - If the inspection procedures reveal the engine block and most engine components are in reusable condition, purchasing individual parts may be the most economical alternative. The block, crankshaft and piston/connecting rod assemblies should all be inspected carefully. Even if the block shows little wear, the cylinder bores should be surface-honed.

**Short-block** - A short-block consists of an engine block with a crankshaft and piston/connecting rod assemblies already installed. All new bearings are incorporated and all clearances will be correct. The existing camshaft, valve train components, cylinder head and external parts can be bolted to the short block with little or no machine shop work necessary.

**Long-block** - A long-block consists of a short block plus an oil pump, oil pan, cylinder head, valve cover, camshaft and valve train components, timing sprockets and a timing belt. All components are installed with new bearings, seals and gaskets incorporated throughout. The installation of manifolds and external parts is all that's necessary. Give careful thought to which alternative is best for you and discuss the situation with local automotive machine shops, auto parts dealers and experienced rebuilders before ordering or purchasing replacement parts.

## 6  Engine removal - methods and precautions

If you've decided the engine must be removed for overhaul or major repair work, several preliminary steps should be taken. Locating a suitable place to work is extremely important. Adequate work space, along with storage space for the vehicle, will be needed.

Cleaning the engine compartment and engine before beginning the removal procedure will help keep tools clean and organized. An engine hoist will also be necessary. Safety is of primary importance, considering the potential hazards involved in removing the engine from this vehicle.

If the engine is being removed by a novice, a helper should be available. Advice and aid from someone more experienced would also be helpful. There are many instances when one person cannot simultaneously perform all of the operations required when lifting or lowering the engine out of the vehicle.

Plan the operation ahead of time. Arrange for or obtain all of the tools and equipment you'll need prior to beginning the job. Some of the equipment necessary to perform engine removal and installation safely and with relative ease (in addition to a hydraulic jack, jackstands and an engine hoist) are a complete set of wrenches and sockets as described in the front of this manual, wooden blocks and plenty of rags and cleaning solvent for mopping up spilled oil, coolant and gasoline.

Plan for the vehicle to be out of use for quite a while. A machine shop will be required to perform some of the work which the do-it-yourselfer can't accomplish without special equipment. These shops often have a busy schedule, so it would be a good idea to consult them before removing the engine in order to accurately estimate the amount of time required to rebuild or repair components that may need work.

Always be extremely careful when removing and installing the engine. Serious injury can result from careless actions. Plan ahead, take your time and a job of this nature, although major, can be accomplished successfully.

➡**Note: Because it may be some time before you reinstall the engine, it is very helpful to make sketches or take photos of various accessory mountings and wiring hookups before removing the engine.**

## 7  Engine - removal and installation

### ❋❋ WARNING 1:

**The models covered by this manual are equipped with airbags. Always disable the airbag system before working in the vicinity of any airbag system component to avoid the possibility of accidental deployment of the airbag(s), which could cause personal injury (see Chapter 12).**

### ❋❋ WARNING 2:

**Gasoline is extremely flammable so take extra precautions when you work on any part of the fuel system. Don't smoke or allow open flames or bare light bulbs near the work area, and don't work in a garage where a gas-type appliance (such as a water heater or a clothes dryer) is present. Since gasoline is carcinogenic, wear latex gloves when there's a possibility of being exposed to fuel, and, if you spill any fuel on your skin, rinse it off immediately with soap and water. Mop up any spills immediately and do not store fuel-soaked rags where they could ignite. The fuel system is under constant pressure, so, if any fuel lines are to be disconnected, the fuel pressure in the system must be relieved first (see Chapter 4). When you perform any kind of work on the fuel system, wear safety glasses and have a Class B type fire extinguisher on hand.**

### ❋❋ WARNING 3:

**The air conditioning system is under high pressure - have a dealer service department or service station evacuate the system and recover the refrigerant before disconnecting any of the hoses or fittings.**

### ❋❋ CAUTION:

**When disassembling the air intake system on turbocharged vehicles, ensure that no foreign material can get into the turbo air intake port. Cover the opening with a sheet of plastic and a rubber band. The turbocharger compressor blades could be severely damaged if debris is allowed to enter.**

### REMOVAL

◗ **Refer to illustrations 7.7, 7.18a, 7.18b and 7.22**

1   Have the air conditioning system properly discharged at a dealership service department or an automotive air conditioning repair facility.

2   Disconnect the cable from the negative terminal of the battery, then disconnect the positive terminal.

### ❋❋ CAUTION:

**Disconnecting the battery can cause severe driveability problems that require a scan tool to rectify. Additionally, disconnecting the battery may cause one or more warning lights on the instrument panel to illuminate, which will also require the use of a scan tool to turn off. Most scan tools available to the public do not have the capacity to perform either of these tasks, which will necessitate taking the vehicle to a dealer service department or other properly equipped repair facility after service work has been performed. See Chapter 5, Section 1 for the use of an auxilliary voltage input ("memory saver") device before disconnecting the battery and for other precautions related to battery disconnection.**

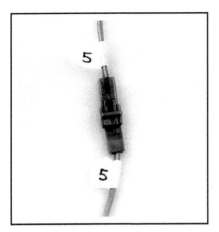

**7.7 Label each wire before unplugging the connector**

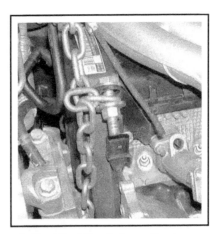

**7.18a Attach the chain or sling to the lifting eyes - all engines are equipped with at least one**

**7.18b On some engines, it will be necessary to attach the sling or chain to the casting protrusion on the cylinder head**

3 Refer to Chapter 11 and remove the radiator support panel.

4 Relieve the fuel system pressure (see Chapter 4).

5 Place protective covers on the fenders and cowl and remove the hood (see Chapter 11).

6 Remove the air cleaner assembly (see Chapter 4). Remove the battery and its mounting bracket (see Chapter 5). On four cylinder models, also disconnect the intercooler hoses from the turbocharger (see Chapter 4).

7 Clearly label, then disconnect all vacuum lines, coolant and emissions hoses, wiring harness connectors, ground straps and fuel lines. Masking tape and/or a touch up paint applicator work well for marking items (see illustration). Take instant photos or sketch the locations of components and brackets.

8 Remove the mechanical cooling fan, and disconnect the radiator hoses and heater hoses from the engine. Also remove the coolant expansion tank (see Chapter 3).

9 Disconnect the throttle cable (if equipped) from the engine (see Chapter 4).

10 Unbolt the power steering pump and the steering pump hoses (see Chapter 10).

11 Disconnect the fuel lines from the fuel rail (see Chapter 4). Plug or cap all open fittings.

12 Raise the vehicle and support it securely on jackstands. Drain the cooling system, transaxle and engine oil and remove the drivebelts (see Chapter 1).

13 Refer to Chapter 3 and unbolt and set aside the air conditioning compressor.

14 Remove the secondary air injection pump and hoses from the engine, if equipped (see Chapter 6).

15 Label and disconnect the main engine electrical harnesses from the starter and the engine.

16 Remove the front section of the exhaust system (see Chapter 4).

17 Support the transaxle with a floor jack. Place a block of wood on the jack head to prevent damage to the transaxle.

18 Attach a lifting sling or chain to the lifting eye (if equipped) on the engine (see illustration). If two lifting eyes are not provided, attach the lifting sling or chain to a safe place such as the side of the cylinder head (see illustration). Position a hoist and connect the sling to it. Take up the slack until there is slight tension on the hoist.

**7.22 Pull the engine forward far enough for the clutch or torque converter to clear the transaxle bellhousing, then raise it high enough to remove it from the engine compartment**

19 Remove the transaxle-to-engine bolts. If equipped with an automatic transaxle, remove the torque converter bolts and push the torque converter away from the driveplate back towards the transaxle (see Chapter 7). This will help keep the torque converter engaged in the transaxle while the engine and driveplate are being separated from the transaxle.

20 Recheck to be sure nothing except the mounts are still connecting the engine/transaxle to the vehicle. Disconnect anything still remaining.

21 Remove the lower engine mount retaining nuts (see Chapter 2A or 2B).

22 Slowly raise the engine out of the vehicle (see illustration).

**✳✳ WARNING:**

**Do not place any part of your body under the engine when it's supported only by a hoist or other lifting device.**

23 Place the engine on the floor or remove the flywheel/driveplate and mount the engine on an engine stand.

## INSTALLATION

24 Install the flywheel/driveplate on the engine (see Chapter 2A or 2B). Check the engine and transaxle mounts. If they're worn or damaged, replace them.

25 If you're working on a vehicle with a manual transaxle, install the clutch and pressure plate onto the flywheel (see Chapter 7A). Now is a good time to install a new clutch.

26 Carefully lower the engine into the engine compartment - make sure the engine mounts line up.

27 If you're working on a vehicle with an automatic transaxle, guide the torque converter into the crankshaft following the procedure outlined in Chapter 7B.

28 If you're working on a vehicle with a manual transaxle, apply a dab of high-temperature grease to the input shaft and guide it into the crankshaft pilot bearing until the bellhousing is flush with the engine block.

➡ **Note: It may be necessary to turn the crankshaft with a socket and breaker bar until the splines on the input shaft align with the splines on the clutch disc.**

29 Install the transaxle-to-engine bolts and tighten them securely.

### ✳ CAUTION:

**DO NOT use the bolts to force the transaxle and engine together!**

30 Reinstall the remaining components in the reverse order of removal.

31 Add coolant, oil, power steering and transaxle fluid as needed.

### ✳ CAUTION:

**Be sure to bleed the air from the cooling system as described in Chapter 1.**

32 Run the engine and check for leaks and proper operation of all accessories, then install the hood and test drive the vehicle.

33 Have the air conditioning system recharged and leak tested.

## 8  Engine overhaul - disassembly sequence

1 It's much easier to disassemble and work on the engine if it's mounted on a portable engine stand. A stand can often be rented quite cheaply from an equipment rental yard. Before it's mounted on a stand, the flywheel/driveplate should be removed from the engine.

2 If a stand isn't available, it's possible to disassemble the engine with it blocked up on the floor. Be extra careful not to tip or drop the engine when working without a stand.

3 If you're going to obtain a rebuilt engine, all external components must come off first, to be transferred to the replacement engine, just as they will if you're doing a complete engine overhaul yourself. These include:

> *Alternator mounting brackets*
> *Emissions control components*
> *Ignition coil/module assembly, spark plug wires and spark plugs*
> *Valve cover*
> *Timing belt covers and timing belt*
> *Water pump*
> *Thermostat and housing cover and coolant supply tubes*
> *Coolant supply tubes (four cylinder engine)*
> *Fuel system components*
> *Turbocharger (if equipped)*
> *Intake/exhaust manifolds*
> *Coolant supply tubes (V6 engines)*
> *Camshaft and crankshaft position sensors*
> *Oil filter and adapter housing/oil cooler*
> *Oil dipstick and tube*
> *Flywheel/driveplate*

➡ **Note: When removing the external components from the engine, pay close attention to details that may be helpful or important during installation. Note the installed position of gaskets, seals, spacers, pins, brackets, washers, bolts and other small items.**

4 If you're obtaining a short-block, then the cylinder heads, oil pan and oil pump will have to be removed as well. See *Engine rebuilding alternatives* for additional information regarding the different possibilities to be considered.

5 If you're planning a complete overhaul, the engine must be disassembled and the internal components removed in the following general order:

> *Camshaft and crankshaft sprockets*
> *Rear timing covers (if equipped)*
> *Camshaft and lifters*
> *Cylinder head*
> *Oil pan, front oil seal housing, oil pump, oil pump drive gear and*
>   *drive gear plug (four cylinder engine)*
> *Lower oil pan (V6 engines)*
> *Oil pick-up cover, lower oil pan gasket, upper oil pan and oil pump/*
>   *front oil seal housing (SOHC and early DOHC V6 engines with*
>   *gear-driven oil pump)*
> *Oil supply tubes, drive chain cover, oil pump drive chain, drive*
>   *chain tensioner, oil pump, upper oil pan and front oil seal housing*
>   *(later DOHC engines with chain driven oil pump)*
> *Oil spray nozzles*
> *Piston/connecting rod assemblies*
> *Rear main oil seal housing*
> *Crankshaft and main bearings*
> *Intermediate shaft (four-cylinder engine)*

6   Before beginning the disassembly and overhaul procedures, make sure the following items are available. Also, refer to *Engine overhaul - reassembly* sequence for a list of tools and materials needed for engine reassembly.

Common hand tools
Small cardboard boxes or plastic bags for storing parts
Gasket scraper
Ridge reamer
Engine balancer puller
Micrometers

Telescoping gauges
Dial indicator set
Valve spring compressor
Cylinder surfacing hone
Piston ring groove-cleaning tool
Electric drill motor
Tap and die set
Wire brushes
Oil gallery brushes
Cleaning solvent

## 9   Cylinder head - disassembly

♦ **Refer to illustrations 9.2 and 9.3**

➡**Note: New and rebuilt cylinder heads are commonly available for most engines at dealerships and auto parts stores. Due to the fact that some specialized tools are necessary for the disassembly and inspection procedures, and replacement parts aren't always readily available, it may be more practical and economical for the home mechanic to purchase a replacement head rather than taking the time to disassemble, inspect and recondition the original.**

1   Cylinder head disassembly involves removal of the intake and exhaust valves and related components. It is already assumed that the camshaft(s) and lifters are removed from the cylinder head. If they're not already removed, label the parts and store them separately so they can be reinstalled in their original locations.

2   Before the valves are removed, arrange to label and store them, along with their related components, so they can be kept separate and reinstalled in their original locations (see illustration).

3   Compress the springs on the first valve with a spring compressor and remove the keepers (see illustration). Carefully release the valve spring compressor and remove the retainer, the spring and the spring seat (if used).

4   Pull the valve out of the head, then remove the oil seal from the guide. If the valve binds in the guide (won't pull through), push it back into the head and deburr the area around the keeper groove with a fine file or whetstone.

5   Repeat the procedure for the remaining valves. Remember to keep all the parts for each valve together so they can be reinstalled in the same locations.

6   Once the valves and related components have been removed and stored in an organized manner, the heads should be thoroughly cleaned and inspected. If a complete engine overhaul is being done, finish the engine disassembly procedures before beginning the cylinder head cleaning and inspection process.

**9.2  A small plastic bag, with an appropriate label, can be used to store the valve train components so they can be kept together and reinstalled in the original positions**

**9.3  You'll need a valve spring compressor with a special adapter to compress the spring and allow removal of the keepers from the valve stem**

## 10  Cylinder head - cleaning and inspection

1   Thorough cleaning of the cylinder head and related valve train components, followed by a detailed inspection, will enable you to decide how much valve service work must be done during the engine overhaul.

➡ **Note: If the engine was severely overheated, the cylinder head is probably warped (see Step 12).**

## CLEANING

2   Scrape all traces of old gasket material and sealant off the head gasket, intake manifold and exhaust manifold mating surfaces. Be very careful not to gouge the cylinder head. Special gasket-removal solvents that soften gaskets and make removal much easier are available at auto parts stores.

3   Remove all built-up scale from the coolant passages.

4   Run a stiff wire brush through the various holes to remove deposits that may have formed in them.

5   Run an appropriate-size tap into each of the threaded holes to remove corrosion and thread sealant that may be present. If compressed air is available, use it to clear the holes of debris produced by this operation.

### ✳✳ WARNING:

**Wear eye protection when using compressed air!**

6   Clean the exhaust manifold and intake manifold stud threads with a wire brush.

7   Clean the cylinder head with solvent and dry it thoroughly.

8   Compressed air will speed the drying process and ensure that all holes and oil passages are clean.

➡ **Note: Decarbonizing chemicals are available and may prove very useful when cleaning cylinder heads and valve train components. They're very caustic and should be used with caution.**

**Be sure to follow the instructions on the container.**

9   Clean all the valve springs, keepers and retainers with solvent and dry them thoroughly. Do the components from one valve at a time to avoid mixing up the parts.

10  Scrape off any heavy deposits that may have formed on the valves, then use a motorized wire brush to remove deposits from the valve heads and stems. Again, make sure the valves don't get mixed up.

## INSPECTION

➡ **Note: Be sure to perform all of the following inspection procedures before concluding machine shop work is required. Make a list of the items that need attention.**

### Cylinder head

▸ **Refer to illustrations 10.11, 10.12a, 10.12b and 10.14**

11  Inspect the head very carefully for cracks, evidence of coolant leakage and other damage. If cracks are found, check with an automotive machine shop concerning repair. If repair isn't possible, a new cylinder head must be obtained (see illustration).

12  Using a straightedge and feeler gauge, check the head gasket mating surface for warpage, then compare the measurement against this Chapter's Specifications (see illustration). If the warpage exceeds the limit, it can be resurfaced at an automotive machine shop as long as the cylinder head is within the specified minimum height listed in this Chapter's Specifications (see illustration).

➡ **Note: If the head is resurfaced, it will be necessary to rework the valve seats the same amount which was removed from the cylinder head or piston-to-valve clearance will be compromised, which may lead to severe engine damage.**

13  Examine the valve seats in each of the combustion chambers. If they're pitted, cracked or burned, the head will require valve service that's beyond the scope of the home mechanic.

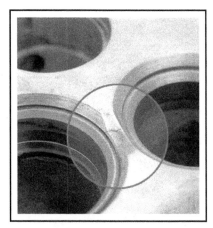

**10.11  Check for cracks between the valve seats**

**10.12a  Check the cylinder head gasket surface for warpage by trying to slip a feeler gauge under the straightedge (see this Chapter's Specifications for the maximum warpage allowed and use a feeler gauge of that thickness)**

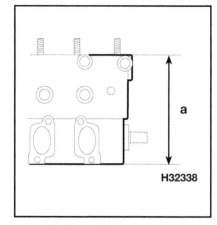

**10.12b  The cylinder head minimum height dimension is measured from the valve cover rail to the deck surface**

14 Check the valve stem-to-guide clearance by measuring the lateral movement of the valve stem with a dial indicator attached securely to the head (see illustration). Install the valve into the guide until the stem is flush with the top of the guide. The total valve stem movement indicated by the gauge needle must be compared to the specifications in this Chapter. After this is done, if there's still some doubt regarding the condition of the valve guides, they should be checked by an automotive machine shop (the cost should be minimal).

## Valves

▶ **Refer to illustrations 10.15 and 10.16**

15 Carefully inspect each valve face for uneven wear, deformation, cracks, pits and burned areas. Check the valve stem for scuffing and galling and the neck for cracks. Rotate the valve and check for any obvious indication that it's bent. Look for pits and excessive wear on the end of the stem. The presence of any of these conditions (see illustration) indicates the need to consult an automotive machine shop.

➡**Note: The manufacturer recommends the valves be replaced, if refacing is necessary.**

16 Also measure the stem diameter at several points along their

lengths (see illustration). Taper should not exceed the limit listed in this Chapter's Specifications.

## Valve components

▶ **Refer to illustrations 10.17 and 10.18**

17 Check each valve spring for wear (on the ends) and pits. Measure the free length of each intake valve spring and compare them with one another (see illustration). Any springs that are shorter have sagged and shouldn't be re-used. Now repeat this check on the exhaust valve springs. If, in either check, any of the springs measures shorter than another (intake-to-intake, exhaust-to-exhaust) replace all of the springs as a set. The tension of all springs should be checked with a special fixture before deciding they're suitable for use in a rebuilt engine (take the springs to an automotive machine shop for this check).

➡**Note: If the engine has accumulated many miles, it's a good idea to replace all of the springs as a matter of course.**

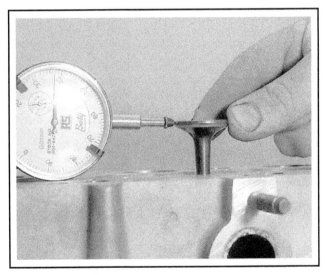

**10.14  A dial indicator can be used to determine the valve stem-to-guide clearance - measure the maximum deflection of the valve in its guide**

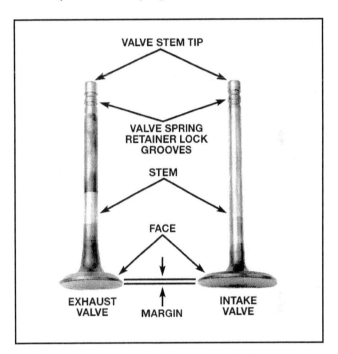

**10.15  Check for valve wear at the points shown here**

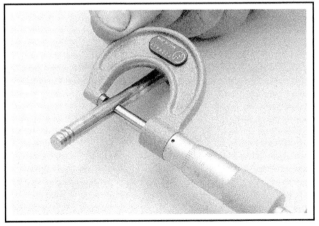

**10.16  Measure the diameter of the valve stems at several points**

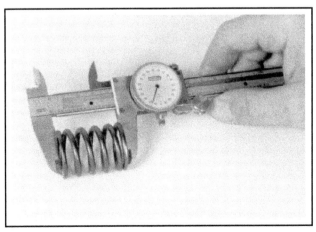

**10.17  Measure the free length of each valve spring with a dial or vernier caliper**

18 Stand each spring on a flat surface and check it for squareness (see illustration). If any of the springs are distorted or sagged, replace all of them with new parts.

19 Check the spring retainers, spring seats and the keepers for obvious wear and cracks. Any questionable parts should be replaced with new ones, as extensive damage will occur if they fail during engine operation.

### Camshaft(s) and lifters

20 Refer to Section 21 of this Chapter for the camshaft and lifter inspection procedures. Be sure to inspect the camshaft bearing journals on the cylinder head before the head is sent to a machine shop to have the valves serviced. If the journals are gouged or scored the cylinder head will have to be replaced regardless of the condition of the valves and related components.

### All components

21 If the inspection process indicates the valve components are in generally poor condition and worn beyond the limits specified, which is usually the case in an engine that's being overhauled, reassemble the valves in the cylinder head (see Section 11 for valve servicing recommendations).

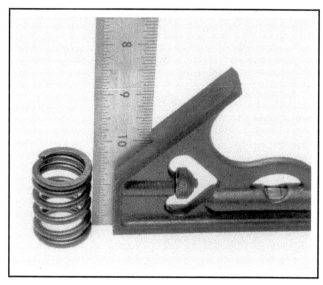

**10.18 Check each valve spring for squareness**

## 11 Valves - servicing

1 Because of the complex nature of the job and the special tools and equipment needed, servicing of the valve seats and the valve guides, commonly known as a valve job, should be done by a professional (the valves themselves aren't serviceable).

2 The home mechanic can remove and disassemble the head, do the initial cleaning and inspection, then reassemble and deliver it to an automotive machine shop for the actual service work. Doing the inspection will enable you to see what condition the head and valvetrain components are in and will ensure that you know what work and new parts are required when dealing with an automotive machine shop.

➡**Note: Be aware that Volkswagen cylinder heads have a maximum valve seat refacing dimension. This is the maximum amount of material that can be removed from the valve seats before cylinder head replacement is required. This measurement will be taken by the automotive machine shop.**

3 The automotive machine shop, will remove the valves and springs, recondition the seats, recondition the valve guides, check and replace the valves, valve springs, spring retainers and keepers (as necessary), replace the valve seals with new ones, reassemble the valve components and make sure the valve stem height is correct. If warped, the cylinder head gasket surface will also be resurfaced as long as the cylinder head is within the specified minimum height listed in this Chapter's Specifications.

4 After the valve job has been performed by a professional, the head will be in like new condition. When the head is returned, be sure to clean it again before installation on the engine to remove any metal particles and abrasive grit that may still be present from the valve service or head resurfacing operations. Use compressed air, if available, to blow out all the oil holes and passages.

## 12 Cylinder head - reassembly

▸ **Refer to illustrations 12.2, 12.3, 12.6, 12.7 and 12.9**

1 Regardless of whether or not the head was sent to an automotive repair shop for valve servicing, make sure it's clean before beginning reassembly.

2 If the head was sent out for valve servicing, the valves and related components will already be in place. Begin the reassembly procedure with Step 9. If the head was not sent out for service, the valves, at the very least should be lapped before reassembly of the cylinder head. Apply a small amount of fine grinding paste on the sealing surface (valve face) of each valve and install them into their appropriate guide in the cylinder head. Attach a valve lapping tool to the valve head. Using a back and forth rotating motion grind the valve head into its seat. Periodically lift the valve and rotate it to redistribute the grinding paste (see illustration). After the lapping process has been completed for each valve it will be necessary to clean the valves and seats of all lapping compound. Be sure to mark the valves before they're removed so they can be installed back into the same valve guide on reassembly.

3 Beginning at one end of the head, lubricate and install the first valve. Apply moly-base grease or clean engine oil to the valve stem (see illustration).

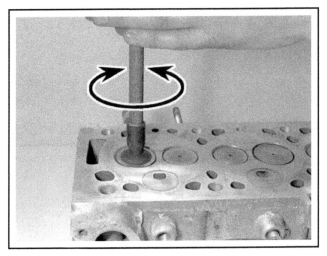

**12.2 Lapping in the valves**

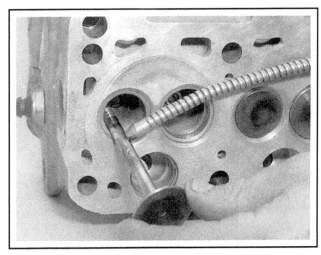

**12.3 Lubricate the valve stem with clean engine oil before installing it into the guide**

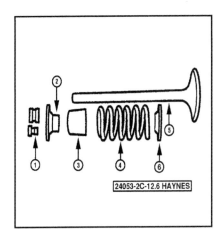

24053-2C-12.6 HAYNES

**12.6 Typical valve components**

| | | | |
|---|---|---|---|
| 1 | Keepers | 5 | Valve |
| 2 | Retainer | 6 | Valve spring |
| 3 | Oil seal | | seat |
| 4 | Spring | | |

**12.7 Apply a small dab of grease to each keeper as shown here before installation - it'll hold them in place on the valve stem as the spring is released**

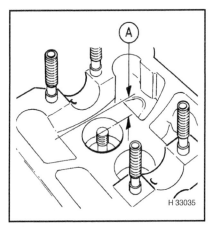

H 33035

**12.9 The valve installed height is measured from the valve cover rail to the top of the valve stem**

4   Install the spring seat and shims, if originally installed, before the valve seals.

5   Install new seals on each of the valve guides. Gently tap each seal into place until it's completely seated on the guide. Many seal sets come with a plastic installer, but use hand pressure. Do not hammer on the seals or they could be driven down too far and subsequently leak. Don't twist or cock the seals during installation or they won't seal properly on the valve stems.

6   The valve components (see illustration) may be installed in the following order:

*Valves*
*Valve spring seat*
*Valve stem seals*
*Valve spring shims (if any)*
*Valve springs*
*Retainers*
*Keepers*

7   Compress the springs with a valve spring compressor and carefully install the keepers in the groove, then slowly release the compressor and make sure the keepers seat properly. Apply a small dab of grease to each keeper to hold it in place if necessary (see illustration). Tap the valve stem tips with a plastic hammer to seat the keepers, if necessary.

8   Repeat the procedure for the remaining valves. Be sure to return the components to their original locations - don't mix them up!

9   Check the installed valve height with a straightedge and a dial or venier caliper. If the head was sent out for service work, the installed height should be correct (but don't automatically assume it is). The measurement is taken from valve cover rail to the top of each valve stem (see illustration). If the height is less than specified in this Chapter, the valve seats have been reworked past their limits and will not allow proper operation of the hydraulic valve lifters. Valve seats that have been reworked past their limits must be replaced with new ones or a new cylinder head is required.

## 13 Pistons and connecting rods - removal

▶ Refer to illustrations 13.1, 13.3a, 13.3b, 13.4 and 13.6

➡Note: Prior to removing the piston/connecting rod assemblies, remove the cylinder head, the oil pan, the oil pump drive chain (if equipped), the oil pump and baffle by referring to the appropriate Sections in Chapter 2 Part A or B. On V6 engines it will also be necessary to remove the upper oil pan before removing the pistons and connecting rods.

1   Use your fingernail to feel if a ridge has formed at the upper limit of ring travel (about 1/4-inch down from the top of each cylinder). If carbon deposits or cylinder wear have produced ridges, they must be completely removed with a special tool (see illustration). Follow the manufacturer's instructions provided with the tool. Failure to remove the ridges before attempting to remove the piston/connecting rod assemblies may result in piston breakage.

2   After the cylinder ridges have been removed, turn the engine upside-down so the crankshaft is facing up.

3   Before the connecting rods are removed, check the endplay (side clearance) with feeler gauges. Slide them between the first connecting rod and the crankshaft throw until the play is removed (see illustration). The endplay is equal to the thickness of the feeler gauge(s). If the endplay exceeds the service limit, new connecting rods will be required. If new rods (or a new crankshaft) are installed, the endplay may fall under the minimum specified in this Chapter (if it does, the rods will have to be machined to restore it - consult an automotive machine shop for advice if necessary). Repeat the procedure for the remaining connecting rods. Where applicable, it will be necessary to remove the oil spray jets from the engine block before removing the piston and connecting rod assemblies (see illustration)

4   Check the connecting rods and caps for identification marks. If they aren't plainly marked, use a small center-punch (see illustration) to make the appropriate number of indentations on each rod and cap (1, 2, 3, etc., depending on the cylinder they're associated with).

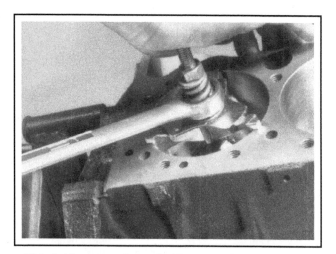

**13.1  A ridge reamer is required to remove the ridge from the top of each cylinder - do this before removing the pistons!**

**13.3a  Check the connecting rod side clearance with a feeler gauge as shown**

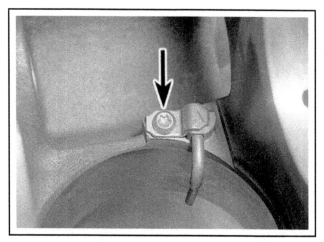

**13.3b  Where applicable, it will be necessary to remove the oil spray nozzles from the block before removing the piston and connecting rod assemblies**

**13.4  Mark the rod bearing caps in order from the front of the engine to the rear (one mark for the front cap, two for the second one and so on)**

5   Loosen each of the connecting rod cap nuts or bolts 1/2-turn at a time until they can be removed by hand. Remove the number one connecting rod cap and bearing insert. Don't drop the bearing insert out of the cap.

6   On 1.8L engines, slip a short length of plastic or rubber hose over each connecting rod cap bolt to protect the crankshaft journal and cylinder wall as the piston is removed (see illustration).

7   Remove the bearing insert and push the connecting rod/piston assembly out through the top of the engine. Use a wooden or plastic hammer handle to push on the upper bearing surface in the connecting rod. If resistance is felt, double-check to make sure all of the ridge was removed from the cylinder.

8   Repeat the procedure for the remaining cylinders.

9   After removal, reassemble the connecting rod caps and bearing inserts in their respective connecting rods and install the cap nuts or bolts finger tight. Leaving the old bearing inserts in place until reassembly will help prevent the connecting rod bearing surfaces from being accidentally nicked or gouged.

10   Don't separate the pistons from the connecting rods.

**13.6  On 1.8L engines, slip sections of rubber or plastic hose over the rod bolts before removing the pistons/rods to prevent damage to the crankshaft journals and cylinder walls**

## 14  Crankshaft - removal

▶ **Refer to illustrations 14.1, 14.3 and 14.4**

➡ **Note: The crankshaft can be removed only after the engine has been removed from the vehicle. It's assumed the flywheel/ driveplate, timing belt, oil pan, oil pump, the front and rear oil seal housings and the piston/connecting rod assemblies have already been removed.**

1   Before the crankshaft is removed, check the endplay. Mount a dial indicator with the stem in line with the crankshaft and touching the snout of the crank (see illustration).

2   Push the crankshaft all the way to the rear and zero the dial indicator. Next, pry the crankshaft to the front as far as possible and check the reading on the dial indicator. The distance it moves is the endplay. If it's greater than listed in this Chapter's Specifications, check the crankshaft thrust surfaces for wear. If no wear is evident, new main bearings should correct the endplay.

3   If a dial indicator isn't available, feeler gauges can be used. Gently pry or push the crankshaft all the way to the front of the engine. Slip feeler gauges between the crankshaft and the front face of the thrust main bearing to determine the clearance (see illustration).

➡ **Note: The thrust bearing is located at the number three main bearing cap on four-cylinder engines and at the number four main bearing cap on V6 engines.**

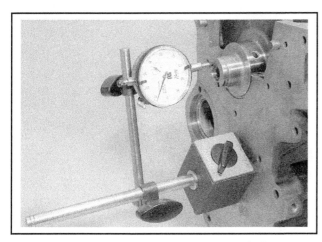

**14.1  Measuring crankshaft endplay using a dial indicator**

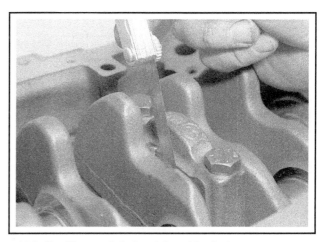

**14.3  Checking crankshaft endplay with a feeler gauge**

4   Check the main bearing caps to see if they're marked to indicate their locations. They should be numbered consecutively from the front of the engine to the rear (see illustration). If they aren't, mark them with number stamping dies or a center-punch. Main bearing caps generally have a cast-in arrow, which points to the front of the engine. Loosen the main bearing cap bolts 1/4-turn at a time each, until they can be removed by hand. Note if any stud bolts are used and make sure they're returned to their original locations when the crankshaft is reinstalled.

5   Gently tap the caps with a soft-face hammer, then separate them from the engine block. If necessary, use the bolts as levers to remove the caps. Try not to drop the bearing inserts if they come out with the caps.

6   Carefully lift the crankshaft straight out of the engine. It may be a good idea to have an assistant available, since the crankshaft is quite heavy. Be careful not to damage the reluctor ring for the crankshaft position sensor. With the bearing inserts in place in the engine block and main bearing caps, return the caps to their respective locations on the engine block and tighten the bolts finger tight.

14.4 Manufacturer's identification markings on the main bearing caps (arrow)

## 15  Intermediate shaft - removal and installation

▶ Refer to illustrations 15.2a, 15.2b, 15.2c, 15.3, 15.4a, 15.4b, 15.4c and 15.4d

## REMOVAL

1   Refer to Chapter 2A and remove the intermediate shaft sprocket.
2   Unbolt the clamp and bearing and remove the oil pump drive gear from the rear of the cylinder block. Use a magnet to lift the gear out. Note which way the gear is installed as it is possible to fit it upside down, causing damage to the gears (see illustrations).

3   Before the shaft is removed, the endplay must be checked. Anchor a dial indicator to the cylinder block with its probe in line with the intermediate shaft center axis. Push the shaft into the cylinder block to the end of its travel, zero the gauge and then draw the shaft out to the opposite end of its travel. Record the maximum deflection and compare

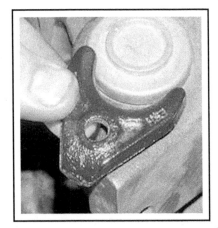

15.2a  Remove the clamp . . .

15.2b  . . . then lift out the bearing . . .

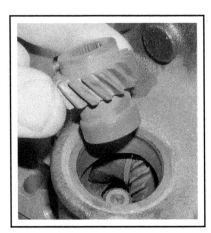

15.2c  . . . and gear

the figure with that listed in this Chapter's Specifications - replace the shaft if the endplay exceeds this limit (see illustration).

4 Loosen the retaining bolts and withdraw the intermediate shaft flange. Recover the O-ring seal, then press out the oil seal (see illustrations).

5 Withdraw the intermediate shaft from the cylinder block and inspect the drive gear at the end of the shaft; if the teeth show signs of excessive wear, or are damaged in any way, the shaft should be replaced.

6 If the oil seal has been leaking, check the shaft mating surface for signs of scoring or damage.

## INSTALLATION

7 Liberally oil the intermediate shaft bearing surfaces and drive gear, then guide the shaft into the cylinder block and engage the journal at the leading end with its support bearing.

8 Press a new shaft oil seal into its housing in the intermediate shaft flange and install a new O-ring seal to the inner sealing surface of the flange.

9 Lubricate the inner lip of the seal with clean engine oil, then slide the flange and seal over the end of the intermediate shaft. Ensure that the O-ring is correctly seated, then install the flange retaining bolts and tighten them to the torque listed in this Chapter's Specifications. Check that the intermediate shaft can rotate freely.

10 Insert the oil pump drive gear in the aperture at the rear of the cylinder block and engage it with the splined oil pump shaft. Make sure the gear is inserted facing the correct direction so that it is fully engaged with the gear on the end of the intermediate shaft. Install the bearing together with the oil seal and secure with the clamp and bolt, tightening the bolt securely.

11 Install the sprocket to the intermediate shaft and tighten the center bolt to the specified torque (see Chapter 2A).

**15.3 Check the intermediate shaft endplay using a dial indicator**

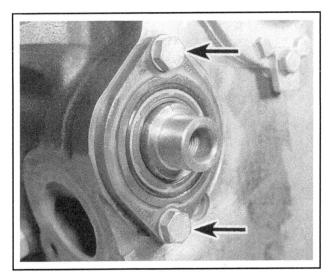

**15.4a Loosen the retaining bolts . . .**

**15.4b . . . and withdraw the intermediate shaft flange**

**15.4c Press out the oil seal . . .**

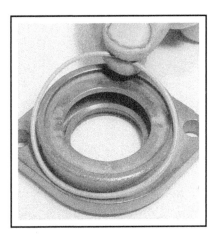

**15.4d . . . then recover the O-ring seal**

## 16 Engine block - cleaning and inspection

### CLEANING

▶ **Refer to illustrations 16.3a, 16.3b, 16.4 and 16.8**

1   Remove the main bearing caps and separate the bearing inserts from the caps and the engine block. Tag the bearings, indicating which cylinder they were removed from and whether they were in the cap or the block, then set them aside.

2   Using a gasket scraper, remove all traces of gasket material from the engine block. Be very careful not to nick or gouge the gasket sealing surfaces.

3   Remove all of the covers and threaded oil gallery plugs from the block. The plugs are usually very tight - they may have to be drilled out and the holes retapped. Use new plugs when the engine is reassembled. On V6 engines remove the crankcase breather chamber, the oil distribution pipe and the oil pressure check valves from the top of the engine block (see illustration).

4   Remove the oil filter adapter (if equipped) the from the engine. Disassemble the components from the adapter housing and inspect them for wear and damage. Look for nicks and scoring especially on the pressure relief valve piston (see illustration). Clean the components and the oil passages in the housing thoroughly with solvent, then dry them with compressed air. If in doubt about the condition of the adapter housing and its components, replace it with a new one.

5   If the engine is extremely dirty, it should be taken to an automotive machine shop to be cleaned.

6   After the block is returned, clean all oil holes and oil galleries one more time. Brushes specifically designed for this purpose are available at most auto parts stores. Flush the passages with warm water until the water runs clear, dry the block thoroughly and wipe all machined surfaces with a light, rust preventive oil. If you have access to compressed air, use it to speed the drying process and blow out all the oil holes and galleries.

### ✳✳ WARNING:

**Wear eye protection when using compressed air!**

7   If the block isn't extremely dirty or sludged up, you can do an adequate cleaning job with hot soapy water and a stiff brush. Take plenty of time and do a thorough job. Regardless of the cleaning method used, be sure to clean all oil holes and galleries very thoroughly, dry the block completely and coat all machined surfaces with light oil.

8   The threaded holes in the block must be clean to ensure accurate torque readings during reassembly. Run the proper size tap into each of the holes to remove rust, corrosion, thread sealant or sludge and restore damaged threads (see illustration). If possible, use compressed air to clear the holes of debris produced by this operation. Now is a good time to clean the threads on the head bolts and the main bearing cap bolts as well.

9   Reinstall the main bearing caps and tighten the bolts finger tight.

10  Apply non-hardening sealant (such as Permatex no. 2 or Teflon pipe sealant) to the new oil gallery plugs and thread them into the holes in the block. Make sure they're tightened securely.

11  If the engine isn't going to be reassembled right away, cover it with a large plastic trash bag to keep it clean.

### INSPECTION

▶ **Refer to illustrations 16.15a, 16.15b and 16.15c**

➡**Note: The manufacturer recommends checking the block deck for warpage and the main bearing bore concentricity and alignment. Since special measuring tools are needed, the checks should be done by an automotive machine shop.**

12  Before the block is inspected, it should be cleaned as described

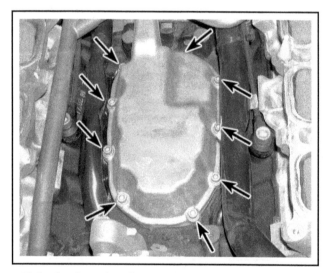

**16.3a  Crankcase breather chamber retaining bolts**

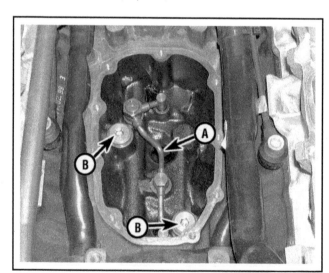

**16.3b  The oil distribution pipe (A) and the oil check valves (B) must be removed before cleaning the engine block - be sure these parts thoroughly cleaned and free of debris before they're reinstalled back onto the engine block**

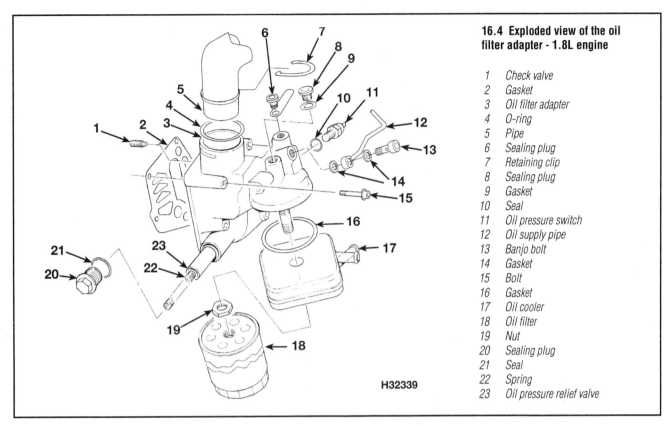

**16.4 Exploded view of the oil filter adapter - 1.8L engine**

1   Check valve
2   Gasket
3   Oil filter adapter
4   O-ring
5   Pipe
6   Sealing plug
7   Retaining clip
8   Sealing plug
9   Gasket
10   Seal
11   Oil pressure switch
12   Oil supply pipe
13   Banjo bolt
14   Gasket
15   Bolt
16   Gasket
17   Oil cooler
18   Oil filter
19   Nut
20   Sealing plug
21   Seal
22   Spring
23   Oil pressure relief valve

H32339

earlier in this Section.

13   Visually check the block for cracks, rust and corrosion. Look for stripped threads in the threaded holes. It's also a good idea to have the block checked for hidden cracks by an automotive machine shop that has the special equipment to do this type of work. If defects are found, have the block repaired, if possible, or replaced.

14   Check the cylinder bores for scuffing and scoring.

➡**Note: The following checks should not be made with the engine block mounted on a stand - the cylinders will be distorted and the measurements will be inaccurate.**

15   Check the cylinders for taper and out-of-round conditions as follows (see illustrations):

16   Measure the diameter of each cylinder at the top (just under the ridge area), center and bottom of the cylinder bore, parallel to the crankshaft axis.

17   Next, measure each cylinder's diameter at the same three locations perpendicular to the crankshaft axis.

18   The taper of each cylinder is the difference between the bore diameter at the top of the cylinder and the diameter at the bottom. The

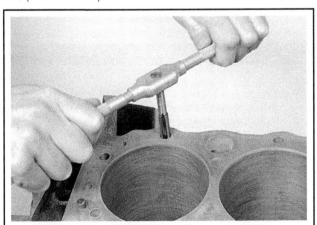

**16.8 All bolt holes in the block - particularly the main bearing cap and head bolt holes - should be cleaned and restored with a tap (be sure to remove debris from the holes after this is done)**

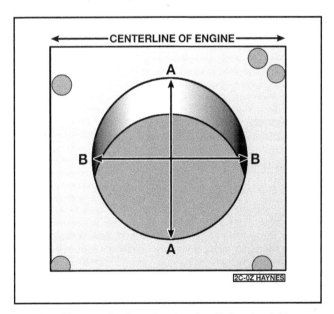

**16.15a Measure the diameter of each cylinder at a right angle to the engine centerline (A), and parallel to the engine centerline (B) - out-of-round is the difference between A and B; taper is the difference between the diameter at the top of the cylinder and the diameter at the bottom of the cylinder**

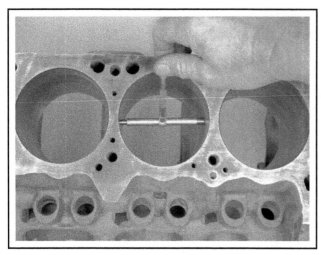

**16.15b The ability to "feel" when the telescoping gauge is at the correct point will be developed over time, so work slowly and repeat the check until you're satisfied the bore measurement is accurate**

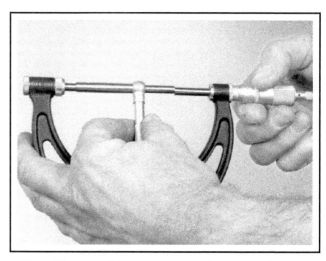

**16.15c The gauge is then measured with a micrometer to determine the bore size**

out-of-round specification of the cylinder bore is the difference between the parallel and perpendicular readings. Compare your results to this Chapter's Specifications.

19 If the cylinder walls are badly scuffed or scored, or if they're out-of-round or tapered beyond the limits given in this Chapter's Specifications, have the engine block rebored and honed at an automotive machine shop.

20 If a rebore is done, oversize pistons and rings will be required.

21 Using a precision straightedge and feeler gauge, check the block

deck (the surface the cylinder heads mate with) for distortion as you did with the cylinder heads (see Section 10). If it's distorted beyond the specified limit, the block decks can be resurfaced by an automotive machine shop, but is not recommended on these engines.

22 If the cylinders are in reasonably good condition and not worn to the outside of the limits, and if the piston-to-cylinder clearances can be maintained properly, they don't have to be rebored. Honing is all that's necessary (see Section 17).

## 17 Cylinder honing

▶ **Refer to illustrations 17.3a and 17.3b**

1 Prior to engine reassembly, the cylinder bores must be honed so the new piston rings will seat correctly and provide the best possible combustion chamber seal.

➡ **Note: If you don't have the tools or don't want to tackle the honing operation, most automotive machine shops will do it for a reasonable fee.**

2 Before honing the cylinders, install the main bearing caps and tighten the bolts to the torque listed in this Chapter's Specifications.

3 Two types of cylinder hones are commonly available - the flex hone or "bottle brush" type and the more traditional surfacing hone with spring-loaded stones. Both will do the job, but for the less experienced mechanic the "bottle brush" hone will probably be easier to use. You'll also need some honing oil (kerosene will work if honing oil isn't avail-

able), rags and an electric drill motor. Proceed as follows:

a) *Mount the hone in the drill motor, compress the stones and slip it into the first cylinder (see illustration). Be sure to wear safety goggles or a face shield!*

b) *Lubricate the cylinder with plenty of honing oil, turn on the drill and move the hone up-and-down in the cylinder at a pace that will produce a fine crosshatch pattern on the cylinder walls, and with the drill square and centered with the bore. Ideally, the crosshatch lines should intersect at approximately a 45 to 60-degree angle (see illustration). Be sure to use plenty of lubricant and don't take off any more material than is absolutely necessary to produce the desired finish.*

➡ **Note: Piston ring manufacturers may specify a different crosshatch angle - read and follow any instructions included with the new rings.**

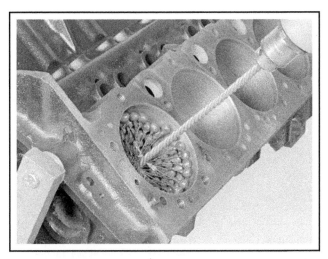

**17.3a  A "bottle brush" hone will produce a better crosshatch pattern when using a drill motor to hone the cylinders**

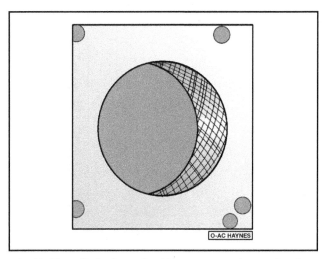

**17.3b  The cylinder hone should leave a smooth, crosshatch pattern with the lines intersecting at approximately a 60-degree angle**

c) *Don't withdraw the hone from the cylinder while it's running. Instead, shut off the drill and continue moving the hone up-and-down in the cylinder until it comes to a complete stop, then compress the stones and withdraw the hone. If you're using a "bottle brush" type hone, stop the drill motor, then turn the chuck in the normal direction of rotation while withdrawing the hone from the cylinder.*

d) *Wipe the oil out of the cylinder and repeat the procedure for the remaining cylinders.*

4   After the honing job is complete, chamfer the top edges of the cylinder bores with a small file so the rings won't catch when the pistons are installed. Be very careful not to nick the cylinder walls with the end of the file.

5   The entire engine block must be washed again very thoroughly with warm, soapy water to remove all traces of the abrasive grit produced during the honing operation.

➡**Note: The bores can be considered clean when a lint-free white cloth - dampened with clean engine oil - used to wipe them out doesn't pick up any more honing residue, which will show up as gray areas on the cloth. Be sure to run a brush through all oil holes and galleries and flush them with running water.**

6   After rinsing, dry the block and apply a coat of light rust preventive oil to all machined surfaces. Wrap the block in a plastic trash bag to keep it clean and set it aside until reassembly.

## 18  Pistons and connecting rods - inspection

▶ **Refer to illustrations 18.4a, 18.4b, 18.10, 18.11, 18.13a and 18.13b**

1   Before the inspection process can be carried out, the piston/connecting rod assemblies must be cleaned and the original piston rings removed from the pistons.

➡**Note: Always use new piston rings when the engine is reassembled.**

2   Using a piston ring installation tool, carefully remove the rings

from the pistons. Be careful not to nick or gouge the pistons in the process.

3   Scrape all traces of carbon from the top of the piston. A hand-held wire brush or a piece of fine emery cloth can be used once the majority of the deposits have been scraped away. Do not, under any circumstances, use a wire brush mounted in a drill motor to remove deposits from the pistons. The piston material is soft and may be eroded away by the wire brush.

4   Use a piston ring groove-cleaning tool to remove carbon deposits

**18.4a The piston ring grooves can be cleaned with a special tool, as shown here . . .**

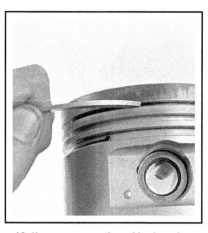

**18.4b . . . or a section of broken ring**

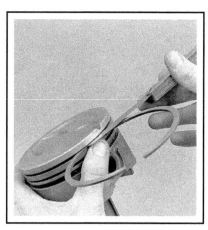

**18.10 Check the ring side clearance with a feeler gauge at several points around the groove**

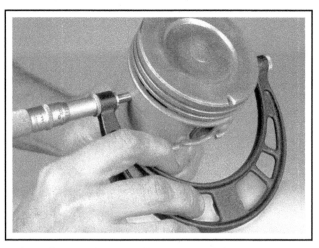

**18.11 Measure the piston diameter at a 90-degree angle to the piston pin and at the specified distance from the bottom of the piston skirt**

from the ring grooves. If a tool isn't available, a piece broken off the old ring will do the job. Be very careful to remove only the carbon deposits - don't remove any metal and do not nick or scratch the sides of the ring grooves (see illustrations).

5   Once the deposits have been removed, clean the piston/rod assemblies with solvent and dry them with compressed air (if available).

### ✳✳ WARNING:

**Wear eye protection. Make sure the oil return holes in the back sides of the ring grooves are clear.**

6   If the pistons and cylinder walls aren't damaged or worn excessively, and if the engine block isn't rebored, new pistons won't be necessary. Normal piston wear appears as even vertical wear on the piston thrust surfaces and slight looseness of the top ring in its groove. New piston rings, however, should always be used when an engine is rebuilt.

7   Carefully inspect each piston for cracks around the skirt, at the pin bosses and at the ring lands.

8   Look for scoring and scuffing on the thrust faces of the skirt,

holes in the piston crown and burned areas at the edge of the crown. If the skirt is scored or scuffed, the engine may have been suffering from overheating and/or abnormal combustion, which caused excessively high operating temperatures. The cooling and lubrication systems should be checked thoroughly. A hole in the piston crown is an indication that abnormal combustion (preignition) was occurring. Burned areas at the edge of the piston crown are usually evidence of spark knock (detonation). If any of the above problems exist, the causes must be corrected or the damage will occur again. The causes may include intake air leaks, incorrect fuel/air mixture, low octane fuel, ignition timing and EGR system malfunctions.

9   Corrosion of the piston, in the form of small pits, indicates coolant is leaking into the combustion chamber and/or the crankcase. Again, the cause must be corrected or the problem may persist in the rebuilt engine.

10   Measure the piston ring side clearance by laying a new piston ring in each ring groove and slipping a feeler gauge in beside it (see illustration). Check the clearance at three or four locations around each groove. Be sure to use the correct ring for each groove - they are different. If the side clearance is greater than specified in this Chapter, new pistons will have to be used.

11   Check the piston-to-bore clearance by measuring the bore (see Section 16) and the piston diameter. Make sure the pistons and bores are correctly matched. Measure the piston across the skirt, at a 90-degree angle to the piston pin (see illustration). The measurement must be taken at a specific point to be accurate: The pistons are measured 0.390 inch (10 mm) from the bottom of the skirt, at right angles to the piston pin. Measure the cylinder bore at three equal places in the bore (top, middle and bottom) and use the average dimension for comparison with the piston measurement.

12   Subtract the piston diameter from the bore diameter to obtain the clearance. If it's greater than specified, the block will have to be rebored and new pistons and rings installed. If the pistons are graphite coated (dark gray in color) an 0.0008 inch (0.02 mm) tighter clearance is acceptable as the piston coating will eventually wear off though normal use.

13   Check the piston pin to the rod and piston clearances by twisting the piston and rod in opposite directions. Any noticeable play indicates excessive wear, which must be corrected (see illustrations). The piston/connecting rod assemblies should be taken to an automotive machine shop to have the pistons and rods re-sized and new pins installed.

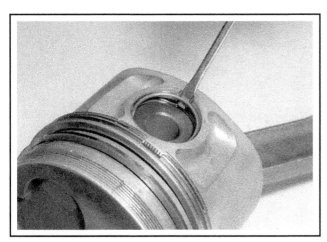

**18.13a  Insert a small screwdriver into the slot and pry out the piston pin retaining clips**

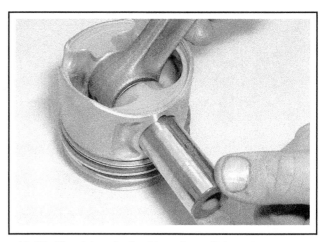

**18.13b  The piston pin should require a slight push to remove and install it - any excessive movement or looseness will require replacement of the rod and/or the piston and pin assembly**

14  If the pistons must be removed from the connecting rods for any reason, they should be taken to an automotive machine shop to also have the connecting rods checked for bend and twist, since automotive machine shops have special equipment for this purpose.

➡**Note: Unless new pistons and/or connecting rods are being installed it is not necessary to completely disassemble the pistons from the connecting rods.**

15  Check the connecting rods for cracks and other damage. Temporarily remove the rod caps, lift out the old bearing inserts, wipe the rod and cap bearing surfaces clean and inspect them for nicks, gouges and scratches. After checking the rods, replace the old bearings, slip the caps into place and tighten the nuts finger tight.

➡**Note: If the engine is being rebuilt because of a connecting rod knock, be sure to install new or remanufactured connecting rods.**

## 19  Crankshaft - inspection

▶ **Refer to illustrations 19.1, 19.2, 19.5 and 19.7**

1   Remove all burrs from the crankshaft oil holes with a stone, file or scraper (see illustration).
2   Clean the crankshaft with solvent and dry it with compressed air (if available).

✳✳ **WARNING:**

**Wear eye protection when using compressed air. Be sure to clean the oil holes with a stiff brush (see illustration) and flush them with solvent.**

**19.1  The oil holes should be chamfered so sharp edges don't gouge or scratch the new bearings**

**19.2  Use a wire or stiff plastic bristle brush to clean the oil passages in the crankshaft**

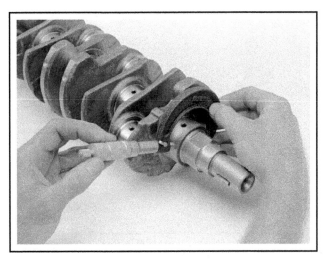

**19.5 Measure the diameter of each crankshaft journal at several points to detect taper and out-of-round conditions**

**19.7 If the seals have worn grooves in the crankshaft journals, or if the seal contact surfaces are nicked or scratched, the new seals will leak**

3   Check the main and connecting rod bearing journals for uneven wear, scoring, pits and cracks.

4   Check the rest of the crankshaft for cracks and other damage. It should be Magnafluxed to reveal hidden cracks - an automotive machine shop will handle the procedure.

5   Using a micrometer, measure the diameter of the main and connecting rod journals and compare the results to this Chapter's Specifications (see illustration). By measuring the diameter at a number of points around each journal's circumference, you'll be able to determine whether or not the journal is out-of-round. Take the measurement at each end of the journal, near the crank throws, to determine if the journal is tapered.

6   If the crankshaft journals are damaged, tapered, out-of-round or

worn beyond the limits given in the Specifications, have the crankshaft reground by an automotive machine shop. Be sure to use the correct-size bearing inserts if the crankshaft is reconditioned.

7   Check the oil seal journals at each end of the crankshaft for wear and damage. If the seal has worn a groove in the journal, or if it's nicked or scratched (see illustration), the new seal may leak when the engine is reassembled. In some cases, an automotive machine shop may be able to repair the journal by pressing on a thin sleeve. If repair isn't feasible, a new or different crankshaft should be installed.

8   Examine the main and rod bearing inserts (see Section 20). Also inspect the crankshaft reluctor ring at the rear of the crankshaft for nicks and damage. Damage to this component may result in severe driveabilty problems.

## 20   Main and connecting rod bearings - inspection and selection

▶ **Refer to illustration 20.1**

1   Even though the main and connecting rod bearings should be replaced with new ones during the engine overhaul, the old bearings should be retained for close examination, as they may reveal valuable information about the condition of the engine (see illustration).

2   Bearing failure occurs because of lack of lubrication, the presence of dirt or other foreign particles, overloading the engine and corrosion. Regardless of the cause of bearing failure, it must be corrected before the engine is reassembled to prevent it from happening again.

3   When examining the bearings, remove them from the engine block, the main bearing caps, the connecting rods and the rod caps and lay them out on a clean surface in the same general position as their location in the engine. This will enable you to match any bearing prob-

lems with the corresponding crankshaft journal.

4   Dirt and other foreign particles get into the engine in a variety of ways. It may be left in the engine during assembly, or it may pass through filters or the PCV system. It may get into the oil, and from there into the bearings. Metal chips from machining operations and normal engine wear are often present. Abrasives are sometimes left in engine components after reconditioning, especially when parts aren't thoroughly cleaned using the proper cleaning methods. Whatever the source, these foreign objects often end up embedded in the soft bearing material and are easily recognized. Large particles won't embed in the bearing and will score or gouge the bearing and journal. The best prevention for this cause of bearing failure is to clean all parts thoroughly and keep everything spotlessly clean during engine assembly. Frequent

and regular engine oil and filter changes are also recommended.

5   Lack of lubrication (or lubrication breakdown) has a number of interrelated causes. Excessive heat (which thins the oil), overloading (which squeezes the oil from the bearing face) and oil leakage or throw off (from excessive bearing clearances, worn oil pump or high engine speeds) all contribute to lubrication breakdown. Blocked oil passages, which usually are the result of misaligned oil holes in a bearing shell, will also oil starve a bearing and destroy it. When lack of lubrication is the cause of bearing failure, the bearing material is wiped or extruded from the steel backing of the bearing. Temperatures may increase to the point where the steel backing turns blue from overheating.

6   Driving habits can have a definite effect on bearing life. Low speed operation in too high a gear (lugging the engine) puts very high loads on bearings, which tends to squeeze out the oil film. These loads cause the bearings to flex, which produces fine cracks in the bearing face (fatigue failure). Eventually the bearing material will loosen in pieces and tear away from the steel backing. Short trip driving leads to corrosion of bearings because insufficient engine heat is produced to drive off the condensed water and corrosive gases. These products collect in the engine oil, forming acid and sludge. As the oil is carried to the engine bearings, the acid attacks and corrodes the bearing material.

7   Incorrect bearing installation during engine assembly will lead to bearing failure as well. Tight-fitting bearings leave insufficient oil clearance and will result in oil starvation. Dirt or foreign particles trapped behind a bearing insert result in high spots on the bearing which lead to failure.

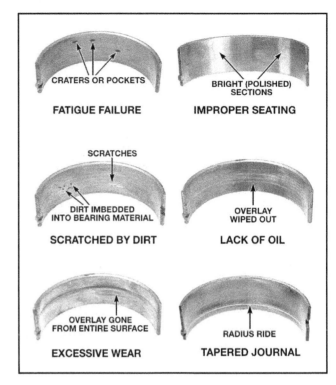

**20.1 Typical bearing failures**

## 21  Camshaft, lifters and bearings - inspection

♦ **Refer to illustrations 21.1, 21.2, 21.4, 21.5, 21.6a, 21.6b and 21.9**

1   Visually check the camshaft bearing surfaces for pitting, score marks, galling and abnormal wear. If the bearing surfaces are damaged, the cylinder head will have to be replaced (see illustration).

2   Measure the outside diameter of each camshaft bearing journal and record your measurements (see illustration). Compare them to the journal outside diameter specified in this Chapter, then measure the inside diameter of each corresponding camshaft bearing and record the measurements. Subtract each cam journal outside diameter from its respective cam bearing bore inside diameter to determine the oil clearance for each bearing. Compare the results to the specified journal-to-bearing clearance. If any of the measurements fall outside the standard specified wear limits in this Chapter, either the camshaft or the cylinder head, or both, must be replaced.

**21.1  Inspect the camshaft bearing surfaces in the cylinder head for pits, score marks and abnormal wear - if damage is noted, the cylinder head must be replaced**

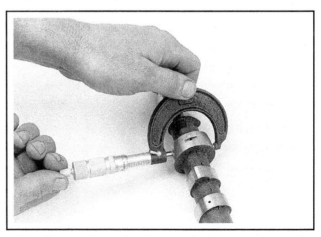

**21.2  Measure the outside diameter of each camshaft journal and the inside diameter of each bearing surface on the cylinder head to determine the oil clearance measurement**

# ENGINE BEARING ANALYSIS

## Debris

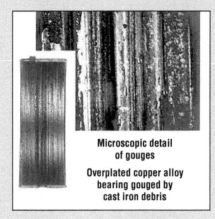

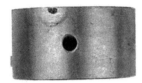

Babbitt bearing embedded with debris from machinings

Microscopic detail of debris

Microscopic detail of gouges

Overplated copper alloy bearing gouged by cast iron debris

Aluminum bearing embedded with glass beads

Microscopic detail of glass beads

Damaged lining caused by dirt left on the bearing back

## Misassembly

Result of a lower half assembled as an upper - blocking the oil flow

Excessive oil clearance is indicated by a short contact arc

Polished and oil-stained backs are a result of a poor fit in the housing bore

Result of a wrong, reversed, or shifted cap

## Overloading

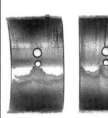

Damage from excessive idling which resulted in an oil film unable to support the load imposed

Damaged upper connecting rod bearings caused by engine lugging; the lower main bearings (not shown) were similarly affected

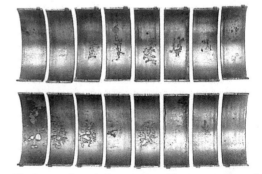

The damage shown in these upper and lower connecting rod bearings was caused by engine operation at a higher-than-rated speed under load

# Misalignment

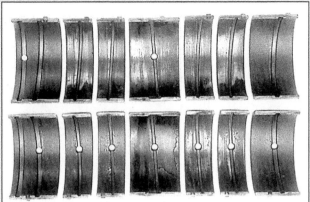

A warped crankshaft caused this pattern of severe wear in the center, diminishing toward the ends

A poorly finished crankshaft caused the equally spaced scoring shown

A bent connecting rod led to the damage in the "V" pattern

A tapered housing bore caused the damage along one edge of this pair

# Corrosion

Microscopic detail of corrosion

Corrosion is an acid attack on the bearing lining generally caused by inadequate maintenance, extremely hot or cold operation, or inferior oils or fuels

# Lubrication

Result of dry start: The bearings on the left, farthest from the oil pump, show more damage

Result of a low oil supply or oil starvation

Severe wear as a result of inadequate oil clearance

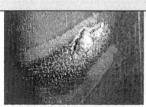

Microscopic detail of cavitation

Example of cavitation - a surface erosion caused by pressure changes in the oil film

Damage from excessive thrust or insufficient axial clearance

Bearing affected by oil dilution caused by excessive blow-by or a rich mixture

© 1986 Federal-Mogul Corporation
Copy and photographs courtesy of Federal Mogul Corporation

**21.4 Checking camshaft endplay with a dial indicator**

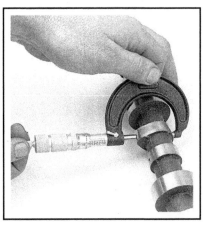

**21.5 Measuring the camshaft lobe height with a micrometer - make sure you move the micrometer to get the highest reading (top of cam lobe)**

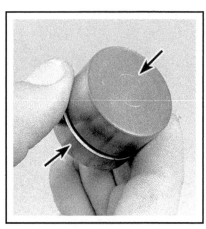

**21.6a Inspect the valve lifters at the area shown (arrows)**

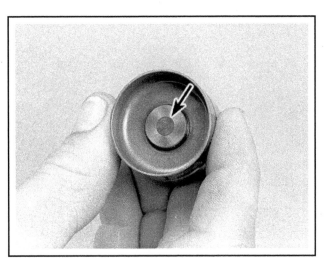

**21.6b Also check the valve stem contact area of the lifter**

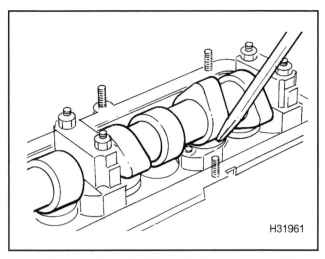

**21.9 When checking noisy lifters with the engine in vehicle, it may be necessary to bleed the lifter down by depressing it with a wooden or plastic tool**

3   Check camshaft runout by placing the camshaft back into the cylinder head and set up a dial indicator on the center journal. Zero the dial indicator. Turn the camshaft slowly and note the dial indicator readings. Record your readings and compare them with the specified runout in this Chapter. If the measured runout exceeds the runout specified in this Chapter, replace the camshaft.

4   Check the camshaft endplay by placing a dial indicator with the stem in line with the camshaft and touching the snout (see illustration). Push the camshaft all the way to the rear and zero the dial indicator. Next, pry the camshaft to the front as far as possible and check the reading on the dial indicator. The distance it moves is the endplay. If it's greater than the Specifications listed in Chapter 2A or 2B, check the bearing caps for wear. If the bearing caps are worn the cylinder head must be replaced.

5   Compare the camshaft lobe height by measuring each lobe with a micrometer (see illustration). Measure of each of the intake lobes and write the measurements and relative positions down on a piece of paper. Then measure of each of the exhaust lobes and record the measure-

ments and relative positions also. This will let you compare all of intake lobes to one another and all of the exhaust lobes to one another. If the difference between the lobes exceeds 0.005 inch the camshaft should be replaced. Do not compare intake lobe heights to exhaust lobe heights as lobe lift may be different. Only compare intake lobes-to-intake lobes and exhaust lobes-to exhaust lobes for this comparison.

6   Inspect the contact and sliding surfaces of each lifter for wear and scratches (see illustrations).

➡**Note: If the lifter pad is worn, it's a good idea to check the corresponding camshaft. Do not lay the lifters on their side or upside down, or air can become trapped inside and the lifter will have to be bled. The lifters can be laid on their side only if they are submerged in a pan of clean engine oil until reassembly.**

7   Check that each lifter moves up and down freely in its bore on the cylinder head. If it doesn't the valve may stick open and cause internal engine damage.

8   In any case make sure all the parts, new or old, have been thoroughly inspected before reassembly.

## HYDRAULIC LIFTERS - IN VEHICLE CHECK

9   Noisy valve lifters can be checked for wear without disassembling the engine by following the procedure outlined below:

a)  *Run the engine until it reaches normal operating temperature.*
b   *Remove the valve cover (see Chapter 2A or 2B).*
c)  *Rotate the engine by hand until the No.1 piston is located at TDC (see Chapter 2A or 2B).*
d)  *Insert a feeler gauge between the camshaft lobe and the lifter to measure the clearance. If the clearance exceeds 0.008 inch (0.2 mm) the lifter and/or the camshaft lobe has worn beyond it limits.*

e)  *If no clearance exists, depress the lifter to let it bleed down and check the clearance again (see illustration).*
f)  *Lifter clearance on the remaining cylinders can be checked by following the firing order sequence and positioning each of the remaining pistons at TDC.*

➡**Note: Lifter clearance can also be checked on any lifter whose cam lobe is pointing upward.**

g)  *If the clearance is beyond the maximum allowed, inspect the camshaft as described in Step 5.*
h)  *If the camshaft is OK, the lifters are faulty and must be replaced.*

## 22  Engine overhaul - reassembly sequence

1   Before beginning engine reassembly, make sure you have all the necessary new parts, gaskets and seals as well as the following items on hand:

*Common hand tools*
*Torque wrench (1/2-inch drive) with angle-torque gauge*
*Piston ring Installation tool*
*Piston ring compressor*
*Short lengths of rubber or plastic hose to fit over connecting rod bolts*
*Plastigage*
*Feeler gauges*
*Fine-tooth file*
*New engine oil*
*Engine assembly lube or moly-base grease*
*Gasket sealant*
*Thread locking compound*

2   In order to save time and avoid problems, engine reassembly must be done in the following general order:

*Crankshaft and main bearings*
*Piston/connecting rod assemblies*

*Oil spray nozzles (if equipped)*
*Front and rear main oil seal housings (this*
*    includes the oil pump/front oil seal housing on SOHC V6 engines)*
*Intermediate shaft (four-cylinder engine)*
*Oil pump, oil pump pick-up tube, oil pump drive gear and drive gear*
*    plug (four-cylinder engine)*
*Upper oil pan (V6 engines)*
*Lower oil pan gasket and oil pick-up cover (SOHC and early DOHC*
*    V6 engines)*
*Oil pump, oil pump drive chain, drive chain tensioner, drive chain*
*    cover and the oil supply tubes (later DOHC engines)*
*Oil pan and baffle (four-cylinder engine)*
*Lower oil pan (V6 engines)*
*Oil filter adapter housing (four-cylinder engines)*
*Cylinder head with lifters and camshaft(s)*
*Timing belt and sprockets*
*Valve cover*
*Flywheel/driveplate*
*Intake and exhaust manifolds*

## 23  Piston rings - installation

▶ **Refer to illustrations 23.3, 23.4, 23.5, 23.9a, 23.9b and 23.12**

1   Before installing the new piston rings, the ring end gaps must be checked. It's assumed the piston ring side clearance has been checked and verified correct (see Section 18).

2   Lay out the piston/connecting rod assemblies and the new ring sets so the ring sets will be matched with the same piston and cylinder during the end gap measurement and engine assembly.

3   Insert the top (number one) ring into the first cylinder and square it up with the cylinder walls by pushing it in with the top of the piston (see illustration). The ring should be near the bottom of the cylinder, at the lower limit of ring travel.

4   To measure the end gap, slip feeler gauges between the ends of the ring until a gauge equal to the gap width is found (see illustration). The feeler gauge should slide between the ring ends with a slight amount of drag. Compare the measurement to this Chapter's Specifications. If the gap is larger or smaller than specified, double-check to make sure you have the correct rings before proceeding.

5   If the gap is too small, it must be enlarged or the ring ends may come in contact with each other during engine operation, which can cause serious engine damage. The end gap can be increased by filing the ring ends very carefully with a fine file. Mount the file in a vise equipped with soft jaws, slip the ring over the file with the ends contacting the file teeth and slowly move the ring to remove material from the

**23.3 When checking piston ring end gap, the ring must be square in the cylinder bore (this is done by pushing the ring down with the top of a piston as shown)**

**23.4 With the ring square in the cylinder, measure the end gap with a feeler gauge**

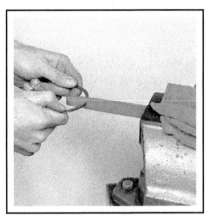

**23.5 If the end gap is too small, clamp a file in a vise and file the ring ends (from the outside in only) to enlarge the gap slightly**

**23.9a Installing the spacer/expander in the oil control ring groove**

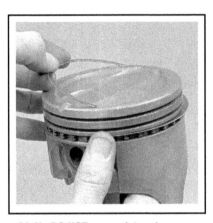

**23.9b DO NOT use a piston ring installation tool when installing the oil ring side rails**

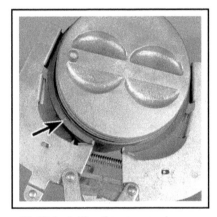

**23.12 Installing the compression rings with a ring expander - the "TOP" mark (arrow) must face up**

ends. When performing this operation, file only from the outside in (see illustration).

➡**Note: When you have the end gap correct, remove any burrs from the filed ends of the rings with a whetstone.**

6   Excess end gap isn't critical unless it's greater than 0.040-inch (1.0 mm). Again, double-check to make sure you have the correct rings for the engine. If the engine block has been bored oversize, necessitating oversize pistons, matching oversize rings are required.

7   Repeat the procedure for each ring that will be installed in the first cylinder and for each ring in the remaining cylinders. Remember to keep rings, pistons and cylinders matched up.

8   Once the ring end gaps have been checked/corrected, the rings can be installed on the pistons.

9   The oil control ring (lowest one on the piston) is usually installed first. Some piston ring manufacturers supply one-piece oil rings - others may supply three-piece oil rings. One-piece rings can be installed as shown in illustration 23.12. If you're installing three-piece oil rings, slip the spacer/expander into the groove (see illustration). If an anti-rotation tang is used, make sure it's inserted into the drilled hole in the ring groove. Next, install the lower side rail. Don't use a piston ring installation tool on the oil ring side rails, as they may be damaged. Instead, place one end of the side rail into the groove between the spacer/expander and the ring land, hold it firmly in place and slide

a finger around the piston while pushing the rail into the groove (see illustration). Next, install the upper side rail in the same manner.

➡**Note: Some engines may have a two piece oil ring. If so, follow the installation instructions that come with the piston rings if they differ from the instructions outlined here.**

10   After the three oil ring components have been installed, check to make sure both the upper and lower side rails can be turned smoothly in the ring groove.

11   The number two (middle) ring is installed next. It's usually stamped with a mark, which must face up, toward the top of the piston.

➡**Note: Always follow the instructions printed on the ring package or box - different manufacturers may require different approaches. Don't mix up the top and middle rings, as they have different cross-sections.**

12   Use a piston ring installation tool and make sure the identification mark is facing the top of the piston, then slip the ring into the middle groove on the piston (see illustration). Don't expand the ring any more than necessary to slide it over the piston.

13   Install the number one (top) ring in the same manner. Make sure the mark is facing up. Be careful not to confuse the number one and number two rings.

14   Repeat the procedure for the remaining pistons and rings.

## 24 Crankshaft - installation and main bearing oil clearance check

1   Crankshaft installation is the first step in engine reassembly. It's assumed at this point that the engine block and crankshaft have been cleaned, inspected and repaired or reconditioned.

2   Position the engine with the bottom facing up.

3   Remove the main bearing cap bolts and lift out the caps. Lay them out in the proper order to ensure correct installation.

4   If they're still in place, remove the original bearing inserts from the block and the main bearing caps. Wipe the bearing surfaces of the block and caps with a clean, lint-free cloth. They must be kept spotlessly clean.

## MAIN BEARING OIL CLEARANCE CHECK

▶ **Refer to illustrations 24.5, 24.6, 24.11 and 24.15**

➡**Note: Don't touch the faces of the new bearing inserts with your fingers. Oil and acids from your skin can etch the bearings.**

5   Clean the back sides of the new main bearing inserts and lay one in each main bearing saddle in the block. If one of the bearing inserts from each set has a large groove in it, make sure the grooved insert is installed in the block (see illustration). Lay the other bearing from each set in the corresponding main bearing cap. Make sure the tab on the bearing insert fits into the recess in the block or cap, neither higher than the cap's edge nor lower.

### ❊❊ CAUTION:

**The oil holes in the block must line up with the oil holes in the bearing inserts. Do not hammer the bearing into place and don't nick or gouge the bearing faces. No lubrication should be used at this time.**

6   Install the crankshaft thrust washers.

➡**Note: On 1.8L four-cylinder engines, two thrust washers must be installed on each side of the number three main cap (see illustration). On V6 engines, two thrust washers must be installed on each side of the number four main journal in the block and two more thrustwashers must be installed on each side of number four bearing cap.**

7   Clean the faces of the bearings in the block and the crankshaft main bearing journals with a clean, lint-free cloth.

8   Check or clean the oil holes in the crankshaft, as any dirt here can go only one way - straight through the new bearings.

9   Once you're certain the crankshaft is clean, carefully lay it in position in the main bearings.

10  Before the crankshaft can be permanently installed, the main bearing oil clearance must be checked.

11  Cut several pieces of the appropriate size Plastigage (they should be slightly shorter than the width of the main bearings) and place one piece on each crankshaft main bearing journal, parallel with the journal axis (see illustration).

12  Clean the faces of the bearings in the caps and install the caps in their original locations (don't mix them up) with the arrows pointing toward the front of the engine. Don't disturb the Plastigage.

13  Starting with the center main and working out toward the ends, tighten the main bearing cap bolts to the first step listed in the torque Specifications when checking the bearing clearances with plastigage. Do not tighten the bolts to the final torque specification or the rotate the crankshaft at any time during the plastigage operation, and do not tighten one cap completely - tighten all caps equally. Before tightening, the main caps should be seated using light taps with a brass or plastic mallet.

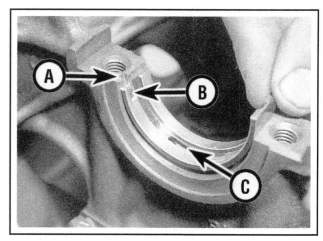

**24.5  Bearing shell correctly installed**

A   Recess in bearing saddle          C   Oil hole
B   Lug on bearing shell

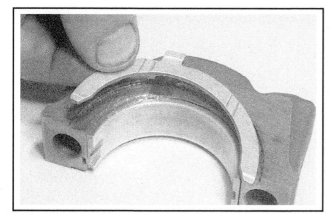

**24.6  Installing the thrust washers on the No. 3 main bearing cap of a four-cylinder engine**

**24.11  Lay the Plastigage strips (arrow) on the main bearing journals, parallel to the crankshaft centerline**

14 Remove the bolts/studs and carefully lift off the main bearing caps. Keep them in order. Don't disturb the Plastigage or rotate the crankshaft. If any of the main bearing caps are difficult to remove, tap them gently from side-to-side with a soft-face hammer to loosen them.

15 Compare the width of the crushed Plastigage on each journal to the scale printed on the Plastigage envelope to obtain the main bearing oil clearance (see illustration). Check the Specifications to make sure it's correct.

16 If the clearance is not as specified, the bearing inserts may be the wrong size (which means different ones will be required). Before deciding different inserts are needed, make sure no dirt or oil was between the bearing inserts and the caps or block when the clearance was measured. If the Plastigage was wider at one end than the other, the journal may be tapered (see Section 19).

17 Carefully scrape all traces of the Plastigage material off the main bearing journals and/or the bearing faces. Use your fingernail or the edge of a credit card - don't nick or scratch the bearing faces.

## FINAL CRANKSHAFT INSTALLATION

18 Carefully lift the crankshaft out of the engine.

19 Clean the bearing faces in the block, then apply a thin, uniform layer of moly-base grease or engine assembly lube to each of the bearing surfaces. Be sure to coat the thrust faces as well as the journal face of the thrust bearing.

20 Make sure the crankshaft journals are clean, then lay the crankshaft back in place in the block.

21 Clean the faces of the bearings in the caps, then apply lubricant to them.

22 Install the caps in their original locations with the arrows (made earlier) pointing toward the front of the engine.

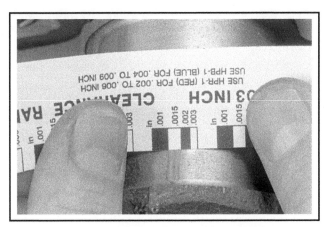

**24.15 Measuring the width of the crushed Plastigage to determine the main bearing oil clearance (be sure to use the correct scale - standard and metric ones are included)**

23 With all caps in place and bolts just started, tap the ends of the crankshaft forward and backward with a lead or brass hammer to line up the main bearing and crankshaft thrust surfaces.

24 Following the procedures outlined in Step 13, retighten all main bearing cap bolts to the final torque listed in this Chapter's Specifications, starting with the center main and working out toward the ends.

25 Rotate the crankshaft a number of times by hand to check for any obvious binding.

26 The final step is to check the crankshaft endplay with feeler gauges or a dial indicator as described in Section 14. The endplay should be correct if the crankshaft thrust faces aren't worn or damaged and new bearings have been installed.

27 If you're working on a four-cylinder engine, install the intermediate shaft now.

## 25  Rear main oil seal - replacement

All models are equipped with a one piece rear main oil seal and housing. The crankshaft must be installed first and the main bearing caps bolted in place before the seal and housing can be installed on the engine block. Refer to Chapter 2A for the rear main seal replacement procedure. Disregard the Steps that do not apply since the engine is out of the vehicle and the oil pan is not installed.

## 26  Pistons and connecting rods - installation and rod bearing oil clearance check

1 Before installing the piston/connecting rod assemblies, the cylinder walls must be perfectly clean, the top edge of each cylinder must be chamfered, and the crankshaft must be in place.

2 Remove the cap from the end of the number one connecting rod (check the marks made during removal). Remove the original bearing inserts and wipe the bearing surfaces of the connecting rod and cap with a clean, lint-free cloth. They must be kept spotlessly clean.

### PISTON INSTALLATION AND ROD BEARING OIL CLEARANCE CHECK

♦ **Refer to illustrations 26.5a, 26.5b, 26.11, 26.13 and 26.17**

3 Clean the back side of the new upper bearing insert, then lay it in

place in the connecting rod. Make sure the tab on the bearing fits into the recess in the rod. Don't hammer the bearing insert into place and be very careful not to nick or gouge the bearing face. Don't lubricate the bearing at this time.

4 Clean the back side of the other bearing insert and install it in the rod cap. Again, make sure the tab on the bearing fits into the recess in the cap, and don't apply any lubricant. It's critically important that the mating surfaces of the bearing and connecting rod are perfectly clean and oil free when they're assembled.

5 Stagger the piston ring gaps around the piston (see illustrations).

6 On four-cylinder engines, slip a section of plastic or rubber hose over each connecting rod cap bolt.

7 Lubricate the piston and rings with clean engine oil and attach a piston ring compressor to the piston. Leave the skirt protruding about

1/4-inch (6 mm) to guide the piston into the cylinder. The rings must be compressed until they're flush with the piston.

8    Rotate the crankshaft until the number one connecting rod journal is at BDC (bottom dead center) and apply a coat of engine oil to the cylinder walls.

9    With the mark or notch on top of the piston facing the front of the engine, gently insert the piston/connecting rod assembly into the number one cylinder bore and rest the bottom edge of the ring compressor on the engine block.

10  Tap the top edge of the ring compressor to make sure it's contacting the block around its entire circumference.

11  Gently tap on the top of the piston with the end of a wooden or plastic hammer handle (see illustration) while guiding the end of the connecting rod into place on the crankshaft journal. The piston rings may try to pop out of the ring compressor just before entering the cylinder bore, so keep some pressure down on the ring compressor. Work slowly, and if any resistance is felt as the piston enters the cylinder, stop immediately. Find out what's hanging up and fix it before proceeding. Do not, for any reason, force the piston into the cylinder - you might break a ring and/or the piston.

12  Once the piston/connecting rod assembly is installed, the connecting rod bearing oil clearance must be checked before the rod cap is permanently bolted in place.

13  Cut a piece of the appropriate size Plastigage slightly shorter than the width of the connecting rod bearing and lay it in place on the number one connecting rod journal, parallel with the journal axis (see illustration).

14  Clean the connecting rod cap bearing face, remove the protective hoses from the connecting rod bolts and install the rod cap. Make sure the mating mark on the cap is on the same side as the mark on the connecting rod.

15  Install the nuts (four-cylinder engine) or bolts (V6 engines) and tighten them to the first step listed in the torque Specifications when checking the bearing clearances with plastigage. Do not tighten the bolts or nuts to the final torque specification or the rotate the crankshaft at any time during the plastigage operation. Work up to it in two steps.

➡**Note: Use a thin-wall socket to avoid erroneous torque readings that can result if the socket is wedged between the rod cap and nut or bolt. If the socket tends to wedge itself between the nut or bolt and the cap, lift up on it slightly until it no longer contacts the cap.**

16  Remove the nuts or bolts and detach the rod cap, being very careful not to disturb the Plastigage.

17  Compare the width of the crushed Plastigage to the scale printed on the Plastigage envelope to obtain the oil clearance (see illustration). Compare it to this Chapter's Specifications to make sure the clearance is correct.

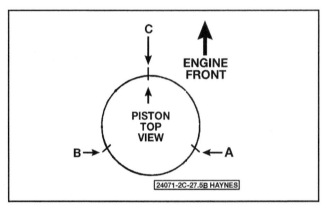

**26.5a  Ring end gap positions (with one-piece oil rings)**

A   Top compression ring gap
B   Second compression ring gap
C   Oil ring gap

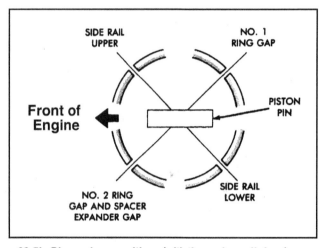

**26.5b  Ring end gap positions (with three-piece oil rings)**

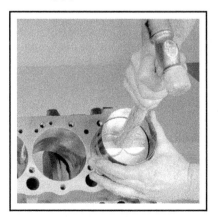

**26.11  Drive the piston into the cylinder bore with the end of a wooden or plastic hammer handle**

**26.13  Lay the Plastigage strips on each rod bearing journal, parallel to the crankshaft centerline**

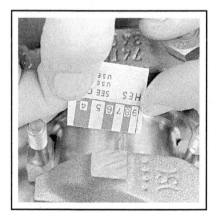

**26.17  Measuring the width of the crushed Plastigage to determine the rod bearing oil clearance (be sure to use the correct scale - standard and metric ones are included)**

18 If the clearance is not as specified, the bearing inserts may be the wrong size (which means different ones will be required). Before deciding different inserts are needed, make sure no dirt or oil was between the bearing inserts and the connecting rod or cap when the clearance was measured. Also, recheck the journal diameter. If the Plastigage was wider at one end than the other, the journal may be tapered (see Section 19).

## FINAL CONNECTING ROD INSTALLATION

19 Carefully scrape all traces of the Plastigage material off the rod journal and/or bearing face. Be very careful not to scratch the bearing - use your fingernail or the edge of a credit card.

20 Make sure the bearing faces are perfectly clean, then apply a uniform layer of clean moly-base grease or engine assembly lube to both of them. You'll have to push the piston into the cylinder to expose the face of the bearing insert in the connecting rod - be sure to slip the protective hoses over the rod bolts first.

21 Slide the connecting rod back into place on the journal, remove the protective hoses (if used) from the rod cap bolts, install the rod cap and tighten the nuts or bolts to the final torque listed in this Chapter's Specifications. Work up to the final torque in three steps.

22 Repeat the entire procedure for the remaining pistons/connecting rods.

23 The important points to remember are:

a) *Keep the back sides of the bearing inserts and the insides of the connecting rods and caps perfectly clean when assembling them.*
b) *Make sure you have the correct piston/rod assembly for each cylinder.*
c) *The arrow or mark on the piston must face the front of the engine.*
d) *Lubricate the cylinder walls with clean oil.*
e) *Lubricate the bearing faces when installing the rod caps after the oil clearance has been checked.*

24 After all the piston/connecting rod assemblies have been properly installed, rotate the crankshaft a number of times by hand to check for any obvious binding.

25 As a final step, the connecting rod endplay must be checked (see Section 13).

26 Compare the measured endplay to this Chapter's Specifications to make sure it's correct. If it was correct before disassembly and the original crankshaft and rods were reinstalled, it should still be right. If new rods or a new crankshaft were installed, the endplay may be inadequate. If so, the rods will have to be removed and taken to an automotive machine shop for re-sizing.

## 27 Initial start-up and break-in after overhaul

### ❊❊ WARNING:

**Have a fire extinguisher handy when starting the engine for the first time.**

➡**Note: It is likely that, after performing major engine repairs, the check engine light on the instrument panel will be illuminated. This will require taking the vehicle to a dealer service department (or other repair shop equipped with the necessary scan tool) to have the cam sensor(s) synchronized and the trouble code(s) cleared.**

1 Once the engine has been installed in the vehicle, double-check the oil and coolant levels.

2 With the spark plugs out of the engine and the fuel and ignition systems disabled (see Section 3) crank the engine until oil pressure registers on the gauge.

3 Install the spark plugs and hook up the spark plug wires (V6 engines) or install the ignition coils (four-cylinder engine). Restore the fuel and ignition system functions.

4 Start the engine. It may take a few moments for the fuel system to build up pressure, but the engine should start without a great deal of effort.

➡**Note: If the engine keeps backfiring, recheck the valve timing and spark plug wire routing.**

5 After the engine starts, it should be allowed to warm up to normal operating temperature. While the engine is warming up, make a thorough check for fuel, oil and coolant leaks.

6 Shut the engine off and recheck the engine oil and coolant levels.

7 Drive the vehicle to an area with no traffic, accelerate from 30 to 50 mph, then allow the vehicle to slow to 30 mph with the throttle closed. Repeat the procedure 10 or 12 times. This will load the piston rings and cause them to seat properly against the cylinder walls. Check again for oil and coolant leaks.

8 Drive the vehicle gently for the first 500 miles (no sustained high speeds) and keep a constant check on the oil level. It isn't unusual for an engine to use oil during the break-in period.

9 At approximately 500 to 600 miles, change the oil and filter.

10 For the next few hundred miles, drive the vehicle normally. Don't pamper it or abuse it.

11 After 2000 miles, change the oil and filter again and consider the engine broken in.

## 28 Engine oil cooler - removal and installation

1 Locate the engine oil cooler between the engine block and the oil filter (V6 engine) or between the oil filter housing and the oil filter (four cylinder engine) (see illustration 2.2a). Remove the oil filter (see Chapter 1).

2 Drain the engine coolant (see Chapter 1).

3 Disconnect the coolant hoses from the engine oil cooler. Be sure to position a pan below the oil filter to catch any residual coolant.

4 Remove the large retaining nut from the housing and separate the oil cooler from the engine block.

5 Be sure to install a new engine oil cooler O-ring gasket.

6 Installation is the reverse of removal. Tighten the oil cooler retaining nut to the Specifications listed in this Chapter.

## Specifications

### General

| | |
|---|---|
| Engine designation | |
| 1.8L turbo | AEB, AUG, AWM and ATW |
| 2.8L SOHC | AFC |
| 2.8L DOHC | AHA, ACK, ALG, APR, ATQ and AQD |
| Displacement | |
| 1.8L turbo | 109 cubic inches (1.8 liters) |
| 2.8L SOHC and DOHC | 171 cubic inches (2.8 liters) |
| Bore and stroke | |
| 1.8L turbo | 3.189 x 3.401 inches (81.0 x 86.4 mm) |
| 2.8L SOHC and DOHC | 3.248 x 3.401 inches (82.5 x 86.4 mm) |
| Cylinder numbers (front to rear) | |
| 1.8L turbo | 1-2-3-4 |
| 2.8L SOHC and DOHC | |
| Left bank | 4-5-6 |
| Right bank | 1-2-3 |
| Firing order | |
| 1.8L turbo | 1-3-4-2 |
| 2.8L SOHC and DOHC | 1-4-3-6-2-5 |
| Cylinder compression pressure | |
| 1.8L turbo | |
| Minimum | 101 psi (7.07 kg/cm2) |
| Maximum variation between cylinders | 30-percent from the highest reading |
| 2.8L SOHC and DOHC | |
| Minimum | 110 psi (7.7 kg/cm2) |
| Maximum variation between cylinders | 30-percent from the highest reading |
| Oil pressure | |
| 1.8L turbo | |
| At 2000 rpm | 29 psi (2.03 kg/cm2) minimum |
| 2.8L SOHC | |
| At idle | 15 to 36 psi (1.05 to 2.52 kg/cm2) |
| At 3,000 rpm | 44 to 73 psi (3.08 to 5.11 kg/cm2) |
| 2.8L DOHC | |
| At 2000 rpm | 29 psi (2.03 kg/cm2) minimum |
| Maximum oil pressure | |
| All engines | 101 psi (7.07 kg/cm2) |

### Cylinder head

| | |
|---|---|
| Resurfacing dimension (minimum height) | |
| 1.8L turbo | 5.480 inches (139.20 mm) |
| SOHC | 5.226 inches (132.75 mm) |
| DOHC | 5.480 inches (139.20 mm) |
| Warpage limit | |
| 1.8L turbo | 0.004 inch (0.10 mm) |
| SOHC | 0.002 inch (0.05 mm) |
| DOHC | 0.004 inch (0.10 mm) |

## Specifications (continued)

### Valves and related components

Valve face angle
    All engines        45-degrees
Valve seat angle
    All engines        45-degrees
Valve margin width
    All engines        N/A
Valve stem diameter
    Intake valves
        1.8L turbo        0.235 inch (5.963 mm)
        2.8L SOHC        N/A
        2.8L DOHC        0.235 inch (5.963 mm)
    Exhaust valves
        1.8L turbo        0.234 inch (5.943 mm)
        2.8L SOHC        N/A
        2.8L DOHC        0.234 inch (5.943 mm)
Valve stem-to-guide clearance (maximum deflection)
    Intake valves
        1.8L turbo        0.031 inch (0.8 mm)
        2.8L SOHC        0.039 inch (1.0 mm)
        2.8L DOHC        0.031 inch (0.8 mm)
    Exhaust valves
        1.8L turbo        0.031 inch (0.8 mm)
        2.8L SOHC        0.051 inch (1.3 mm)
        2.8L DOHC        0.031 inch (0.8 mm)
Valve spring
    Free length        N/A
    Installed height        N/A
Valve installed height (minimum dimension)*
    Intake valves
        1.8L turbo
            Outer valves        1.339 inch (34.0 mm)
            Inner valve        1.327 inch (33.7 mm)
        2.8L SOHC        1.331 inch (33.8 mm)
        2.8L DOHC
            Outer valves        1.339 inch (34.0 mm)
            Inner valve        1.327 inch (33.7 mm)
    Exhaust valves
        1.8L turbo        1.354 inch (34.4 mm)
        2.8L SOHC        1.343 inch (34.1 mm)
        2.8L DOHC        1.354 inch (34.4 mm)

*Measured from tip of valve stem to top of the valve cover rail.

### Crankshaft

Endplay
    Standard        0.0028 to 0.0091 inch (0.07 to 0.23 mm)
    Service limit        0.0118 inch (0.30 mm)
Runout        N/A

Main bearing journal diameters
    1.8L turbo
        Standard                    2.1260 inches (54.0 mm)
        1st undersize            2.1161 inches (53.75 mm)
        2nd undersize          2.1063 inches (53.50 mm)
        3rd undersize          2.0965 inches (53.25 mm)
        Tolerance               -0.0008 to -0.00165 inch (-0.022 to -0.042 mm)
    2.8L SOHC and DOHC
        Standard                    2.559 inches (65.0 mm)
        1st undersize            2.549 inches (64.75 mm)
        2nd undersize          2.539 inches (64.50 mm)
        3rd undersize          2.530 inches (64.25 mm)
        Tolerance               -0.0008 to -0.00165 inch (-0.022 to -0.042 mm)
    All engines
        Out-of-round limit      0.0002 inch (0.005 mm)
        Taper limit             0.0003 inch (0.007 mm)
Main bearing oil clearance
    1.8L turbo
        Standard                    0.0008 to 0.0023 inch (0.02 to 0.06 mm)
        Service limit            0.0059 inch (0.15 mm)
    2.8L SOHC and DOHC
        Standard                    0.0007 to 0.0018 inch (0.018 to 0.045 mm)
        Service limit            0.0039 inch (0.10 mm)
Intermediate shaft endplay - 1.8L turbo
    (service limit)               0.01 inch (0.25 mm)

## Connecting rods

Connecting rod bearing journal diameters
    1.8L turbo
        Standard                    1.8819 inches (47.80 mm)
        1st undersize            1.8720 inches (47.55 mm)
        2nd undersize          1.8622 inches (47.30 mm)
        3rd undersize          1.8524 inches (47.05 mm)
        Tolerance               -0.0008 to -0.0016 inch (-0.022 to -0.042 mm)
    2.8L SOHC and DOHC
        Standard                    2.1260 inches (54.0 mm)
        1st undersize            2.1161 inches (53.75 mm)
        2nd undersize          2.1063 inches (53.50 mm)
        3rd undersize          2.0965 inches (53.25 mm)
        Tolerance               -0.0008 to -0.00165 inch (-0.022 to -0.042 mm)
    All engines
        Out-of-round limit      0.0002 inch (0.005 mm)
        Taper limit             0.0003 inch (0.007 mm)
Connecting rod bearing oil clearance
    1.8L turbo
        Standard                    0.0004 to 0.0024 inch (0.01 to 0.06 mm)
        Service limit            0.0047 inch (0.12 mm)
    2.8L SOHC and DOHC
        Standard                    0.0006 to 0.00244 inch (0.015 to 0.062 mm)
        Service limit            0.0047 inch (0.12 mm)
Connecting rod side clearance (endplay)
    All engines                0.0020 to 0.0122 inch (0.05 to 0.31 mm)

### Engine block

Cylinder bore diameter
    1.8L turbo
        Standard                        3.189 inches (81.01 mm)
        1st oversize               3.209 inches (81.51 mm)
    2.8L SOHC and DOHC
        Standard                        3.2484 inches (82.51 mm)
        1st oversize               3.2583 inches (82.76 mm)
        2nd oversize              3.2681 inches (83.01 mm)
Cylinder bore diameter
    All engines
        Out-of-round limit        0.0031 inch (0.08 mm)
        Taper limit                0.0031 inch (0.08 mm)
Block deck warpage limit             0.004 inch (0.10 mm)

### Pistons and rings

Piston diameter*
    1.8L turbo
        Standard                        3.1875 inches (80.965 mm)
        1st oversize               3.2073 inches (81.465 mm)
    2.8L SOHC and DOHC
        Standard                        3.2472 inches (82.48 mm)
        1st oversize               3.2575 inches (82.74 mm)
        2nd oversize              3.2669 inches (82.98 mm)
Piston ring end gap
    1.8L turbo
        Top compression ring       0.0079 to 0.0157 inch (0.20 to 0.40 mm)
        Second compression ring   0.0079 to 0.0157 inch (0.20 to 0.40 mm)
        Oil control ring               0.0098 to 0.0197 inch (0.25 to 0.50 mm)
    2.8L SOHC and DOHC
        Top compression ring       0.014 to 0.020 inch (0.35 to 0.50 mm)
        Second compression ring   0.020 to 0.028 inch (0.50 to 0.70 mm)
        Oil control ring               0.0098 to 0.0197 inch (0.25 to 0.50 mm)
Piston ring side clearance
    1.8L turbo
        Top compression ring       0.0023 to 0.0035 inch (0.06 to 0.09 mm)
        Second compression ring   0.0023 to 0.0035 inch (0.06 to 0.09 mm)
        Oil control ring               0.0012 to 0.0024 inch (0.03 to 0.06 mm)
    1.9 liter
        Top compression ring       0.0024 to 0.0035 inch (0.06 to 0.09 mm)
        Second compression ring   0.0020 to 0.0031 inch (0.05 to 0.08 mm)
        Oil control ring               0.0012 to 0.0024 inch (0.03 to 0.06 mm)
    2.8L SOHC and DOHC
        Top compression ring       0.001 to 0.003 inch (0.02 to 0.08 mm)
        Second compression ring   0.001 to 0.003 inch (0.02 to 0.08 mm)
        Oil control ring               0.001 to 0.003 inch (0.02 to 0.08 mm)

*Dimension without graphite coating (the graphite coating, which is dark gray, will add approximately 0.0008 inch [0.02 mm] to the diameter).*

| Torque specifications*** | Ft-lbs (unless otherwise indicated) | Nm |
|---|---|---|

➡Note: One foot-pound (ft-lb) of torque is equivalent to 12 inch-pounds (in-lbs) of torque. Torque values below approximately 15 ft-lbs are expressed in inch-pounds, since most foot-pound torque wrenches are not accurate at these smaller values.

| | | |
|---|---|---|
| Main bearing cap side bolts** | | |
| V6 engines | 18 | 25 |
| Main bearing cap bolts (always replace) | | |
| 1.8L turbo | | |
| Step 1 | 48 | 65 |
| Step 2 | Tighten an additional 90-degrees | |
| 2.8L SOHC | | |
| Step 1 | 44 | 60 |
| Step 2 | Tighten an additional 180-degrees | |
| 2.8L DOHC | | |
| Step 1 | 44 | 60 |
| Step 2 | Tighten an additional 90-degrees | |
| Connecting rod cap nuts or bolts (always replace) | | |
| All engines | | |
| Step 1 | 22 | 30 |
| Step 2 | Tighten an additional 90-degrees | |
| Intermediate shaft flange bolts | 25 | 18 |
| Engine oil cooler retaining nut | | |
| Four cylinder engines | 18 | 25 |
| V6 engines | 22 | 30 |
| Oil filter housing mounting bolts (four cylinder models) (always replace) | | |
| Step 1 | 132 in-lbs | 15 |
| Step 2 | Tighten an additional 90-degrees | |
| Oil spray nozzles | | |
| 1.8L turbo | 20 | 27 |
| 2.8L SOHC and early DOHC | 84 in-lbs | 10 |
| Oil distribution pipe | | |
| V6 engines | 133 in-lbs | 15 |
| Oil retention valves | | |
| V6 engines | 18 | 25 |

** Always tighten the main bearing cap side bolts by hand before tightening the main bearing cap bolts.
***Note: Refer to Chapter 2A or 2B for additional torque specifications.

**NOTES**

**Section**

**Reference to other Chapters**

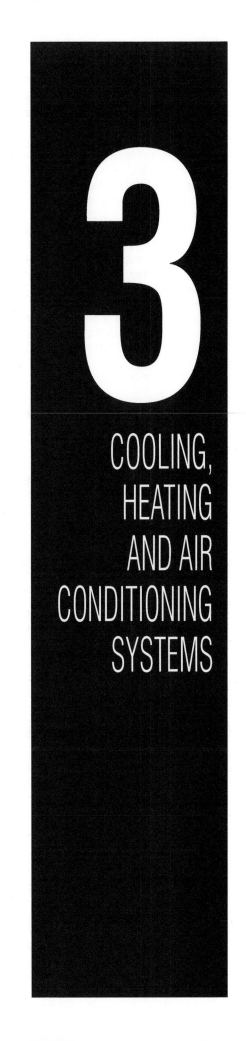

# 3

# COOLING, HEATING AND AIR CONDITIONING SYSTEMS

## 1   General information

All vehicles covered by this manual employ a pressurized engine cooling system with thermostatically controlled coolant circulation. Coolant is drawn from the radiator by an impeller-type water pump mounted at the front of the block. The coolant is then circulated through the engine block and the cylinder head before it's redirected back into the radiator.

A wax pellet type thermostat is located in the thermostat housing on the engine. During warm up, the closed thermostat prevents coolant from circulating through the radiator. When the engine reaches normal operating temperature, the thermostat opens and allows hot coolant to travel through the  radiator, where it is cooled before returning to the engine.

The cooling system is pressurized by a spring-loaded expansion tank cap, which, by maintaining pressure, increases the boiling point of the coolant. If the coolant temperature goes above this increased boiling point, the extra pressure in the system forces the reservoir cap valve off its seat and allows the coolant to escape through the overflow tube into the expansion tank. When the system cools, the excess coolant is automatically drawn from the reservoir tank back into the radiator.

The expansion tank serves as both the point at which fresh coolant is added to the cooling system to maintain the proper fluid level and as a holding tank for overheated coolant.

The heating system works by directing air through the heater core mounted in the dash and then to the interior of the vehicle by a system of ducts. Temperature is controlled by mixing heated air with fresh air, using a system of doors in the ducts, and a blower motor.

The air conditioning system consists of an evaporator core located under the dash, a condenser in front of the radiator, a accumulator in the engine compartment and a belt-driven compressor mounted at the front of the engine.

## 2   Antifreeze - general information

▶ **Refer to illustration 2.4**

### ❋❋ WARNING:

**Do not allow antifreeze to come in contact with your skin or painted surfaces of the vehicle. Rinse off spills immediately with plenty of water. Antifreeze is highly toxic if ingested. Never leave antifreeze lying around in an open container or in puddles on the floor; children and pets are attracted by it's sweet smell and may drink it. Check with local authorities about disposing of used antifreeze. Many communities have collection centers which will see that antifreeze is disposed of safely. Never dump used antifreeze on the ground or pour it into drains.**

### ❋❋ CAUTION:

**Refer to Chapter 1 Specifications to find out what type of coolant your vehicle uses. Typically, on the models covered in this manual, the proper coolant designation is molded into the top of the expansion tank. Never mix green-colored ethylene glycol anti-freeze and red-colored phosphate-free coolant because doing so will destroy the efficiency of the antifreeze.**

The cooling system should be filled with the proper antifreeze solution which will prevent freezing down to at least -20-degrees F (even lower in cold climates). It also provides protection against corrosion and increases the coolant boiling point.

The cooling system should be drained, flushed and refilled at least every other year (see Chapter 1). The use of antifreeze solutions for periods of longer than two years is likely to cause damage and encourage the formation of rust and scale in the system. However, most of these models are filled with a new, long-life phosphate-free coolant which the manufacturer claims is good for the lifetime of the vehicle.

Before adding antifreeze to the system, check all hose connections. Antifreeze can leak through very minute openings.

The exact mixture of antifreeze to water which you should use depends on the relative weather conditions. The mixture should contain at least 50-percent antifreeze, but should never contain more than 70-percent anti-freeze. Consult the mixture ratio chart on the antifreeze container before adding coolant. Hydrometers are available at most auto parts stores to test the coolant (see illustration). Always use antifreeze which meets the vehicle manufacturer's specifications.

**2.4 An inexpensive hydrometer can be used to test the condition of your coolant**

## 3   Thermostat - check and replacement

**✳✳ WARNING:**

**The engine must be completely cool when this procedure is performed.**

**✳✳ CAUTION:**

**Don't drive the vehicle without a thermostat! The computer may stay in open loop mode and emissions and fuel economy will suffer.**

## CHECK

1   Before assuming the thermostat is to blame for a cooling system problem, check the coolant level, drivebelt tension (see Chapter 1) and temperature gauge (or light) operation.

2   If the engine seems to be taking a long time to warm up (based on heater output or temperature gauge operation), the thermostat is probably stuck open. Replace the thermostat with a new one.

3   If the engine runs hot, use your hand to check the temperature of the upper radiator hose. If the hose isn't hot, but the engine is, the thermostat is probably stuck closed, preventing the coolant inside the engine from escaping to the radiator. Replace the thermostat.

4   If the upper radiator hose is hot, it means the coolant is flowing and the thermostat is open. Consult the *Troubleshooting* Section at the front of this manual for cooling system diagnosis.

## REPLACEMENT

▶ **Refer to illustrations 3.11, 3.13a, 3.13b and 3.17**

5   Raise the front of the vehicle and support it securely on jackstands.

6   Remove the splash shield from below the engine.

7   Drain the coolant from the radiator (see Chapter 1). If the coolant is relatively new or in good condition, save it and reuse it. If it is to be replaced, see Section 2 for cautions about proper handling of used anti-freeze.

8   On models equipped with V6 engines, place the radiator support panel in the service position (see Chapter 11).

9   Follow the lower radiator hose to the engine to locate the thermostat housing cover. The thermostat on four-cylinder engines is located on the driver's side of the engine at the bottom of the water pump housing. The thermostat on V6 engines is located on the front of the engine block.

10   Squeeze the tabs on the hose clamp and pull the hose clamp back over the hose.

11   Detach the hose from the thermostat housing cover (see illustration). If the hose sticks, grasp it near the end with a pair of adjustable pliers and twist it to break the seal, then pull it off. If the hose is old or deteriorated, cut it off and install a new one.

**✳✳ CAUTION:**

**Try not to get coolant on the timing belt. It's a good idea to cover the belt with rags to prevent this from happening.**

12   If the outer surface of the cover fitting that mates with the hose is deteriorated (corroded, pitted, etc.) it may be damaged further by hose removal. If it is, the thermostat housing cover will have to be replaced.

13   Remove the bolts/nuts and detach the thermostat cover (see illustrations). If the cover is stuck, tap it with a soft-face hammer to jar it loose. Be prepared for some coolant to spill as the gasket seal is broken.

14   Note how it's installed (which end is facing up or out), then remove the thermostat and the cover O-ring.

15   Clean the mating surfaces of the engine block and the thermostat housing cover.

**3.11  Squeeze the hose clamp with pliers and position it back over the hose**

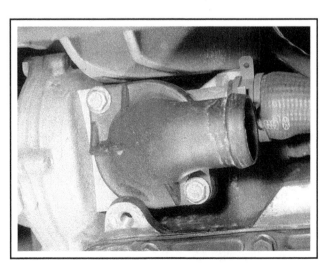

**3.13a  Thermostat housing cover bolts (four-cylinder engine)**

**3.13b Thermostat housing cover bolts (arrows) (V6 engines)**

**3.17 Correct installation of the thermostat and the cover O-ring - on V6 engines the jiggle valve (arrow) must face up**

16  Install the thermostat and make sure the correct end faces out - the spring is directed toward the engine.

17  Install a new O-ring over the thermostat and reattach the thermostat housing cover to the engine block (see illustration). Tighten the bolts to the torque listed in this Chapter's Specifications. Now may be a good time to check and replace the hoses and clamps (see Chapter 1).

18  The remainder of installation is the reverse of the removal procedure.

19  Refer to Chapter 1 and refill and bleed the system, then run the engine and check carefully for leaks.

## 4   Engine cooling fans - check and replacement

### ❄❄ WARNING:

**Keep hands, tools and clothing away from the fan. To avoid injury or damage DO NOT operate the engine with a damaged fan. Do not attempt to repair fan blades - Always replace a damaged fan with a new one.**

## CHECK

### Mechanical cooling fan

1  Rock the fan back and forth by hand to check for excessive bearing play.

2  With the engine cold (and not running), turn the fan blades by hand. The fan should turn freely.

3  Visually inspect for substantial fluid leakage from the clutch assembly. If problems are noted, replace the clutch assembly.

4  With the engine completely warmed up, turn off the ignition switch. Turn the fan by hand. Some drag should be evident. If the fan turns easily, replace the fan clutch.

### Auxiliary electric cooling fan

▶ Refer to illustration 4.5

5  If the fans do not operate, check the heating and cooling system fuses in the engine compartment fuse holder and in the passenger compartment fuse box. If the fuses are OK, test the fan motor for proper operation by unplugging the electrical connector at the motor and use fused jumper wires to connect battery power and ground directly to the fan (see illustration). If the fan doesn't operate, replace the motor.

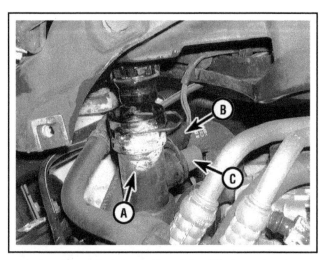

**4.5 Location of the lower radiator hose (A), the cooling fan electrical connectors (B) and the cooling fan thermo switch (C) on a V6 engine**

6   If the fan motor tests OK, reconnect the fan connector and check the cooling fan thermo switch. The switch is located at the bottom of the radiator on the left-hand side on four-cylinder models and on V6 engines it's located in the lower radiator hose on the right hand side of the engine compartment (see illustration 4.5).

7   Referring to the wiring diagrams at the end of this manual, bridge the appropriate terminals on the switch with the ground terminals on the switch connector, then turn On the ignition switch. The fan should run at low speed when one of the terminals is connected to ground and in high speed when the other terminal is connected to ground.

➡️**Note: Some models use a single speed fan which runs at high speed only when the two terminals are bridged.**

8   If the fan runs as described in Step 7, it indicates that the auxiliary cooling fan and circuit are operating properly and the thermo switch is faulty.

9   If the fan does not operate with the thermo switch bypassed, check the cooling fan relays for proper operation. The auxiliary cooling fan relays are located in the instrument panel fuse/relay box (see Chapter 12).

10   If the relays are OK, the problem lies in the wiring harness, the fan control module (automatic A/C only) or the A/C high pressure switch.

11   Refer to the wiring schematics at the end of Chapter 12 and check the wiring for open or short circuits. The fan control module on automatic A/C systems can only be diagnosed as faulty through process of elimination.

## REPLACEMENT

### Mechanical cooling fan

▶ **Refer to illustration 4.12**

12   Loosen the fan retaining nut.

➡️**Note: A spanner wrench, obtainable at most auto parts stores, is required to loosen the fan. The fan attaches to the idler pulley with a large central nut that is part of the fan drive (see illustration). The spanner wrench holds the idler pulley while a large wrench is used to loosen the large drive nut. Sometimes it is possible to hold the idler pulley by applying considerable hand pressure to the serpentine belt while the large nut is turned clockwise (left-hand threads), but it may require the tool if the fan has been installed for years.**

13   Lift the fan/clutch assembly past the shroud and out of the engine compartment.

14   Carefully inspect the fan blades for damage and defects. Replace it if necessary.

15   At this point, the fan may be unbolted from the clutch, if necessary. If the fan clutch is stored, position it with the radiator side facing down.

16   Installation is the reverse of removal. Be sure to tighten the fan and clutch mounting nuts evenly and securely.

### Auxiliary electric cooling fan

▶ **Refer to illustration 4.20**

17   Raise the vehicle and support it securely on jackstands. Remove the lower engine cover and disconnect the electrical connector(s) from the cooling fan(s).

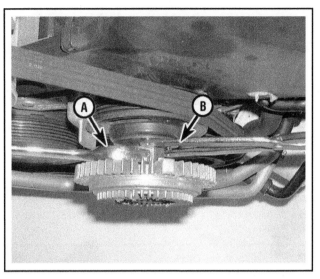

**4.12  Removing the mechanical cooling fan/clutch assembly with a large wrench (A) while holding the idler pulley with a spanner wrench (B) (fan blade removed for clarity)**

**4.20  Auxiliary cooling fan mounting nuts (arrows)**

18   Drain the cooling system (see Chapter 1) and remove the radiator (see Section 5).

19   Remove the forward mounted air intake duct from the radiator support panel and the air cleaner housing.

20   Working in between the radiator support panel and the engine, remove the auxiliary cooling fan mounting nuts (see illustration).

21   Pull the auxiliary cooling fan out toward the front of the vehicle. On models with air conditioning it will be necessary to squeeze the fan out between the condenser and the front of the radiator support panel. Be careful not to contact the condenser cooling fins.

22   If the fan blades or the fan motor are damaged, they can be replaced by removing the fan blade from the fan motor.

23   Installation is the reverse of removal.

## 5 Radiator and expansion tank - removal and installation

### ✳✳ WARNING 1:

**The air conditioning system is under high pressure. DO NOT loosen any fittings or remove any components until after the system has been discharged. Air conditioning refrigerant should be properly discharged into an EPA-approved container at a dealership service department or an automotive air conditioning repair facility. Always wear eye protection when disconnecting air conditioning system fittings.**

### ✳✳ WARNING 2:

**The engine must be completely cool when this procedure is performed.**

### RADIATOR

▶ **Refer to illustrations 5.3 and 5.6**

1   Raise the front of the vehicle and support it securely on jackstands. Drain the cooling system as described in Chapter 1. Refer to the coolant **Warning** in Section 2.

2   Place the radiator support panel in the service position (see Chapter 11).

3   Disconnect the upper and lower coolant hoses (see illustration) and the electrical connector from the fan thermo switch.

4   If the vehicle is equipped with an automatic transaxle, detach the transmission oil cooler lines from the bottom of the radiator.

5   If the vehicle is equipped with air conditioning, disconnect the retaining clamps for the air conditioning lines. Remove the condenser mounting bolts and separate the condenser from the radiator without disconnecting the air conditioning lines (see Section 15). Pull the condenser forward as far as possible and secure the condenser to the body with a piece of wire so it does not hang on the refrigerant lines.

6   Detach the radiator mounting clips and remove the radiator from the front of the support panel (see illustration).

7   Prior to installation of the radiator, replace any damaged hose clips and/or radiator hoses and fittings. If leaks have been noticed or there have been cooling problems, have the radiator cleaned and tested at a radiator shop.

8   Radiator installation is the reverse of removal.

9   After installation, fill the system with the proper mixture of anti-freeze and bleed the air from the cooling system as described in Chapter 1.

### EXPANSION TANK

▶ **Refer to illustration 5.11**

10  Drain the cooling system as described in Chapter 1 until the expansion tank is empty. Refer to the coolant **Warning** in Section 2.

11  Remove the coolant recovery hoses from the expansion tank (see illustration).

12  Detach the reservoir mounting bolts, then lift the expansion tank from the firewall and disconnect the coolant level sensor connector.

13  Remove the expansion tank from the engine compartment.

14  Prior to installation make sure the reservoir is clean and free of debris which could be drawn into the radiator (wash it with soapy water and a brush if necessary, then rinse thoroughly).

15  Installation is the reverse of removal.

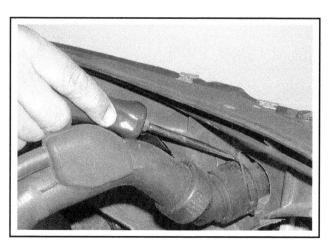

**5.3  Removing the upper radiator hose - pry the clip outward and pull the hose off the radiator**

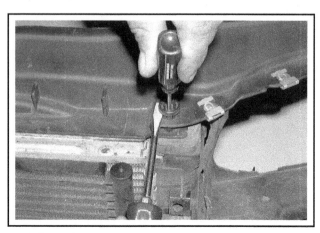

**5.6  Removing a radiator mounting clip**

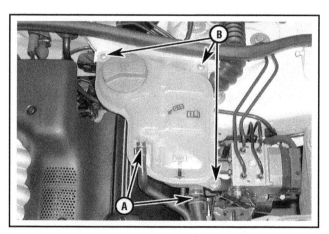

**5.11  Remove the coolant recovery hoses (A) and the expansion tank mounting bolts (B)**

**6    Water pump and after-run coolant pump - removal and installation**

### ✳✳ WARNING:

**Wait until the engine is completely cool before starting this procedure.**

## WATER PUMP

1    Raise the vehicle and support it securely on jackstands.
2    Drain the coolant (see Chapter 1).
3    Place the radiator support panel in the service position (see Chapter 11).
4    Remove the serpentine drivebelt (see Chapter 1).
5    Remove the mechanical cooling fan/clutch assembly as described in Section 5.

### Four-cylinder engine

▸ **Refer to illustrations 6.11, 6.13 and 6.14**

6    Counterhold the power steering pump pulley by passing a screwdriver through the hole in the pulley and bracing it against the pump mounting bracket. Unbolt the outer section of the water pump pulley assembly and remove the V-belt.
7    Remove the alternator with reference to Chapter 5.
8    Remove the power steering pump from its mounting bracket, as described in Chapter 10, noting that there is no need to disconnect the hydraulic pipe/hose(s) from the pump. Secure the pump clear of its mounting bracket using cable ties or wire.
9    Remove the nuts/bolts (as applicable) and remove the alternator/power steering pump bracket.
10    Release the hose clamps and disconnect the coolant hoses from the back of the water pump housing and the thermostat housing cover.
11    Unscrew the retaining nut, then remove hammer head bolt securing the water pump to the lower timing cover and remove the water pump housing from the engine (see illustration).
12    Remove the O-ring located between the housing and block and discard it; a new one should be used on installation.

13    With the assembly on a bench, unscrew the retaining bolts and remove the pump from the housing (see illustration). Discard the gasket; a new one must be used on installation. Note it is not possible to overhaul the pump. If it is faulty, the unit must be replaced with a new one.

➡**Note the fitted position of the hammer head bolt in the coolant pump housing.**

14    Installation is the reverse of removal. Be sure to install a new gasket between the pump and the housing and a new O-ring between the water pump housing and the engine block (see illustration).
15    Add coolant to the specified level (see Chapter 1). Start the engine and check for the proper coolant level and the water pump and hoses for leaks. Bleed the cooling system of air as described in Chapter 1.

**6.11  Removing the water pump housing (four-cylinder engine)**

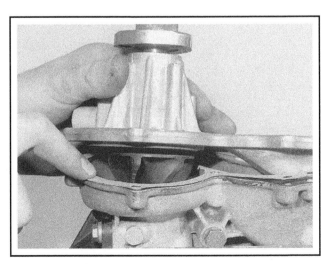

**6.13  Unscrew the retaining bolts and remove the water pump from the housing**

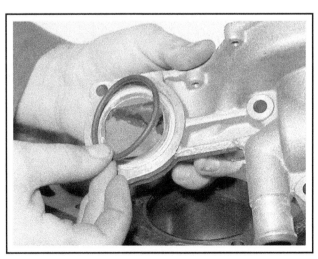

**6.14  Installing a new water pump housing-to-engine block O-ring**

### V6 engines

▶ **Refer to illustrations 6.17 and 6.19**

16  Remove the timing belt (see Chapter 2B).

17  Remove the water pump mounting bolts and the water pump from the engine block (see illustration).

18  Remove all traces of gasket material from the mating surfaces on the water pump and the engine block. Wipe the mating surfaces clean with lacquer thinner or acetone.

19  Apply a small amount of RTV sealant to the water pump gasket surface to help keep the gasket in place as the water pump is being installed. Install the gasket on the water pump, then install the water pump onto the engine block (see illustration).

20  Install the water pump bolts and tighten them to the torque listed in this Chapter's Specifications.

21  The remainder of the installation procedure is the reverse of removal. Be sure to properly install the timing belt. Add coolant to the specified level and bleed the cooling system (see Chapter 1). Start the engine and check for the proper coolant level and the water pump and hoses for leaks.

## AFTER-RUN COOLANT PUMP

▶ **Refer to illustration 6.26**

22  Raise the vehicle and support it securely on jackstands.

23  Drain the coolant (see Chapter 1).

24  Remove the lower radiator hose.

25  Remove the inlet and outlet hoses from the after-run coolant pump.

26  Disconnect the electrical connector from the pump, then remove the retaining bracket bolt and detach the after-run coolant pump from the engine (see illustration).

27  Installation is the reverse of removal. Refill and bleed the cooling system (see Chapter 1).

**6.17  Water pump mounting bolts (arrows) (V6 engines)**

**6.19  Use a small amount of RTV sealant to hold the gasket to the water pump**

**6.26  After-run coolant pump mounting bracket bolt**

## 7   Coolant temperature gauge sending unit - check and replacement

### CHECK

1   The coolant temperature indicator system is composed of a temperature gauge or warning light mounted in the dash and a coolant temperature sensor mounted on the engine. This coolant temperature sensor doubles as an information sensor for the fuel and emissions systems (see Chapter 6) and as a sending unit for the temperature gauge.

2   If an overheating indication occurs, check the coolant level in the system and then make sure the wiring between the gauge and the sending unit is secure and all fuses are intact.

3   Sensor operation is explained in Chapter 6. If the sensor is defective, replace it with a new part of the same specification.

4   If the coolant temperature sensor is good, have the temperature gauge checked by a dealer service department. This test will require a scan tool to access the information as it is processed by the Engine Control Module.

### REPLACEMENT

5   Refer to Chapter 6 for the engine coolant temperature sensor replacement procedure.

## 8   Blower motor and circuit - check

▶ **Refer to illustration 8.4**

### ❄❄ WARNING:

**These models have airbags. Always disable the airbag system before working in the vicinity of any airbag system component to avoid the possibility of accidental deployment of the airbag(s), which could cause personal injury (see Chapter 12).**

➡**Note: This procedure applies to vehicles equipped with manual heating and air conditioning systems only. Vehicles equipped with automatic heating and air conditioning systems are very complex and considered beyond the scope of the home mechanic. Vehicles equipped with automatic heating and air conditioning systems should be taken to dealer service department or other qualified repair facility.**

1   Check the fuses and all connections in the circuit for looseness and corrosion. Make sure the battery is fully charged.

2   Place the transaxle in Park (automatic) or Neutral (manual) and set the parking brake securely.

3   Remove the glove box (see Chapter 11).

4   Turn the ignition switch to the Run position (it isn't necessary to start the vehicle). Backprobe the blower motor electrical connector and connect a voltmeter to the two terminals in the blower motor connector (see illustration).

➡**Note: Refer to Chapter 12 for additional information on backprobing a connector.**

5   Move the blower switch through each of its positions and note the voltage readings. Changes in voltage indicate that the motor speeds will also vary as the switch is moved to the different positions.

6   If there is voltage present, but the blower motor does not operate, follow the blower motor ground wire from the motor to the chassis and check the ground terminal for continuity to ground against the chassis metal.

7   If the ground wire is OK, the blower motor is probably faulty. Disconnect the blower motor connector, then hook one side of the blower motor terminals to a chassis ground and the other to a fused source of battery voltage. If the blower doesn't operate, it is faulty.

8   If there was no voltage present at the blower motor at one or more speeds, and the motor itself tested OK, check the blower motor resistor.

9   Disconnect the electrical connector from the blower motor resistor (see illustration). With the ignition On, check for voltage at each of the terminals in the connector as the blower speed switch is moved to the different positions. If the voltmeter does not respond correctly to the switch and the blower is known to be good then the resistor is probably faulty. If there is no voltage present from the switch, then the switch, control panel or related wiring is probably faulty.

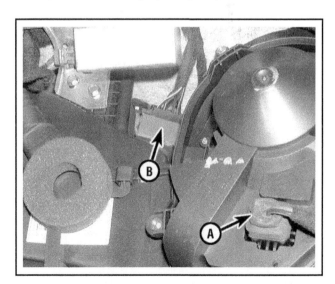

**8.4   The blower motor connector (A) and the blower motor resistor (B) are located directly behind the instrument panel glove box**

## 9   Blower motor - removal and installation

▶ **Refer to illustration 9.2**

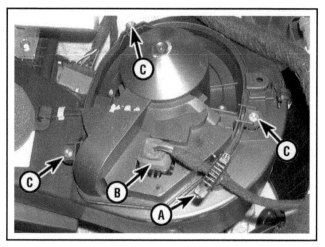

### ☀☀ WARNING:

**These models have airbags. Always disable the airbag system before working in the vicinity of any airbag system component to avoid the possibility of accidental deployment of the airbag(s), which could cause personal injury (see Chapter 12).**

1   Remove the glove box and the glove box support brace (see Chapter 11).

2   Disconnect the electrical connector and the wiring harness from the plastic retainer.

3   Remove the blower motor retaining screws and withdraw the blower motor through the glove box opening (see illustration).

4   Installation is the reverse of removal.

**9.2  Detach the wire retainer (A), unplug the electrical connector (B) and remove the blower motor retaining screws (C)**

## 10   Heater and air conditioning control assembly - removal and installation

### ☀☀ WARNING:

**These models have airbags. Always disable the airbag system before working in the vicinity of any airbag system component to avoid the possibility of accidental deployment of the airbag(s), which could cause personal injury (see Chapter 12).**

### REMOVAL

▶ **Refer to illustration 10.2**

1   Remove the instrument panel center trim panel to allow access to the heater/air conditioning control unit (see Chapter 11).

2   Once the center trim panel is removed, simply pull the control unit outward several inches away from the dash (see illustration).

3   Disconnect the electrical connections from the rear of the control head.

4   Use a small screwdriver to release the clips and detach the cables from the actuating arms. Note the color and location of the cables as the cables are removed.

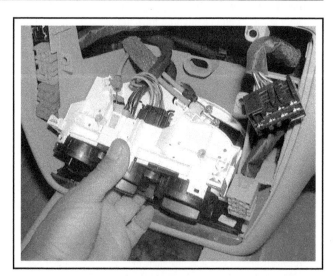

**10.2  With the center trim bezel removed, the heater control unit can easily be pulled outward from the instrument panel to access the electrical connections and cables at the rear**

## INSTALLATION

**◆ Refer to illustration 10.5**

5   To install the control assembly, attach the cables to the actuating arms first, then snap the cable retaining clip in place.

➡️**Note: When reconnecting cables to the control assembly, be sure to attach the correct color cable with the correct actuating arm (see illustration).**

6   The remainder of the installation is the reverse the removal.

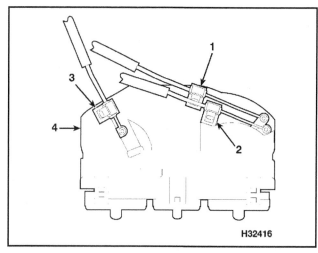

H32416

**10.5  Heater/air conditioning control cable installation details**

1   *Air distribution control knob (black cable to center duct)*
2   *Air distribution control knob (white cable to defroster duct)*
3   *Temperature control knob (red cable to temperature duct)*
4   *Control unit*

## 11  Heater core - removal and installation

### ✳️✳️ WARNING 1:

**These models have airbags. Always disable the airbag system before working in the vicinity of any airbag system component to avoid the possibility of accidental deployment of the airbag(s), which could cause personal injury (see Chapter 12).**

### ✳️✳️ WARNING 2:

**The air conditioning system is under high pressure. DO NOT loosen any fittings or remove any components until after the system has been discharged. Air conditioning refrigerant should be properly discharged into an EPA-approved container at a dealership service department or an automotive air conditioning facility. Always wear eye protection when disconnecting air conditioning system fittings.**

➡️**Note: This procedure requires removal of the entire instrument panel to access the heater core. Refer to Chapter 11 and read through the entire instrument panel removal procedure before attempting to remove the heater core. The instrument panel removal procedure is quite lengthy and can be particularly difficult for a beginner.**

## REMOVAL

**◆ Refer to illustrations 11.3, 11.4, 11.7a, 11.7b, 11.7c, 11.8a, 11.8b, 11.9a, 11.9b, 11.11, 11.12a, 11.12b, 11.13, 11.14a and 11.14b**

1   Have the air conditioning system discharged by a dealership service department or an automotive air conditioning facility.

2   Drain the cooling system (see Chapter 1) Refer to the coolant **Warning** in Section 2. Disconnect the cable from the negative terminal of the battery, then disconnect the positive terminal.

### ✳️✳️ CAUTION:

**Disconnecting the battery can cause severe driveability problems that require a scan tool to rectify. Additionally, disconnecting the battery may cause one or more warning lights on the instrument panel to illuminate, which will also require the use of a scan tool to turn off. Most scan tools available to the public do not have the capacity to perform either of these tasks, which will necessitate taking the vehicle to a dealer service department or other properly equipped repair facility after service work has been performed. See Chapter 5, Section 1 for other precautions related to battery disconnection.**

**11.3 Disconnect the heater core hoses (arrows) at the engine compartment firewall**

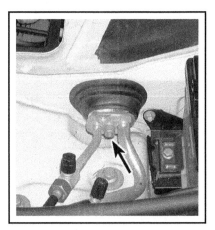

**11.4 Disconnect the air conditioning lines at the firewall (if equipped) - the arrow indicates the refrigerant line flange retaining bolt**

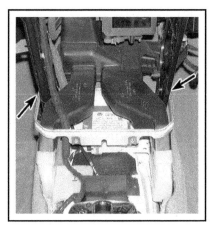

**11.7a Center console support brace retaining bolts (arrows)**

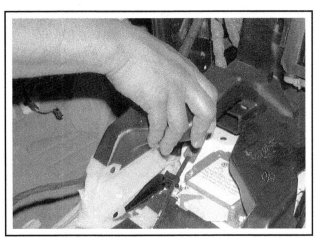

**11.7b Detach the lower air ducts . . .**

**11.7c . . . and the upper air duct from the heater/air conditioning unit**

3   Disconnect the heater hoses at the heater core inlet and outlet on the engine side of the firewall (see illustration) and plug the open fittings. If the hoses are stuck to the pipes, cut them off and replace them with new ones upon installation.

4   Remove the air conditioning lines from the evaporator core fittings at the firewall (see illustration).

5   Remove the instrument panel (see Chapter 11).

6   Once the instrument panel is removed from the vehicle, remove the passenger side airbag (see Chapter 12).

7   Remove the center console support brace (see illustration). Remove the upper and lower air ducts from the heating/air conditioning unit (see illustrations).

8   Disconnect the electrical connections and the ground connections at the left side of the instrument panel (see illustrations).

9   Disconnect the electrical connectors at the right side of the instrument panel (see illustrations).

10  Refer to Chapter 10 and remove the steering column lower pinch bolt.

11  Remove the bolts securing the cross beam center support braces (see illustration).

12  Remove the mounting nuts/bolts securing the instrument panel cross beam to the door pillars (see illustrations).

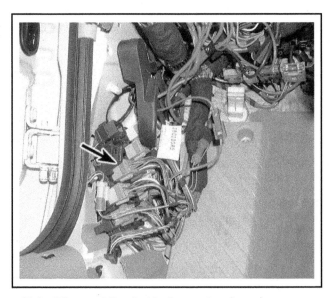

**11.8a Disconnect the electrical connectors (arrow) . . .**

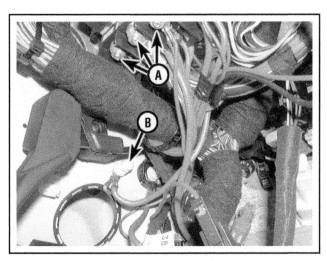

**11.8b** . . . the power supply wires (A) and the ground wire (B) from the left side of the vehicle

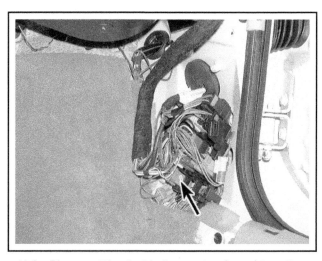

**11.9a** Disconnect the electrical connectors (arrow) from the right side of the vehicle

**11.9b** Also disconnect this single connector from the right side of the vehicle

**11.11** Cross beam center support brace retaining bolts

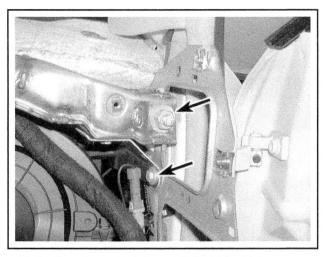

**11.12a** Cross beam retaining nuts/bolts (right side)

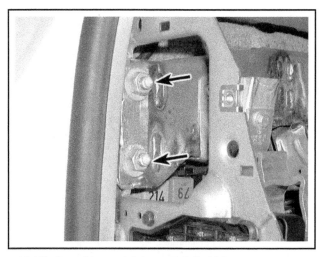

**11.12b** Cross beam retaining nuts (left side)

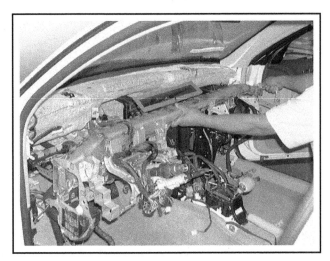

**11.13 Removing the instrument panel cross beam with the heating/air conditioning unit and the steering column attached**

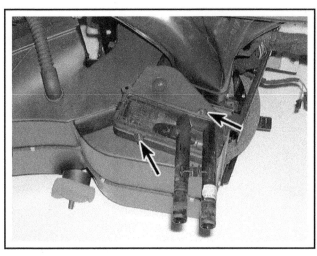

**11.14a Detach the heater core retaining screws (arrows) . . .**

13 Once the cross beam is unbolted from the door pillars it can be removed from the vehicle and set on a workbench with the heating/air conditioning unit and the steering column attached (see illustration).

14 Remove the heater core retaining screws and carefully remove the heater core from the heating/air conditioning unit (see illustrations).

## INSTALLATION

15 Installation is the reverse of removal.

➡**Note: When reinstalling the heater core, make sure any original insulating/sealing materials are in place around the heater core pipes and around the core.**

16 Refill and bleed the cooling system (see Chapter 1).

17 Start the engine and check for proper operation.

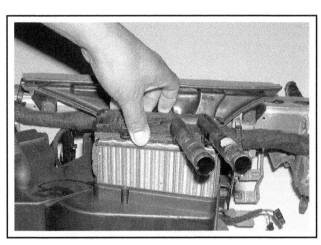

**11.14b . . . and remove the heater core from the housing**

## 12 Air conditioning and heating system - check and maintenance
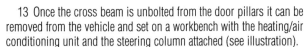

### AIR CONDITIONING SYSTEM

▶ **Refer to illustration 12.1**

**❊❊ WARNING:**

The air conditioning system is under high pressure. Do not loosen any hose fittings or remove any components until after the system has been discharged. Air conditioning refrigerant should be properly discharged into an EPA-approved recovery/recycling unit at a dealer service department or an automotive

air conditioning repair facility. Always wear eye protection when disconnecting air conditioning system fittings.

**❊❊ CAUTION 1:**

All models covered by this manual use environmentally friendly R-134a. This refrigerant (and its appropriate refrigerant oils) are not compatible R-12 refrigerant system components and must never be mixed or the components will be damaged.

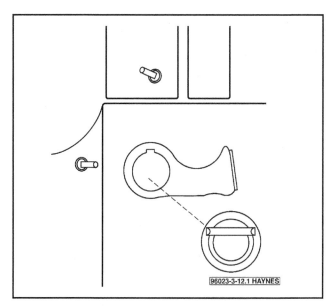

**12.1 Remove the right-side catalytic converter heat shield and check to make sure the evaporator housing drain valve is clear of any blockage**

**12.9 Insert a thermometer in the center duct while operating the air conditioning system - the output air should be 35-40 degrees F less than the ambient temperature, depending on humidity (but not lower than 40-degrees F)**

## ❋❋ CAUTION 2:

**When replacing entire components, additional refrigerant oil should be added equal to the amount that is removed with the component being replaced. Be sure to read the can before adding any oil to the system, to make sure it is compatible with the R-134a system.**

1   The following maintenance checks should be performed on a regular basis to ensure that the air conditioning continues to operate at peak efficiency.

    *a) Inspect the condition of the compressor drivebelt. If it is worn or deteriorated, replace it (see Chapter 1).*

    *b) Check the drivebelt tension and, if necessary, adjust it (see Chapter 1).*

    *c) Inspect the system hoses. Look for cracks, bubbles, hardening and deterioration. Inspect the hoses and all fittings for oil bubbles or seepage. If there is any evidence of wear, damage or leakage, replace the hose(s).*

    *d) Inspect the condenser fins for leaves, bugs and any other foreign material that may have embedded itself in the fins. Use a "fin comb" or compressed air to remove debris from the condenser.*

    *e) Make sure the system has the correct refrigerant charge.*

    *f) If you hear water sloshing around in the dash area or have water dripping on the carpet, remove the right-side catalytic converter heat shield and check the evaporator housing drain valve on the firewall (see illustration). Insert a piece of wire into the opening to check for blockage.*

➡**Note: This valve is very difficult to access on V6 models.**

2   It's a good idea to operate the system for about ten minutes at least once a month. This is particularly important during the winter months because long term non-use can cause hardening, and subsequent failure, of the seals. Note that using the Defrost function operates the compressor.

3   If the air conditioning system is not working properly, proceed to Step 6 and perform the general checks outlined below.

4   Because of the complexity of the air conditioning system and the special equipment necessary to service it, in-depth troubleshooting and repairs beyond checking the refrigerant charge and the compressor clutch operation are not included in this manual. However, simple checks and component replacement procedures are provided in this Chapter.

5   The most common cause of poor cooling is simply a low system refrigerant charge. If a noticeable drop in system cooling ability occurs, one of the following quick checks will help you determine whether the refrigerant level is low. Should the system lose its cooling ability, the following procedure will help you pinpoint the cause.

### Check

▶ **Refer to illustration 12.9**

6   Warm the engine up to normal operating temperature.

7   Place the air conditioning temperature selector at the coldest setting and put the blower at the highest setting. Open the doors (to make sure the air conditioning system doesn't cycle off as soon as it cools the passenger compartment).

8   After the system reaches operating temperature, feel the two pipes connected to the evaporator at the firewall.

9   The pipe (thinner tubing) leading from the condenser outlet to the evaporator should be cold, and the evaporator outlet line (the thicker tubing that leads back to the compressor) should be slightly colder (3 to 10 degrees F colder). If the evaporator outlet is considerably warmer than the inlet, the system needs a charge. Insert a thermometer in the center air distribution duct (see illustration) while operating the air conditioning system at its maximum setting - the temperature of the output air should be 35 to 40 degrees F below the ambient air temperature (down to approximately 40 degrees F). If the ambient (outside) air temperature is very high, say 110 degrees F, the duct air temperature may be as high as 60 degrees F, but generally the air conditioning is 35 to

40 degrees F cooler than the ambient air.

10 If the air isn't as cold as it used to be, the system probably needs a charge.

11 If the air is warm and the system doesn't seem to be operating properly check the operation of the compressor clutch.

12 Have an assistant switch the air conditioning On while you observe the front of the compressor. The clutch will make an audible click and the center of the clutch should rotate.

13 If the clutch does not operate, check the appropriate fuses. Inspect the fuses in the interior fuse panel.

14 If the clutch doesn't respond, refer to the wiring diagrams at the end of this manual and check for battery voltage at the compressor clutch connector. There should be battery voltage with the air conditioning switched On at one of the terminals.

15 Check for continuity to ground on the opposite terminal of the compressor clutch connector.

➡ **Note: On some models there is a single (power) wire leading to the compressor only. On these models the ground circuit, grounds the compressor body directly to the engine block with no external ground wire. Therefore it will be necessary to check for continuity to ground from the center of the compressor clutch to the chassis.**

16 If power and ground are available and the clutch doesn't operate when connected, the compressor clutch is defective.

17 Further inspection or testing of the system is beyond the scope of the home mechanic and should be left to a professional.

### Adding refrigerant

▶ **Refer to illustrations 12.18 and 12.21**

### ✳✳ CAUTION:

**Make sure any refrigerant, refrigerant oil or replacement component your purchase is designated as compatible with environmentally friendly R-134a systems.**

18 Purchase a R-134a automotive charging kit at an auto parts store (see illustration). A charging kit includes a 12-ounce can of refrigerant, a tap valve and a short section of hose that can be attached between the tap valve and the system low side service valve.

### ✳✳ WARNING:

**Never add more than one can of refrigerant to the system. If more than one can is required, the system should be evacuated, leak tested, and recharged by a licensed air conditioning technician.**

19 Hook up the charging kit by following the manufacturer's instructions.

### ✳✳ WARNING:

**DO NOT hook the charging kit hose to the system high side! The fittings on the charging kit are designed to fit only on the low side of the system.**

**12.18 A basic charging kit for 134a systems is available at most auto parts stores - it must say 134a (not R-12) and so should the can of refrigerant**

**12.21 Attach the refrigerant kit to the low-side charging port**

20 Back off the valve handle on the charging kit and screw the kit onto the refrigerant can, making sure first that the O-ring or rubber seal inside the threaded portion of the kit is in place.

### ✳✳ WARNING:

**Wear protective eyewear when dealing with pressurized refrigerant cans.**

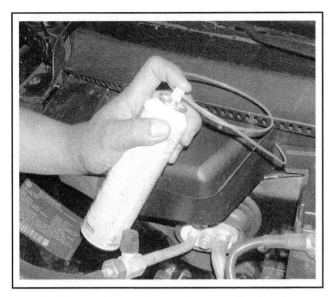

**12.32 With the nozzle inserted in the pollen filter housing, spray the disinfectant at the evaporator core**

21 Remove the dust cap from the low-side charging and attach the quick-connect fitting on the kit hose (see illustration).

22 Warm up the engine and turn On the air conditioning. Keep the charging kit hose away from the fan and other moving parts.

➡**Note: The charging process requires the compressor to be running. If the clutch cycles off, you can put the air conditioning switch on High and leave the car doors open to keep the clutch on and compressor working.**

23 Turn the valve handle on the kit until the stem pierces the can, then back the handle out to release the refrigerant. You should be able to hear the rush of gas. Add refrigerant to the low side of the system, keeping the can upright at all times, but shaking it occasionally. Allow stabilization time between each addition.

➡**Note: The charging process will go faster if you wrap the can with a hot-water-soaked shop rag to keep the can from freezing up.**

24 If you have an accurate thermometer, you can place it in the center air conditioning duct inside the vehicle and keep track of the output air temperature (see illustration 12.9). A charged system that is working properly should cool down to approximately 40-degrees F. If the ambient (outside) air temperature is very high, say 110 degrees F, the duct air temperature may be as high as 60 degrees F, but generally the air conditioning is 30-40 degrees F cooler than the ambient air.

25 When the can is empty, turn the valve handle to the closed position and release the connection from the low-side port. Replace the dust cap.

26 Remove the charging kit from the can and store the kit for future use with the piercing valve in the UP position, to prevent inadvertently piercing the can on the next use.

## HEATING SYSTEMS

27 If the carpet under the heater core is damp, or if antifreeze vapor or steam is coming through the vents, the heater core is leaking. Remove it (see Section 11) and install a new unit (most radiator shops will not repair a leaking heater core).

28 If the air coming out of the heater vents isn't hot, the problem could stem from any of the following causes:

 a) *The thermostat is stuck open, preventing the engine coolant from warming up enough to carry heat to the heater core. Replace the thermostat (see Section 3).*

 b) *There is a blockage in the system, preventing the flow of coolant through the heater core. Feel both heater hoses at the firewall. They should be hot. If one of them is cold, there is an obstruction in one of the hoses or in the heater core, or the heater control valve is shut. Detach the hoses and back flush the heater core with a water hose. If the heater core is clear but circulation is impeded, remove the two hoses and flush them out with a water hose.*

 c) *If flushing fails to remove the blockage from the heater core, the core must be replaced (see Section 11).*

## ELIMINATING AIR CONDITIONING ODORS

**Refer to illustration 12.32**

29 Unpleasant odors that often develop in air conditioning systems are caused by the growth of a fungus, usually on the surface of the evaporator core. The warm, humid environment there is a perfect breeding ground for mildew to develop.

30 The evaporator core on most vehicles is difficult to access, and factory dealerships have a lengthy, expensive process for eliminating the fungus by opening up the evaporator case and using a powerful disinfectant and rinse on the core until the fungus is gone. You can service your own system at home, but it takes something much stronger than basic household germ-killers or deodorizers.

31 Aerosol disinfectants for automotive air conditioning systems are available in most auto parts stores, but remember when shopping for them that the most effective treatments are also the most expensive. The basic procedure for using these sprays is to start by running the system in the RECIRC mode for ten minutes with the blower on its highest speed. Use the highest heat mode to dry out the system and keep the compressor from engaging by disconnecting the wiring connector at the compressor (see Section 14).

32 The disinfectant can usually comes with a long spray hose. Point the nozzle inside the hole on the pollen filter housing towards the evaporator core, and spray according to the manufacturer's recommendations (see illustration). Try to cover the whole surface of the evaporator core, by aiming the spray up, down and sideways. Follow the manufacturer's recommendations for the length of spray and waiting time between applications.

33 Once the evaporator has been cleaned, the best way to prevent the mildew from coming back again is to make sure your evaporator housing drain tube is clear (see illustration 12.1).

## 13 Air conditioning accumulator - removal and installation

### REMOVAL

▶ Refer to illustration 13.4

**✳✳ WARNING:**

**The air conditioning system is under high pressure. DO NOT loosen any fittings or remove any components until after the system has been discharged. Air conditioning refrigerant should be properly discharged into an EPA-approved container at a dealership service department or an automotive air conditioning repair facility. Always wear eye protection when disconnecting air conditioning system fittings.**

1  Have the air conditioning system discharged (see **Warning** above).

2  Raise the vehicle and support it securely on jackstands.

3  Place the radiator support panel in the service position (see Chapter 11).

4  Disconnect the refrigerant lines from the accumulator (see illustration). Cap or plug the open lines immediately to prevent the entry of dirt or moisture.

5  Lift the accumulator out of its mounting bracket.

### INSTALLATION

6  If you are replacing the accumulator with a new one, add fresh

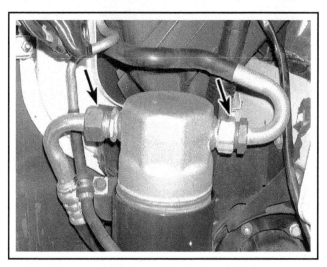

**13.4 Disconnect the refrigerant lines (arrows) from the accumulator - view here is from the front of the vehicle**

refrigerant oil to the new unit following the directions included with the accumulator (the oil must be R-134a compatible).

7  Place the new accumulator into position in the bracket.

8  Install the inlet and outlet lines, using clean refrigerant oil on the new O-rings. Tighten the refrigerant lines securely.

9  Have the system evacuated, recharged and leak tested by a dealership service department or an automotive air conditioning repair facility.

## 14 Air conditioning compressor - removal and installation

➡**Note 1: Whenever the compressor is replaced because of internal damage, the orifice tube should also be replaced (see Section 16).**

➡**Note 2: The accumulator (see Section 13) should be replaced whenever the compressor is replaced.**

### REMOVAL

▶ Refer to illustrations 14.7 and 14.8

**✳✳ WARNING:**

**The air conditioning system is under high pressure. DO NOT loosen any fittings or remove any components until after the system has been discharged. Air conditioning refrigerant should be properly discharged into an EPA-approved container at a dealership service department or an automotive air conditioning repair facility. Always wear eye protection when disconnecting air conditioning system fittings.**

1  Have the air conditioning system discharged (see **Warning** above).

2  Raise the vehicle and support it securely on jackstands.

3  Remove the splash shield from below the engine (if equipped). Place the radiator support panel in the service position (see Chapter 11).

4  Remove the A/C compressor drivebelt (see Chapter 1). On V6 engines it will also be necessary to remove the oil filter to allow access to the compressor.

5  Clean the compressor thoroughly around the refrigerant line fittings.

6  Disconnect the electrical connector from the air conditioning compressor.

➡**Note: Some A/C compressors have compressor speed sensor which also must be disconnected.**

7  Disconnect the suction and discharge lines from the compressor (see illustration).

**Note: On certain models, the compressor lines are mounted to the back of the compressor with a single bolt. Plug the open fittings to prevent the entry of dirt and moisture, and discard the seals between the plate and compressor**

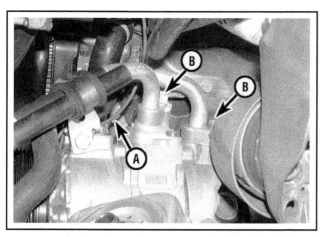

**14.7 Disconnect the wiring (A) from the compressor clutch and the retaining bolts (B) securing the refrigerant lines to the compressor (V6 engine shown)**

8   Remove the compressor mounting bolts (see illustration). Detach the compressor-from the mounting bracket and remove the compressor from the engine compartment.

## INSTALLATION

9   If a new compressor is being installed, pour the oil from the old compressor into a graduated container and add that exact amount of new refrigerant oil to the new compressor. Also follow any directions included with the new compressor.

➡**Note: Some replacement compressors come with refrigerant oil in them. Follow the directions with the compressor regarding the draining of excess oil prior to installation.**

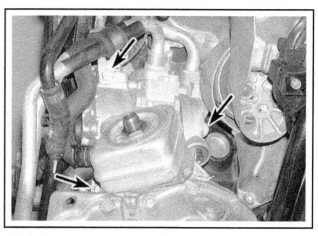

**14.8 Compressor mounting bolts (arrows) (V6 engine shown)**

❄❄ **CAUTION:**

**The oil used must be labeled as compatible with R-134a refrigerant systems.**

10   Installation is the reverse of the disassembly. When installing the line fitting bolt to the compressor, use new seals lubricated with clean refrigerant oil, and tighten the bolt securely.

11   Have the system evacuated, recharged and leak tested by a dealership service department or an automotive air conditioning repair facility.

## 15  Air conditioning condenser - removal and installation

▶ Refer to illustrations 15.5a, 15.5b, 15.5c, 15.6a, 15.6b, 15.7a and 15.7b

❄❄ **WARNING:**

**The air conditioning system is under high pressure. DO NOT loosen any fittings or remove any components until after the system has been discharged. Air conditioning refrigerant should be properly discharged into an EPA-approved container at a dealership service department or an automotive air conditioning repair facility. Always wear eye protection when disconnecting air conditioning system fittings.**

1   Have the air conditioning system discharged (see **Warning** above).
2   Raise the vehicle and support it securely on jackstands.
3   Remove the splash shield from below the engine (if equipped).
4   Remove the front bumper assembly (see Chapter 11).
5   Detach the power steering oil cooler line from the radiator support panel and set it aside (see illustrations).

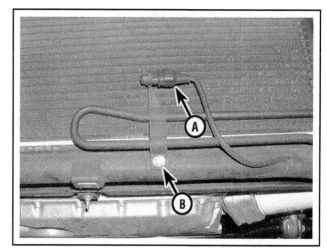

**15.5a  Unclip the ambient temperature sensor (A) from the bracket, then remove the bolt (B) securing the bracket to the radiator support panel**

15.5b Remove the screw (A) and detach the plastic air deflector panel (B) from the left side of the condenser

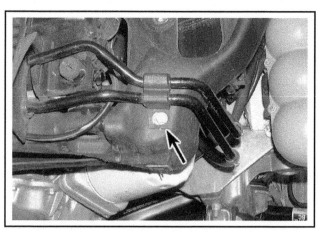

15.5c Detach the remaining power steering oil cooler line clamp and position the line aside

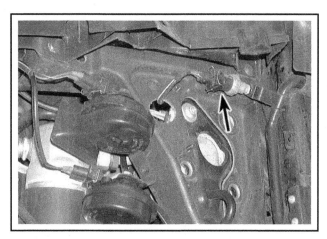

15.6a Disconnect the connector from the A/C pressure switch (arrow)

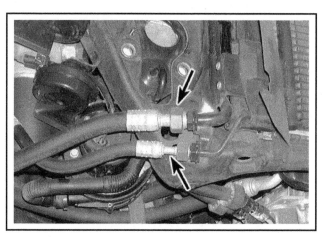

15.6b Condenser inlet and outlet line fittings (arrows)

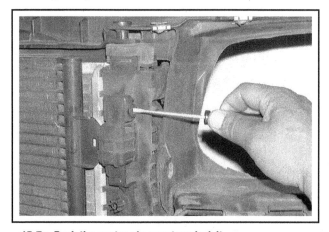

15.7a Push the center pin over to unlock it . . .

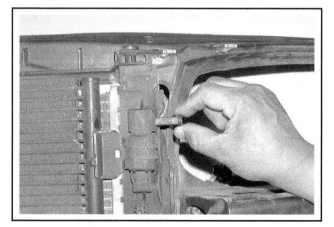

15.7b . . . then pull the retaining pin outward to release the condenser - there is one retaining pin on each side securing the condenser to the radiator

6  Disconnect the A/C pressure switch and condenser inlet and outlet lines (see illustrations).

7  Remove the condenser retaining pins and separate the condenser from the radiator (see illustrations).

8  Installation is the reverse of removal. Always use new O-rings on air conditioning system fittings. If you are replacing the condenser with a new one, add fresh refrigerant oil to the new unit following the directions included with the new condenser (the oil must be R-134a compatible).

9  Have the system evacuated, recharged and leak tested by a dealership service department or an automotive air conditioning repair facility.

## 16 Air conditioning expansion (orifice) tube - removal and installation

♦ **Refer to illustrations 16.2 and 16.4**

### ✳✳ WARNING:

**The air conditioning system is under high pressure. DO NOT loosen any fittings or remove any components until after the system has been discharged. Air conditioning refrigerant should be properly discharged into an EPA-approved container at a dealership service department or an automotive air conditioning repair facility. Always wear eye protection when disconnecting air conditioning system fittings.**

1   Have the air conditioning system discharged and the refrigerant recovered (see **Warning** above).

2   Disconnect the refrigerant high-pressure line at the orifice tube (see illustration).

3   The expansion tube is a tube with a fixed-diameter orifice and a mesh filter at each end. When you separate the pipe at the fitting you will see one end of the orifice tube inside the pipe leading to the evapo-rator. Use needle-nose pliers to remove the orifice tube.

4   The orifice tube acts to meter the refrigerant, changing it from high-pressure liquid to low-pressure liquid/vapor. It is possible to reuse the orifice tube if (see illustration):

    *a) The screens aren't plugged with grit or foreign material*
    *b) Neither screen is torn*
    *c) The plastic housing over the screens is intact*
    *d) The brass orifice inside the plastic housing is unrestricted*

5   Installation is the reverse of removal. Be sure to insert the expansion tube with the shorter end in first, toward the evaporator.

### ✳✳ CAUTION:

**Always use a new O-ring when installing the expansion (orifice) tube.**

6   Retighten the fitting and refrigerant line, then have the system evacuated, recharged and leak-tested by the shop that discharged it.

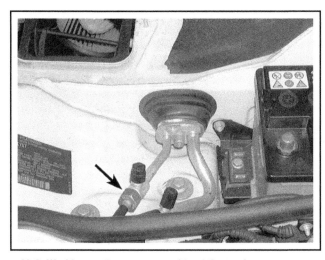

**16.2  Working on the passenger side of the engine compartment, use a pair of wrenches to detach the pipe fitting (arrow)**

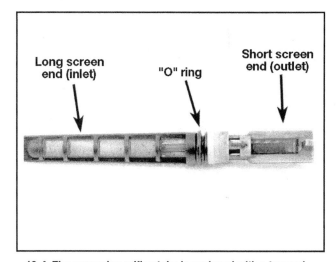

**16.4  The expansion orifice tube is equipped with a tapered mesh screen that must be clean and not have any holes or damage**

## Specifications

### General

| | |
|---|---|
| Coolant capacity | See Chapter 1 |
| Expansion tank pressure cap rating | 15 psi |
| Thermostat opening temperature | 185 to 190-degrees F (85 to 87-degrees C) |
| Refrigerant type | R-134a |
| Refrigerant capacity | 24.5 ounces (700 grams) |

## Torque specifications

➡**Note: One foot-pound (ft-lb) of torque is equivalent to 12 inch-pounds (in-lbs) of torque. Torque values below approximately 15 ft-lbs are expressed in inch-pounds, since most foot-pound torque wrenches are not accurate at these smaller values.**

| | |
|---|---|
| Thermostat housing nuts/bolts | 84 in-lbs (10 Nm) |
| Coolant outlet pipe-to-cylinder head | 84 in-lbs (10 Nm) |
| Water pump attaching bolts | 84 in-lbs (10 Nm) |

**Section**

**Reference to other Chapters**

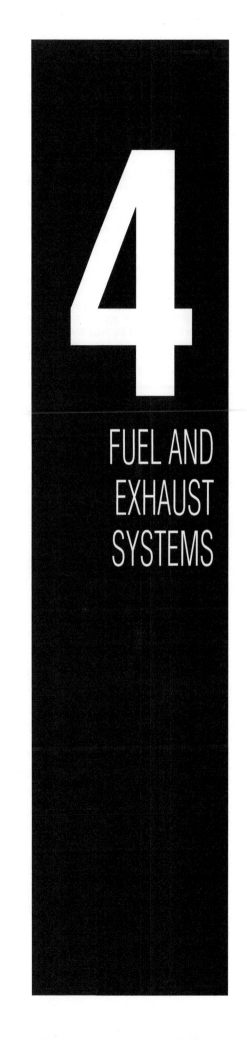

# 4

# FUEL AND EXHAUST SYSTEMS

## 1   General information

▶ **Refer to illustrations 1.1a and 1.1b**

❊❊ **WARNING:**

Gasoline is extremely flammable, so take extra precautions when you work on any part of the fuel system. Don't smoke or allow open flames or bare light bulbs near the work area, and don't work in a garage where a gas-type appliance (such as a water heater or a clothes dryer) is present. Since gasoline is carcinogenic, wear fuel-resistant gloves when there's a possibility of being exposed to fuel, and, if you spill any fuel on your skin, rinse it off immediately with soap and water. Mop up any spills immediately and do not store fuel-soaked rags where they could ignite. The fuel system is under constant pressure, so, if any fuel lines are to be disconnected, the fuel pressure in the system must be relieved first. When you perform any kind of work on the fuel system, wear safety glasses and have a Class B type fire extinguisher on hand.

All models covered by this manual are equipped with a Bosch Motronic fuel injection system (see illustrations). This system uses timed impulses to sequentially inject the fuel directly into the intake ports of each cylinder. The injectors are controlled by the Engine Control Module (ECM). The ECM monitors various engine parameters and delivers the exact amount of fuel, in the correct sequence, into the intake ports. This Chapter's information pertains to the air and fuel delivery components of the system only. Refer to Section 12 for additional general information regarding the fuel injection system. Refer to Chapter 6 for information regarding the electronic control system.

All models are equipped with an electric fuel pump, mounted in the fuel tank. Access to the fuel pump is provided through an access hole under the rear seat cushion. The fuel level sending unit is an integral component of the fuel pump module and it must be removed from the fuel tank in the same manner.

The exhaust system consists of exhaust manifolds, catalytic converters, exhaust pipes and mufflers. Each of these components is replaceable. For further information regarding the catalytic converter, refer to Chapter 6.

1.8L four-cylinder models are equipped with a turbocharger and intercooler. The turbocharger increases power by using an exhaust gas driven turbine to pressurize the intake charge before it enters the combustion chambers. The amount of intake manifold pressure (boost) is regulated by an exhaust by-pass valve (wastegate). The wastegate is controlled by the ECM. The heated compressed air is routed through an air-to-air radiator (intercooler). The intercooler removes excess heat from the compressed air, increasing its density and allowing for more boost pressure.

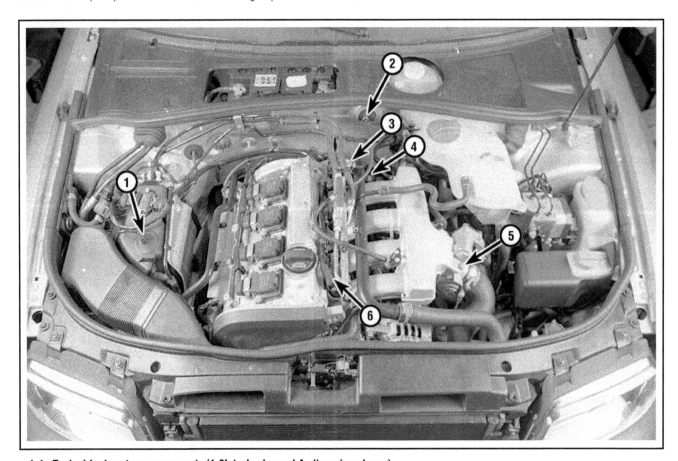

**1.1a Typical fuel system components (1.8L turbocharged Audi engine shown)**

| | | | | | |
|---|---|---|---|---|---|
| *1* | *Air filter housing* | *3* | *Fuel pressure regulator* | *5* | *Throttle body/throttle control module* |
| *2* | *Accelerator cable* | *4* | *Fuel feed and return lines* | *6* | *Fuel rail and injectors* |

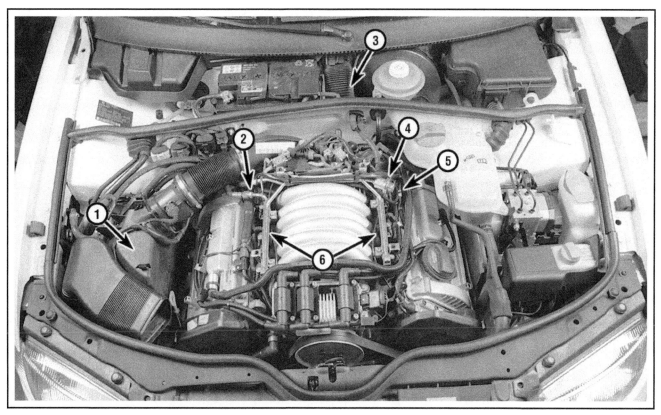

**1.1b Typical fuel system components (2.8L VW engine shown)**

1   Air filter housing
2   Fuel feed and return lines
3   Accelerator cable

4   Fuel pressure regulator
5   Throttle body (located between the intake
    manifold and the engine compartment
    firewall)

6   Fuel rail and injectors

## 2   Fuel pressure relief procedure

▶ Refer to illustration 2.3

### ❊❊ WARNING:

See the Warning in Section 1.

➡Note: After the fuel pressure has been relieved, it's a good idea to lay a shop towel over any fuel connection to be disassembled, to absorb the residual fuel that may leak out when servicing the fuel system.

1   Before servicing any fuel system component, you must relieve the fuel pressure to minimize the risk of fire or personal injury.

2   Remove the fuel filler cap - this will relieve any pressure built up in the tank.

3   Remove the fuel pump fuse (no. 28) from the fuse box (see illustration).

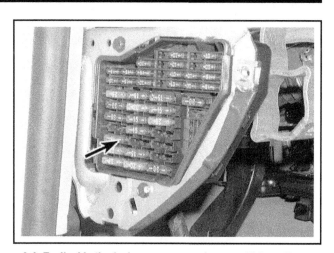

**2.3 To disable the fuel pump, remove fuse no. 28 from the fuse box**

4   Attempt to start the engine. The engine should immediately stall. Continue to crank the engine for approximately three seconds.

5   Turn the ignition Off and remove the key, then disconnect the cable from the negative terminal of the battery.

### ✳✳ CAUTION:

**Disconnecting the battery can cause driveability problems that require a scan tool to rectify. Additionally, disconnecting the battery may cause one or more warning lights on the instrument panel to illuminate, which will also require the use of a scan tool to turn off. Most scan tools available to the public do not**

**have the capability to perform either of these tasks, which will necessitate taking the vehicle to a dealer service department or other properly equipped repair facility after service work has been performed. See Chapter 5, Section 1 for the use of an auxiliary voltage input device ("memory saver") before disconnecting the battery and for other precautions related to battery disconnection.**

6   Place shop towels around the fuel fitting to be disconnected to absorb any residual fuel that may spill out.

7   After completing the service or repair, install the fuel pump fuse.

## 3   Fuel pump/fuel pressure - check

### ✳✳ WARNING:

**See the Warning in Section 1.**

## PRELIMINARY CHECK

▶ **Refer to illustration 3.3**

1   If you suspect insufficient fuel delivery check the following items first:

    a) *Check the battery and make sure it's fully charged (see Chapter 5).*

    b) *Check the fuel pump fuse (see Section 2).*

    c) *Check the fuel filter for restriction.*

    d) *Inspect all fuel lines to ensure that the problem is not simply a leak in a line.*

2   Place the transmission in Park (automatic) or neutral (manual) and apply the parking brake. Have an assistant cycle the ignition key On and Off several times (or attempt to start the engine, if the engine does not start) while you listen for the sound of the fuel pump operating inside the fuel tank. Remove the rear seat cushion and the fuel pump access cover and listen at, or feel the top of the fuel pump module, if necessary. You should hear a "whirring" sound indicating the fuel pump is operating. If the fuel pump is operating, proceed to the pressure check.

3   If there is no sound, remove the fuel pump access cover and disconnect the fuel pump electrical connector. Connect a test light or voltmeter to terminals 1 and 4 of the fuel pump harness connector (see illustration). Cycle the ignition key On and Off several times - battery voltage should be indicated. If battery voltage is not indicated, check the fuel pump circuit, referring to Chapter 12 and the wiring diagrams.

Check the related fuses, the fuel pump relay and the related wiring to ensure power is reaching the fuel pump connector. Check the ground circuit for continuity.

➡**Note: The fuel pump relay is located in Position 6 in the relay box near the driver's side kick panel.**

4   If the power and ground circuits are good and the fuel pump does not operate, remove the fuel pump and check for open circuits in the fuel pump module wiring and connectors. If the wiring and connectors are good, replace the fuel pump (see Section 7).

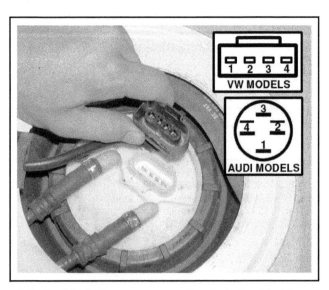

**3.3  Check for battery power across terminals 1 and 4 of the fuel pump connector with the ignition key On**

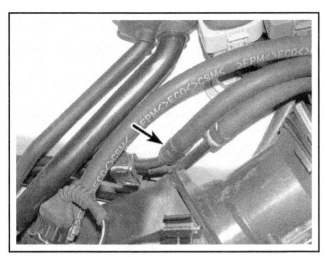

**3.6a  To test the fuel pressure, remove the fuel supply line (arrow) from the fuel rail . . .**

**3.6b  . . . and attach a fuel pressure gauge between the fuel line and fuel rail with a T-fitting**

## PRESSURE CHECK

♦ **Refer to illustrations 3.6a and 3.6b**

➡**Note: In order to perform the fuel pressure test, you will need a fuel pressure gauge capable of measuring high fuel pressure. The fuel gauge must be equipped with the proper fittings or adapters required to attach it to the fuel line and fuel rail.**

5   Relieve the fuel pressure (see Section 2). Remove the engine cover (if equipped).

6   Disconnect the fuel supply line from the fuel rail (see illustration). Connect the pressure gauge with a T-fitting and adapter hose to the fuel line and fuel rail (see illustration).

7   Cycle the ignition key On and Off several times (or attempt to start the engine, if the engine does not start). Note the pressure indicated on the gauge and compare your reading with the pressure listed in this Chapter's Specifications.

8   If the fuel pressure is lower than specified, pinch-off the fuel return line.

### ✶✶ CAUTION:

**Use special pliers designed specifically for pinching a rubber fuel line (available at most auto parts stores). Use of any other type pliers may damage the fuel line.**

Cycle the ignition key On and Off several times and note the fuel pressure.

### ✶✶ CAUTION:

**Do not allow the fuel pressure to rise above 85 psi (586 kPa) or damage to the fuel pressure regulator may occur.**

If the fuel pressure is now above the specified pressure, replace the fuel pressure regulator (see Section 15). If the fuel pressure is still lower than specified, check the fuel lines and the fuel filter for restric-

tions. If no restriction is found, remove the fuel pump module (see Section 7) and check the fuel strainer for restrictions, check the fuel pipe for leaks and check the fuel pump wiring for high resistance. If no problems are found, replace the fuel pump.

9   If the fuel pressure recorded in Step 7 is higher than specified, check the fuel return line for restrictions. If no restrictions are found, replace the fuel pressure regulator (see Section 15).

10  If the fuel pressure is within specifications, start the engine.

### ✶✶ WARNING:

**Make sure the fuel pressure gauge hose is positioned away from the engine drivebelt before starting the engine.**

With the engine running, the fuel pressure should be 5 to 10 psi (34 to 69 kPa) below the pressure recorded in Step 7. If it isn't, remove the vacuum hose from the fuel pressure regulator and verify there is 12 to 14 in-Hg (305 to 356 mm-Hg) of vacuum present at the hose. If vacuum is not present at the hose, check the hose for a restriction or a break. If vacuum is present, reconnect the hose to the fuel pressure regulator. If the fuel pressure regulator does not decrease the fuel pressure with vacuum applied, replace the fuel pressure regulator.

11  Now check the fuel system hold pressure. Turn the engine off and monitor the fuel pressure for ten minutes; the fuel pressure should not drop below 32 psi (220 kPa) within ten minutes. If it does, there is a leak in the fuel line, a fuel injector is leaking, the fuel pump check valve is defective or the fuel pressure regulator is defective. To determine the source of the leak, cycle the ignition key On and Off several times to obtain the highest fuel pressure reading, then immediately pinch-off the fuel supply hose between the fuel gauge T-fitting and the fuel rail. If the pressure drops below 32 psi (220 kPa) within ten minutes, the main fuel line is leaking or the fuel pump is defective. If the pressure holds, remove the clamp from the supply line, pressurize the system and clamp off the return line. If the pressure drops below 32 psi (220 kPa) within ten minutes, an injector is probably leaking (or the fuel rail is leaking, but such a leak should be very apparent). If the pressure holds, remove the clamp from the return line. If the pressure now drops, the fuel pressure regulator is defective.

## 4  Fuel lines and fittings - repair and replacement

▶ **Refer to illustration 4.2**

### ✴✴ WARNING:

**See the Warning in Section 1.**

1    Always relieve the fuel pressure before servicing fuel lines or fittings (see Section 2).

2    Special fuel supply, return and vapor lines extend from the fuel tank to the engine compartment. The lines are secured to the underbody with retainers (see illustration). Rubber hose completes the connection from the engine compartment junction block to the fuel rail. All fuel lines must be occasionally inspected for leaks or damage.

3    If evidence of contamination is found in the system or fuel filter during disassembly, the line should be disconnected and blown out. Check the fuel strainer on the fuel pump for damage and deterioration.

4    Don't route fuel line or hose within four inches of any part of the exhaust system or within ten inches of the catalytic converter. Fuel line must never be allowed to chafe against the engine, body or frame. A minimum of 1/4-inch clearance must be maintained around a fuel line.

5    Because fuel lines used on fuel-injected vehicles are under high pressure, they require special consideration.

6    In the event of fuel line damage, it is necessary to replace the damaged lines with factory replacement parts. Others may fail from the high pressures of this system.

7    When replacing a fuel line, remove all fasteners attaching the fuel line to the vehicle body and route the new line exactly as originally installed.

8    When replacing rubber hose, always use hose specifically designated as fuel hose and replace the hose clamp with a new one.

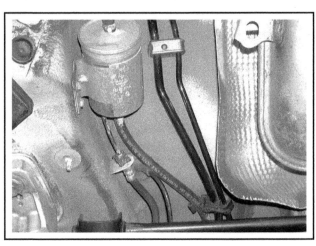

**4.2  The fuel lines are secured to the underbody with plastic retainers**

## 5  Fuel tank - removal and installation

▶ **Refer to illustrations 5.4, 5.7, 5.8. 5.9 and 5.10**

### ✴✴ WARNING:

**See the Warning in Section 1.**

1    Remove the fuel tank filler cap to relieve fuel tank pressure.

2    Relieve the fuel system pressure (see Section 2).

3    Using a siphoning kit (available at most auto parts stores), siphon the fuel into an approved gasoline container.

### ✴✴ WARNING:

**Do not start the siphoning action by mouth!**

4    Open the fuel filler door, remove the screws and remove the filler door and splash shield assembly (see illustration).

5    Loosen the right rear wheel bolts, then raise the vehicle and support it securely on jackstands. Remove the wheel.

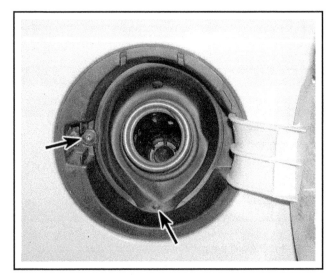

**5.4  Remove the fuel filler door splash shield mounting bolts (arrows)**

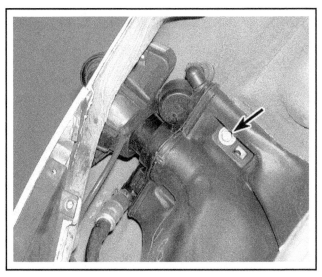

**5.7 Loosen the hose clamps and disconnect the vent hoses from the filler neck - remove the filler neck mounting bolt (arrow)**

**5.8 Disconnect the fuel supply line from the fuel filter**

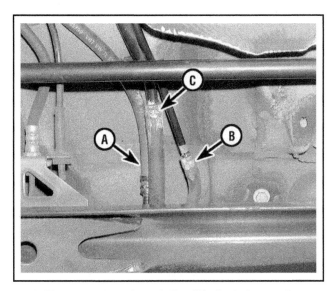

**5.9 Disconnect the fuel supply line (A), the fuel return line (B) and the vent hose (C)**

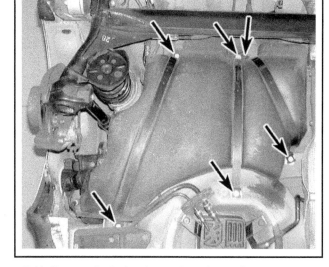

**5.10 Remove the fuel tank strap bolts (arrows) and remove the straps**

6   Remove the right rear wheel well liner (see Chapter 11). On all-wheel drive models, remove the protective plates from the filler neck and fuel tank.

7   Remove the fuel tank filler neck mounting bolts (see illustration).

8   Disconnect the fuel supply line from the fuel filter (see illustration).

9   Disconnect the fuel return hose and vent hose from the metal fuel lines at the body (see illustration).

10   Position a transmission jack under the fuel tank and support the tank. Remove the fuel tank strap bolts and remove the straps (see illustration).

11   Lower the jack slightly and disconnect the electrical connector from the fuel pump module.

12   Lower the jack and remove the tank from the vehicle.

13   Installation is the reverse of removal with the following additions:

   *a) Make sure the fuel hoses are not kinked and seated on the metal lines properly.*

   *b) Use new clamps specifically designed for fuel injection systems on the fuel hoses.*

   *c) Turn the ignition key on and off several times and check for leaks before returning the vehicle to service.*

   *d) Tighten the wheel bolts to the torque listed in the Chapter 1 Specifications.*

## 6 Fuel tank cleaning and repair - general information

1  The fuel tank installed in the vehicles covered by this manual are not repairable. If the fuel tank becomes damaged, it must be replaced.

2  Cleaning the fuel tank (due to fuel contamination) should be performed by a professional with the proper training to carry out this critical and potentially dangerous work. Even after cleaning and flushing, explosive fumes may remain inside the fuel tank.

3  If the fuel tank is removed from the vehicle, it should not be placed in an area where sparks or open flames could ignite the fumes coming out of the tank. Be especially careful inside a garage where a gas-type appliance is located.

## 7 Fuel pump - removal and installation

▶ **Refer to illustrations 7.2, 7.3a, 7.3b, 7.4, 7.5, 7.6, 7.7a, 7.7b, 7.8 and 7.12**

### ✳✳ WARNING:

**See the Warning in Section 1.**

1  Relieve the fuel system pressure (see Section 2).

2  On front-wheel drive models, remove the carpet from the luggage compartment. On all-wheel drive models, remove the rear seat cushion (see Chapter 11). Remove the fuel pump access cover (see illustration).

3  Disconnect the fuel pump electrical connector. Disconnect the fuel supply and return lines from the fuel tank flange (see illustrations).

➡**Note: The fuel supply line is identified with a black mark. The fuel return line is identified with a blue mark.**

4  Apply alignment marks on the fuel tank and flange (if none exist) so the fuel tank flange can be installed in the original position (see illustration).

5  Loosen the fuel tank flange retaining ring (see illustration).

6  Pull the flange up and disconnect the electrical connector for the fuel level sending unit from the underside of the flange. Unlatch the fuel return hose from the flange (see illustration).

7  Using a special tool (see illustration), rotate the fuel pump 15-degrees counterclockwise and withdraw the fuel pump from the fuel reservoir (see illustration). The inner section of the reservoir will come out with the fuel pump. Place the fuel pump and flange in a clean shallow pan as some gasoline will remain in the fuel pump.

8  Remove the wiring terminal nuts and disconnect the fuel pump wiring from the fuel pump. Loosen the clamp and disconnect the fuel hose from the fuel pump (see illustration). Note the routing of the wiring and fuel hose so they can be installed in their original locations.

9  Remove the fuel pump from the inner section of the reservoir.

10  Assemble the new fuel pump, inner reservoir section and flange in the reverse order of removal. Make sure the toothed washers are used under the wiring terminal nuts and use a new clamp to retain the fuel hose.

11  Clean the fuel tank sealing surface and install a new seal on the fuel tank.

12  Install the fuel pump and inner reservoir section into the fuel tank reservoir, aligning the notch in the inner reservoir with the right-hand mark on the reservoir (see illustration). Press the fuel pump and inner reservoir section into the reservoir until seated. Using the spanner, rotate the fuel pump and inner reservoir section 15-degrees clockwise to engage the clips. The notch on the inner section should now align with the left-hand mark on the reservoir.

13  Connect the fuel level sending unit connector and return hose and install the fuel tank flange aligning the marks made in Step 4.

14  Press the flange down until seated and tighten the retaining ring.

15  The remainder of installation is the reverse of removal.

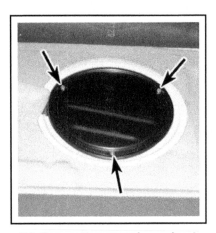

**7.2  Remove the screws (arrows) and the fuel pump access cover**

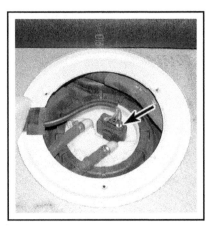

**7.3a  Unplug the electrical connector from the fuel pump module**

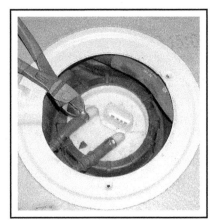

**7.3b  Remove the fuel pump lines from the module using a wire cutter to slice the clamp – use new clamps when reassembling**

7.4 Alignment marks on a 2-wheel drive model – all-wheel drive models position the alignment marks opposite

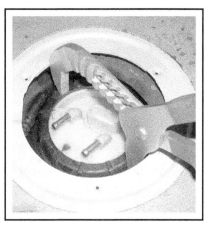

7.5 Using an appropriate tool, loosen the fuel pump module retaining ring

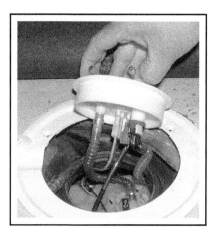

7.6 Lift the fuel pump flange up, disconnect the harness electrical connector and fuel lines and remove the assembly

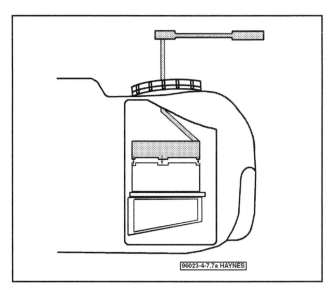

7.7a Use a special tool to extend into the fuel tank and release fuel pump module

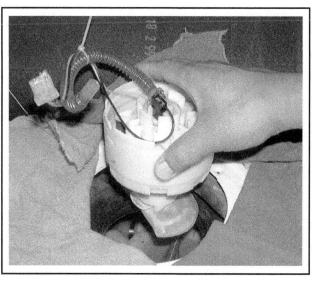

7.7b Carefully remove the fuel pump module from the tank by rotating the module 15-degrees counterclockwise

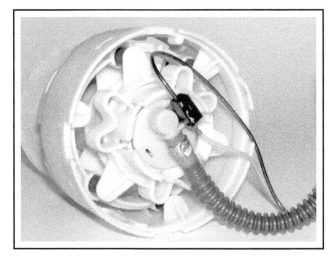

7.8 Remove the fuel line from the fuel pump

7.12 Be sure to align the tab with the mating mark (arrow)

## 8 Fuel level sending unit - check and replacement

### ✳✳ WARNING:

**See the Warning in Section 1.**

➡Note: Some all-wheel drive models are equipped with two additional fuel level sending units. One is mounted adjacent to the fuel pump/fuel level sending unit module while the other is mounted to the rear, accessible through a removable panel in the luggage compartment floor.

### CHECK

▶ **Refer to illustrations 8.2 and 8.3**

1    Remove the fuel pump access cover (see Section 7).
2    Disconnect the electrical connector from the fuel tank flange and connect the probes of an ohmmeter to the two fuel level sensor terminals (2 and 3) (see illustration). If the tank is full, the sending unit resistance should be low (approximately 40 ohms), if the tank is empty the resistance will be higher (approximately 280 ohms).

3    Further testing of the sending unit may be performed by removing the unit and connecting the ohmmeter to the two terminals of the sending unit electrical connector (see illustration). Position the float in the down (empty) position and note the reading on the ohmmeter. Move the float up to the full position while watching the meter. If the fuel level sending unit resistance does not decrease smoothly as the float travels from empty to full, replace the fuel level sending unit assembly.

### REPLACEMENT

▶ **Refer to illustration 8.9**

4    Relieve the fuel system pressure (see Section 2).
5    Using a siphoning kit (available at most auto parts stores), siphon the fuel into an approved gasoline container.

### ✳✳ WARNING:

**Do not start the siphoning action by mouth!**

6    Remove the fuel pump access cover, disconnect the fuel pump electrical connector and remove the fuel lines from the flange (see Section 7).
7    Loosen the fuel tank flange retaining ring (see Section 7).
8    Pull the flange up and disconnect the electrical connector for the fuel level sending unit from the underside of the flange (see Section 7). Position the flange aside with the hoses attached.
9    Reach through the fuel tank opening and depress the locking tab on the sending unit (see illustration). Pull the sending unit off the fuel reservoir and carefully withdraw the sending unit from the tank. Be careful not to damage the float arm.
10   Installation is the reverse of removal with the following additions:
   a)  *Press the sending unit onto the fuel reservoir and engage the latch.*
   b)  *Make sure the float arm is positioned properly and not binding.*
   c)  *Make sure the sending unit wiring is routed properly before connecting it to the flange.*

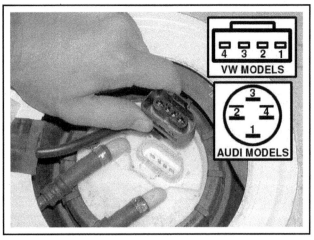

**8.2  Check the resistance of the fuel level sending unit across terminals 2 and 3 (VW model shown)**

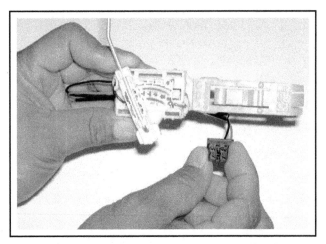

**8.3  Check the fuel level sending unit resistance directly at the electrical connector**

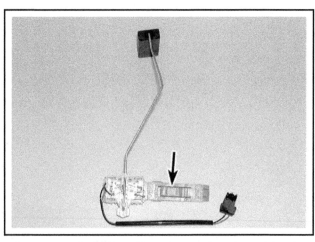

**8.9  Depress the locking tab on the sending unit to release it from the tank assembly (sending unit removed for clarity)**

## 9   Air filter housing - removal and installation

▶ **Refer to illustrations 9.1, 9.2, 9.4, 9.5 and 9.6**

1   Remove intake duct resonator screws and remove the resonator (see illustration).

2   Disconnect the electrical connectors from the Mass Airflow sensor, power transistor and EVAP purge valve, as required (see illustration). Detach the wiring harness from the air filter cover and position the harness aside.

3   On DOHC V6 models, detach the purge valve from the cover and position the purge valve and hoses aside.

4   Loosen the hose clamp and detach air intake duct from the Mass Airflow sensor (see illustration).

5   Disconnect the secondary air injection pump hose from the air filter cover (if equipped). Loosen the latches and remove the cover with the Mass Airflow sensor attached (see illustration). Remove the air filter element.

6   Remove the mounting bolt, pull the housing up and detach the housing from grommets on the inner fender (see illustration). Remove the assembly from the engine compartment.

7   Installation is the reverse of removal.

**9.1  Remove the intake air resonator screws (arrows) and the air filter cover**

**9.2  Disconnect the electrical connectors from the mass airflow sensor, the intake air temperature sensor and the power transistor (arrows)**

**9.4  Remove the hose clamp (arrow) on the intake duct at the MAF sensor**

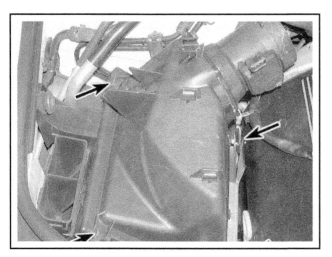

**9.5  Unlatch the air filter housing clips (arrows)**

**9.6  Location of the air filter housing mounting bolt (arrow)**

## 10 Accelerator cable - replacement

▸ **Refer to illustrations 10.1, 10.3, 10.4, 10.5, 10.6, 10.7 and 10.8**

1 Remove the trim panel from under the dash and detach the cable from the accelerator pedal (see illustration).

2 Remove the engine cover.

3 On V6 models equipped with cruise control, disconnect the vacuum hose from the servo unit and remove the bracket mounting bolts (see illustration). Disconnect the throttle rod and remove the servo unit.

4 Rotate the throttle lever and separate the accelerator cable end from the throttle lever (see illustration).

5 Remove the clip and detach the cable grommet from the accelerator cable bracket (see illustration).

6 On vehicles equipped with an automatic transaxle, disconnect the electrical connector from the kick-down switch (see illustration).

7 Remove the cable retaining clip at the bulkhead (see illustration).

8 Detach the cable from the firewall by rotating the cable 90-degrees counterclockwise (using the tabs provided) (see illustra-

tion). Pull the cable and the firewall grommet through the firewall and into the engine compartment.

9 Remove the cable from the engine compartment.

10 Installation is the reverse of removal, but before installing the accelerator cable bracket clip, adjust the cable as follows:

   a) Install the cable grommet onto the accelerator cable bracket.
   b) Have an assistant fully depress the accelerator pedal.
   c) Pull the cable out of the grommet until the throttle plate reaches wide open throttle.
   d) Install the retaining clip in the nearest notch the bracket.
   e) Have your assistant release and depress the accelerator pedal several times. Check the throttle lever and make sure it contacts both the closed throttle and wide open throttle stops, if it doesn't, readjust the cable.

11 On vehicles equipped with an automatic transaxle, check the operation of the kick-down switch (see Chapter 7B).

**10.1 Remove the cable end (arrow) from the accelerator pedal**

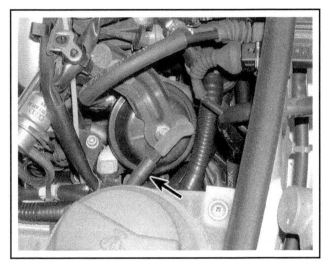

**10.3 Remove the vacuum line (arrow) from the servo**

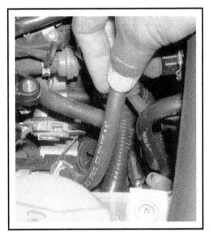

**10.4 Rotate the throttle lever and pass the cable through the slot in the lever**

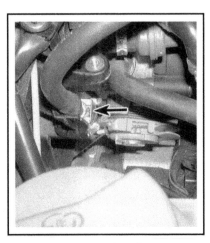

**10.5 Remove the cable clip from the bracket**

**10.6 Disconnect the kickdown switch harness connector**

**10.7 Remove the cable clip at the bulkhead using a screwdriver**

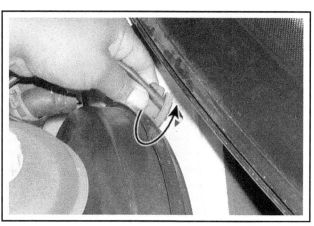

**10.8 Rotate the cable 90-degrees to unlock it from the bulkhead**

## 11 Electronic accelerator pedal module (E-Gas) - replacement

♦ **Refer to illustration 11.4**

1   Some models may be equipped with an electronically controlled accelerator system (also known as E-Gas). The system consists of the accelerator pedal module, the throttle control module and the ECM. The system does not use the traditional accelerator cable. The ECM controls the throttle position based on the voltage signal received from the accelerator pedal module. Refer to Chapter 6 for more information on the electronic accelerator system.

2   Remove the insulation panel in the driver's footwell.

3   Disconnect the electrical connector from the pedal module.

4   Remove the mounting nuts and remove the module from the footwell (see illustration).

5   Installation is the reverse of removal.

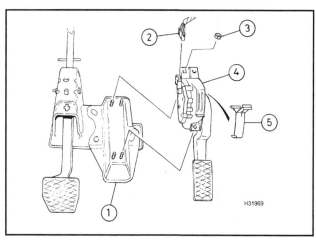

**11.4 Electronic accelerator pedal module and related components (1.8L turbo)**

| | | | |
|---|---|---|---|
| 1 | Pedal bracket | 4 | Module |
| 2 | Pedal module electrical | 5 | Footwell panel bracket |
| 3 | Mounting nut | | |

## 12 Fuel injection system - general information

The fuel injection system consists of three sub-systems: air intake, engine control and fuel delivery. The system uses an Engine Control Module (ECM) along with the sensors (coolant temperature sensor, throttle position sensor, mass airflow sensor, oxygen sensor, etc.) to determine the proper air/fuel ratio under all operating conditions.

The fuel injection system and the engine control system are closely linked in function and design. For additional information, refer to Chapter 6.

### AIR INTAKE SYSTEM

The air intake system consists of the air filter, the air intake ducts, the throttle body, the air intake plenum and the intake manifold.

When the engine is idling, the air/fuel ratio is controlled by the idle control system, which consists of the Engine Control Module (ECM) and the throttle position actuator. The throttle position actuator is an electric motor contained within the throttle control module and controlled by the ECM. The ECM commands the throttle position actuator to open or close the throttle plate depending upon the running conditions of the engine (air conditioning system, power steering, cold and warm running etc.). The ECM receives information from the sensors (vehicle speed, coolant temperature, air conditioning, power steering mode etc.) and adjusts the idle according to the demands of the engine and driver. Refer to Chapter 6 for information on the throttle control module.

## EMISSIONS AND ENGINE CONTROL SYSTEM

The emissions and engine control system is described in detail in Chapter 6.

## FUEL DELIVERY SYSTEM

The fuel delivery system consists of these components: the fuel pump, the fuel pressure regulator, the fuel rail, the fuel injectors and the associated hoses and lines.

The fuel pump is an electric type located in the fuel tank. Fuel is drawn through an inlet screen into the pump, flows through the one-way valve, passes through the fuel filter and is delivered to the fuel rail and injectors. The pressure regulator maintains a constant fuel pressure to the injectors. Excess fuel is routed back to the fuel tank through the fuel pressure regulator.

The injectors are solenoid-actuated pintle type consisting of a solenoid, plunger, needle valve and housing. When current is applied to the solenoid coil, the needle valve raises and pressurized fuel sprays out the nozzle. The injection quantity is determined by the length of time the valve is open (the length of time during which current is supplied to the solenoid coils).

The fuel pump relay is located in the relay panel under the left (driver's) side of the instrument panel. The fuel pump relay connects battery voltage to the fuel pump. The ECM controls the fuel pump relay. If the ECM senses there is NO signal from the engine speed sensor (as with the engine not running or cranking), the ECM will de-energize the relay.

## 13  Fuel injection system - check

♦ **Refer to illustrations 13.7 and 13.8**

➡**Note: The following procedure is based on the assumption that the fuel pressure is adequate (see Section 3).**

1   Check all electrical connectors that are related to the system. Check the ground wire connections for tightness. Loose connectors and poor grounds can cause many problems that resemble more serious malfunctions.

2   Check to see that the battery is fully charged, as the control unit and sensors depend on an accurate supply voltage in order to properly meter the fuel.

3   Check the air filter element - a dirty or partially blocked filter will severely impede performance and economy (see Chapter 1).

4   Check the related fuses. If a blown fuse is found, replace it and see if it blows again. If it does, search for a wire shorted to ground in the harness.

5   Check the air intake duct to the intake manifold for leaks, which will result in an excessively lean mixture. Also check the condition of all vacuum hoses connected to the intake manifold and/or throttle body.

6   Remove the air intake duct from the throttle body and check for dirt, carbon or other residue build-up. If it's dirty, clean it with carburetor cleaner spray, a toothbrush and a shop towel.

7   With the engine running, place an automotive stethoscope against each injector, one at a time, and listen for a clicking sound, indicating operation (see illustration). If you don't have a stethoscope, place the tip of a screwdriver against the injector and listen through the handle. If you hear the injectors operating but there is a misfire condition present, the electrical circuits are functioning, but the injectors may be dirty or fouled from carbon deposits - commercial cleaning products may help or they may require replacement.

8   If you can't hear an injector operating, disconnect the injector electrical connector and measure the resistance of the injector (see illustration). Compare the measurement with the resistance value listed in this Chapter's Specifications. Replace any injector whose resistance value does not fall within specifications.

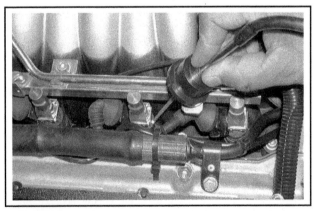

**13.7  Use a stethoscope to determine if the injectors are working properly - they should make a steady clicking sound that rises and falls with engine speed changes**

**13.8  Measure the resistance of each injector across the two terminals of the injector**

## 14  Throttle body - removal and installation

❋❋ **WARNING:**

**The engine must be completely cool before beginning this procedure.**

### FOUR-CYLINDER ENGINE

▶ **Refer to illustration 14.5**

1   Remove the air intake duct between the Mass Airflow sensor and the throttle body.

2   Disconnect the electrical connector from the throttle control module.

3   Drain the coolant (see Chapter 1). Detach the hoses from the throttle body.

4   Detach the accelerator cable from the throttle body (see Section 10).

5   Remove the mounting bolts and remove the throttle body (see illustration).

6   Install the throttle body with a new gasket and tighten the bolts to the torque listed in this Chapter's Specifications.

7   The remainder of installation is the reverse of removal. Refill and bleed the cooling system (see Chapter 1).

### V6 ENGINES

▶ **Refer to illustrations 14.11 and 14.16**

8   Remove the engine cover.

9   Remove the air intake duct between the Mass Airflow sensor and the intake air elbow.

10  On models equipped with cruise control, disconnect the vacuum hose from the servo unit and remove the bracket mounting bolts. Disconnect the throttle rod and remove the servo unit.

11  Disconnect the electrical connectors from the intake air temperature sensor, secondary air injection solenoid valve and intake manifold tuning solenoid valve. Carefully label and detach the vacuum hoses from the valves and remove the valve bracket assembly (see illustration).

12  Remove the air intake elbow from the throttle body.

13  Disconnect the electrical connector from the throttle control module.

14  Drain the coolant (see Chapter 1). Detach the hoses from the throttle body.

15  Detach the accelerator cable from the throttle body (see Section 10).

16  Remove the mounting bolts and remove the throttle body (see illustration).

17  Inspect the O-ring seal and replace it if necessary.

18  Install the throttle body and tighten the bolts to the torque listed in this Chapter's Specifications.

19  The remainder of installation is the reverse of removal. Refill and bleed the cooling system (see Chapter 1)

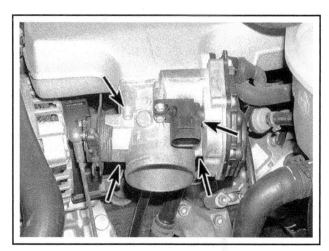

**14.5  Throttle body mounting bolts (arrows) (1.8L turbocharged engine shown)**

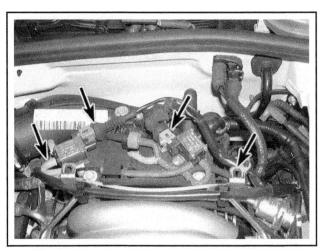

**14.11  Disconnect the electrical connectors to the IAT sensor, the secondary air injection solenoid and the intake manifold tuning valve (arrows)**

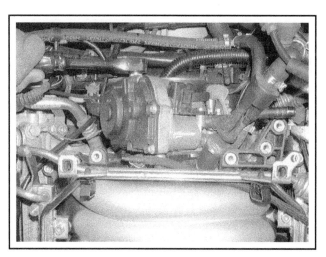

**14.16  Removing the throttle body mounting bolts from the 2.8L V6 engine**

## 15 Fuel pressure regulator - replacement

▶ **Refer to illustration 15.4 and 15.5**

### ✳✳ WARNING:

**See the Warning in Section 1.**

1  Relieve the fuel system pressure (see Section 2).
2  Remove the engine cover (if equipped).
3  Disconnect the vacuum hose from the port on the regulator.

4  Remove the fuel pressure regulator retaining clip and withdraw the fuel pressure regulator from the fuel rail (see illustration).
5  Be sure to replace the O-ring seals, lubricating them with a light film of engine oil (see illustration).
➡**Note: The small O-ring may remain in the fuel rail; recover and replace it.**
6  Press the fuel pressure regulator into the fuel rail until fully seated and install the retaining clip.
7  The remainder of installation is the reverse of removal.

**15.4  Remove the fuel pressure regulator retaining clip (arrow) and withdraw the regulator from the fuel rail**

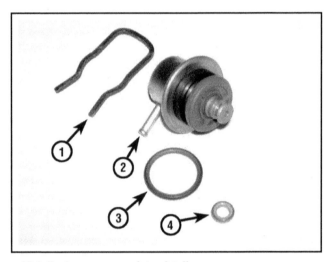

**15.5  Fuel pressure regulator details**

| | | | |
|---|---|---|---|
| 1 | Retaining clip | 3 | Large O-ring |
| 2 | Fuel pressure regulator | 4 | Small O-ring |

## 16 Fuel rail and injectors - removal and installation

### ✳✳ WARNING:

**See the Warning in Section 1.**

## REMOVAL

▶ **Refer to illustrations 16.5, 16.6, 16.7, 16.8a and 16.8b**

1  Relieve the fuel pressure (see Section 2).
2  Remove the engine cover (if equipped).
3  Clearly label and remove any vacuum hoses or electrical wiring that will interfere with the fuel rail removal.
4  Disconnect the vacuum hose from the fuel pressure regulator.
5  Disconnect the fuel inlet and return lines from the fuel rail (see illustration).

6  Disconnect the fuel injector electrical connectors. Detach the wiring harness retainers from the fuel rail and position the harness aside (see illustration).
➡**Note: Apply a numbered tag to each connector with the corresponding cylinder number.**

7  Clean any debris from around the injectors. Remove the fuel rail mounting bolts (see illustration). Gently rock the fuel rail and injectors to loosen the injectors. Remove the fuel rail and fuel injectors as an assembly.
8  Remove the retaining clip and remove the injector(s) from the fuel rail assembly (see illustrations). Remove and discard the O-rings and seals.

➡**Note: Whether you're replacing an injector or a leaking O-ring, it's a good idea to remove all the injectors from the fuel rail and replace all the O-rings.**

**16.5 Disconnect the fuel supply line and return line from the fuel rail**

**16.6 Disconnect the fuel injector electrical connectors (arrows) (2.8L V6 engine shown)**

**16.7 Remove the fuel rail mounting bolts (arrows) (2.8L V6 engine shown)**

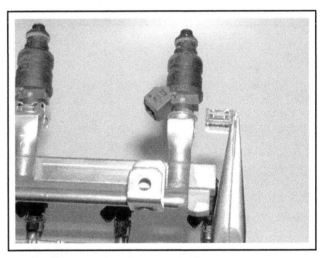

**16.8a Remove the injector retaining clip and pull the injector off the fuel rail**

## INSTALLATION

9   Coat the new O-rings with clean engine oil and install them on the injector(s), then insert each injector into its corresponding bore in the fuel rail. Install the injector retaining clips.

10  Install the injector and fuel rail assembly on the intake manifold. Fully seat the injectors, then tighten the fuel rail mounting nuts to the torque listed in this Chapter's Specifications.

11  Connect the fuel lines and make sure they're securely installed.

12  Connect the electrical connectors to each injector, referring to the numbered tags.

13  The remainder of installation is the reverse of removal.

14  After the injector/fuel rail assembly installation is complete, turn the ignition switch to On, but don't operate the starter (this activates the fuel pump for about two seconds, which builds up fuel pressure in the fuel lines and the fuel rail). Repeat this about two or three times, then check the fuel lines, fuel rail and injectors for fuel leakage.

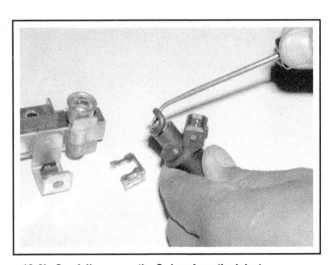

**16.8b Carefully remove the O-rings from the injector**

## 17 Turbocharger and intercooler (four-cylinder engine) - check and replacement

### CHECK

1 The turbocharger is a precision component which can be severely damaged by a lack of lubrication or from foreign material entering the air intake duct. Turbocharger failure may be indicated by poor engine performance, blue/gray exhaust smoke or unusual noises from the turbocharger. If a turbocharger failure is suspected, check the following areas:

a) *Check the intake air duct for looseness or damage. Make sure there are no restrictions in the air intake system, or a dirty air filter element or damaged intercooler.*

b) *Check the system vacuum hoses for restrictions or damage.*

c) *Check the system wiring for damage and electrical connectors for looseness or corrosion.*

d) *Make sure the wastegate actuator linkage is not binding.*

e) *Check the exhaust system for damage or restrictions.*

f) *Check the lubricating oil supply and drainback lines for damage or restrictions.*

g) *Check the coolant supply and return lines for damage and restrictions.*

h) *If the turbocharger requires replacement due to failure, be sure to change the engine oil and filter (see Chapter 1).*

2 Complete diagnosis of the turbocharger and control system require special techniques and equipment. If the previous checks fail to identify the problem, take the vehicle to a dealership service department or other properly equipped repair facility for diagnosis.

### REPLACEMENT

#### Turbocharger

▶ Refer to illustrations 17.8a, 17.8b, 17.9 and 17.10

### ✳✳ WARNING:

**Wait until the engine is completely cool before beginning this procedure.**

3 Drain the cooling system (see Chapter 1).

4 Raise the vehicle and support it securely on jackstands.

5 Remove the engine compartment undercover.

6 Remove the air conditioning compressor from the mounting bracket and position the compressor aside without disconnecting the refrigerant hoses.

7 Remove the turbocharger support bracket.

8 Loosen the hose clamps and disconnect the air inlet and outlet ducts from the turbocharger (see illustrations).

9 Remove the bolts and disconnect the oil return pipe from the turbocharger (see illustration).

10 Loosen the oil supply union bolt from the turbocharger (see illustration). Remove the bracket bolt and disconnect the oil supply pipe from the turbocharger.

11 Disconnect the vacuum hose from the wastegate actuator and remove the coolant line bracket bolt.

**17.8a Disconnect the boost pressure recirculation valve hose from the air ducts**

**17.8b Remove the inlet duct (1) and the outlet duct (2) from the turbocharger**

**17.9 Location of the oil return pipe (arrow) – follow this pipe up into the engine compartment to locate the junction at the turbocharger**

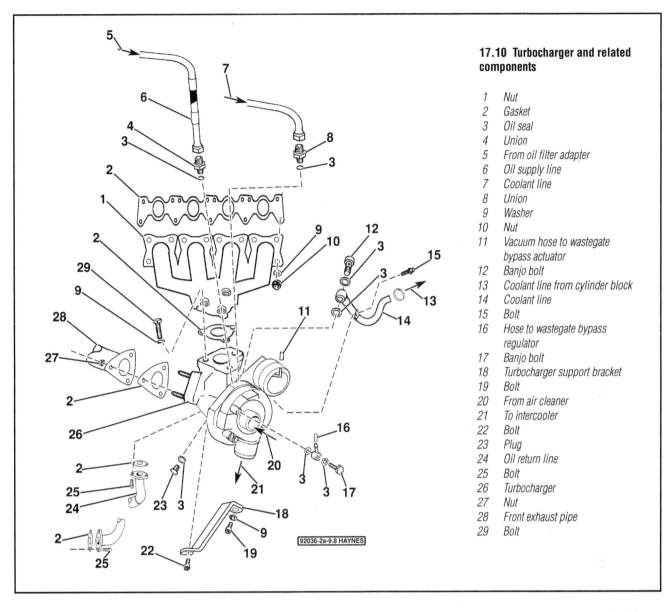

**17.10 Turbocharger and related components**

| | |
|---|---|
| 1 | Nut |
| 2 | Gasket |
| 3 | Oil seal |
| 4 | Union |
| 5 | From oil filter adapter |
| 6 | Oil supply line |
| 7 | Coolant line |
| 8 | Union |
| 9 | Washer |
| 10 | Nut |
| 11 | Vacuum hose to wastegate bypass actuator |
| 12 | Banjo bolt |
| 13 | Coolant line from cylinder block |
| 14 | Coolant line |
| 15 | Bolt |
| 16 | Hose to wastegate bypass regulator |
| 17 | Banjo bolt |
| 18 | Turbocharger support bracket |
| 19 | Bolt |
| 20 | From air cleaner |
| 21 | To intercooler |
| 22 | Bolt |
| 23 | Plug |
| 24 | Oil return line |
| 25 | Bolt |
| 26 | Turbocharger |
| 27 | Nut |
| 28 | Front exhaust pipe |
| 29 | Bolt |

*92036-2a-9.8 HAYNES*

12  Disconnect the required hoses and electrical connectors at the air filter assembly and remove the air filter housing and the duct from the air filter housing to the turbocharger.

13  Disconnect the crankcase ventilation hose from the valve cover and heat shield. Remove the oil line bracket bolts and remove the heat shield.

14  Loosen the clamps, remove the bracket bolts and remove the turbocharger outlet pipe and elbow.

15  Loosen the coolant return pipe union bolt at the turbocharger. Remove the bracket bolt and disconnect the coolant return pipe from the turbocharger. Be sure to recover the sealing washers.

16  Remove the nuts securing the exhaust pipe to the turbocharger. Remove the exhaust support bolts, as necessary and lower the exhaust pipe.

17  Remove the turbocharger-to-exhaust manifold bolts. Lower the turbocharger, tilt it to the side, loosen the coolant supply pipe union bolt and disconnect the coolant supply pipe from the turbocharger.

18  Remove the turbocharger from the engine compartment.

19  Installation is the reverse of removal with the following additions:

a) Replace all gaskets, seals, union bolt washers and self-locking nuts.

b) Tighten the turbocharger mounting bolts to the torque listed in this Chapter's Specifications.

c) Change the oil and filter (see Chapter 1).

d) Before starting the engine, remove the fuel pump fuse (see Section 2) from the fuse box and crank the engine over until oil pressure builds.

## Intercooler

20  Raise the vehicle and support it securely on jackstands.

21  Remove the engine compartment undercover.

22  Detach the air deflector from the intercooler.

23  Loosen the hose clamp and disconnect the upper air duct from the intercooler.

24  Loosen the hose clamp and disconnect the lower air duct from the intercooler.

25  Remove the mounting bolts and remove the intercooler.

26  Installation is the reverse of removal.

## 18 Exhaust system servicing - general information

### ✳✳ WARNING:

**Inspection and repair of exhaust system components should be done only after enough time has elapsed after driving the vehicle to allow the system components to cool completely. Also, when working under the vehicle, make sure it is securely supported on jackstands.**

1   The exhaust system consists of the exhaust manifold, catalytic converter, muffler, resonators, the tailpipe and all connecting pipes, brackets, hangers and clamps. The exhaust system is attached to the body with mounting brackets and rubber hangers. If any of the parts are improperly installed, excessive noise and vibration will be transmitted to the body.

### MUFFLER AND PIPES

▶ **Refer to illustrations 18.2a through 18.2e**

2   Conduct regular inspections of the exhaust system to keep it safe and quiet. Look for any damaged or bent parts, open seams, holes, loose connections, excessive corrosion or other defects which could allow exhaust fumes to enter the vehicle (see illustrations). Also check the catalytic converter when you inspect the exhaust system (see below). Deteriorated exhaust system components should not be repaired; they should be replaced with new parts.

3   If the exhaust system components are extremely corroded or rusted together, welding equipment will probably be required to remove them. The convenient way to accomplish this is to have a muffler repair shop remove the corroded sections with a cutting torch. If, however, you want to save money by doing it yourself (and you don't have a welding outfit with a cutting torch), simply cut the exhaust pipes with a hacksaw at the separation point. If you do decide to tackle the job at home, be sure to wear safety goggles to protect your eyes from metal chips and work gloves to protect your hands.

4   Here are some simple guidelines to follow when repairing the exhaust system:

a)   Work from the back to the front when removing exhaust system components.
b)   Apply penetrating oil to the exhaust system component fasteners to make them easier to remove.
c)   Use new gaskets, hangers and clamps when installing exhaust system components.
d)   Apply anti-seize compound to the threads of all exhaust system fasteners during reassembly.
e)   Be sure to allow sufficient clearance between newly installed parts and all points on the underbody to avoid overheating the floor pan and possibly damaging the interior carpet and insulation. Pay particularly close attention to the catalytic converter and heat shield.

### CATALYTIC CONVERTER

### ✳✳ WARNING:

**The converter gets very hot during operation. Make sure it has cooled down before you touch it.**

➡**Note: See Chapter 6 for additional information on the catalytic converter.**

5   Periodically inspect the heat shield for cracks, dents and loose or missing fasteners.
6   Inspect the converter for cracks or other damage.
7   If the catalytic converter requires replacement, refer to Chapter 6.

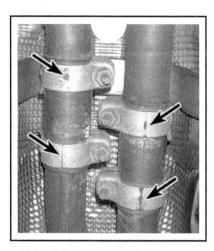

**18.2a  Exhaust pipe slip-joint clamps (arrows)**

**18.2b  Exhaust pipe (arrow) - front portion**

**18.2c  Exhaust system resonator**

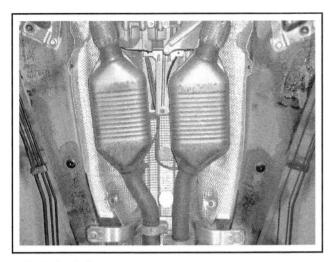

**18.2d  Catalytic converters**

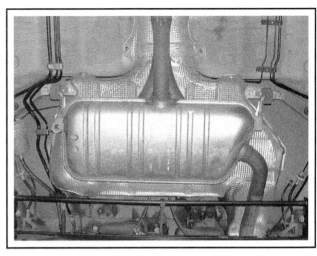

**18.2e  Exhaust system muffler**

## Specifications

### General

Fuel pressure
 Key On, engine Off       55 to 61 psi (379 to 421 kPa)
 Engine running        46 to 55 psi (317 to 379 kPa)
Fuel injector resistance (at room temperature)
 Four-cylinder engine      11.0 to 13.0 ohms
 V6 engines
  Siemans injectors      13.5 to 15.5 ohms
  Bosch injectors       15 to 17 ohms

| Torque specifications | Ft-lbs (unless otherwise indicated) | Nm |
| --- | --- | --- |
| Fuel rail mounting bolts | 84 in-lbs | 10 |
| Fuel tank mounting strap bolts | 18 | 25 |
| Throttle body mounting bolts | 84 in-lbs | 10 |
| Turbocharger-to-exhaust manifold bolts | 84 in-lbs | 35 |
| Turbocharger-to-exhaust pipe nuts | 18 | 25 |

**Notes**

## Section

### Reference to other Chapters

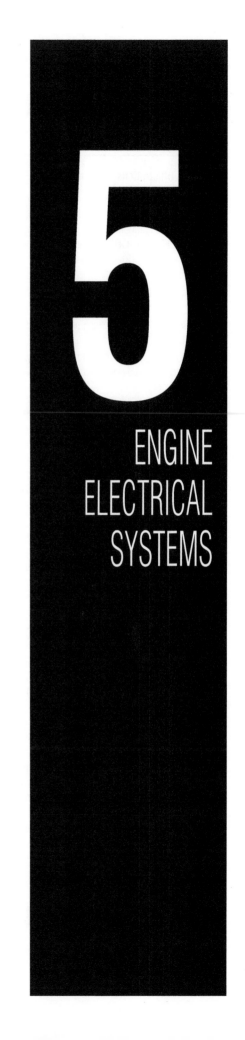

# 5

# ENGINE ELECTRICAL SYSTEMS

## 1  General information, precautions and battery disconnection

### GENERAL INFORMATION

▶ **Refer to illustrations 1.1a and 1.1b**

The engine electrical systems include all ignition, charging and starting components (see Illustrations). Because of their engine-related functions, these components are discussed separately from body electrical devices such as the lights, the instruments, etc. (which are included in Chapter 12).

### PRECAUTIONS

Always observe the following precautions when working on the electrical system:

a) *Be extremely careful when servicing engine electrical components. They are easily damaged if checked, connected or handled improperly.*
b) *Never leave the ignition switched on for long periods of time when the engine is not running.*
c) *Never disconnect the battery cables while the engine is running.*
d) *Maintain correct polarity when connecting battery cables from another vehicle during jump starting - see the "Booster battery (jump) starting" section at the front of this manual.*
e) *Always disconnect the negative battery cable from the battery before working on the electrical system, but read the following battery disconnection information first.*

It's also a good idea to review the safety-related information regarding the engine electrical systems located in the "Safety first!" section at the front of this manual, before beginning any operation included in this Chapter.

### BATTERY DISCONNECTION

Several systems on the vehicle require battery power to be available at all times, either to ensure their continued operation (such as the radio, alarm system, power door locks, windows, etc.) or to maintain control unit memories (such as that in the engine management system's Engine Control Module [ECM]) which would be lost if the battery were to be disconnected. Therefore, whenever the battery is to be disconnected, first note the following to ensure that there are no unforeseen consequences of this action:

a) *These models are equipped with an anti-theft radio. Before performing a procedure that requires disconnecting the battery, make sure you have the proper activation code.*

b) *The engine management system's ECM (and, on models equipped with an automatic transaxle, the Transmission Control Module [TCM]) will lose the information stored in its memory when the battery is disconnected. This includes idling and operating values, any fault codes detected and system monitors required for emissions testing. Whenever the battery is disconnected, the information relating to idle speed control and other operating values will have to be re-programmed into the unit's memory using a scan tool. This will require taking the vehicle to a dealer service department or other properly equipped repair facility, and may involve considerable expense.*
c) *On any vehicle with power door locks, it is a wise precaution to remove the key from the ignition and to keep it with you, so that it does not get locked inside if the power door locks should engage accidentally when the battery is reconnected!*

Devices known as "memory savers" can be used to avoid some of the above problems. Precise details vary according to the device used. Typically, it is plugged into the cigarette lighter and is connected by its own wires to a spare battery; the vehicle's own battery is then disconnected from the electrical system, leaving the "memory saver" to pass sufficient current to maintain audio unit security codes and ECM (and TCM) memory values, and also to run permanently live circuits such as the clock and radio memory, all the while isolating the battery in the event of a short-circuit occurring while work is carried out.

### ❋❋ WARNING 1:

**Some of these devices allow a considerable amount of current to pass, which can mean that many of the vehicle's systems are still operational when the main battery is disconnected. If a "memory saver" is used, ensure that the circuit concerned is actually "dead" before carrying out any work on it!**

### ❋❋ WARNING 2:

**If work is to be performed around any of the airbag system components, the battery must be disconnected and no "memory saver" can be used. If a memory-saver device is used, power will be supplied to the airbag and personal injury may result if the airbag is accidentally deployed.**

The battery on these vehicles is located under the cowl at the rear of the engine compartment (see Section 1). To disconnect the battery for service procedures requiring power to be cut from the vehicle, lift the battery cover lid, peel back the insulator, loosen the negative cable clamp nut and detach the negative cable from the negative battery post (see Section 3). Isolate the cable end to prevent it from accidentally coming into contact with the battery post.

## 2  Battery - emergency jump starting

Refer to the *Booster battery (jump) starting* procedure at the front of this manual.

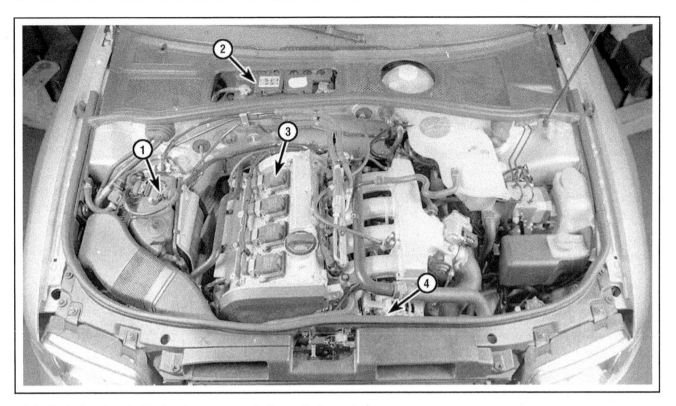

**1.1a  Typical engine electrical system components (four-cylinder engine)**

| | | | |
|---|---|---|---|
| 1 | Power transistor | 3 | Ignition coil/spark plug assembly |
| 2 | Battery | 4 | Alternator |

**1.1b  Typical engine electrical system components (V6 DOHC engine)**

| | | | | | |
|---|---|---|---|---|---|
| 1 | Battery | 3 | Coil harness connector | 5 | Ignition wires (right bank) |
| 2 | Ignition wires (left bank) | 4 | Ignition coils | | |

## 3  Battery - check and replacement

### ✳✳ WARNING:

Hydrogen gas is produced by the battery, so keep open flames and lighted cigarettes away from it at all times. Always wear eye protection when working around a battery. Rinse off spilled electrolyte immediately with large amounts of water.

### ✳✳ CAUTION 1:

These models are equipped with an anti-theft radio. Before performing a procedure that requires disconnecting the battery, make sure you have the activation code.

### ✳✳ CAUTION 2:

Disconnecting the battery can cause driveability problems that require a scan tool to rectify. Additionally, disconnecting the battery may cause one or more warning lights on the instrument panel to illuminate, which will also require the use of a scan tool to turn off. Most scan tools available to the public do not have the capability to perform either of these tasks, which will necessitate taking the vehicle to a dealer service department or other properly equipped repair facility after service work has been performed. See Section 1 for the use of an auxiliary voltage input device ("memory saver") before disconnecting the battery and for other precautions related to battery disconnection.

## CHECK

▶ **Refer to illustrations 3.1a, 3.1b and 3.1c**

1    A battery cannot be accurately tested until it is at or near a fully charged state. Disconnect the negative battery cable from the battery and perform the following tests:

a)  **Battery state of charge test** - *Visually inspect the indicator eye (if equipped) on the top of the battery. If the indicator eye is dark in color, charge the battery as described in Chapter 1. If the battery is equipped with removable caps, check the battery electrolyte. The electrolyte level should be above the upper edge of the plates. If the level is low, add distilled water. DO NOT OVERFILL. The excess electrolyte may spill over during periods of heavy charging. Test the specific gravity of the electrolyte using a hydrometer (see illustration). Remove the caps and extract a sample of the electrolyte and observe the float inside the barrel of the hydrometer. Follow the instructions from the tool manufacturer and determine the specific gravity of the electrolyte for each cell. A fully charged battery will indicate approximately 1.270 (green zone) at 68-degrees F (20-degrees C). If the specific gravity of the electrolyte is low (red zone), charge the battery as described in Chapter 1.*

b)  **Open circuit voltage test** - *Using a digital voltmeter, perform an open circuit voltage test (see illustration). Connect the negative probe of the voltmeter to the negative battery post and the positive probe to the positive battery post. The battery voltage should be greater than 12.5 volts. If the battery is less than the specified*

*voltage, charge the battery before proceeding to the next test. Do not proceed with the battery load test until the battery is fully charged.*

c)  **Battery load test** - *An accurate check of the battery condition can only be performed with a load tester (available at most auto parts stores). This test evaluates the ability of the battery to operate the starter and other accessories during periods of heavy amperage draw (load). Install a special battery load testing tool onto the battery terminals (see illustration). Load test the battery according to the tool manufacturer's instructions. This tool utilizes a carbon pile to increase the load demand (amperage draw) on the battery. Maintain the load on the battery for 15 seconds and observe that the battery voltage does not drop below 9.6 volts. If the battery condition is weak or defective, the tool will indicate this condition immediately.*

➡Note: Cold temperatures will cause the minimum voltage requirements to drop slightly. Follow the chart given in the tool manufacturer's instructions to compensate for cold climates. Minimum load voltage for freezing temperatures (32 degrees F/0-degrees C) should be approximately 9.1 volts.

d)  **Battery drain test** - *This test will indicate whether there's a constant drain on the vehicle's electrical system that can cause the battery to discharge. Make sure all accessories are turned Off. If the vehicle has an underhood light, verify it's working properly, then disconnect it. Connect one lead of a digital ammeter to the disconnected negative battery cable clamp and the other lead to the negative battery post. A drain of approximately 100 milliamps or less is considered normal (due to the engine control computer, digital clocks, digital radios and other components which normally cause a key-off battery drain). An excessive drain (approximately 500 milliamps or more) will cause the battery to discharge. The problem circuit or component can be located by removing the fuses, one at a time, until the excessive drain stops and normal drain is indicated on the meter.*

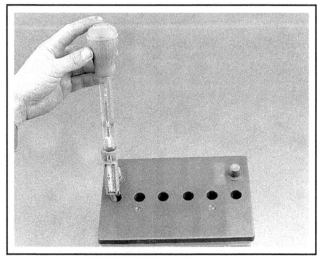

**3.1a  Use a battery hydrometer to draw electrolyte from the battery cell - this hydrometer is equipped with a thermometer to make temperature corrections**

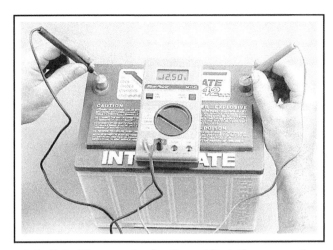

**3.1b** To test the open circuit voltage of the battery, connect the black probe of the voltmeter to the negative terminal and the red probe to the positive terminal of the battery - a fully charged battery should indicate approximately 12.5 volts depending on the outside air temperature

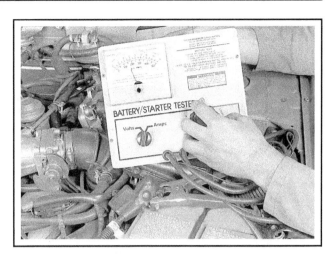

**3.1c** Some battery load testers are equipped with an ammeter which enables the battery load to be precisely dialed in, as shown - less expensive testers have a load switch and a voltmeter only

## REPLACEMENT

▶ **Refer to illustrations 3.4 and 3.5**

### ✳✳ CAUTION:

**Always disconnect the negative cable first and hook it up last or the battery may be shorted by the tool being used to loosen the cable clamps.**

2   Loosen the cable clamp nut and remove the negative battery cable from the negative battery post. Isolate the cable end to prevent it from accidentally coming into contact with the battery post.

3   Loosen the cable clamp nut and remove the positive battery cable from the positive battery post.

4   Remove the battery hold-down clamp (see illustration).

5   Disconnect the vent hose (see illustration) and lift out the battery. Be careful - it's heavy.

➡**Note: Battery straps and handlers are available at most auto parts stores for a reasonable price. They make it easier to remove and carry the battery.**

6   While the battery is out, inspect the battery tray for corrosion. If corrosion exists, clean the deposits with a mixture of baking soda and water to prevent further corrosion. Flush the area with plenty of clean water and dry thoroughly.

7   If you are replacing the battery, make sure you replace it with a battery with the identical dimensions, amperage rating, cold cranking rating, etc.

8   When installing the battery, make sure the center notch in the battery foot is aligned with the hold-down clamp hole in the battery tray. Install the hold-down clamp and tighten the bolt to the torque listed in this Chapter's Specifications. Do not over-tighten the bolt.

9   The remainder of installation is the reverse of removal.

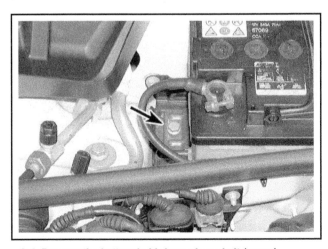

**3.4  Remove the battery hold-down clamp bolt (arrow)**

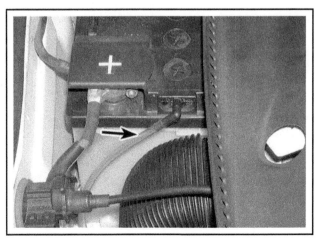

**3.5  Disconnect the battery vent hose (arrow)**

## 4    Battery cables - replacement

♦ Refer to illustrations 4.4a and 4.4b

### ✳✳ CAUTION 1:

These models are equipped with an anti-theft radio. Before performing a procedure that requires disconnecting the battery, make sure you have the activation code.

### ✳✳ CAUTION 2:

Disconnecting the battery can cause driveability problems that require a scan tool to rectify. Additionally, disconnecting the battery may cause one or more warning lights on the instrument panel to illuminate, which will also require the use of a scan tool to turn off. Most scan tools available to the public do not have the capability to perform either of these tasks, which will necessitate taking the vehicle to a dealer service department or other properly equipped repair facility after service work has been performed. See Section 1 for the use of an auxiliary voltage input device ("memory saver") before disconnecting the battery and for other precautions related to battery disconnection.

1    Periodically inspect the entire length of each battery cable for damage, cracked or burned insulation and corrosion. Poor battery cable connections can cause starting problems and decreased engine performance.

2    Check the cable-to-terminal connections at the ends of the cables for cracks, loose wire strands and corrosion. The presence of white, fluffy deposits under the insulation at the cable terminal connection is a sign that the cable is corroded and should be replaced. Check the terminals for distortion, missing mounting bolts and corrosion.

3    When removing the cables, always disconnect the negative cable from the negative battery post first and hook it up last or the battery may be shorted by the tool used to loosen the cable clamps. Even if only the positive cable is being replaced, be sure to disconnect the negative cable from the negative battery post first (see Chapter 1 for further information regarding battery cable maintenance).

4    Disconnect the old cables from the battery, then disconnect them from the opposite end. Detach the cables from the starter solenoid, underhood fuse box and ground terminals, as necessary (see illustrations). Note the routing of each cable to ensure correct installation.

5    If you are replacing either or both of the battery cables, take them with you when buying new cables. It is vitally important that you replace the cables with identical parts. Cables have characteristics that make them easy to identify: Positive cables are usually red and larger in cross-section; ground cables are usually black and smaller in cross-section.

6    Clean the threads of the starter solenoid or ground connection with a wire brush to remove rust and corrosion. Apply a light coat of battery terminal corrosion inhibitor or petroleum jelly to the threads to prevent future corrosion.

7    Attach the cable to the terminal and tighten the mounting nut/bolt securely.

8    Before connecting a new cable to the battery, make sure that it reaches the battery post without having to be stretched.

9    After installing the cables, connect the negative cable to the negative battery post.

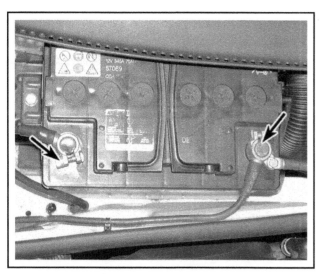

4.4a  Remove the negative (first) and positive (second) battery cables (arrows)

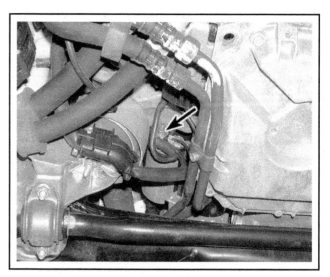

4.4b  Remove the battery cable from the starter solenoid (arrow)

## 5   Ignition system - general information

All models are equipped with a distributorless ignition system. The ignition system consists of the battery, ignition coils, spark plugs, camshaft position sensor, engine speed sensor and the Engine Control Module (ECM). The ECM controls the ignition timing and spark advance characteristics for the engine. The ignition timing is not adjustable.

The engine speed sensor and camshaft position sensor generate pulses that are input to the Engine Control Module. The ECM determines piston position and engine speed from these two sensors. The ECM calculates injector sequence and ignition timing from the piston position. Refer to Chapter 6 for replacement procedures for the engine speed sensor and camshaft position sensor.

The V6 engines utilize a coil pack consisting of two ignition coils and a power unit. This type of ignition system uses a "waste spark" method of spark distribution. Each cylinder is paired with its opposing cylinder in the firing order (1-6, 2-4, 3-5) so one cylinder under compression fires simultaneously with its opposing cylinder, where the piston is on the exhaust stroke. Since the cylinder on the exhaust stroke requires very little of the available voltage to fire its plug, most of the voltage is used to fire the plug of the cylinder on the compression stroke. In a conventional ignition system, one end of the ignition coil

secondary winding is connected to engine ground. In a waste spark system, neither end of the secondary winding is grounded - instead, one end of the coil secondary winding is directly attached to the spark plug and the other end is attached to the spark plug of the companion cylinder.

Four-cylinder models utilize an individual ignition coil/power unit for each cylinder. The unit is positioned directly over each spark plug. The ECM fires each coil sequentially in the firing order sequence.

The ECM controls the ignition system by opening and closing the primary ignition coil control circuit. The computerized ignition system provides complete control of the ignition timing by determining the optimum timing in response to engine speed, coolant temperature, throttle position and engine load. These parameters are relayed to the ECM by the camshaft position sensor, engine speed sensor, throttle position sensor, coolant temperature sensor and mass airflow sensor. Refer to Chapter 6 for additional information on the various sensors.

The ignition system is also integrated with a knock sensor system. The system uses two knock sensors in conjunction with the ECM to control spark timing. The knock sensor system allows the engine to use maximum spark advance without spark knock, which improves driveability and fuel economy.

## 6   Ignition system - check

▶ **Refer to illustrations 6.2, 6.4a, 6.4b, 6.5a, 6.5b and 6.9**

### ❋❋ WARNING:

**Because of the high voltage generated by the ignition system, extreme care should be taken whenever an operation is performed involving ignition components. This not only includes the ignition coil, but related components and test equipment.**

➡**Note: The ignition system components on these models are expensive and difficult to diagnose. In the event of ignition system failure, if the checks do not clearly indicate the source of the ignition system problem, have the vehicle tested by a dealer service department or other qualified auto repair facility.**

1   If a malfunction occurs and the vehicle won't start, do not immediately assume that the ignition system is causing the problem. First, check the following items:
  a) *Make sure the battery cable clamps, where they connect to the battery, are clean and tight.*
  b) *Test the condition of the battery (see Section 3). If it does not pass all the tests, replace it with a new battery.*
  c) *Check the external ignition coil wiring and connections.*
  d) *Check the related fuses inside the fuse box (see Chapter 12). If they're burned, determine the cause and repair the circuit.*

2   If the engine turns over but won't start, make sure there is sufficient secondary ignition voltage to fire the spark plug. On V6 models, disconnect a spark plug wire from one of the spark plugs and attach a calibrated ignition system tester (available at most auto parts stores) to the spark plug boot. Connect the clip on the tester to a bolt or metal

bracket on the engine (see illustration). On four-cylinder models, remove an ignition coil (see Section 7) and attach the calibrated ignition system tester to the spark plug boot. Reconnect the electrical connector to the coil and clip the tester to a good ground. Crank the engine and watch the end of the tester to see if a bright blue, well-defined spark occurs (weak spark or intermittent spark is the same as no spark).

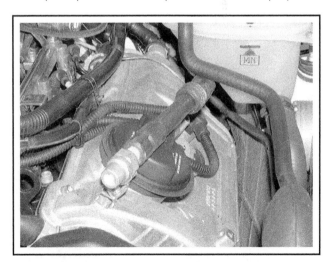

**6.2 To use a calibrated ignition tester, disconnect a wire from a spark plug, connect the tester to the spark plug boot, clip the tester to a convenient ground and crank the engine over - if there's enough power to fire the plug, bright blue sparks will be visible between the electrode tip and the tester body (weak sparks or intermittent sparks are the same as no sparks)**

3   If sparks occur, sufficient voltage is reaching the plug to fire it (repeat the check at the remaining spark plug wires or ignition coils to verify that the spark plug wires, connectors and ignition coils are good). If the ignition system is operating properly the problem lies elsewhere; i.e. a mechanical or fuel system problem. However, the plugs themselves may be fouled, so remove and check them as described in Chapter 1.

4   If no spark occurs, remove the spark plug wire or boot from the suspected ignition coil and check the terminals for damage. Using an ohmmeter (see illustration), check the wire or boot for an open or high resistance (compare your measurement with the values listed in this Chapter's Specifications). Disconnect the spark plug wires from all the ignition coil towers and check the ignition coil secondary resistance across each pair of coil towers (see illustration). Compare your measurement with the values listed in this Chapter's Specifications. Replace the ignition coil assembly if any coil is not within specifications.

➡**Note: The ignition wire and coil resistance checks are only applicable to the 2.8L V6 engines. The spark plug boot on the four-cylinder engines can be checked for resistance.**

5   Check for battery voltage to the ignition coil with the ignition key On (engine not running). Disconnect the coil electrical connector and check for power at the corresponding terminals of the coil connector (see illustrations). Battery voltage should be available with the ignition key On. If there is no battery voltage present, check the wiring and/or circuit between the fuse box and ignition coil (don't forget to check the fuses). Also check the ground circuit for continuity.

➡**Note: Refer to the wiring diagrams at the end of Chapter 12 for wire color identification for testing.**

6   If battery voltage is available to the ignition coil, check the ignition coil control circuits as follows:

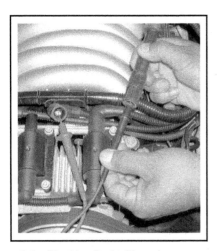

**6.4a  Checking the ignition wire resistance on a V6 engine**

**6.4b  Checking the ignition coil secondary resistance on a V6 engine**

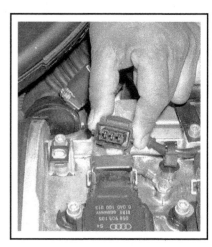

**6.5a  Check for battery voltage to the ignition coil assembly on the red/green wire of the harness connector (2000 1.8L engine shown)**

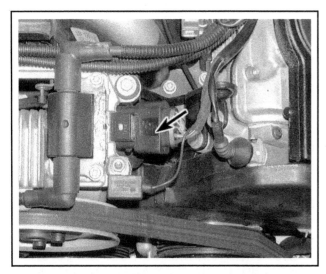

**6.5b  Location of the ignition coil harness connector (arrow) on a V6 engine**

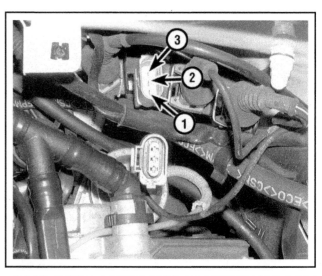

**6.9  Check the speed sensor resistance across terminal number 1 and number 2**

**✳✳ CAUTION:**

**Use only an LED test light to avoid damaging the ECM.**

a) Remove the fuel injector supply fuse from the fuse box (this disables the fuel injectors so the engine will not start or flood).

   1) On VW models, remove fuse number 34 (1998 through 2000, 2005) or fuse number 32 (2001 through 2004)

   2) On Audi models, remove fuse number 34 (1996 through 1999) or fuse number 32 (2000 and 2001)

b) Connect an LED test light onto the trigger signal wire(s) of the coil harness connector. Refer to the wiring diagrams at the end of Chapter 12 for the wire color designations for each year and model. For example, on 1999 VW 2.8L V6 models, use the gray/yel, grn/wht and the grn/gray wires to check for the trigger signal from the ECM. Perform the remainder of the test, then connect the LED test light to each of the other wire terminals in turn and perform the test. Be sure to note that 1997 through 1999 1.8L and 1996 and 1997 SOHC 2.8L engines incorporate an ignition power transistor in the circuit (see Step 8).

c) Crank the engine with the starter and confirm that the LED test light flashes as the engine rotates. A flashing LED test light indicates the ECM and power transistor (if equipped), camshaft position sensor and engine speed sensor are functioning properly.

7  If battery voltage, ground and a control signal exist at the ignition coil and there is no spark, replace the ignition coil.

8  1997 through 1999 1.8L and 1996 and 1997 V6 models are equipped with a power transistor in the control circuit. If a control signal is not present at the ignition coil, check the power transistor for battery power, ground and a control signal in the same manner as the ignition coils. See illustration 1.1a for location of the power transistor on the 1.8L turbocharged engine.

➡**Note: Refer to the wiring diagrams at the end of Chapter 12 for wire color identification for testing.**

9  If a control signal is not present at the ignition coil or power transistor (if equipped), check the engine speed sensor as follows (the engine will not start if the engine speed sensor is defective) (see illustration):

a) Disconnect the electrical connector from the engine speed sensor (see Chapter 6).

b) Using an ohmmeter, measure the resistance across terminals 1 and 2 of the engine speed sensor. Engine speed sensor resistance should be 400 to 1,000 ohms (VW models) or 1 ohm (Audi models).

c) Check for a short between terminals 1 and 3, and terminals 2 and 3. There should be infinite resistance.

d) Replace the sensor if defective (remove the sensor and check the sensor wheel for damage before condemning the sensor).

10  If the engine speed sensor is good and there is no control signal, have the ECM checked by a dealer service department or other qualified repair shop.

<br>

## 7  Ignition coils - replacement

▶ **Refer to illustrations 7.3 and 7.4**

1  Remove the engine cover.
2  Disconnect the electrical connector from the ignition coil.
3  On V6 models, label and detach the spark plug wires. Remove the mounting bolts and remove the coil assembly (see illustration).
4  On four-cylinder models, remove the ignition coil mounting screws and pull the coil straight up and out of the cylinder head (see Illustration).
5  Installation is the reverse of removal.

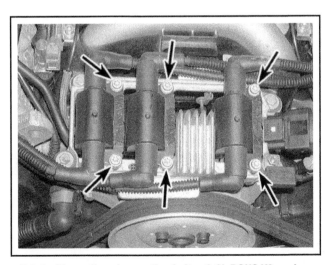

**7.3  Ignition coil pack mounting bolts - 2.8L DOHC V6 engine**

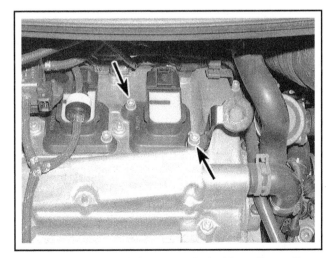

**7.4  On 1.8L turbo models, remove the ignition coil mounting screws (arrows) and pull the coil straight up**

## 8 Power transistor (1997 through 1999 four-cylinder and SOHC V6 engines) - replacement

♦ Refer to illustration 8.2

1   Disconnect the electrical connector from the power transistor.
2   Remove the mounting bolts and remove the power transistor (see illustration).
3   Installation is the reverse of removal.

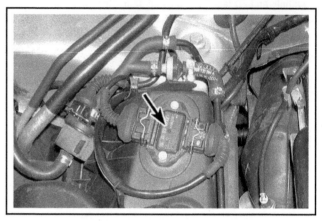

**8.2 Location of the power transistor on the 1.8L engine**

## 9 Charging system - general information and precautions

The charging system includes the alternator, a charge indicator light, the battery and the wiring between all the components. The battery supplies electrical power for the vehicle electrical system. The charging system maintains the battery in a fully charged condition. The alternator generates DC voltage to charge the battery and is driven by a serpentine drivebelt at the front of the engine.

The alternator voltage regulator regulates the alternator voltage output. The voltage regulator limits the charging voltage to a maximum preset value. This prevents overcharging the battery. The voltage regulator and brush assembly may be replaced in the event of failure.

The charging system doesn't ordinarily require periodic maintenance. However, the drivebelt, battery, battery cables, wiring and connections should be inspected at the intervals outlined in Chapter 1.

The dashboard warning light should come ON when the ignition key is turned to ON, but it should go off immediately after the engine is started. If it remains on, there is a malfunction in the charging system. Some vehicles are also equipped with a voltmeter. If the voltmeter indicates abnormally high or low voltage, check the charging system (see Section 10).

Be very careful when making electrical circuit connections to a vehicle equipped with an alternator and note the following:

a)  *When reconnecting wires to the alternator from the battery, be sure to note the polarity.*
b)  *Before using arc welding equipment to repair any part of the vehicle, disconnect the wires from the alternator and the battery terminals.*
c)  *Never start the engine with a battery charger connected.*
d)  *Always disconnect both battery cables before using a battery charger.*
e)  *The alternator is turned by an engine drivebelt which could cause serious injury if your hands, hair or clothes become entangled in it with the engine running.*
f)  *Because the alternator is connected directly to the battery, it could arc or cause a fire if overloaded or shorted out.*
g)  *Wrap a plastic bag over the alternator and secure it with rubber bands before steam-cleaning the engine.*

## 10 Charging system - check

♦ Refer to illustration 10.6

1   If a malfunction occurs in the charging circuit, do not immediately assume that the alternator is causing the problem. First check the following items:

a)  *The battery cables where they connect to the battery. Make sure the connections are clean and tight.*
b)  *Check the battery as described in Section 3. If the battery is defective, replace the battery.*
c)  *Check the external alternator wiring and connections.*
d)  *Check the drivebelt condition and tension (see Chapter 1).*
e)  *Check the alternator mounting bolts for tightness.*
f)  *Run the engine and check the alternator for abnormal noise.*

2   The charging system warning light on the instrument cluster should illuminate when the ignition key is switched on and go off when the engine is running.

3   If the warning light does not illuminate when the ignition key is

switched on, switch the ignition off and disconnect the blue wire from the alternator (do not disconnect the large output wire). Connect the blue wire terminal to a good engine ground point using a jumper wire and switch the ignition on. The warning light should illuminate - if it doesn't, there is an open circuit in the blue wire between the alternator and the instrument cluster, or the instrument cluster is defective.

4　If the warning light illuminates with the connector grounded but not when connected to the alternator, remove the voltage regulator brush assembly and check the brushes (see Section 11). Measure the brush length and compare it with the value listed in this Chapter's Specifications. Check the brush holder and springs for damage. If the brushes are worn or defective, replace the voltage regulator/brush assembly. If the brushes are good, replace the alternator.

5　If the warning light is illuminated with the engine running, switch the engine off and disconnect the blue wire from the alternator (do not disconnect the large output wire). Switch the ignition on and check the warning light. If the warning light is still on, the blue wire from the alternator to the instrument cluster is grounded or the instrument cluster is defective. If the warning light went out, switch the ignition off, reconnect the blue wire and proceed with the charging system check.

6　Connect a voltmeter to the positive and negative battery terminals (see Illustration). Check the battery voltage with the engine off. It should be approximately 12.4 to 12.6 volts if the battery is fully charged.

7　Start the engine and check the battery voltage again. It should now be greater than the voltage recorded in Step 2, but not more than 14.5 volts. Turn On all the vehicle accessories (air conditioning, rear window defogger, blower motor, etc.) and increase the engine speed to 2000 rpm - the voltage should not drop below the voltage recorded in Step 2.

8　If the indicated voltage is greater than the specified charging voltage, replace the voltage regulator (see Section 11).

9　If the indicated voltage reading is less than the specified charging voltage, the alternator is probably defective. Have the charging system checked at a dealer service department or other properly equipped repair facility.

➡**Note: Many auto parts stores will bench test an alternator off the vehicle. Refer to your local auto parts store regarding their policy, many will perform this service free of charge.**

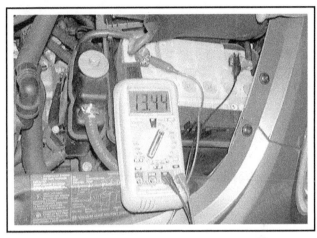

**10.6  To measure battery voltage, attach the voltmeter leads to the remote battery terminals (engine OFF) - to measure charging voltage, start the engine**

## 11  Alternator - removal and installation

### ✳✳ CAUTION 1:

**These models are equipped with an anti-theft radio. Before performing a procedure that requires disconnecting the battery, make sure you have the activation code.**

### ✳✳ CAUTION 2:

**Disconnecting the battery can cause driveability problems that require a scan tool to rectify. Additionally, disconnecting the battery may cause one or more warning lights on the instrument panel to illuminate, which will also require the use of a scan tool to turn off. Most scan tools available to the public do not have the capability to perform either of these tasks, which will necessitate taking the vehicle to a dealer service department or other properly equipped repair facility after service work has been performed. See Section 1 for the use of an auxiliary volt-**age input device ("memory saver") before disconnecting the battery and for other precautions related to battery disconnection.

1　Disconnect the negative battery cable from the battery.
2　Remove the engine cover from above and the splash shield from below the front of the engine.

### FOUR-CYLINDER MODELS

▸ **Refer to illustrations 11.5 and 11.6**

3　Remove the air intake duct from the intercooler to the throttle body and remove the alternator drivebelt (see Chapter 1).
4　Remove the engine cooling fan (see Chapter 3).

**11.5 Disconnect the alternator electrical connections (arrows) (1.8L engine)**

**11.6 Remove the alternator mounting bolts - upper bolt shown (arrow) (1.8L engine)**

5   Disconnect the wire terminals from the alternator (see illustration).

6   Remove the mounting bolts and remove the alternator from the engine (see illustration).

## V6 MODELS

▶ **Refer to illustrations 11.11 and 11.12**

7   Position the front end/radiator support in the service position (see Chapter 11).

8   Remove the alternator drivebelt (see Chapter 1).

9   Detach the refrigerant line clamp from below the alternator.

10   Detach the starter/alternator wiring harness from the bracket.

11   Remove the alternator cooling duct and disconnect the wire terminals from the alternator (see illustration).

12   Remove the mounting bolts and remove the alternator from the engine (see illustration).

## ALL MODELS

13   If you are replacing the alternator, take the old one with you when purchasing a replacement unit. Make sure the new/rebuilt unit looks identical to the old alternator. Look at the terminals - they should be the same in number, size and location as the terminals on the old alternator. Finally, look at the identification numbers - they will be stamped into the housing or printed on a tag attached to the housing. Make sure the numbers are the same on both alternators.

14   Many new/rebuilt alternators do not have a pulley installed, so you may have to switch the pulley from the old unit to the new/rebuilt one. When buying an alternator, find out the shop's policy regarding pulleys; some shops will perform this service free of charge.

15   Installation is the reverse of removal. Tighten the mounting bolts to the torque listed in this Chapter's Specifications.

16   Install the drivebelt (see Chapter 1).

17   Check the charging voltage to verify proper operation of the alternator (see Section 10).

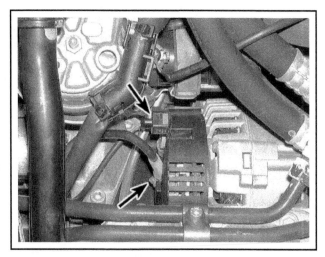

**11.11 Location of the alternator electrical connections (arrows) on the V6 engine**

**11.12 Alternator mounting bolts on the 2.8L V6 engine**

## 12  Voltage regulator and brushes - replacement

♦ **Refer to illustrations 12.3a, 12.3b, 12.3c, 12.4a, 12.4b, 12.5 and 12.6**

1  Remove the alternator (see Section 11).

2  Place the alternator on a clean work surface, with the pulley facing down.

3  Remove the retaining screws, pry open the clips and remove the plastic cover from the rear of the alternator (see illustrations).

4  Remove the voltage regulator/brush assembly screws and remove the unit from the alternator (see illustrations).

5  Measure the brush contact free length and compare your measurement with the value listed in this Chapter's Specifications (see illustration). Replace the unit if the brushes are worn below the minimum value.

6  Inspect the brush contact surface of the slip rings (see illustration). Minor imperfections may be cleaned with crocus cloth. If they are excessively worn, burnt or pitted, replace the alternator.

7  Installation is the reverse of removal.

8  Install the alternator and check the charging voltage as described in Section 10.

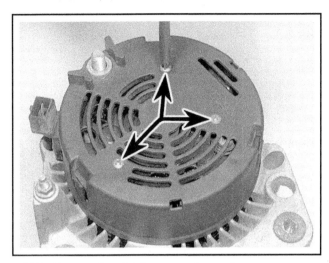

**12.3a  Remove the retaining screws (arrows) . . .**

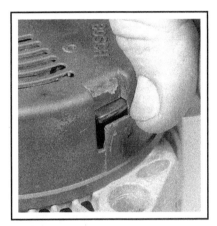

**12.3b . . . then pry open the clips . . .**

**12.3c . . . and remove the plastic cover from the rear of the alternator**

**12.4a  Unscrew the voltage regulator/brush assembly screws . . .**

**12.4b . . . and remove the assembly from the alternator**

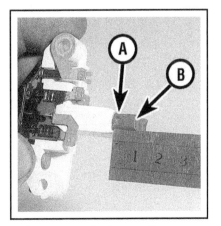

**12.5  Measure the free length of the brush contacts - take the measurement from the manufacturers emblem (A) etched in the side of the brush contact, to the shallowest part of the curved end face of the brush (B)**

**12.6  Inspect the surfaces of the slip rings (arrows), at the end of the alternator shaft**

## 13  Starting system - general information and precautions

The starting system consists of the battery, starter circuit, starter motor assembly and the wiring connecting the components.

The starter motor assembly is bolted to the front of the transaxle bellhousing.

When the ignition key is turned to the START position, the starter solenoid is actuated through the starter control circuit which includes a starter relay located in the relay panel. The starter solenoid then connects the battery to the starter motor. The battery supplies the electrical energy to the starter motor, which does the actual work of cranking the engine.

Always observe the following precautions when working on the starting system:

a) *Excessive cranking of the starter motor can overheat it and cause serious damage. Never operate the starter motor for more than 15 seconds at a time without pausing to allow it to cool for at least two minutes.*

b) *The starter is connected directly to the battery and could arc or cause a fire if mishandled, overloaded or shorted.*

c) *Always detach the cable from the negative terminal of the battery before working on the starting system (see Section 1).*

## 14  Starter motor and circuit - check

▶ **Refer to illustration 14.4**

1    If a malfunction occurs in the starting circuit, do not immediately assume that the starter is causing the problem. First, check the following items:

a) *Make sure the battery cable clamps, where they connect to the battery, are clean and tight.*

b) *Check the condition of the battery cables (see Section 4). Replace any defective battery cables with new parts.*

c) *Test the condition of the battery (see Section 3). If it does not pass all the tests, replace it with a new battery.*

d) *Check the starter motor wiring and connections.*

e) *Check the starter motor mounting bolts for tightness.*

f) *Check the related fuses in the fuse box (see Chapter 12). If they're blown, determine the cause and repair the circuit.*

g) *Check the ignition switch circuit for correct operation (see Chapter 12).*

h) *Check the starter relay (see Chapter 12).*

i) *Check the multi-function transmission range switch.*

2    If the starter does not activate when the ignition switch is turned to the start position, check for battery voltage to the starter solenoid. This will determine if the solenoid is receiving the correct voltage from the ignition switch. Install a 12-volt test light or a voltmeter to the starter solenoid terminal. While an assistant turns the ignition switch to the start position, observe the test light or voltmeter. The test light should shine brightly or battery voltage should be indicated on the voltmeter. If voltage is not available to the starter solenoid, refer to the wiring diagrams in Chapter 12 and check the fuses, switches and starter relay in series with the starting system. If voltage is available but there is no movement from the starter motor, remove the starter from the engine (see Section 15) and bench test the starter (see Step 4).

3    If the starter turns over slowly, check the starter cranking voltage and the current draw from the battery. This test must be performed with the starter assembly on the engine. Crank the engine over (for 10 seconds or less) and observe the battery voltage. It should not drop below 8.5 volts. Also, observe the current draw using an ammeter. Typically a starter amperage draw should not exceed 300 amps. If the starter motor

amperage draw is excessive, have it tested by a dealer service department or other qualified repair shop. There are several conditions that may affect the starter cranking potential. The battery must be in good condition and the battery cold-cranking rating must not be underrated for the particular application. Be sure to check the battery specifications carefully. The battery terminals and cables must be clean and not corroded. Also, in cases of extreme cold temperatures, make sure the battery and/or engine block is warmed before performing the tests.

4    If the starter is receiving voltage but does not activate, remove and check the starter motor assembly on the bench. Most likely the starter motor or solenoid is defective. In some rare cases, the engine may be seized so be sure to try and rotate the crankshaft pulley (see Chapter 2A or 2B) before proceeding. With the starter assembly mounted in a vise on the bench, install one jumper cable from the posi-

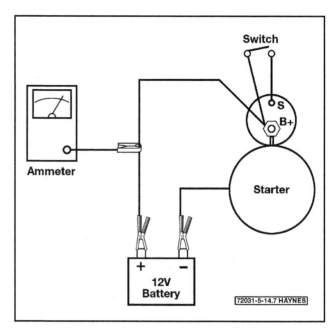

**14.4  Starter motor bench testing details**

tive terminal of a test battery to the B+ terminal on the starter. Install another jumper cable from the negative terminal of the battery to the body of the starter (see illustration). Install a starter switch and apply battery voltage to the solenoid S terminal (for 10 seconds or less) and observe the solenoid plunger, shift lever and overrunning clutch extend and rotate the pinion drive. If the pinion drive extends but does not rotate, the solenoid is operating but the starter motor is defective. If there is no movement but the solenoid clicks, the solenoid and/or the starter motor is defective. If the solenoid plunger extends and rotates the pinion drive, the starter assembly is operating properly.

## 15  Starter motor - removal and installation

▶ Refer to illustrations 15.7 and 15.8

### ✳✳ CAUTION 1:

**These models are equipped with an anti-theft radio. Before performing a procedure that requires disconnecting the battery, make sure you have the activation code. See Section 1 for the use of an auxiliary voltage input device before disconnecting the battery.**

### ✳✳ CAUTION 2:

**Disconnecting the battery can cause driveability problems that require a scan tool to rectify. Additionally, disconnecting the battery may cause one or more warning lights on the instrument panel to illuminate, which will also require the use of a scan tool to turn off. Most scan tools available to the public do not have the capability to perform either of these tasks, which will necessitate taking the vehicle to a dealer service department or other properly equipped repair facility after service work has been performed. See Section 1 for the use of an auxiliary voltage input device ("memory saver") before disconnecting the battery and for other precautions related to battery disconnection.**

1   Disconnect the negative battery cable from the battery.
2   Raise the vehicle and support it securely on jackstands.
3   Remove the splash shield from under the engine.
4   On four-cylinder models, remove the air conditioning compressor from the mounting bracket and position it aside without disconnecting the refrigerant lines (see Chapter 3).
5   On V6 models, remove the alternator (see Section 11).
6   Remove the starter heat shield (if equipped).
7   Disconnect the wires from the terminals on the starter motor solenoid (see illustration).
8   Remove the starter mounting bolts from the transaxle bellhousing and remove the starter (see illustration).
9   Installation is the reverse of removal.

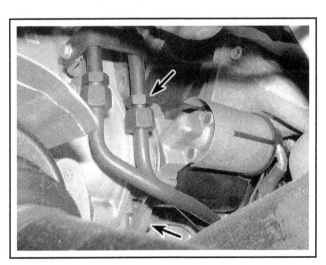

**15.7  Remove the nut (A) and disconnect the battery cable from the starter motor - disconnect the connector (B) from the starter solenoid**

**15.8  Remove the starter mounting bolts (arrows) (DOHC V6 engine shown)**

## Specifications

### General

| | |
|---|---|
| Alternator brush wear limit | 0.19 inch (5.0 mm) |
| Battery voltage | |
|     Engine off | 12.0 to 12.5 volts |
|     Engine running | 13.5 to 14.5 volts |
| Ignition coil secondary resistance (V6 engines) | 8,000 to 14,000 ohms |
| Spark plug wire resistance (V6 engines) | 4,000 to 6,000 ohms |
| Spark plug boot resistance (four-cylinder engine) | 2,000 ohms |

### Torque specifications

| | Ft-lbs | Nm |
|---|---|---|
| Alternator mounting bolts | | |
|     Four-cylinder engine | 22 | 30 |
|     V6 engines | 33 | 45 |
| Battery hold-down clamp bolt | 16 | 22 |
| Starter mounting bolts | 48 | 65 |

## Section

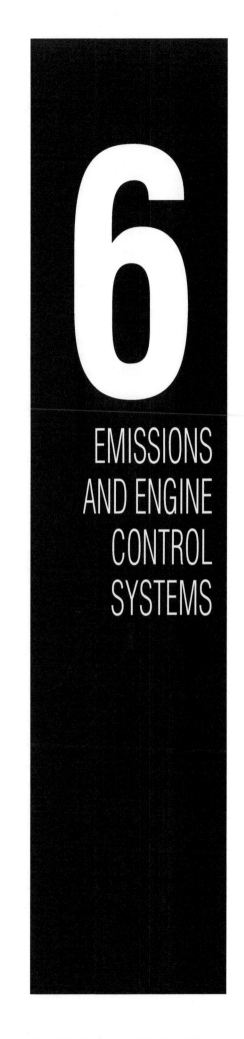

# 6

# EMISSIONS AND ENGINE CONTROL SYSTEMS

## 1   General information

♦ **Refer to illustrations 1.1a, 1.1b, 1.7a and 1.7b**

To prevent pollution of the atmosphere from incompletely burned and evaporating gases, and to maintain good driveability and fuel economy, a number of emission control systems are incorporated (see illustrations). They include the:

Electronic engine control system
Crankcase ventilation system
Evaporative emissions control system
Secondary air injection system
Exhaust gas recirculation system
Catalytic converter

All of these systems are linked, directly or indirectly, to the emission control system.

The Sections in this Chapter include general descriptions, checking procedures within the scope of the home mechanic (when possible) and component replacement procedures for each of the systems listed above.

Before assuming that an emissions control system is malfunctioning, check the fuel and ignition systems carefully. The diagnosis of some emission control devices requires specialized tools, equipment and training. If checking and servicing become too difficult or if a procedure is beyond your ability, consult a dealer service department

or other properly equipped repair facility. Remember, the most frequent cause of emissions problems is simply a loose or broken vacuum hose or wire, so always check the hose and wiring connections first.

A scan tool is required for complete diagnosis of the system control circuits. If a scan tool is not available, have the system checked at a dealer service department or other properly equipped repair facility.

➡**Note: Because of a Federally mandated warranty which covers the emission control system components, check with your dealer about warranty coverage before working on any emissions-related systems. Once the warranty has expired, you may wish to perform some of the replacement procedures in this Chapter to save money.**

Pay close attention to any special precautions outlined in this Chapter. It should be noted that the illustrations of the various systems may not exactly match the system installed on the vehicle you're working on because of changes made by the manufacturer during production or from year-to-year.

A Vehicle Emissions Control Information (VECI) label is located in the engine compartment (see illustration). This label contains important emissions specifications and other information. When servicing the engine or emissions systems, the VECI label in your particular vehicle should always be checked for up-to-date information. A vacuum hose schematic is located adjacent to the VECI label (see illustration).

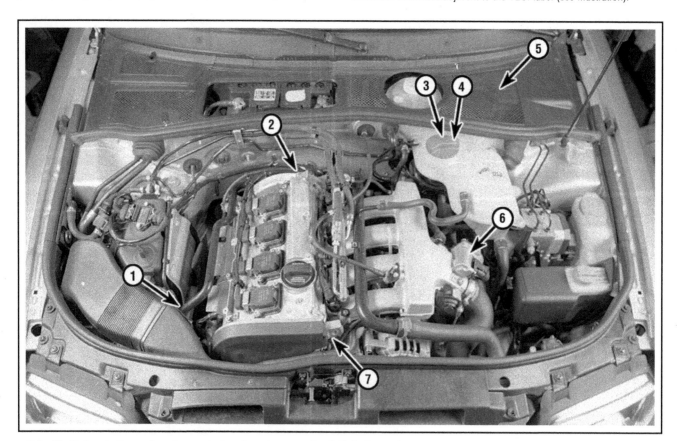

**1.1a  Typical emission and engine control system components - 1.8L turbo engine**

| | | | |
|---|---|---|---|
| 1 | Mass airflow sensor | 3 | Oxygen sensor connectors (below coolant reservoir) |
| 2 | Engine coolant temperature sensor connector | 4 | Engine speed sensor connector (below coolant reservoir) |
| | | 5 | Engine control module (ECM) |
| | | 6 | Throttle control module |
| | | 7 | Camshaft position sensor (behind camshaft sprocket) |

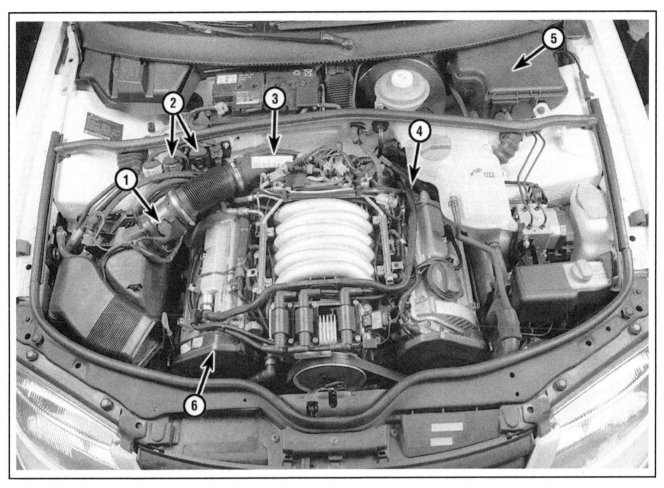

**1.1b  Typical emission and engine control system components - 2.8L DOHC engine**

1   Mass airflow sensor
2   Oxygen sensor connectors
3   Engine coolant temperature sensor

4   Camshaft sensor - left bank (behind camshaft sprocket)
5   Engine control module (ECM)

6   Camshaft sensor - right bank (behind camshaft sprocket)

**1.7a  The Vehicle Emission Control Information (VECI) label is located in the engine compartment and contains information on the emission devices on your vehicle**

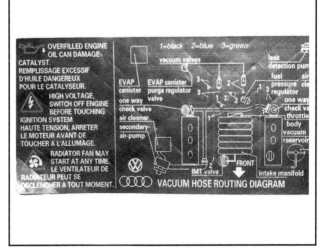

**1.7b  Vacuum hose routing diagram of the emissions system vacuum hoses**

## 2   On-Board Diagnostic (OBD) system and trouble codes

## DIAGNOSTIC TOOL INFORMATION

▶ **Refer to illustration 2.1**

1    A digital multimeter is necessary for checking fuel injection and emission related components (see illustration). A digital volt-ohmmeter is preferred over the older style analog multimeter for several reasons. The analog multimeter cannot display the volts-ohms or amps measurement in hundredths and thousandths increments. When working with electronic circuits which are often very low voltage, this accurate reading is most important. Another good reason for the digital multimeter is the high impedance circuit. The digital multimeter is equipped with a high resistance internal circuitry (10 million ohms). Because a voltmeter is hooked up in parallel with the circuit when testing, it is vital that none of the voltage being measured should be allowed to travel the parallel path set up by the meter itself. This dilemma does not show itself when measuring larger amounts of voltage (9 to 12 volt circuits) but if you are measuring a low voltage circuit such as the oxygen sensor signal voltage, a fraction of a volt may be a significant amount when diagnosing a problem. However, there are several exceptions where using an analog voltmeter may be necessary to test certain sensors.

2    Hand-held scanners are the most powerful and versatile tools for analyzing engine management systems used on later model vehicles. Several tool manufacturers have released OBD-II scan tools for the home mechanic. Unfortunately, at the time of writing, only a couple of tools on the market are compatible with the OBD system on these vehicles. They are very expensive and generally only available to dealer service departments and professional mechanics.

## ON-BOARD DIAGNOSTIC SYSTEM GENERAL DESCRIPTION

3    All models described in this manual are equipped with the second generation On-Board Diagnostic (OBD-II) system. The system

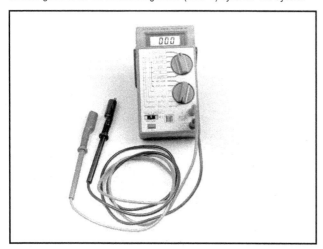

**2.1  Digital multimeters can be used for testing all types of circuits; because of their high impedance, they are much more accurate than analog meters for measuring low-voltage computer circuits**

consists of an on-board computer, known as the Engine Control Module (ECM), information sensors and output actuators.

4    The information sensors monitor various functions of the engine and send data to the ECM. Based on the data and the information programmed into the computer's memory, the ECM generates output signals to control various engine functions via control relays, solenoids and other output actuators. The ECM is specifically calibrated to optimize the emissions, fuel economy and driveability of the vehicle.

5    Because of a Federally mandated warranty which covers the emissions system components and because any owner-induced damage to the ECM, the sensors and/or the control devices may void the warranty, it isn't a good idea to attempt diagnosis or replacement of the ECM at home while the vehicle is under warranty. Take the vehicle to a dealer service department if the ECM or a system component malfunctions.

## INFORMATION SENSORS

6    **Camshaft position sensor** - The camshaft position sensor provides information on camshaft position. The ECM uses this information, along with the engine speed sensor information, to control the ignition timing and fuel injection synchronization.

7    **Engine speed sensor** - The engine speed sensor senses crankshaft position (TDC) during each engine revolution. The ECM uses this information to control the ignition system.

8    **Engine coolant temperature sensor** - The engine coolant temperature sensor senses engine coolant temperature. The ECM uses this information to control fuel injection duration and ignition timing.

9    **Intake air temperature sensor** - The intake air temperature senses the temperature of the air entering the intake manifold. The ECM uses this information to control fuel injection duration.

10   **Knock sensor** - The knock sensor is a piezoelectric element that detects the sound of engine detonation, or "pinging." The ECM uses the input signal from the knock sensor to recognize detonation and retard spark advance to avoid engine damage.

11   **Mass airflow sensor** - The mass airflow sensor measures the amount of air passing through the sensor body and ultimately entering the engine. The ECM uses this information to control fuel delivery.

12   **Oxygen sensor** - The oxygen sensors generate a voltage signal that varies with the difference between the oxygen content of the exhaust and the oxygen in the surrounding air. The ECM uses this information to determine if the fuel system is running rich or lean.

13   **Throttle position sensor** - The throttle position sensor senses throttle movement and position. This signal enables the ECM to determine when the throttle is closed, in a cruise position, or wide open. The ECM uses this information to control fuel delivery and ignition timing. The throttle position sensor is a component of the throttle control module.

14   **Closed throttle position switch** - The closed throttle position switch signals the ECM when the throttle is in the fully closed position. The ECM uses this information to control the engine idle speed. The closed throttle position switch is a component of the throttle control module.

15   **Vehicle speed sensor** - The vehicle speed sensor provides information to the ECM to indicate vehicle speed.

16   **Miscellaneous ECM inputs** - In addition to the various sen-

sors, the ECM monitors various switches and circuits to determine vehicle operating conditions. The switches and circuits include:

a) *Accelerator pedal position (E-Gas system)*
b) *Air conditioning system*
c) *Antilock brake system*
d) *Barometric pressure sensor (inside ECM)*
e) *Battery voltage*
f) *Brake switch*
g) *Cruise control system*
h) *EVAP system*
i) *Transmission Range switch*
j) *Power steering pressure switch*
k) *Turbocharger boost pressure (1.8L turbo engine)*
l) *Sensor signal and ground circuits*
m) *Transaxle control module*

## OUTPUT ACTUATORS

17 **Check Engine light** - The ECM will illuminate the Check Engine light if a malfunction in the electronic engine control system occurs.

18 **EVAP canister purge valve solenoid** - The evaporative emission canister purge valve solenoid is operated by the ECM to purge the fuel vapor canister and route fuel vapor to the intake manifold for combustion.

19 **EVAP system leak detection pump** - The EVAP system is equipped with a self-diagnostic leak detection system. The ECM controls the operation of the leak detection pump.

20 **Secondary air injection pump and vacuum valve/solenoid** - The ECM operates the secondary air injection pump and opens the vacuum valve to inject fresh air into the exhaust stream, lowering emission levels under certain operating conditions.

21 **Fuel injectors** - The ECM opens the fuel injectors individually in firing order sequence. The ECM also controls the time the injector is held open (pulse width). The pulse width of the injector (measured in milliseconds) determines the amount of fuel delivered. For more information on the fuel delivery system and the fuel injectors, including injector replacement, refer to Chapter 4.

22 **Fuel pump relay** - The fuel pump relay is activated by the ECM with the ignition switch in the Start or Run position. When the ignition switch is turned on, the relay is activated to supply initial line pressure to the system. For more information on fuel pump check and replacement, refer to Chapter 4.

23 **Throttle valve actuator (except E-Gas system)** - The throttle valve actuator (mechanical) controls the throttle plate. The more the throttle plate is opened, the higher the idle speed. The throttle plate opening and the resulting idle speed is controlled by accelerator cable.

24 **Throttle valve actuator (E-Gas system)** - The throttle valve actuator (electronic) controls the throttle plate at all engine speeds. The more the throttle plate is opened, the higher the engine speed. The throttle plate opening and the resulting engine speed is controlled by the ECM. The throttle valve actuator is a component of the throttle control module.

25 **Ignition coils** - The ECM controls spark delivery and ignition timing depending on engine operation conditions. Refer to Chapter 5 for more information on the ignition system.

26 **Oxygen sensor heaters** - Each oxygen sensor is equipped with a heating element. Heating the oxygen sensor allows it to reach operating temperature quickly. The ECM controls the oxygen sensor heaters.

27 **Turbocharger boost control (1.8L turbo engine)** - The ECM monitors intake manifold pressure and controls the turbocharger wastegate with the boost pressure control valve. The engine control system calculates the engine torque needed depending on driver demand and engine operating conditions, the ECM will then adjust the boost pressure to meet the demands.

## OBTAINING DIAGNOSTIC TROUBLE CODES

▶ **Refer to illustration 2.29**

**✳✳ CAUTION:**

**Disconnecting the battery can cause driveability problems that require a scan tool to rectify. Additionally, disconnecting the battery may cause one or more warning lights on the instrument panel to illuminate, which will also require the use of a scan tool to turn off. Most scan tools available to the public do not have the capability to perform either of these tasks, which will necessitate taking the vehicle to a dealer service department or other properly equipped repair facility after service work has been performed. See Chapter 5, Section 1 for the use of an auxiliary voltage input device (memory saver) before disconnecting the battery and for other precautions related to battery disconnection.**

➡**Note: The diagnostic trouble codes on all models can only be extracted from the Engine Control Module (ECM) using a specialized scan tool. Have the vehicle diagnosed by a dealer service department or other qualified automotive repair facility if the proper scan tool is not available.**

28 The ECM will illuminate the CHECK ENGINE light (also known as the Malfunction Indicator Lamp) on the dash if it recognizes a fault in the system. The light will remain illuminated until the problem is repaired and the code is cleared or the ECM does not detect any malfunction for several consecutive drive cycles.

29 The diagnostic codes for the On-Board Diagnostic (OBD) system can only be extracted from the ECM using a scan tool. The scan tool is programmed to interface with the OBD system by plugging into the diagnostic connector (see illustration). When used, the scan tool has the ability to diagnose in-depth driveability problems and it allows

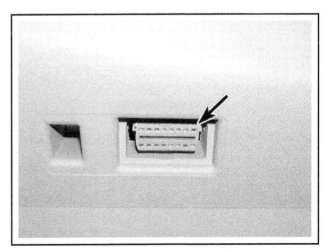

**2.29 The diagnostic connector is typically located under the instrument panel**

freeze frame data to be retrieved from the ECM stored memory. Freeze frame data is an OBD II ECM feature that records all related sensor and actuator activity on the ECM data stream whenever an engine control or emissions fault is detected and a trouble code is set. This ability to look at the circuit conditions and values when the malfunction occurs provides a valuable tool when trying to diagnose intermittent driveability problems. If the tool is not available and intermittent driveability problems exist, have the vehicle checked at a dealer service department or other qualified repair shop.

➡**Note: The diagnostic connector on 1996 models with 2.8L SOHC engines is located next to the rear ashtray receptacle.**

## CLEARING DIAGNOSTIC TROUBLE CODES

30 After the system has been repaired, the codes must be cleared from the ECM memory using a scan tool. Do not attempt to clear the codes by disconnecting battery power. If battery power is disconnected from the ECM, the ECM will lose the current engine operating parameters and driveability will suffer until the ECM is programmed with a scan tool.

31 Always clear the codes from the ECM before starting the engine after a new electronic emission control component is installed onto the engine. The ECM stores the operating parameters of each sensor. The ECM may set a trouble code if a new sensor is allowed to operate before the parameters from the old sensor have been erased.

## DIAGNOSTIC TROUBLE CODE IDENTIFICATION

32 The accompanying list of diagnostic trouble codes is a compilation of most of the codes that may be encountered. Additional trouble codes are available with the use of the manufacturer specific scan tool. Not all codes pertain to all models and not all codes will illuminate the Check Engine light when set. All models require a scan tool to access the diagnostic trouble codes.

## TROUBLE CODES

| Code | Code Identification |
| --- | --- |
| P0102 | Mass air flow sensor circuit, low input |
| P0103 | Mass air flow sensor circuit, high input |
| P0112 | Intake air temperature circuit, low input |
| P0113 | Intake air temperature circuit, high input |
| P0116 | Engine coolant temperature circuit, range or performance problem |
| P0117 | Engine coolant temperature circuit, low input |
| P0118 | Engine coolant temperature circuit, high input |
| P0120 | Throttle position sensor circuit, malfunction |
| P0121 | Throttle position sensor circuit, range or performance problem |
| P0122 | Throttle position sensor circuit, low input |
| P0123 | Throttle position sensor circuit, high input |
| P0130 | Oxygen sensor circuit malfunction (pre-converter sensor, bank 1) |
| P0131 | Oxygen sensor circuit, low voltage (pre-converter sensor, bank 1) |
| P0132 | Oxygen sensor circuit, high voltage (pre-converter sensor, bank 1) |
| P0133 | Oxygen sensor circuit, slow response (pre-converter sensor, bank 1) |
| P0134 | Oxygen sensor circuit - no activity detected (pre-converter sensor, bank 1) |

| Code | Code Identification |
| --- | --- |
| P0135 | Oxygen sensor heater circuit malfunction (pre-converter sensor, bank 1) |
| P0136 | Oxygen sensor circuit malfunction (post-converter sensor, bank 1) |
| P0137 | Oxygen sensor circuit, low voltage (post-converter sensor, bank 1) |
| P0138 | Oxygen sensor circuit, high voltage (post-converter sensor, bank 1) |
| P0140 | Oxygen sensor circuit - no activity detected (post-converter sensor, bank 1) |
| P0141 | Oxygen sensor heater circuit malfunction (post-converter sensor, bank 1) |
| P0150 | Oxygen sensor circuit malfunction (pre-converter sensor, bank 2) |
| P0151 | Oxygen sensor circuit, low voltage (pre-converter sensor, bank 2) |
| P0152 | Oxygen sensor circuit, high voltage (pre-converter sensor, bank 2) |
| P0153 | Oxygen sensor circuit, slow response (pre-converter sensor, bank 2) |
| P0154 | Oxygen sensor circuit - no activity detected (pre-converter sensor, bank 2) |
| P0155 | Oxygen sensor heater circuit malfunction (pre-converter sensor, bank 2) |
| P0156 | Oxygen sensor circuit malfunction (post-converter sensor, bank 2) |
| P0157 | Oxygen sensor circuit, low voltage (post-converter sensor, bank 2) |
| P0158 | Oxygen sensor circuit, high voltage (post-converter sensor, bank 2) |
| P0160 | Oxygen sensor circuit - no activity detected (post-converter sensor, bank 2) |
| P0161 | Oxygen sensor heater circuit malfunction (post-converter sensor, bank 2) |
| P0170 | Fuel trim malfunction (bank 1) |
| P0171 | System too lean (bank 1) |
| P0172 | System too rich (bank 1) |
| P0173 | Fuel trim malfunction (bank 2) |
| P0174 | System too lean (bank 2) |
| P0175 | System too rich (bank 2) |
| P0300 | Random/multiple cylinder misfire detected |
| P0301 | Cylinder no. 1 misfire detected |
| P0302 | Cylinder no. 2 misfire detected |
| P0303 | Cylinder no. 3 misfire detected |

## TROUBLE CODES (CONTINUED)

| Code | Code Identification |
| --- | --- |
| P0304 | Cylinder no. 4 misfire detected |
| P0305 | Cylinder no. 5 misfire detected |
| P0306 | Cylinder no. 6 misfire detected |
| P0321 | Engine speed sensor, range or performance problem |
| P0322 | Engine speed sensor, no input |
| P0327 | Knock sensor no. 1 circuit, low input |
| P0328 | Knock sensor no. 1 circuit, high input |
| P0332 | Knock sensor no. 2 circuit, low input |
| P0333 | Knock sensor no.2 circuit, high input |
| P0337 | Crankshaft position sensor, low input |
| P0401 | Exhaust gas recirculation flow insufficient |
| P0402 | Exhaust gas recirculation flow excessive |
| P0411 | Secondary air injection system, incorrect flow |
| P0422 | Catalyst system efficiency below threshold (bank 1) |
| P0432 | Catalyst system efficiency below threshold (bank 2) |
| P0440 | Evaporative emission control system malfunction |
| P0441 | Evaporative emission control system, incorrect purge flow |
| P0442 | Evaporative emission control system, small leak detected |
| P0452 | Evaporative emission control system pressure sensor, low input |
| P0453 | Evaporative emission control system pressure sensor, high input |
| P0455 | Evaporative emission control system, large leak detected |
| P0461 | Fuel level sensor, range or performance problem |
| P0501 | Vehicle speed sensor circuit, range or performance problem |
| P0506 | Idle control system, rpm lower than expected |
| P0507 | Idle control system, rpm higher than expected |
| P0560 | System voltage malfunction |

| Code | Code Identification |
|------|---------------------|
| P0562 | System voltage low |
| P0563 | System voltage high |
| P0601 | Engine Control Module, programming error |
| P0604 | Engine Control Module, memory error (RAM) |
| P0605 | Engine Control Module, memory error (ROM) |
| P0706 | Transmission range sensor circuit range or performance problem |
| P0707 | Transmission range sensor circuit, low input |
| P0708 | Transmission range sensor circuit, high input |
| P1101 | Oxygen sensor low voltage (pre-converter sensor, bank 1) |
| P1102 | Oxygen sensor heater circuit, short to B+ (pre-converter sensor, bank 1) |
| P1104 | Oxygen sensor low voltage (post-converter sensor, bank 1) |
| P1105 | Oxygen sensor heater circuit, short to B+ (post-converter sensor, bank 1) |
| P1106 | Oxygen sensor low voltage (pre-converter sensor, bank 2) |
| P1107 | Oxygen sensor heater circuit, short to B+ (pre-converter sensor, bank 2) |
| P1109 | Oxygen sensor low voltage (post-converter sensor, bank 2) |
| P1110 | Oxygen sensor heater circuit, short to B+ (post-converter sensor, bank 2) |
| P1127 | Long term fuel trim too rich (bank 1) |
| P1128 | Long term fuel trim too lean (bank 1) |
| P1129 | Long term fuel trim too rich (bank 2) |
| P1130 | Long term fuel trim too lean (bank 2) |
| P1136 | Fuel trim too rich at idle (bank 1) |
| P1137 | Fuel trim too lean at idle (bank 1) |
| P1138 | Fuel trim too rich at idle (bank 2) |
| P1139 | Fuel trim too lean at idle (bank 2) |
| P1141 | Load calculation cross check range or performance problem |
| P1176 | Oxygen sensor, correction limit attained (bank 1) |
| P1177 | Oxygen sensor, correction limit attained (bank 2) |

## TROUBLE CODES (CONTINUED)

| Code | Code Identification |
|---|---|
| P1196 | Oxygen sensor heater electrical malfunction (pre-converter sensor, bank 1) |
| P1197 | Oxygen sensor heater electrical malfunction (pre-converter sensor, bank 2) |
| P1198 | Oxygen sensor heater electrical malfunction (post-converter sensor, bank 1) |
| P1199 | Oxygen sensor heater electrical malfunction (post-converter sensor, bank 2) |
| P1201 | Fuel injector circuit electrical malfunction (cylinder no. 1) |
| P1202 | Fuel injector circuit electrical malfunction (cylinder no. 2) |
| P1203 | Fuel injector circuit electrical malfunction (cylinder no. 3) |
| P1204 | Fuel injector circuit electrical malfunction (cylinder no. 4) |
| P1205 | Fuel injector circuit electrical malfunction (cylinder no. 5) |
| P1206 | Fuel injector circuit electrical malfunction (cylinder no. 6) |
| P1213 | Fuel injector circuit, short to B+ (cylinder no. 1) |
| P1214 | Fuel injector circuit, short to B+ (cylinder no. 2) |
| P1215 | Fuel injector circuit, short to B+ (cylinder no. 3) |
| P1216 | Fuel injector circuit, short to B+ (cylinder no. 4) |
| P1217 | Fuel injector circuit, short to B+ (cylinder no. 5) |
| P1218 | Fuel injector circuit, short to B+ (cylinder no. 6) |
| P1225 | Fuel injector circuit, short to ground (cylinder no. 1) |
| P1226 | Fuel injector circuit, short to ground (cylinder no. 2) |
| P1227 | Fuel injector circuit, short to ground (cylinder no. 3) |
| P1228 | Fuel injector circuit, short to ground (cylinder no. 4) |
| P1229 | Fuel injector circuit, short to ground (cylinder no. 5) |
| P1230 | Fuel injector circuit, short to ground (cylinder no. 6) |
| P1237 | Fuel injector circuit, open (cylinder no. 1) |
| P1238 | Fuel injector circuit, open (cylinder no. 2) |
| P1239 | Fuel injector circuit, open (cylinder no. 3) |

| Code | Code Identification |
|------|---------------------|
| P1240 | Fuel injector circuit, open (cylinder no. 4) |
| P1241 | Fuel injector circuit, open (cylinder no. 5) |
| P1242 | Fuel injector circuit, open (cylinder no. 6) |
| P1250 | Fuel level low |
| P1325 | Knock sensor limit attained (cylinder no. 1) |
| P1326 | Knock sensor limit attained (cylinder no. 2) |
| P1327 | Knock sensor limit attained (cylinder no. 3) |
| P1328 | Knock sensor limit attained (cylinder no. 4) |
| P1329 | Knock sensor limit attained (cylinder no. 5) |
| P1330 | Knock sensor limit attained (cylinder no. 6) |
| P1337 | Camshaft position sensor, short to ground (bank 1) |
| P1338 | Camshaft position sensor, open circuit or short to B+ (bank 1) |
| P1339 | Camshaft position sensor and engine speed sensor cross connected |
| P1340 | Camshaft position sensor out of sequence |
| P1341 | Ignition coil power output stage 1, short to ground |
| P1343 | Ignition coil power output stage 2, short to ground |
| P1345 | Ignition coil power output stage 3, short to ground |
| P1386 | Knock sensor circuit error (internal Engine Control Module) |
| P1391 | Camshaft position sensor, short to ground (bank 2) |
| P1392 | Camshaft position sensor, open circuit or short to B+ (bank 2) |
| P1393 | Ignition coil power output stage 1, electrical malfunction |
| P1394 | Ignition coil power output stage 2, electrical malfunction |
| P1395 | Ignition coil power output stage 3, electrical malfunction |
| P1396 | Engine speed sensor gear (starter ring gear) missing teeth |
| P1400 | Exhaust gas recirculation valve circuit, electrical malfunction |
| P1402 | Exhaust gas recirculation valve circuit, short to B+ |
| P1407 | Exhaust gas recirculation temperature sensor, signal low |
| P1408 | Exhaust gas recirculation temperature sensor, signal high |

## TROUBLE CODES (CONTINUED)

| Code | Code Identification |
| --- | --- |
| P1409 | Fuel tank vent valve electrical malfunction |
| P1410 | Fuel tank vent valve, short to B+ |
| P1421 | Secondary air injection valve circuit, short to ground |
| P1422 | Secondary air injection valve circuit, Short to B+ |
| P1425 | Fuel tank vent valve circuit, short to ground |
| P1426 | Fuel tank vent valve circuit, open |
| P1432 | Secondary air injection valve open |
| P1433 | Secondary air injection system, pump relay circuit open |
| P1434 | Secondary air injection system, pump relay circuit short to B+ |
| P1435 | Secondary air injection system, pump relay circuit short to ground |
| P1436 | Secondary air injection system, pump relay circuit, electrical malfunction |
| P1471 | EVAP system leak detection pump circuit, short to B+ |
| P1472 | EVAP system leak detection pump circuit, short to ground |
| P1473 | EVAP system leak detection pump circuit, open |
| P1476 | EVAP system leak detection pump, low vacuum |
| P1477 | EVAP system leak detection pump system malfunction |
| P1500 | Fuel pump relay circuit malfunction |
| P1501 | Fuel pump relay circuit, short to ground |
| P1502 | Fuel pump relay circuit, short to B+ |
| P1504 | Intake air bypass system, intake air leak |
| P1505 | Closed throttle position switch, open circuit |
| P1506 | Closed throttle position switch, short to ground |
| P1509 | Idle air control circuit malfunction |
| P1510 | Idle air control circuit, short to B+ |
| P1511 | Intake manifold changeover valve malfunction |
| P1512 | Intake manifold changeover valve, short to B+ |
| P1515 | Intake manifold changeover valve, short to ground |

| Code | Code Identification |
|------|---------------------|
| P1516 | Intake manifold changeover valve, open circuit |
| P1519 | Intake camshaft controller, malfunction (bank 1) |
| P1522 | Intake camshaft controller, malfunction (bank 2) |
| P1543 | Throttle actuator sensor, signal low |
| P1544 | Throttle actuator sensor, signal high |
| P1545 | Throttle position controller malfunction |
| P1546 | Turbocharger boost pressure control valve circuit, short to B+ |
| P1547 | Turbocharger boost pressure control valve circuit, short to ground |
| P1548 | Turbocharger boost pressure control valve circuit, open |
| P1555 | Turbocharger boost pressure control valve, upper limit exceeded |
| P1556 | Turbocharger boost pressure control valve, negative deviation |
| P1557 | Turbocharger boost pressure control valve, positive deviation |
| P1558 | Throttle valve actuator, electrical malfunction |
| P1559 | Throttle control module to Engine Control Module adaptation error |
| P1560 | Maximum engine speed exceeded |
| P1564 | Idle speed control system, low system voltage |
| P1565 | Idle speed control system, lower limit not attained |
| P1600 | Engine Control Module, power supply low voltage |
| P1602 | Engine Control Module, power supply low voltage |
| P1606 | Antilock Brake Control Module malfunction |
| P1611 | Transmission control module, short to ground |
| P1612 | Engine Control Module, incorrect programming |
| P1613 | Transmission control module, short to B+ |
| P1624 | Transmission control module, Check Engine light malfunction |
| P1626 | Engine Control Module, no data from Transmission control module |
| P1640 | Engine control module, EPROM error |
| P1681 | Engine Control Module, programming not finished |
| P1690 | Check Engine light circuit malfunction |

## 3   Engine Control Module (ECM) - removal and installation

▶ Refer to illustrations 3.2, 3.3 and 3.4

### ✳✳ CAUTION 1:

Disconnecting the battery can cause driveability problems that require a scan tool to rectify. Additionally, disconnecting the battery may cause one or more warning lights on the instrument panel to illuminate, which will also require the use of a scan tool to turn off. Most scan tools available to the public do not have the capability to perform either of these tasks, which will necessitate taking the vehicle to a dealer service department or other properly equipped repair facility after service work has been performed. See Chapter 5, Section 1 for the use of an auxiliary voltage input device (memory saver) before disconnecting the battery and for other precautions related to battery disconnection.

### ✳✳ CAUTION 2:

Avoid static electricity damage to the Engine Control Module (ECM) by grounding yourself to the body of the vehicle before touching the ECM and using a special anti-static pad to store the ECM on, once it is removed.

➡Note: Anytime the ECM is replaced with a new unit it must be programmed by a dealership service department or other repair facility equipped with a V.A.G. 1551 scan tool. The following procedure pertains to removal and installation of the original ECM only. If the ECM must be replaced with a new unit, take the vehicle to a dealership service department or other repair facility equipped with the proper tool.

1   Disconnect the cable from the negative battery terminal.

### ✳✳ CAUTION:

These models are equipped with an anti-theft radio. Before performing a procedure that requires disconnecting the battery, make sure you have the activation code.

2   The ECM is located in the plenum, at the base of the windshield on the drivers side. Remove the cover from the ECM housing (see illustration).

3   Carefully pry off the ECM retainer clips (see illustration). Remove the ECM from the housing.

4   Disconnect the electrical connectors from the ECM (see illustration).

5   Installation is the reverse of removal.

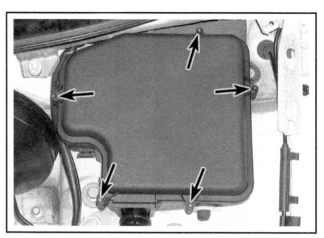

3.2  Remove the screws (arrows) and lift the cover off the ECM housing

3.3  Remove the ECM retaining clip

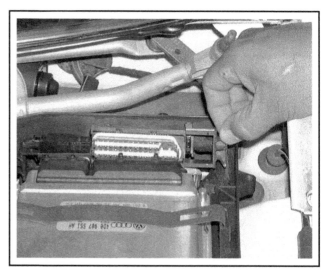

3.4  Disconnect the electrical connectors from the ECM

## 4   Throttle position sensor (SOHC V6 engine) - removal and installation

▶ **Refer to illustration 4.5**

1   The Throttle Position Sensor (TPS) is a variable-resistance poten-tiometer, mounted on the side of the throttle body and connected to the throttle shaft. By monitoring the output voltage from the TPS, the PCM can determine fuel delivery based on throttle valve angle (driver

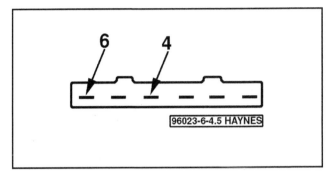

**4.5  The throttle switch should read continuity, then infinity when the throttle valve is slightly rotated from the idle position**

demand). A broken or loose TPS can cause intermittent bursts of fuel from the injector and an unstable idle because the PCM thinks the throttle is moving. A problem with the TPS or circuit will set a diagnostic trouble code. A closed throttle switch is incorporated into the throttle position sensor. The closed throttle switch closes the particular throttle circuit in response to the position of the throttle valve (open or closed).

2   Remove the air intake duct (see Chapter 4) and disconnect the TPS electrical connector.

3   Remove the two retaining screws and separate the TPS from the throttle body.

4   Install the new TPS leaving the mounting screws loose.

5   Adjust the TPS as follows (see illustration):

   a) *Rotate the TPS clockwise in the screw slots, then back counterclockwise until the closed throttle switch stop can just be felt. Tighten the screws.*

   b) *Connect an ohmmeter to terminals 4 and 6 of the TPS. The ohmmeter should indicate continuity (zero ohms)*

   c) *Open the throttle slightly; the ohmmeter should now indicate no continuity (infinity). If it doesn't, readjust the switch.*

6   The remainder of installation is the reverse of removal.

## 5   Throttle control module - removal and installation

▶ **Refer to illustration 5.1**

### ✳✳ CAUTION:

**If the throttle control module is replaced, the ECM must be programmed with a scan tool to accept the new throttle control module.**

1   The throttle control module is connected to the end of the throttle shaft on the throttle body (see illustration). The throttle control module contains four major components of the engine control system; the throttle position sensor, the closed throttle switch, the throttle valve actuator and a feedback sensor for the throttle valve actuator. The ECM uses the throttle position sensor to control fuel delivery based on driver demand. The throttle valve actuator controls the idle speed and cruise control functions. Models equipped with the electronic accelerator system (E-Gas) are not equipped with a throttle cable, the throttle valve actuator controls the throttle plate under all driving conditions.

2   The throttle control module is calibrated to the throttle body during manufacture, therefore if the throttle control module requires replacement, replace the throttle body assembly (see Chapter 4).

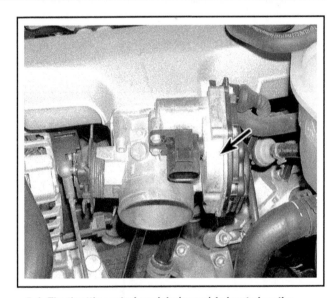

**5.1  The throttle control module (arrow) is located on the side of the throttle body - 1.8L turbo engine**

## 6 Mass airflow sensor - removal and installation

♦ **Refer to illustration 6.2**

1   The mass airflow sensor measures the amount of air passing through the sensor body and ultimately entering the engine through the throttle body. The ECM uses this information to control fuel delivery - the more air entering the engine (acceleration), the more fuel required.

2   Loosen the hose clamp and disconnect the air duct from the mass airflow sensor (see illustration).

3   Disconnect the electrical connector from the mass airflow sensor.

4   Remove the air filter housing cover (see Chapter 4).

5   Remove the screws retaining the mass airflow sensor to the air filter cover and remove the sensor.

### ✳✳ CAUTION:

**Handle the mass airflow sensor with care. Damage to this sensor will affect the operation of the entire fuel injection system.**

6   Installation is the reverse of removal.

**6.2  The mass airflow sensor is located in the air intake duct attached to the air filter housing cover**

## 7 Intake air temperature sensor - removal and installation

♦ **Refer to illustrations 7.1, 7.2a and 7.2b**

1   The intake air temperature sensor is a thermistor (a resistor which varies the value of its resistance in accordance with temperature changes). The change in the resistance values will directly affect the voltage signal from the sensor to the ECM (see illustration). As the sensor temperature INCREASES, the resistance values will DECREASE.

As the sensor temperature DECREASES, the resistance values will INCREASE.

2   Disconnect the electrical connector from the sensor, remove the mounting bolt and withdraw the sensor from the intake manifold (see illustrations).

3   Installation is the reverse of removal.

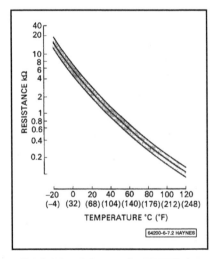

**7.1  Intake air temperature sensor and engine coolant temperature sensor approximate temperature vs. resistance values**

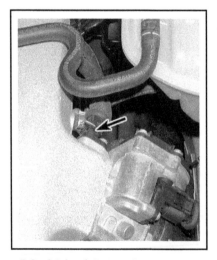

**7.2a  Intake air temperature sensor location - 1.8L turbo engine**

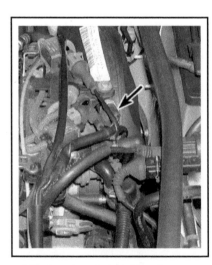

**7.2b  Location of the intake air temperature sensor harness connector (arrow) on the 2.8L DOHC engine**

## 8  Engine coolant temperature sensor - removal and installation

▶ **Refer to illustrations 8.4a and 8.4b**

### ✳✳ WARNING:

**Wait until the engine is completely cool before beginning this procedure.**

1    The engine coolant temperature sensor is a thermistor (a resistor which varies the value of its resistance in accordance with temperature changes). The change in the resistance values will directly affect the voltage signal from the sensor to the ECM (see illustration 7.1). As the sensor temperature INCREASES, the resistance values will DECREASE. As the sensor temperature DECREASES, the resistance values will INCREASE.

2    Partially drain the cooling system (see Chapter 1).

3    Disconnect the electrical connector from the sensor.

4    Carefully pry the retaining clip out and withdraw the sensor from the coolant pipe (see illustrations).

5    Before installing the new sensor, replace the O-ring.

6    Installation is the reverse of removal.

7    Refill and bleed the cooling system (see chapter 1).

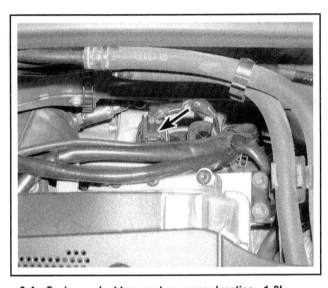

8.4a  Engine coolant temperature sensor location - 1.8L turbo engine

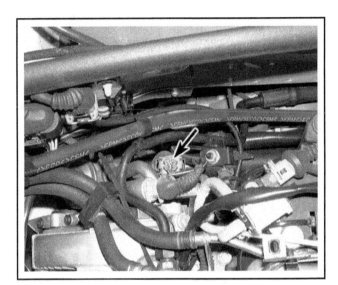

8.4b  Engine coolant temperature sensor location - 2.8L DOHC engine

## 9  Crankshaft position sensor (SOHC V6 engine) - removal and installation

1    The crankshaft position sensor is mounted at the lower left side of the engine near the engine oil drain plug. It detects gaps in the sensor ring corresponding to TDC of each cylinder (120-degree signal). The sensor uses a permanent magnet, core and coil. The changing gap causes the magnetic field near the sensor to change, this in turn varies the voltage signal to the ECM. The sensor signals are used by the ECM for ignition timing, fuel synchronization.

2    Disconnect the electrical connector from the sensor.

3    Remove the crankshaft sensor retaining bolt and remove the sensor.

4    Replace the O-ring and lightly lubricate it with clean engine oil.

5    Installation is the reverse of removal.

## 10 Engine speed sensor - removal and installation

▶ **Refer to illustrations 10.3a and 10.3b**

1   The engine sensor provides the ECM with a crankshaft position signal. The ECM uses the signal to determine a crankshaft reference point (Top Dead Center) and calculate engine speed (RPM). The signal is also used by the On-board Diagnostic system for misfire detection. The ignition system will not operate if the ECM does not receive an engine speed sensor input.

2   Disconnect the engine speed sensor electrical connector.

3   Remove the engine speed sensor mounting bolt and withdraw the sensor from the bellhousing (see illustrations).

4   Replace the O-ring and lightly lubricate it with clean engine oil.

5   Installation is the reverse of removal.

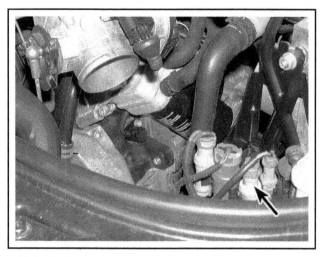

10.3a  Engine speed sensor electrical connector location (arrow) - 1.8L turbo engine

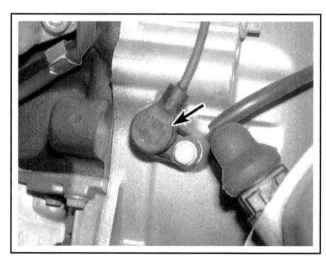

10.3b  Engine speed sensor location (arrow) - 2.8L DOHC engine

## 11 Camshaft position sensor - removal and installation

▶ **Refer to illustrations 11.4a, 11.4b and 11.4c**

1   The camshaft position sensor, in conjunction with the crankshaft position sensor, determines the ignition timing and fuel injection synchronization on each cylinder. The sensor is a magnetic pick-up device triggered from a reluctor wheel on the end of the camshaft. The camshaft sensor on the 1.8L engine is located under the timing belt cover near the camshaft sprocket. The 2.8L DOHC V6 engine is equipped with two camshaft sensors; one on the right bank on the front of the cylinder head and the other sensor on the left bank, at the rear of the cylinder head.

2   Disconnect the camshaft position sensor electrical connector.

3   Remove the timing belt upper cover (see Chapter 2), timing belt and sprocket.

4   Remove the camshaft position sensor mounting bolts and remove the sensor from the cylinder head (see illustrations).

5   Installation is the reverse of removal.

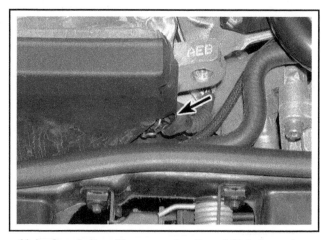

11.4a  Camshaft position sensor location - 1.8L turbo engine

**11.4b  Camshaft position sensor location on the right cylinder bank - 2.8L DOHC engine**

**11.4c  Camshaft position sensor location on the left cylinder bank - 2.8L DOHC engine**

## 12  Oxygen sensor - removal and installation

♦ **Refer to illustrations 12.9a, 12.9b, 12.10a and 12.10b**

➡**Note: Four-cylinder models are equipped with two oxygen sensors; one pre-converter oxygen sensor and one post-converter oxygen sensor. V6 models are equipped with four oxygen sensors; two pre-converter oxygen sensors and two post-converter oxygen sensors.**

1  The oxygen in the exhaust reacts with the elements inside the oxygen sensor to produce a voltage output that varies from 0.1 volt (high oxygen, lean mixture) to 0.9 volt (low oxygen, rich mixture). The pre-converter oxygen sensor (mounted in the exhaust system before the catalytic converter) provides a feedback signal to the ECM that indicates the amount of oxygen remaining in the exhaust gas after combustion. The ECM monitors this variable voltage continuously to determine the required fuel injector pulse width and to control the engine air/fuel ratio. A mixture ratio of 14.7 parts air to 1 part fuel is the ideal ratio for minimum exhaust emissions, as well as the best combination of fuel economy and engine performance. Based on oxygen sensor signals, the ECM tries to maintain this air/fuel ratio of 14.7:1 at all times.

2  The post-converter oxygen sensor (mounted in the exhaust system after the catalytic converter) has no effect on ECM control of the air/fuel ratio. However, the post-converter sensor is identical to the pre-converter sensor and operates in the same way. The ECM uses the post-converter signal to monitor the efficiency of the catalytic converter. A post-converter oxygen sensor will produce a slower fluctuating voltage signal that reflects the lower oxygen content in the post-catalyst exhaust.

3  An oxygen sensor produces no voltage when it is below its normal operating temperature of about 600-degrees F. During this warm-up period, the ECM operates in an open-loop fuel control mode. It does not use the oxygen sensor signal as a feedback indication of residual oxygen in the exhaust. Instead, the ECM controls fuel metering based on the inputs of other sensors and its own programs.

4  Proper operation of an oxygen sensor depends on four conditions:

a)  **Electrical** - *The low voltages generated by the sensor require good, clean connections which should be checked whenever a sensor problem is suspected or indicated.*

b)  **Outside air supply** - *The sensor needs air circulation to the internal portion of the sensor. Whenever the sensor is installed, make sure the air passages are not restricted.*

c)  **Proper operating temperature** - *The ECM will not react to the sensor signal until the sensor reaches approximately 600-degrees F. This factor must be considered when evaluating the performance of the sensor.*

d)  **Unleaded fuel** - *Unleaded fuel is essential for proper operation of the sensor.*

5  The ECM can detect several different oxygen sensor problems and set diagnostic trouble codes to indicate the specific fault (see Section 2). When an oxygen sensor fault occurs, the ECM will disregard the oxygen sensor signal voltage and revert to open-loop fuel control as described previously.

6  The exhaust pipe contracts when cool, so the oxygen sensor may be hard to loosen when the engine is cold. To make sensor removal

easier, start and run the engine for a minute or two, then shut it off. Be careful not to burn yourself during the following procedure. Also observe these guidelines when replacing an oxygen sensor.

a) *The sensor has a permanently attached pigtail and electrical connector which should not be removed from the sensor. Damage or removal of the pigtail or electrical connector can harm operation of the sensor.*

b) *Keep grease, dirt and other contaminants away from the electrical connector and the louvered end of the sensor.*

c) *Do not use cleaning solvents of any kind on the oxygen sensor.*

d) *Do not drop or roughly handle the sensor.*

7   If you're replacing the pre-converter oxygen sensor, remove the necessary components to gain access to the sensor.

8   If you're replacing the post-converter oxygen sensor, raise the vehicle and place it securely on jackstands.

9   Remove the cover (if equipped) and disconnect the electrical connector from the sensor (remote connectors are used on most oxygen sensors; follow the pigtail from the oxygen sensor to the connector) (see illustrations).

10  Using a suitable wrench or specialized oxygen sensor socket, unscrew the sensor from the exhaust manifold (see illustrations).

11  Anti-seize compound must be used on the threads of the sensor to aid future removal. The threads of most new sensors will be coated with this compound. If not, be sure to apply anti-seize compound before installing the sensor.

12  Install the sensor and tighten it securely.

13  Reconnect the electrical connector to the sensor.

14  The remainder of installation is the reverse of removal.

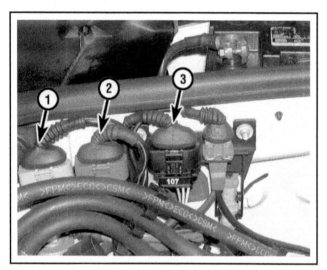

**12.9a  Oxygen sensor harness connector locations on right side of vehicle - 2.8L DOHC engine**

1   *Post-converter O2 sensor, right cylinder bank*
2   *Post-converter O2 sensor, left cylinder bank*
3   *Pre-converter O2 sensor, right cylinder bank*

**12.9b  Oxygen sensor harness connector locations on left side of vehicle - 2.8L DOHC engine**

4   *Pre-converter O2 sensor, left cylinder bank*

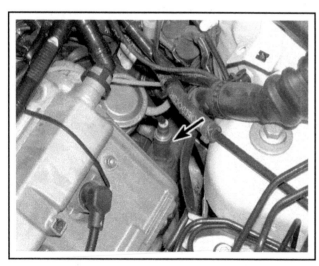

**12.10a  The pre-converter oxygen sensor, left bank, is located in the exhaust pipe before the catalytic converter**

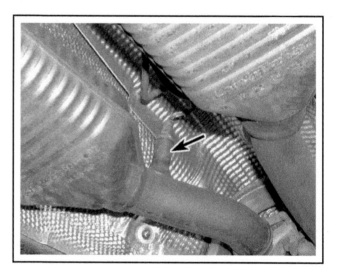

**12.10b  The post-converter oxygen sensor is located in the exhaust pipe after the catalytic converter**

## 13 Knock sensor - removal and installation

▶ **Refer to illustrations 13.1 and 13.4**

➡**Note: All models are equipped with two knock sensors. On four-cylinder models, the knock sensors are located on the left side of the engine block, below the intake manifold. On V6 models, one knock sensor is located on each side of the engine block underneath the intake manifold.**

1   The knock sensor detects abnormal vibration (spark knock or pinging) in the engine (see illustration). The knock control system is designed to reduce spark knock during periods of heavy detonation. This allows the engine to use maximum spark advance to improve driveability. Knock sensors produce AC output voltage which increases with the severity of the knock. The signal is fed into the ECM and the timing is retarded to eliminate for the detonation.

2   Remove the necessary components to gain access to the knock sensor.

3   Disconnect the knock sensor electrical connector (remote connectors are used on most knock sensors, follow the pigtail from the knock sensor to the connector).

4   Unscrew the bolt and remove the sensor from the engine block (see illustration).

5   Installation is the reverse of removal. Tighten the knock sensor bolt to 15 ft-lbs (20 Nm).

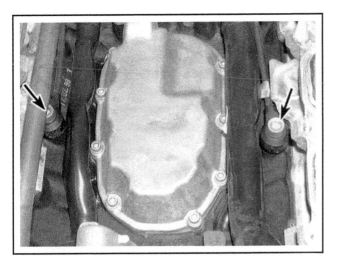

**13.1  Knock sensor locations (arrows) on the 2.8L DOHC engine**

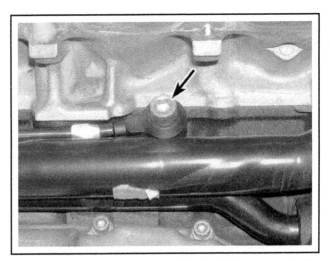

**13.4  Remove the bolt (arrow) and lift the knock sensor out of the engine block**

## 14 Vehicle speed sensor - removal and installation

▶ **Refer to illustration 14.4**

1   The Vehicle Speed Sensor (VSS) is mounted on the transaxle. The sensor is triggered by a toothed rotor on the transaxle output shaft. As the output shaft rotates, the sensor produces a fluctuating voltage, the frequency of which is proportional to vehicle speed. The ECM uses the sensor input signal for several different engine and transmission control functions. The VSS signal also drives the speedometer on the instrument panel. A defective VSS can cause various driveability and transaxle problems.

2   Raise the vehicle and support it securely on jackstands.

3   Disconnect the electrical connector from the VSS.

4   Remove mounting bolt and withdraw the VSS from the transaxle case (see illustration).

5   Replace the sensor O-ring.

6   Installation is the reverse of removal.

**14.4  Vehicle speed sensor location (arrow)**

## 15  Idle air control valve (SOHC V6 engine) - removal and installation

1  The Idle Air Control (IAC) valve controls the amount of air that bypasses the throttle valve, which controls the engine idle speed. The IAC valve is a stepper motor type actuator mounted on the upper intake plenum (manifold) and controlled by voltage pulses from the ECM. The IAC valve pintle moves in or out, allowing more or less intake air into the system. To increase idle speed, the ECM commands the stepper motor to pull the IAC valve pintle from the seat, allowing more air to bypass the throttle bore. To decrease idle speed, the ECM commands the IAC valve pintle towards the seat, reducing the air flow.

2  Remove the necessary components to gain access to the IAC valve.

3  Disconnect the electrical connector from the IAC valve.

4  Remove the mounting bolts and separate the IAC valve from the intake manifold.

5  Remove the gasket and clean the sealing surface on the intake manifold and the IAC housing to ensure a good seal.

### ✳✳ CAUTION:

**The IAC valve itself is an electrical component and must not be soaked in any liquid cleaner, as damage may result.**

6  Install the IAC valve using a new gasket and tighten the screws securely.

## 16  Electronic accelerator control system (E-Gas)

Some models may be equipped with an electronic accelerator control system (also known as E-Gas). The system does not use an accelerator cable; the throttle valve is operated by the throttle valve actuator contained within the throttle control module. The throttle valve actuator is controlled by the ECM. The accelerator pedal module contains a pedal position sensor. The pedal position signal is fed to the ECM and the ECM commands the throttle valve actuator to open the throttle accordingly.

With the engine Off and the ignition key On, the ECM opens the throttle valve in direct relationship to the accelerator pedal input. But when the engine is running, under load, the ECM operates the throttle valve independently of the accelerator pedal. The throttle valve may be opened significantly farther than the driver may be demanding. This allows the ECM to maintain the engine at peak operating efficiency.

The system operation is monitored by the On Board Diagnosis system, but uses a separate warning light. The Electronic Power Control (EPC) warning light is located near the top of the speedometer. The EPC light will illuminate with the ignition key on and should go out when the engine is started. If the EPC warning light remains on, or comes on, with the engine running, a problem with the system has been identified and a Diagnostic Trouble Code is retained in the ECM memory. If there is a complete failure with the system the ECM will limit the engine speed to approximately 1200 rpm.

A quick check of the system can be performed by removing the air intake duct and observing the throttle valve as an assistant depresses the accelerator pedal. When the accelerator pedal is depressed half-way, the throttle valve should open to half throttle. When the accelerator pedal is depressed fully, the throttle valve should open to full throttle. The manufacturer's scan tool is required to properly diagnose the system. If a fault is suspected with the system or if the EPC warning light is on, take the vehicle to a dealership service department or other properly equipped repair facility as soon as possible. For additional information, refer to Chapter 4 for information on the electronic accelerator pedal module and Section 5 of this Chapter for information on the throttle control module.

## 17  Turbocharger boost control system (four-cylinder engine)

▶ **Refer to illustrations 17.1a, 17.1b, 17.1c and 17.1d**

The turbocharger boost control system consists of the boost pressure sensor, the boost pressure (wastegate) control valve, the recirculation air valve and solenoid, the ECM and the connecting hoses and wiring (see illustrations).

The ECM monitors the boost pressure sensor signal and controls the vacuum supplied to the turbocharger wastegate actuator with the boost pressure control valve. The engine control system calculates the engine torque needed depending on driver demand and engine operating conditions, the ECM will then adjust the boost pressure to meet the demands. The recirculation air valve returns air to the intake duct when the throttle is closed minimizing turbo lag. The ECM controls the recirculation air valve with a vacuum valve/solenoid.

If a problem is suspected with the system, check the wiring, connectors, vacuum hoses and connecting pipes. Using a hand-held vacuum pump, apply vacuum to the wastegate actuator (with the engine not running) and check the actuator arm and lever for freedom of movement. Disconnect the electrical connector from the boost pressure control valve and using an ohmmeter, check the solenoid windings for an open circuit. The boost pressure control valve should measure approximately 25 to 35 ohms, if an open circuit or excessive resistance is indicated, replace the control valve.

A scan tool is required for complete diagnosis of the system control circuits. If a scan tool is not available, have the system checked at a dealer service department or other properly equipped repair facility.

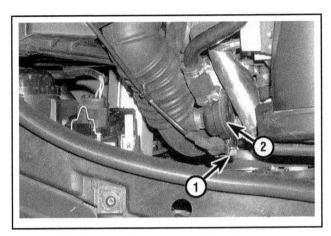

**17.1a  Turbocharger system component locations**

1   *Boost pressure sensor*
2   *Pressure regulating valve*

**17.1b Turbocharger system component locations**

1   *Turbocharger*
2   *Boost pressure recirculation valve*

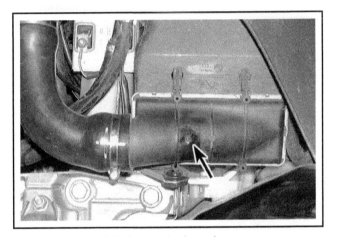

**17.1c  Location of the intercooler (arrow)**

**17.1d  Location of the boost pressure recirculation valve connecting hose**

## 18  Intake manifold tuning control system (V6 engines)

♦ **Refer to illustration 18.2**

1   V6 models are equipped with an intake manifold tuning control system.

2   The intake manifold tuning control system consists of the change-over valve (located inside the intake manifold), the change-over valve actuator, the change-over valve control solenoid and the ECM (see illustration). The change-over valve diverts the path of the incoming air through the intake manifold, one path being longer than the other. At low engine speeds vacuum is applied to the change-over valve actuator, the change-over valve is closed and air is diverted through the longer path to enhance maximum torque at low speed. At a preset engine speed the ECM de-energizes the solenoid, vacuum is vented from the actuator and the change-over valve opens. Air is then drawn through the shorter path enhancing high speed power.

3   To check the system, place the transmission in Park or Neutral, apply the parking brake and block the rear wheels. Start the engine

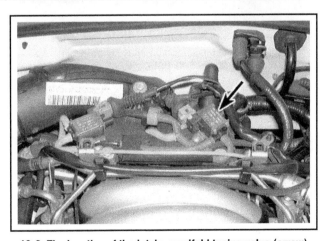

**18.2  The location of the intake manifold tuning valve (arrow)**

and rev the engine above 4,500 rpm several times while watching the change-over valve actuator. The actuator rod should move when engine speed exceeds 4,500 rpm. If the actuator is not responding, stop the engine and continue testing. Make sure the ignition key remains off.

4   Check the vacuum hoses from the vacuum source to the actuator for leaks or damage. Check the vacuum tank to make sure it holds vacuum. Check the one-way valve - it should allow air to pass one way but not the other.

5   Unplug the electrical connector from the solenoid and remove the vacuum lines from the change-over valve control solenoid ports. Using a pair of fused jumper wires, connect battery voltage and ground to the two terminals of the change-over valve control solenoid. The solenoid should click and air should be allowed to pass between the two ports. Remove the jumpers, no air should pass between the two ports. If the solenoid does not operate as described, replace the change-over valve control solenoid.

6   Remove the vacuum hose from the change-over valve actuator. Connect a hand-held vacuum pump to the actuator and apply vacuum. The rod should move and the actuator should hold vacuum. If it doesn't, replace the actuator.

7   If all the above tests are good, have the ECM diagnosed by a dealer service department or other qualified repair facility.

## 19  Camshaft timing control system (DOHC V6 engine)

▶ **Refer to illustrations 19.1a and 19.1b**

1   DOHC V6 models are equipped with a camshaft timing control system. The system consists of the actuators (incorporated into each intake camshaft sprocket), two camshaft timing control solenoids, the two camshaft position sensors and the ECM (see illustrations). The system advances the intake camshaft timing by directing engine oil pressure to the intake camshaft actuator/sprockets, rotating the sprocket slightly on the camshaft when under pressure. Advancing the intake camshaft timing during high-speed operation increases the mid-to-high range power output, while retarding the intake camshaft during low engine speeds increases engine torque. The camshaft timing control solenoid is controlled by the ECM. The ECM sends a ground signal to the camshaft timing control solenoid which in turn directs oil flow to the actuator/sprocket, advancing the intake camshaft timing when engine speed and load reach the preprogrammed level (over 4,000 rpm under load).

2   To check the camshaft timing control solenoid, turn off the ignition key, unplug the electrical connector and use an ohmmeter to measure the resistance across the two terminals of the solenoid. There should be approximately 10.0 to 18.0 ohms resistance at 68-degrees F (20-degrees C) across the terminals. If the resistance is incorrect, replace the camshaft timing control solenoid.

3   If camshaft timing control solenoid tests good, have the ECM diagnosed by a dealer service department or other qualified repair facility.

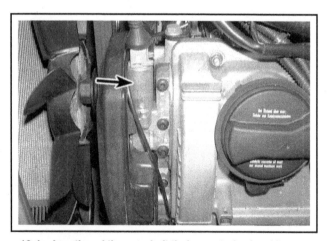

19.1a  Location of the camshaft timing control solenoid on the left cylinder bank - 2.8L DOHC engine

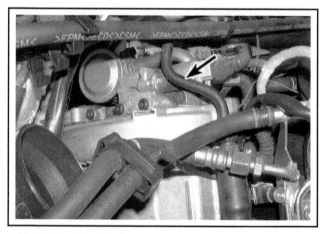

19.1b  Location of the camshaft timing control solenoid on the right cylinder bank - 2.8L DOHC engine

## 20  Crankcase ventilation system

▶ **Refer to illustration 20.2**

1   When the engine is running, a certain amount of the gasses produced during combustion escapes past the piston rings into the crankcase as blow-by gasses. The crankcase ventilation system is designed to reduce the resulting hydrocarbon emissions (HC) by routing the gasses and vapors from the crankcase into the intake manifold and combustion chambers, where they are consumed during engine operation.

2   Crankcase vapors pass through a hose connected from the valve cover to the air intake duct (see illustration). The oil/air separator at the valve cover separates the oil suspended in the blow-by gases and

allows the oil to drain back into the crankcase. The crankcase vapors are drawn from the oil/air separator through a hose connected to the air intake duct where they mix with the incoming air and are burned during the normal combustion process.

3   A plugged breather, valve or hose will cause excessive crankcase pressures resulting in oil leaks and sludge build-up in the crankcase. Check the components for restrictions and clean or replace the components as necessary. Be sure to check the basic mechanical condition of the engine before condemning the crankcase ventilation system (see Chapter 2C).

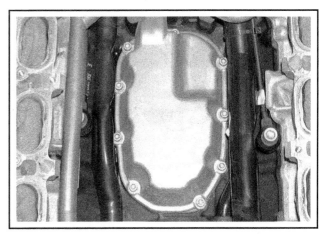

**20.2  Crankcase ventilation system/oil separator on the 2.8L DOHC engine**

## 21  Evaporative emissions control system

1   The fuel evaporative emissions control (EVAP) system absorbs fuel vapors from the fuel tank and, during engine operation, releases them into the engine intake system where they mix with the incoming air/fuel mixture. The main components of the evaporative emissions system are the canister (filled with activated charcoal to absorb fuel vapors), the purge control valve, the leak detection pump, the fuel tank and the vapor and purge lines.

2   After passing through a check valve, fuel tank vapor is carried through the vapor hose to the charcoal canister. The activated charcoal in the canister absorbs and stores the vapors. When a programmed set of conditions are met (engine running, warmed to a pre-set temperature, etc.), the ECM opens the purge valve. Fuel vapors from the canister are then drawn through the purge hose by intake manifold vacuum into the intake manifold and combustion chamber where they are consumed during normal engine operation.

3   The ECM regulates the rate of vapor flow from the canister to the intake manifold by controlling the duty cycle of the EVAP purge control valve solenoid. During cold running conditions and hot start time delay, the ECM does not energize the solenoid. After the engine has warmed up to the correct operating temperature, the ECM purges the vapors into the intake manifold according to the running conditions of the engine. The ECM will cycle (ON then OFF) the purge control valve solenoid about 5 to 10 times per second. The flow rate will be controlled by the pulse width, or length of time, the solenoid is allowed to be energized.

4   The EVAP system is equipped with a leak detection monitor system. The system is a self-diagnostic system designed to detect a leak in the EVAP system. Each time the engine is started cold, the PCM energizes the leak detection pump. The pump pressurizes the EVAP system then shuts off. The PCM is able to detect a leak if the pump continues to run, unable to pressurize the system. If a leak is detected, the PCM will trigger a diagnostic trouble code (see Section 2).

## CHECK

➡**Note: The evaporative emissions control system, like all emission control systems, is protected by a Federally-mandated warranty. The EVAP system probably won't fail during the service life of the vehicle; however, if it does, the hoses or charcoal canister are usually to blame.**

5   Always check the hoses first. A disconnected, damaged or missing hose is the most likely cause of a malfunctioning EVAP system. Repair any damaged hoses or replace any missing hoses as necessary.

6   Check the related fuses and wiring to the purge valve. Refer to the wiring diagrams at the end of Chapter 12, if necessary. The purge valve is normally closed - no vapors will pass through the ports. When the ECM energizes the solenoid (by completing the circuit to ground), the valve opens and vapors flow through.

7   A scan tool is required to thoroughly check the system. If the above checks fail to identify the problem area, have the system diagnosed by a dealer service department or other qualified repair shop.

## COMPONENT REPLACEMENT

▶ **Refer to illustrations 21.9, 21.11, 21.15, 21.16 and 21.18**

### EVAP canister

8   The EVAP canister is located under a cover below the spare tire well.

9   Raise the vehicle and support it securely on jackstands. Remove the cover (see illustration).

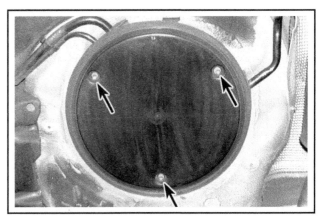

**21.9  Remove the protective plate bolts (arrows) from the bottom of the spare tire carrier**

10 Label and remove the hoses from the canister. Many of the EVAP system hoses are equipped with quick-connect fittings. Disconnect the fitting as follows:

a) *Clean the area around the fitting.*

b) *Twist the fitting back-and-forth several times to loosen the seal.*

c) *Depress the locking tabs and pull the fitting straight off the nipple.*

11 Remove the mounting nuts and remove the canister (see illustration).

12 Installation is the reverse of removal.

### Leak detection pump

13 The leak detection pump is located behind the left rear wheelwell.

14 Loosen the wheel bolts. Raise the vehicle and support it securely on jackstands. Remove the left rear wheel and the wheelwell liner.

15 Label and remove the hoses from the leak detection pump (see illustration).

16 Remove the mounting nuts and remove the pump (see illustration).

17 Installation is the reverse of removal.

### Purge control valve

18 The purge control valve is mounted on the air filter housing, above the right front wheelwell (see Illustration).

19 Disconnect the electrical connector. Label and remove the hoses from the purge control valve.

20 Remove the mounting bolts and remove the purge control valve from the bracket.

21 Installation is the reverse of removal.

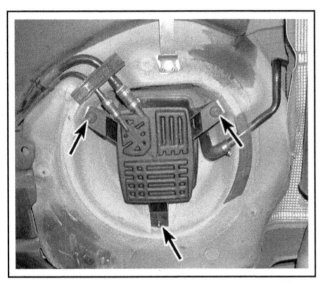

**21.11  Remove the nuts (arrows) that mount the charcoal canister**

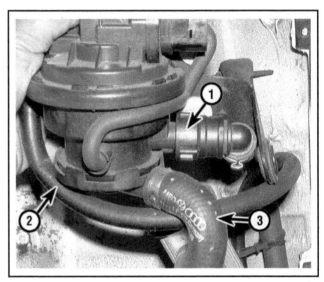

**21.15  Leak detection pump details**

| 1 | *Quick connect fitting* | 3 | *Main vent hose* |
| 2 | *Vent line* | | |

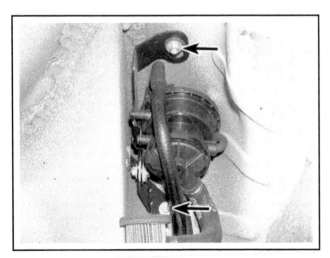

**21.16  Leak detection pump mounting bolts (arrows)**

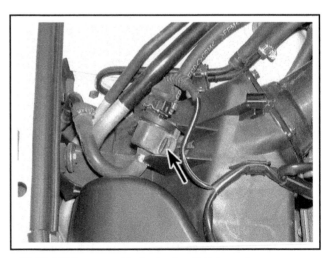

**21.18  EVAP purge control valve location (arrow)**

## 22 Secondary air injection system

1   The secondary air injection system is used to reduce tailpipe emissions on initial engine start-up. The system uses an electric motor/pump assembly, vacuum valve/solenoid, combination air shut-off/check valve and tubing to inject fresh air directly into the exhaust manifolds. The fresh air (oxygen) reacts with the exhaust gas in the catalytic converter to reduce HC and CO levels. The air pump and solenoid are controlled by the ECM. During initial start-up, when the coolant temperature is cold, the ECM will energize the vacuum valve/solenoid, opening the check valve and operate the air pump. During normal operation, the check valve is closed to prevent exhaust backflow into the system.

## CHECK

2   Check the air pump hoses and the vacuum hoses. Repair any damaged hoses or replace any missing hoses as necessary. Check the vacuum source to the vacuum valve. Intake manifold vacuum should be present with the engine running.

3   Check the related fuses and wiring to the air pump and vacuum valve/solenoid. Refer to the wiring diagrams at the end of Chapter 12, if necessary. The vacuum valve is normally closed - no vacuum is applied to the check valve. When the ECM energizes the solenoid (by completing the circuit to ground), the valve opens, vacuum is applied to the check valve, the check valve opens and air flows through the tube into the exhaust pipes.

4   A scan tool is required to thoroughly check the system. If the above checks fail to identify the problem area, have the system diagnosed by a dealer service department or other qualified repair shop.

## COMPONENT REPLACEMENT

▶ **Refer to illustrations 22.7, 22.10 and 22.16**

### Air pump

5   Remove the splash shield from under the front of the engine. Remove the front bumper and place the radiator support assembly in the service position (see Chapter 11).

6   Remove the alternator cooling duct.

7   Disconnect the electrical connector and remove the hoses from the air pump (see illustration).

8   Remove the mounting bolts and remove the air pump assembly.

9   Installation is the reverse of removal.

### Vacuum valve

10  Disconnect the electrical connector from the vacuum valve (see illustration).

11  Label and disconnect the vacuum hoses from the valve.

12  Remove the mounting nut and remove the vacuum valve.

13  Installation is the reverse of removal.

### Check valves

14  To remove the right-side check valve, remove the air intake duct between the mass airflow sensor and throttle body. Remove the air filter housing (see Chapter 4).

15  To remove the left-side check valve, remove the coolant expansion tank (see Chapter 3).

16  Remove the vacuum hose and the air hose from the check valve (see illustration).

17  Remove the mounting bolts and remove the check valve.

18  Installation is the reverse of removal.

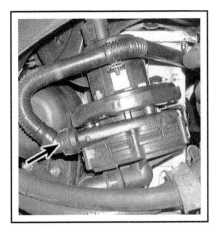

**22.7  Location of the secondary air injection pump (arrow)**

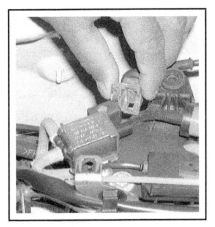

**22.10  Disconnect the harness connector from the vacuum valve**

**22.16  Remove the air hose from the secondary air injection check valve (arrow)**

## 23 Exhaust gas recirculation system (SOHC V6 engine)

1 The Exhaust Gas Recirculation (EGR) system is used to lower NOx (oxides of nitrogen) emission levels caused by high combustion temperatures. The EGR valve recirculates a small amount of exhaust gases into the intake manifold. The additional mixture lowers the temperature of combustion thereby reducing the formation of NOx compounds.

2 The EGR system is equipped with an EGR valve, an EGR control solenoid valve which receives manifold vacuum and an EGR temperature sensor. The operation of the system is controlled by the ECM which operates the EGR control solenoid.

3 The EGR temperature sensor is used to inform the ECM of temperature changes in the EGR passageway. This helps the ECM determine the EGR On/Off time.

### System check

4 Check all hoses for cracks, kinks, broken sections and proper connection. Inspect all system connections for damage, cracks and leaks.

5 To check the EGR system operation, bring the engine up to operating temperature and, with the transmission in Neutral (parking brake set and tires blocked to prevent movement), allow it to idle. Open the throttle so the engine speed is between 2,000 and 4,000 rpm and then allow it to close. The EGR valve stem should move if the control system is working properly. The test should be repeated several times. Movement of the stem indicates the control system is functioning correctly. If the EGR valve stem does not move, check the vacuum signal to the EGR valve. Remove the hose from the valve, place your finger over the end of the hose and perform the procedure again. If no vacuum is present at the end of the hose check the hose connections to make sure they are not leaking or clogged, then check the EGR control solenoid (see Step 11).

### Component checks

#### EGR Valve

6 With the engine Off, disconnect the vacuum hose and apply ten inches of vacuum with a hand-held vacuum pump to the EGR valve. If the valve opens, measure the valve travel to make sure it is approximately 1/8-inch. If the stem does not move, replace the EGR valve with a new one.

7 Next apply vacuum with the pump and then clamp the hose shut. The valve should stay open for 30 seconds or longer. If it does not, the diaphragm is leaking and the valve should be replaced with a new one.

8 Start the engine and apply vacuum to the valve. The engine should idle roughly when the valve is open. If it doesn't, the passages in the manifold are probably clogged. If there's no difference in idle quality with the valve open or closed the EGR valve is probably not closing all the way. Remove the EGR valve and inspect the poppet and seat area for deposits.

9 If the deposits are more than a thin film of carbon, the valve should be cleaned. To clean the valve, apply solvent and allow it to penetrate and soften the deposits, making sure that none gets on the valve diaphragm, as it could be damaged.

10 Use a vacuum pump to hold the valve open and carefully scrape the deposits from the seat and poppet area with a tool. Inspect the poppet and stem for wear and replace the valve with a new one if wear is found.

#### EGR control solenoid

11 First check for a vacuum signal to the solenoid with the engine running, If no vacuum is present check the hose to and from the solenoid for cracks and clogging. If the hose is OK, check the throttle body port for clogging.

12 To check the operation of this EGR control solenoid, make sure the ignition key is turned off, then disconnect the electrical connector and connect a ground wire to one of its terminals and fused battery voltage to the other terminal.

13 It should be possible to blow air through the two vacuum ports that are adjacent to one another when there is battery power applied, but impossible to do so when it is not energized with battery voltage.

#### EGR temperature sensor

14 Remove the EGR temperature sensor. Submerge the tip of the sensor in a container of water. Heat the water on the stove while monitoring the resistance of the sensor. Resistance should decrease as temperature increases.

### Component replacement

#### EGR valve

15 Remove the air intake duct from the throttle body and air filter housing (see Chapter 4).

16 Detach the vacuum line from the EGR valve. Disconnect the EGR pipe nut.

17 Remove the EGR valve mounting fasteners.

18 Remove the EGR valve and gasket from the manifold. Discard the gasket.

19 With a wire wheel, buff the exhaust deposits from the EGR valve mounting surface on the manifold and, if you plan to use the same valve, the mounting surface of the valve itself. Look for exhaust deposits in the valve outlet. Remove deposit build-up with a screwdriver.

### ✳✳ CAUTION:

**Never immerse the valve in solvents or degreaser - both agents will permanently damage the diaphragm. Sandblasting is also not recommended because it will affect the operation of the valve.**

20 If the EGR passage contains an excessive build-up of deposits, clean it out with a wire wheel. Make sure that all loose particles are completely removed to prevent them from clogging the EGR valve or from being ingested into the engine.

21 If there are large amounts of deposits within the EGR valve, remove the EGR pipe from the exhaust manifold and clean out the deposits inside the tube.

➡**Note: Remove the EGR control backpressure transducer valve and pipe and clean any deposits from the passages.**

22 Installation is the reverse of removal.

#### EGR control solenoid

23 Unplug the electrical connector from the solenoid.

24 Clearly label and detach the vacuum hoses.

25 Remove the solenoid mounting nut and remove the solenoid.

26 Installation is the reverse of removal.

**EGR temperature sensor**

27 Disconnect the electrical connector and unscrew the sensor from the pipe.

28 Apply anti-seize to the threads of the sensor before installing it.

29 Installation is the reverse of removal.

## 24 Catalytic converter

➡**Note: Because of a Federally mandated warranty which covers emissions-related components such as the catalytic converter, check with a dealer service department before replacing the converter at your own expense.**

1 The catalytic converter is an emission control device added to the exhaust system to reduce pollutants from the exhaust gas stream. A three-way (reduction) catalyst design is used. The catalytic coating on the three-way catalyst contains platinum and rhodium, which lowers the levels of oxides of nitrogen (NOx) as well as hydrocarbons (HC) and carbon monoxide (CO).

2 The test equipment for a catalytic converter is expensive and highly sophisticated. If you suspect that the converter on your vehicle is malfunctioning, take it to a dealer or authorized emissions inspection facility for diagnosis and repair.

## CHECK

3 Whenever the vehicle is raised for servicing of underbody components, check the converter for leaks, corrosion, dents and other damage. Check the welds/flange bolts that attach the front and rear ends of the converter to the exhaust system. If damage is discovered, the converter should be replaced.

4 A catalytic converter may become plugged. The easiest way to check for a restricted converter is to use a vacuum gauge to diagnose the effect of a blocked exhaust on intake vacuum.

   a) *Connect a vacuum gauge to an intake manifold vacuum source.*
   b) *Warm the engine to operating temperature, place the transmission in Park and apply the parking brake.*
   c) *Note and record the vacuum reading at idle.*

   d) *Open the throttle until the engine speed is about 2000 rpm.*
   e) *Release the throttle quickly and record the vacuum reading.*
   f) *Perform the test three more times, recording the reading after each test.*
   g) *If the reading after the fourth test is more than one in-Hg lower than the reading recorded at idle, the catalytic converter, muffler or exhaust pipes may be plugged or restricted.*

## REPLACEMENT

➡**Note: Refer to the exhaust system servicing section in Chapter 4 for additional information and illustrations.**

5 Raise the vehicle and support it securely on jackstands.

6 Disconnect the electrical connectors from the oxygen sensors.

7 Remove the exhaust pipe-to-exhaust manifold flange bolts and separate the exhaust pipe from the exhaust manifold. Support the exhaust pipe.

## ✳✳ CAUTION:

**Do not allow the flex pipe (if equipped) to bend more than 10-degrees or damage to the pipe may occur.**

8 Loosen the clamp bolts and detach the catalytic converter and header pipe from the exhaust system.

9 Clean the carbon deposits from the mounting flanges and install new gaskets.

10 Installation is the reverse of removal, but before tightening the clamps, make sure the system is aligned properly and not under stress.

**Notes**

**Section**

**Reference to other Chapters**

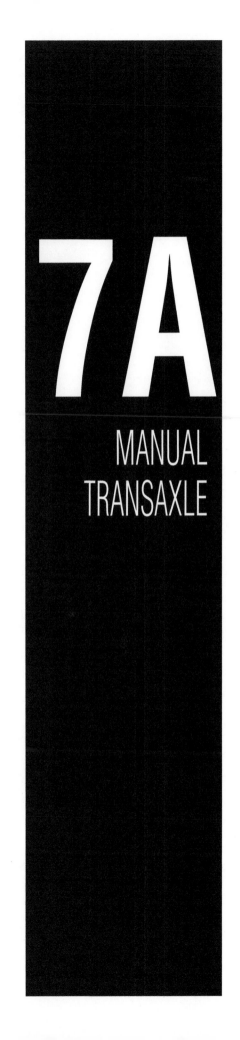

# 7A

## MANUAL TRANSAXLE

## 1    General information

The five-speed manual transaxle is bolted to the rear of the engine. The transaxle on front-wheel drive models transmits the power to a differential unit located at the front of the transaxle, through driveaxles, to the front wheels. On all-wheel drive models, power is also transmitted through a transfer gearbox to a driveshaft, which turns the rear differential, driveaxles and wheels. All gears including reverse incorporate a synchromesh engagement.

Gearshift is by a floor-mounted lever. A single shift rod connects the bottom of the lever to the transaxle, and the shift rod is connected to a rod which protrudes from the rear of the transaxle.

## 2    Shift linkage - adjustment

### STANDARD GEARSHIFT MODELS

▶ **Refer to illustrations 2.3 and 2.7**

1    Apply the parking brake, then raise the front of the vehicle and support it securely on jackstands. Place the shifter in Neutral.

2    Working under the car, remove the heatshield from the underbody for access to the bottom of the shifter assembly. For improved access, remove the exhaust front downpipe and catalytic converter (see Chapter 4).

3    Loosen the clamp bolt attaching the shift rod to the adjustment fork on the bottom of the shift lever (see illustration). Do not remove the bolt.

4    Working inside the car, unscrew and remove the knob from the top of the shifter, then unclip and remove the boot.

5    Remove the noise insulation around the bottom of the shifter housing.

6    Check that the ball housing is horizontal. If not, loosen the two nuts, adjust the position of the housing, and re-tighten the nuts.

7    Have an assistant hold the shift lever in a vertical position so that the distance between the ends of the curved ball stop are the same on both sides with the lever positioned slightly to the rear. The gear lever is now in the 3rd/4th neutral position (see illustration).

8    Make sure the transaxle gear selector rod is positioned in neutral, then tighten the adjustment bolt beneath the car.

9    Check that all gear positions can be selected without difficulty.

10  Install the underbody heatshield together with the exhaust downpipe and catalytic converter (where removed). Lower the car to the ground.

11  Install the noise insulation, boot and shift knob.

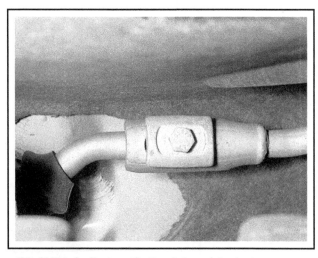

**2.3  Shift rod adjustment bolt and clamp (standard gearshift models)**

**2.7  The distance between the curved ball stop and the shifter base must be the same on both sides**

## SHORT-TRAVEL GEARSHIFT MODELS

**◗ Refer to illustration 2.14**

12 Working inside the car, unscrew and remove the knob from the top of the shifter, then unclip and remove the boot.

13 Remove the noise insulation around the bottom of the shifter housing.

14 Measure the distance between the rear pushrod and body as shown (see illustration). If it is not 1-29/64 inches (37 mm), loosen the pushrod bolt, re-position the pushrod, and tighten the bolt.

15 Working through the shifter aperture, loosen the clamp bolt attaching the shift rod to the adjustment fork on the shift lever. Do not remove the bolt.

16 Check that the ball housing is horizontal. If not, loosen the two nuts, adjust the position of the housing, and re-tighten the nuts.

17 Have an assistant hold the shift lever in a vertical position so that the distance between the ends of the curved ball stop are the same on both sides with the lever positioned slightly to the rear (see illustration 2.7). The shift lever is now in the 3rd/4th neutral position.

18 Make sure the transaxle gear selector rod is positioned in neutral, then tighten the adjustment bolt. Lower the car to the ground.

19 Install the noise insulation, boot and shift knob.

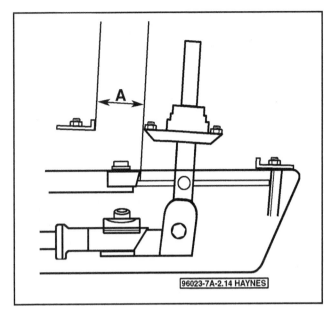

96023-7A-2.14 HAYNES

**2.14  On the short-travel gear shifter, the distance shown must be 1-29/64 inches (37 mm)**

---

## 3   Manual transaxle - removal and installation

**◗ Refer to illustration 3.16**

### ✳ CAUTION 1:

**These models are equipped with an anti-theft radio. Before performing a procedure that requires disconnecting the battery, make sure you have the proper activation code.**

### ✳ CAUTION 2:

**Disconnecting the battery can cause driveability problems that require a scan tool to remedy. See Chapter 5, Section 1 for the use of an auxiliary voltage input device before disconnecting the battery.**

### REMOVAL

1   Select a solid, level surface to park the vehicle upon. Give yourself enough space to move around it easily. Apply the parking brake and chock the rear wheels.

2   Remove the transaxle-to-engine bolts accessible from above.

3   Raise the front of the vehicle and support it securely on jackstands. Remove the splash guard from under the engine compartment.

4   Disconnect the cable from the negative terminal of the battery.

5   Remove the air cleaner assembly as described in Chapter 4.

6   On turbocharged models, remove the screws and lift the coolant expansion tank from its location and move it to one side without disconnecting the coolant hoses. Disconnect the oxygen sensor wiring on the firewall.

7   Remove the exhaust front pipe and catalytic converter (see Chapter 4). Take care not to excessively bend the flexible section of the front pipe.

8   Unbolt the splash guard support bracket from under the front of the engine compartment.

9   Using an Allen key, unbolt the heat shields from over the inner end of the right-hand driveaxle.

10  Detach the driveaxles from the transaxle flanges (see Chapter 8). Rest the driveshafts on the suspension links.

11  If you are working on an all-wheel drive model, remove the driveshaft (see Chapter 8).

12  Support the transaxle on a jack, preferably one made for this purpose. Secure the transaxle to the jack with a safety chain. Unbolt the

right-hand transaxle mount complete with rubber bushing.

13 Disconnect the wiring from the Vehicle Speed Sensor, and from the multi-function switch on the transaxle.

14 Check that all wiring has been disconnected from the transaxle and transaxle-to-engine bolts.

15 Remove the starter motor (see Chapter 5). If preferred, the wiring may be left connected, and the starter supported to one side.

16 Unscrew the bolt and disconnect the shift rod from the rear of the transaxle (see illustration).

17 On models with a short-travel gear-change, unscrew the bolt and detach the steady bar.

18 Make sure that the transaxle is adequately supported, then unscrew the remaining bolts securing the transaxle to the engine.

19 Unscrew the bolt securing the left-hand transaxle mount to the rubber mounting.

20 With the help of an assistant, withdraw the transaxle from the locating dowels on the rear of the engine, making sure that the input shaft does not hang on the clutch. Lower the transaxle sufficient to gain access to the clutch release cylinder. Make sure that the driveshafts are supported clear of the transaxle.

## ✳✳ WARNING:

**Make sure that the transaxle remains steady on the jack head. Keep the transaxle level until the input shaft is fully withdrawn from the clutch friction plate.**

21 Unbolt the release cylinder from the transaxle, and tie it to one side.

➡**Note: Do not depress the clutch pedal with the release cylinder removed.**

22 Lower the transaxle to the ground.

## INSTALLATION

23 Before installing the transaxle, make sure that the location dowels are correctly positioned in the engine cylinder block rear face. Also make sure that the starter motor lower mounting bolt is positioned in the transaxle, as it cannot be inserted with the transaxle in its normal position.

24 Installation of the transaxle is a reversal of the removal procedure, but note the following points:

 a) *Check the rear rubber mounts and replace them if necessary.*
 b) *Apply a little high-melting-point grease to the splines of the transaxle input shaft.*
 c) *Tighten all nuts and bolts to the specified torque where given.*
 d) *On completion, refer to Section 2 and check the shift linkage adjustment.*

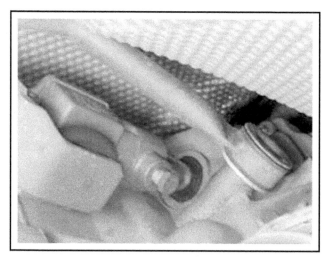

**3.16 Shift rod connection at the rear of the transaxle**

## 4    Manual transaxle overhaul - general information

1    Overhauling a manual transaxle unit is a difficult and involved job for the home mechanic. In addition to dismantling and reassembling many small parts, clearances must be precisely measured and, if necessary, changed by selecting shims and spacers. Internal transaxle components are also often difficult to obtain and in many instances, extremely expensive. Because of this, if the transaxle develops a fault or becomes noisy, the best course of action is to have the unit overhauled by a transmission specialist or to obtain an exchange reconditioned unit.

2    Nevertheless, it is not impossible for the more experienced mechanic to overhaul the transaxle if the special tools are available and the job is carried out in a deliberate step-by-step manner, to ensure that nothing is overlooked.

3    The tools necessary for an overhaul include internal and external snap-ring pliers, bearing pullers, a slide hammer, a set of pin punches, a dial test indicator and possibly a hydraulic press. In addition, a large, sturdy workbench and a vise will be required.

4    During dismantling of the transaxle, make careful notes of how each component is fitted to make reassembly easier and accurate.

5    Before disassembling the transaxle, it will help if you have some idea of where the problem lies. Certain problems can be closely related to specific areas in the transaxle which can make component examination and replacement easier. Refer to the *Troubleshooting* Section in this manual for more information.

## 5  Multi-function switch - removal and installation

▶ **Refer to illustration 5.1**

### REMOVAL

1   The multi-function switch is located on top of the transaxle (see illustration).
2   Apply the parking brake, then raise the front of the vehicle and support it on jackstands.
3   Disconnect the electrical connector, then unscrew the bolts securing the switch lead to the top of the transaxle
4   Note the installed position of the switch, then unscrew the bolt and remove the switch retainer plate.
5   Withdraw the multi-function switch from the transaxle. Recover the O-ring seal.

### INSTALLATION

6   To install the switch, first clean the switch location in the transaxle. Install a new O-ring seal, then insert the switch in the previously noted position.

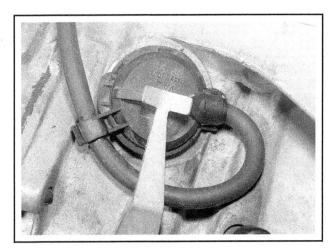

**5.1  Multi-function switch**

7   Install the retainer plate and tighten the bolt.
8   Secure the lead to the top of the transaxle and tighten the bolts.
9   Reconnect the wiring, then lower the vehicle to the ground.

## 6  Oil seals - replacement

### DRIVE FLANGE OIL SEALS

1   Loosen the wheel bolts. Apply the parking brake, then raise the front of the vehicle and support it on jackstands. Remove the wheel.
2   Refer to Chapter 8 and unbolt the heatshield, then unscrew the bolts and detach the relevant driveshaft from the transaxle drive flange. Tie the driveaxle away from the transaxle, and wrap the inner joint in a plastic bag in order to prevent entry of dust and dirt. Turn the steering as necessary to move the driveshaft away from the flange.
3   Position a suitable container beneath the transaxle to catch spilled oil.
4   The drive flange is held in position by a snap-ring, and in order to remove the flange, it is necessary to release the snap-ring. To do this, locate a suitable distance piece (such as a chisel) between the flange and the final drive cover or transaxle casing, then screw a bolt through the flange onto the distance piece. As the bolt is tightened, the flange will be forced outwards and the snap-ring released from its groove. If the flange is tight, turn it 180-degrees and repeat the removal procedure.
5   With the flange out, note the installed depth of the oil seal in the housing, then pry it out using a large flat-bladed screwdriver.
6   Clean all traces of dirt from the area around the oil seal aperture, then apply a smear of grease to the lips of the new oil seal.

7   Ensure the seal is correctly positioned, with its sealing lip facing inwards, and tap it squarely into position, using a suitable tubular drift (such as a socket) which bears only on the hard outer edge of the seal. If the surface of the flange is good, make sure the seal is installed at the same depth in its housing as originally noted; it should be 7/32-inch (5.5 mm) below the outer edge of the transaxle. If the surface of the flange is worn, install the oil seal at a depth of 1/4-inch (6.5 mm).
8   Clean the oil seal and apply a smear of multi-purpose grease to its lips.
9   It is recommended that the snap-ring on the inner end of the drive flange is replaced whenever the flange is removed. To do this, mount the flange in a soft-jawed vise, then pry off the old snap-ring and install the new one. Lightly grease the snap-ring.
10  Insert the drive flange through the oil seal and engage it with the differential gear. Using a suitable drift, drive the flange fully into the gear until the snap-ring is felt to engage.
11  Install the driveshaft (see Chapter 8).
12  Install the wheel, then lower the vehicle to the ground. Check and if necessary top up the transaxle oil level.

### INPUT SHAFT OIL SEAL

13  The transaxle must be removed for access to the input shaft oil

seal. Refer to Section 3 of this Chapter.

14  Remove the clutch release bearing and lever (see Chapter 8).

15  Unscrew the bolts and remove the guide sleeve from inside the bellhousing. Recover the gasket. Do not disturb any shims located on the input shaft.

16  Note the installed depth of the oil seal in the transaxle housing, then use a screwdriver to pry it out taking care not to damage the input shaft.

17  Wipe clean the oil seal seating and input shaft.

18  Smear a little multi-purpose grease on the lips of the new oil seal, then locate the seal over the input shaft with its sealing lip facing inwards. Tap the oil seal squarely into position, using a suitable tubular drift which bears only on the hard outer edge of the seal. Make sure the seal is fitted at the same depth in its housing as originally noted; it should be 3/16-inch (4.5 mm) below the guide sleeve mounting surface.

19  Install the guide sleeve together with a new gasket and new bolts, and tighten the bolts to the torque listed in this Chapter's Specifications.

20  Install the clutch release bearing and lever (see Chapter 8).

21  Install the transaxle (see Section 3).

## SELECTOR SHAFT OIL SEAL

22  Apply the parking brake, then raise the front of the vehicle and support it on jackstands.

23  Unscrew the locking bolt and slide the gearshift coupling from the transaxle selector shaft.

24  Using a small screwdriver, carefully pry the oil seal from the transaxle housing taking care not to damage the surface of the selector shaft or housing.

25  Wipe clean the oil seal seating and selector shaft, then smear a little multi-purpose grease on the new oil seal lips and locate the seal over the end of the shaft. Make sure the closed side of the seal faces outwards. To prevent damage to the oil seal, temporarily wrap some adhesive tape around the end of the shaft.

26  Tap the oil seal squarely into position, using a suitable tubular drift which bears only on the hard outer edge of the seal. The seal should be inserted until it is 1/32-inch (1.0 mm) below the surface of the transaxle.

27  Install the gearshift coupling and tighten the locking bolt.

28  Lower the vehicle to the ground.

## Specifications

### General

| Lubricant type and capacity | See Chapter 1 |
| --- | --- |

| Torque specifications | Ft-lbs (unless otherwise indicated) | Nm |
| --- | --- | --- |

➡Note: One foot-pound (ft-lb) of torque is equivalent to 12 inch-pounds (in-lbs) of torque. Torque values below approximately 15 ft-lbs are expressed in inch-pounds, since most foot-pound torque wrenches are not accurate at these smaller values.

| | Ft-lbs | Nm |
| --- | --- | --- |
| Transaxle-to-engine bolts | | |
| M10 | 33 | 45 |
| M12 | 48 | 65 |
| Shift linkage adjustment bolt | 17 | 23 |
| Guide sleeve bolt | | |
| Aluminum casing | 26 | 35 |
| Magnesium casing* | 18 | 25 |
| Multi-function switch | | |
| Aluminum casing | 18 | 25 |
| Magnesium casing* | 132 in-lbs | 15 |
| Lubricant level check/fill plug | 18 | 25 |
| Lubricant drain plug | 26 | 35 |

* On magnesium casings the code MgAl9Zn1 appears just in front of the left-hand driveaxle, and on the bottom of the casing behind the lefthand driveaxle.

**Section**

**Reference to other Chapters**

# 7B

## AUTOMATIC TRANSAXLE

## 1  General information

The automatic transaxle is a four or five speed unit, incorporating a hydrodynamic torque converter and a planetary gearbox.

The overall operation of the transaxle is managed by the Engine Control Module (ECM) and the Transmission Control Module (TCM), and as a result there are no manual adjustments. Comprehensive troubleshooting can therefore only be carried out using dedicated electronic test equipment.

Due to the complexity of the transaxle and its control system, major repairs and overhaul operations should be left to a dealer service department or other qualified repair facility, who will be equipped to carry out troubleshooting and repair. The information in this Chapter is therefore limited to a description of the removal and installation of the transaxle as a complete unit. The removal, installation and adjustment of the shift cable and key interlock cable is also described.

## 2  Diagnosis - general

Automatic transaxle malfunctions may be caused by five general conditions:

a) *Poor engine performance*
b) *Improper adjustments*
c) *Hydraulic malfunctions*
d) *Mechanical malfunctions*
e) *Malfunctions in the computer or its signal network*

Diagnosis of these problems should always begin with a check of the easily repaired items: fluid level and condition (see Chapter 1) and shift cable adjustment. Next, perform a road test to determine if the problem has been corrected or if more diagnosis is necessary. If the problem persists after the preliminary tests and corrections are completed, additional diagnosis should be done by a dealer service department or transmission repair shop. Refer to the *Troubleshooting* Section at the front of this manual for information on symptoms of transaxle problems.

### PRELIMINARY CHECKS

1   Drive the vehicle to warm the transaxle to normal operating temperature.

2   Check the fluid level as described in Chapter 1.

a) *If the fluid level is unusually low, add enough fluid to bring it up to the proper level, then check for external leaks (see below).*

b) *If the fluid level is abnormally high, drain off the excess, then check the drained fluid for contamination by coolant. The presence of engine coolant in the automatic transmission fluid indicates that a failure has occurred in the transaxle fluid cooler.*

c) *If the fluid is foaming, drain it and refill the transaxle, then check for coolant in the fluid.*

3   Look for a CHECK ENGINE light glowing on the instrument panel (see Chapter 6 for information).

➡**Note: If the engine or its control network is malfunctioning, do not proceed with the preliminary checks until it has been repaired and runs normally.**

4   Inspect the shift cable (see Section 3). Make sure that it's properly adjusted and that it operates smoothly.

### FLUID LEAK DIAGNOSIS

5   Most fluid leaks are easy to locate visually. Repair usually consists of replacing a seal or gasket. If a leak is difficult to find, the following procedure may help.

6   Identify the fluid. Make sure it's transmission fluid and not engine oil or brake fluid (the automatic transmission fluid in these models is a transparent yellow color).

7   Try to pinpoint the source of the leak. Drive the vehicle several miles, then park it over a large sheet of cardboard. After a minute or two, you should be able to locate the leak by determining the source of the fluid dripping onto the cardboard.

8   Make a careful visual inspection of the suspected component and the area immediately around it. Pay particular attention to gasket mating surfaces. A mirror is often helpful for finding leaks in areas that are hard to see.

9   If the leak still cannot be found, clean the suspected area thoroughly with a degreaser or solvent, then dry it.

10  Drive the vehicle for several miles at normal operating temperature and varying speeds. After driving the vehicle, visually inspect the suspected component again.

11  Once the leak has been located, the cause must be determined before it can be properly repaired. If a gasket is replaced but the sealing flange is bent, the new gasket will not stop the leak. The bent flange must be straightened.

12  Before attempting to repair a leak, check to make sure that the following conditions are corrected or they may cause another leak.

➡**Note: Some of the following conditions cannot be fixed without highly specialized tools and expertise. Such problems must be referred to a transmission shop or a dealer service department.**

## Gasket leaks

13 Check the pan periodically. Make sure the bolts are tight, no bolts are missing, the gasket is in good condition and the pan is flat (dents in the pan may indicate damage to the valve body inside).

14 If the pan gasket is leaking, the fluid level may be too high, the vent may be plugged, the pan bolts may be too tight, the pan sealing flange may be warped, the sealing surface of the transaxle housing may be damaged, the gasket may be damaged or the transaxle casting may be cracked or porous. If sealant instead of gasket material has been used to form a seal between the pan and the transaxle housing, it may be the wrong sealant.

## Seal leaks

15 If a transaxle seal is leaking, the fluid level may be too high, the vent may be plugged, the seal bore may be damaged, the seal itself may be damaged or improperly installed, the surface of the shaft protruding through the seal may be damaged or a loose bearing may be causing excessive shaft movement.

16 Make sure the filler tube seal is in good condition and the tube is properly seated. If transmission fluid is evident, check the O-ring for damage.

## Case leaks

17 If the case itself appears to be leaking, the casting is porous and will have to be repaired or replaced.

18 Make sure the oil cooler hose fittings are tight and in good condition.

## Fluid comes out vent pipe or fill tube

19 If this condition occurs, the transaxle is overfilled, there is coolant in the fluid, the vent is plugged or the drain-back holes are plugged.

---

## 3  Shift cable - removal, installation and adjustment

## REMOVAL

1  Move the shift lever to the P position

2  Apply the parking brake, then raise the front of the vehicle and support it on jackstands.

3  Working under the vehicle, remove the screws and lower the heatshield from the shifter mounting bracket onto the exhaust. Slide the heatshield forwards.

4  Release the cover from the bottom of the shift lever bracket by pressing the fastener to the rear.

5  Disconnect the inner cable by squeezing the clip to release the cable from the shift lever. Pull the cable casing from the support, then pull out the locking element and remove the cover and cable from the bottom of the shift lever assembly. Take care not to bend the cable excessively.

6  At the transmission end of the cable, use a screwdriver to pry the end of the inner cable up from the transmission lever.

7  Unscrew the bolt and detach the support bracket and cable from the side of the transmission.

8  Loosen the locknuts and detach the cable from the bracket. Withdraw the cable from under the car.

## INSTALLATION

9  Installation is a reversal of removal, but lightly grease the cable end fittings. Before lowering the car to the ground and before reconnecting the cable to the transmission lever, adjust the cable as follows.

## ADJUSTMENT

10  Move the shift lever inside the car to the P position.

11  Move the shift lever on the transmission to the P position, which is the rear stop. Make sure that both front wheels are locked by attempting to turn them in the same direction at the same time.

➡**Note: Even though the transmission is locked, it will still be possible to turn the front wheels in opposite directions, since the differential gears are able move in relation to each other.**

12  Loosen the bolt(s) securing the shift cable to the transmission.

13  Press the cable end onto the shift lever, then check that the cable is free of any stress by moving it side to side several times. Now tighten the cable securing bolt.

14  Check the adjustment by selecting P. With the brake pedal released, check that the selector lever cannot be moved out of the P position with the lever button pressed. Now depress the brake pedal and check that the lock solenoid releases enabling the selector lever to be moved to any position with the lever button pressed. Check that the display agrees with the actual position of the lever.

15  Select position N. With the brake pedal released, check that the selector lever is locked. Depress the pedal and check that the selector lever can be moved to any position. Note that it is only possible to select R with the button pressed.

16  Check that it is only possible to operate the starter motor in positions P and N with the button released.

17  Lower the car to the ground.

## 4   Shift interlock system - description, check and component replacement

### ✳✳ WARNING 1:

These models are equipped with airbags. Always disable the airbag system before working in the vicinity of any airbag system component to avoid the possibility of accidental deployment of the airbag(s), which could cause personal injury (see Chapter 12).

### ✳✳ WARNING 2:

Do not use a memory saving device to preserve the ECM's memory when working on or near airbag system components.

### ✳✳ CAUTION 1:

These models are equipped with an anti-theft radio. Before performing a procedure that requires disconnecting the battery, make sure you have the proper activation code.

### ✳✳ CAUTION 2:

Disconnecting the battery can cause driveability problems that require a scan tool to rectify. Additionally, disconnecting the battery may cause one or more warning lights on the instrument panel to illuminate, which will also require the use of a scan tool to turn off. Most scan tools available to the public do not have the capability to perform either of these tasks, which will necessitate taking the vehicle to a dealer service department or other properly equipped repair facility after service work has been performed. See Chapter 5, Section 1 for the use of an auxiliary voltage input device ("memory saver") before disconnecting the battery and for other precautions related to battery disconnection.

## DESCRIPTION

1   The shift lock system prevents the shift lever from being shifted out of Park or Neutral until the brake pedal is applied and the button on the lever is pushed in. It also prevents the ignition key from being turned to the Lock position until the shift lever has been placed in the Park position.

## SOLENOID CHECK

2   Remove the center console (see Chapter 11).
3   Follow the wiring harness from the shift lock solenoid back to the electrical connector, then unplug the connector. Using a pair of jumper wires, momentarily apply battery voltage and ground to the solenoid terminals and verify that there's an audible "click."

### ✳✳ CAUTION:

Don't apply battery voltage any longer than necessary to perform this check.

4   If the shift lock solenoid doesn't click when energized, replace it.

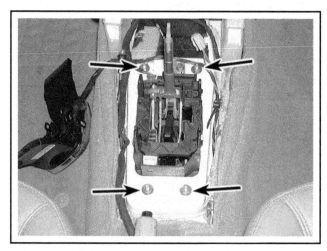

**4.11  The shifter housing is secured to the floorpan with four nuts**

## COMPONENT REPLACEMENT

### Shift lock solenoid

#### 01N transaxle

5   Remove the center console (see Chapter 11).
6   Rotate the solenoid 90-degrees counterclockwise to remove it. Unplug the electrical connector.
7   Installation is the reverse of removal.

#### 01V transaxle

▶ **Refer to illustration 4.11**

8   Remove the center console (see Chapter 11).
9   Detach the key interlock cable from the shift lever (see below).
10  Unplug all electrical connectors from the shift lever housing.
11  Remove the four nuts retaining the shift lever housing to the floorpan (see illustration).
12  Raise the vehicle and support it securely on jackstands.
13  Remove the center and rear portions of the exhaust system (see Chapter 4).
14  Remove the heatshield, then lower the shift lever housing from the floorpan.
15  Remove the clips and detach the shift lever. Drive out the pin from the locking pawl, then remove the shift lock solenoid and pawl as an assembly.
16  Reassembly and installation is the reverse of removal. Be sure to adjust the key interlock cable.

### Key interlock cable

#### Replacement

▶ **Refer to illustrations 4.22a, 4.22b, 4.23, 4.25 and 4.26**

17  Disconnect the cable from the negative terminal of the battery.
18  Remove the steering wheel (see Chapter 10).
19  Remove the steering column cover (see Chapter 11).
20  Remove the shift lever handle.

21 Remove the center console and the trim panel from around the radio (see Chapter 11).

22 Remove the shift knob. To do this, push down the sleeve directly below the knob as far as it will go, then pull out the button on the side of the knob approximately 7/32-inch (5.5 mm) and lift it off the shift lever (see illustrations).

23 Disengage the locking tabs and remove the cover and gear position indicator from the shift lever housing (see illustration).

24 Turn the ignition key to the On position and place the shift lever in the Park position.

25 Pull out the clip that retains the cable to the key lock cylinder housing, then detach the cable from the housing (see illustration).

26 Remove the clamp bolt, then pull up on the retaining spring and detach the interlock cable from the locking lever at the shifter housing.

27 Remove the cable from the instrument panel, noting how it's routed. To ease installation of the new cable, it's a good idea to attach a length of wire to the end of the old cable; as you remove the old cable, the length of wire will occupy its place. Then attach the wire to the new cable and use it to help you pull the new cable into place.

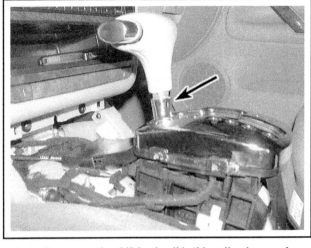

**4.22a  To remove the shift knob, slide this collar down as far as it will go (you may have to pry it down from the knob with a screwdriver) . . .**

**4.22b  . . . then pull the button on the knob out and slide the knob off the lever**

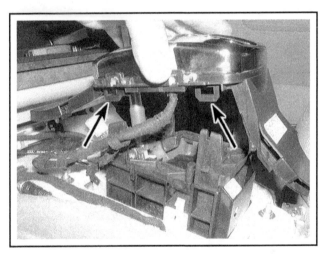

**4.23  The shift housing cover/gear position indicator is secured by two locking tabs on each side**

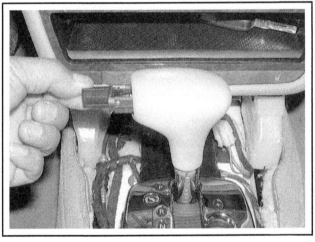

**4.25  Remove this clip to detach the key interlock cable from the lock cylinder housing**

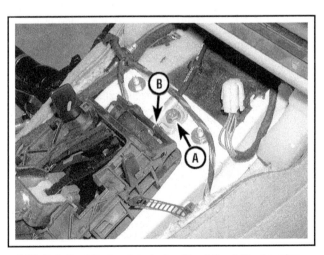

**4.26  Details of the key interlock cable at the shifter housing**

*A  Clamp bolt*              *B  Retaining spring*

28 To install the cable, reverse the removal procedure, then adjust the cable as described in the next Step.

➡Note: When installing the shift knob, pull the button outward, place it on the lever with the button pointing toward the front of the vehicle, push the knob down and rotate it so that the button is pointing toward the driver's seat. Push the button in until it clicks, then slide the collar up to the bottom of the knob.

### Adjustment

29 Remove the center console, if not already done (see Chapter 11).
30 Place the shift lever in the Park position, then turn the ignition

key to Lock and remove it from the lock cylinder.
31 Loosen the cable clamp bolt at the shifter housing (see illustration 4.26).
32 Place the steering column to the lowest position.
33 Prevent the cable end from moving, then pull the cable casing away from the shift housing to take out any slack. Tighten the bolt securely.
34 Turn the ignition key on, depress the brake pedal and move the shift lever to D, then back to P. Make sure it is now possible to remove the ignition key.

---

## 5  Automatic transaxle - removal and installation

### ✳✳ WARNING 1:

These models are equipped with airbags. Always disable the airbag system before working in the vicinity of any airbag system component to avoid the possibility of accidental deployment of the airbag(s), which could cause personal injury (see Chapter 12).

### ✳✳ WARNING 2:

Do not use a memory saving device to preserve the ECM's memory when working on or near airbag system components.

### ✳✳ CAUTION 1:

These models are equipped with an anti-theft radio. Before performing a procedure that requires disconnecting the battery, make sure you have the proper activation code.

### ✳✳ CAUTION 2:

Disconnecting the battery can cause driveability problems that require a scan tool to rectify. Additionally, disconnecting the battery may cause one or more warning lights on the instrument panel to illuminate, which will also require the use of a scan tool to turn off. Most scan tools available to the public do not have the capability to perform either of these tasks, which will necessitate taking the vehicle to a dealer service department or other properly equipped repair facility after service work has been performed. See Chapter 5, Section 1 for the use of an auxiliary voltage input device ("memory saver") before disconnecting the battery and for other precautions related to battery disconnection.

### REMOVAL

▶ Refer to illustrations 5.19a and 5.19b

1  Select a solid, level surface to park the vehicle upon. Give yourself enough space to move around it easily. Apply the parking brake and chock the rear wheels.

2  Disconnect the cable from the negative terminal of the battery.
3  Support the engine with a hoist or support bar located on the front fender inner channels. If necessary, remove the hood as described in Chapter 11 in order to position the hoist over the engine. The engine should be supported using both the front and rear lifting eyes. Depending on the engine, temporarily remove components as necessary to attach the hoist.
4  Unscrew and remove the transaxle-to-engine mounting bolts accessible from the engine compartment.
5  Loosen the front wheel bolts, then raise the front of the vehicle and support it on jackstands. Remove both front wheels.
6  Remove the under-vehicle splash shield and its bracket.
7  Where necessary, unscrew the nuts and remove the top cover from the engine.
8  Remove the exhaust system (see Chapter 4).
9  Disconnect the wiring from the Vehicle Speed Sensor (see Chapter 6).
10 Identify the wiring connections on the transaxle, then unplug them. Loosen and detach the wiring support, and position the wiring to one side.
11 With the selector lever in position P, carefully disconnect the inner cable from the shift lever, then unbolt the support bracket. Position the cable to one side.
12 Using an Allen key, unbolt the heat shields from over the inner end of the right-hand driveaxle.
13 Refer to Chapter 8 and detach the driveaxles from the transaxle flanges. Tie the driveaxles away from the transaxle. If you are working on an all-wheel drive model, remove the driveshaft.
14 Unbolt the right-hand transaxle mounting complete with rubber bushing and shield.
15 Position a suitable container beneath the transaxle to collect spilled hydraulic fluid.
16 Detach the hydraulic lines from the transaxle, and recover the sealing rings. Plug the openings in the transaxle housing to prevent entry of dust and dirt.
17 Remove the starter motor (see Chapter 5).
18 Turn the engine to locate one of the torque converter-to-driveplate nuts or bolts in the starter motor aperture. Unscrew and remove the nuts or bolts while preventing the engine from turning using a wide-bladed screwdriver engaged with the ring gear teeth on the driveplate. Unscrew the remaining two fasteners, turning the engine a third of a turn at a

time to locate them. Mark the relationship of the torque converter to the driveplate so balance will be preserved when reinstalling the transaxle.

19 Unscrew the transaxle-to-engine mounting bolts accessible from under the car (see illustrations).

20 Support the transaxle on a jack, preferably one made for this purpose. Secure the transaxle to the jack with a safety chain. Unbolt the right-hand transaxle mount complete with rubber bushing.

21 On four-cylinder models, mark the location of the subframe beneath the engine compartment, then loosen the front subframe bolts. Support the rear of the subframe with a floor jack. Remove the remaining subframe bolts and lower the rear of the subframe.

➡Note: It is important that the subframe is reinstalled in its correct position otherwise the handling of the car will be affected and excessive wear will occur.

22 With the help of an assistant, withdraw the transaxle from the locating dowels on the rear of the engine, making sure that the torque converter remains fully engaged with the transaxle input shaft. If necessary, use a lever to release the torque converter from the driveplate.

23 When the locating dowels are clear of their mounting holes, lower the transaxle to the ground using the jack. Strap a restraining bar across the front of the bellhousing to keep the torque converter in position.

**✳✳ WARNING:**

**Make sure that the transaxle remains steady on the jack head. Take care to prevent the torque converter from falling out as the transaxle is removed.**

24 Where necessary, remove the intermediate plate from the locating dowels.

## INSTALLATION

25 Installation of the transaxle is a reversal of the removal procedure, but note the following points:

a) *As the torque converter is reinstalled, ensure that the drive pins at the center of the torque converter hub engage with the recesses in the automatic transaxle fluid pump inner wheel.*

b) *When installing the transaxle, make sure the marked stud on the torque converter engages with the marked hole on the driveplate.*

c) *Tighten the bellhousing bolts and torque converter-to-driveplate nuts to the specified torque. Always replace self-locking nuts and bolts.*

d) *Replace the O-ring seals on the fluid pipes and filler tube attached to the transaxle casing.*

e) *Tighten the transaxle mounting bolts to the correct torque.*

f) *Tighten the driveaxle inner CV joint bolts and, on all-wheel drive models, the driveshaft-to-center differential flange bolts, to the torque values listed in the Chapter 8 Specifications.*

g) *Check the final drive oil level and transaxle fluid level as described in Chapter 1.*

h) *On completion, refer to Section 4 and check the shift cable adjustment.*

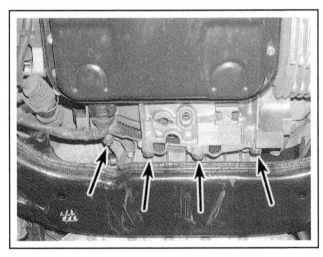

**5.19a From below, these four transaxle mounting fasteners can be removed (V6 engine shown)**

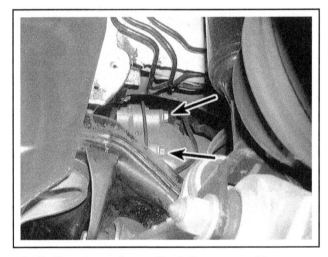

**5.19b These transaxle mounting bolts are accessible through the left wheel well**

## 6 Automatic transaxle overhaul - general information

In the event of a fault occurring, it will be necessary to establish whether the fault is electrical, mechanical or hydraulic in nature, before repair work can be contemplated. Diagnosis requires detailed knowledge of the transaxle's operation and construction, as well as access to specialized test equipment, and so is deemed to be beyond the scope of this manual. It is therefore essential that problems with the automatic transaxle are referred to a dealer service department or other qualified repair facility for assessment.

Note that a faulty transaxle should not be removed before the vehicle has been assessed by a knowledgeable technician equipped with the proper tools, as troubleshooting must be performed with the transaxle installed in the vehicle.

## 7   Transaxle mounts - check and replacement

▶ **Refer to illustration 7.5**

### CHECK

1   Raise the vehicle and support it securely on jackstands.
2   Insert a large screwdriver or prybar between the transaxle and the subframe and try to pry the transaxle up slightly.
3   The transmission should not move much at all - if the mount is cracked or torn, replace it.

### REPLACEMENT

4   Support the transaxle with a floor jack. Place a block of wood on the jack head to act as a cushion.
5   Remove the bolts and nuts attaching the mount to the subframe and transaxle (see illustration).
6   Raise the transaxle slightly with the jack and remove the mount.
7   Installation is the reverse of the removal procedure. Be sure to tighten all fasteners securely.

**7.5  Transaxle mount fasteners**

## Specifications

### General

| Designation | |
|---|---|
| Four-speed | 01N |
| Five speed | 01V |
| Automatic transaxle fluid type and capacity | See Chapter 1 |

| Torque specifications | Ft-lbs (unless otherwise indicated) | Nm |
|---|---|---|

➡**Note: One foot-pound (ft-lb) of torque is equivalent to 12 inch-pounds (in-lbs) of torque. Torque values below approximately 15 ft-lbs are expressed in inch-pounds, since most foot-pound torque wrenches are not accurate at these smaller values.**

| | Ft-lbs | Nm |
|---|---|---|
| Automatic transaxle selector cable support bolt | 17 | 23 |
| Bracket for fluid pipe | 84 in-lbs | 10 |
| Torque converter-to-driveplate bolts | 63 | 85 |
| Fluid pipes-to-transaxle bolt | 15 | 20 |
| Fluid pipe unions | 22 | 30 |
| Transaxle bellhousing-to-engine bolts | | |
| M10 | 33 | 45 |
| M12 | 48 | 65 |
| Transaxle mount | | |
| Center bolt | 30 | 40 |
| To subframe | 17 | 23 |
| To transaxle | 30 | 40 |

**Section**

**Reference to other Chapters**

# 8

# CLUTCH AND DRIVELINE

## 1 General information

The information in this Chapter deals with the components from the rear of the engine to the drive wheels, except for the transaxle, which is dealt with in Chapter 7A and 7B. Included in this Chapter is service information on the clutch and its release system, driveaxles, and, on all-wheel drive models, the driveshaft and rear differential. Information on the center differential on all-wheel drive models can be found in Chapter 7.

Since nearly all the procedures covered in this Chapter involve working under the vehicle, make sure it's securely supported on sturdy jackstands or a hoist where the vehicle can be easily raised and lowered.

## 2 Clutch - description and check

1   All models with a manual transaxle use a single dry plate, diaphragm spring type clutch. The clutch disc has a splined hub which allows it to slide along the splines of the transaxle input shaft. The clutch and pressure plate are held in contact by spring pressure exerted by the diaphragm in the pressure plate.

2   The clutch release system is hydraulically operated. The release system consists of the clutch pedal, the clutch master cylinder, the clutch release cylinder, the hydraulic line between the master cylinder and release cylinder, the clutch release bearing and the clutch release lever.

3   When pressure is applied to the clutch pedal to release the clutch, the clutch master cylinder transmits this movement to the clutch release cylinder, which moves the clutch release lever. As the lever pivots, the release bearing pushes against the fingers of the diaphragm spring of the pressure plate assembly, which in turn releases the clutch plate.

4   Terminology can be a problem regarding the clutch components because common names have in some cases changed from that used by the manufacturer. For example, the clutch release cylinder is sometimes referred to as a slave cylinder, the driven plate is also called the clutch plate or disc, the pressure plate assembly is also known as the clutch cover, and the clutch release bearing is sometimes called a throw-out bearing.

5   Other than replacing components that have obvious damage, some preliminary checks should be performed to diagnose a clutch system failure.

*a) To check clutch "spin down" time, run the engine at normal idle speed with the transaxle in Neutral (clutch pedal up, engaged). Disengage the clutch (pedal down), wait several seconds and shift the transaxle into Reverse. No grinding noise should be heard. A grinding noise would most likely indicate a problem in the pressure plate or the clutch disc.*

*b) To check for complete clutch release, run the engine (with the parking brake applied to prevent movement) and hold the clutch pedal approximately 1/2-inch from the floor. Shift the transaxle between 1st gear and Reverse several times. If the shift is not smooth, component failure is indicated.*

*c) Visually inspect the clutch pedal pivot at the top of the clutch pedal to make sure there is no sticking or excessive wear.*

*d) Make sure that the hydraulic lines aren't leaking at either the master cylinder or the release cylinder (see Sections 3 and 4). Bleed the system if necessary (see Section 5).*

## 3 Clutch master cylinder - removal and installation

▶ **Refer to illustrations 3.4, 3.7 and 3.8**

### REMOVAL

1   Place the shift lever in Neutral and remove the key from the ignition switch.

2   If necessary for access to the hydraulic lines, remove the electronics box from the cowl chamber (see Chapter 6).

3   Place rags on the floor under the clutch pedal to absorb any brake fluid that may spill.

4   Working in the cowl chamber, detach the fluid supply hose from the clutch master cylinder (see illustration). Have a plug ready and immediately plug the line.

5   Detach the pressure line fitting from the clutch master cylinder. On some models a quick-connect fitting is used; pull out the clip to separate the line from the master cylinder (see illustration 3.4). On other models a conventional hydraulic fitting is used; unscrew the fitting nut using a flare nut wrench to avoid rounding off the corners of the fitting. Plug the line to prevent excessive fluid loss and the entry of contaminants.

6   Working inside the vehicle, remove the left-side under-dash panel.

7   Unclip and slide out the pin securing the clutch master cylinder to the clutch pedal (see illustration).

8   Unscrew the mounting bolts and detach the master cylinder from the firewall (see illustration). Be careful not to spill fluid on the carpet.

## INSTALLATION

9   Installation is the reverse of removal, with the following points:

   a) *If you're installing a new cylinder, measure the distance from the mounting flange (firewall side) to the center of the mounting hole on the pushrod. It should be approximately 6-1/2 inches (165*

*mm). If necessary, loosen the locknut and turn the pushrod end on the pushrod to achieve this dimension.*

   b) *Tighten all fasteners to the torque values listed in this Chapter's Specifications.*

   c) *Bleed the clutch hydraulic system (see Section 5).*

   d) *Check the brake fluid level in the brake fluid reservoir, adding as necessary to bring it to the appropriate level (see Chapter 1).*

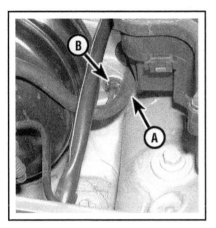

**3.4  Clutch master cylinder feed hose (A) and pressure line fitting (B)**

**3.7  Unclip the pushrod pin and slide it out**

**3.8  Clutch master cylinder mounting bolts**

## 4   Clutch release cylinder - removal and installation

▶ **Refer to illustration 4.3**

### REMOVAL

1   Place the shift lever in Neutral and remove the key from the ignition switch.

2   Apply the parking brake and block the rear wheels to prevent the vehicle from rolling. Raise the front of the vehicle and support it securely on jackstands.

3   Detach the pressure line from the release cylinder. On some models a quick-connect fitting is used; pull out the clip to separate the line from the master cylinder (see illustration). On other models a conventional hydraulic fitting is used; unscrew the fitting nut using a flare nut wrench to avoid rounding off the corners of the fitting. Have some rags handy, as some fluid will be spilled when the line is disconnected. Plug the line to prevent excessive fluid loss and the entry of contaminants.

4   Remove the release cylinder mounting bolt and pull the cylinder straight out of the transaxle.

### INSTALLATION

5   Lubricate the end of the pushrod with copper grease. Apply lithium-base grease to the area of the boot that seats in the bore in the transaxle. Install the release cylinder, inserting it straight into its bore (otherwise the pushrod may not seat in its pocket in the release lever).

6   Attach the bracket for the hydraulic line (if equipped), install the release cylinder mounting bolt and tighten it to the torque listed in this

Chapter's Specifications.

7   Connect the pressure line to the cylinder. If equipped with a quick-connect fitting insert the clip and make sure the line is completely attached and won't pull off. If equipped with a conventional hydraulic fitting, tighten the fitting securely with a flare-nut wrench.

8   Bleed the system (see Section 5).

9   Check the fluid level in the brake fluid reservoir, adding as necessary to bring it to the appropriate level (see Chapter 1).

10  Wash off any spilled brake fluid with water.

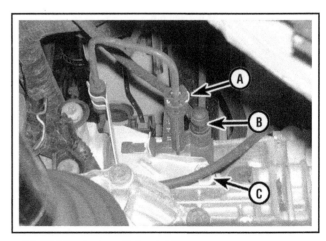

**4.3  Clutch release cylinder details**

| A | Hydraulic line fitting | C | Mounting bolt |
|---|---|---|---|
| B | Bleeder screw | | |

## 5   Clutch hydraulic system - bleeding

1   The hydraulic system should be bled of all air whenever any part of the system has been removed or if the fluid level has been allowed to fall so low that air has been drawn into the master cylinder. The procedure is similar to bleeding a brake system.

2   Fill the brake master cylinder with new brake fluid conforming to DOT 4 specifications.

### ✳✳ WARNING:

**Do not re-use any of the fluid coming from the system during the bleeding operation or use fluid which has been inside an open container for an extended period of time.**

3   Apply the parking brake and block the rear wheels to prevent the vehicle from rolling. Raise the front of the vehicle and support it securely on jackstands. Locate the bleeder screw on the clutch release cylinder (see illustration 4.3). Remove the dust cap from the bleeder screw and push a length of snug-fitting (preferably clear) hose over the screw. Place the other end of the hose into a clear container with about two inches of brake fluid in it. The hose end must be submerged in the fluid.

4   Have an assistant depress the clutch pedal and hold it. Open the bleeder screw on the release cylinder, allowing fluid to flow through the hose. Close the bleeder screw when fluid stops flowing from the hose. Once closed, have your assistant release the pedal.

5   Continue this process until all air is evacuated from the system, indicated by a full, solid stream of fluid being ejected from the bleeder screw each time and no air bubbles in the hose or container. Keep a close watch on the fluid level inside the brake master cylinder reservoir; if the level drops too low, air will be sucked back into the system and the process will have to be started over again.

6   Install the dust cap on the bleeder screw. Check carefully for proper operation before placing the vehicle in normal service.

7   Lower the vehicle.

8   Recheck the brake fluid level.

## 6   Clutch components - removal, inspection and installation

### ✳✳ WARNING:

**Dust produced by clutch wear and deposited on clutch components is hazardous to your health. DO NOT blow it out with compressed air and DO NOT inhale it. DO NOT use gasoline or petroleum-based solvents to remove the dust. Brake system cleaner should be used to flush the dust into a drain pan. After the clutch components are wiped clean with a rag, dispose of the contaminated rags and cleaner in a labeled, covered container.**

### REMOVAL

▶ **Refer to illustrations 6.4 and 6.6**

1   Access to the clutch components is normally accomplished by removing the transaxle, leaving the engine in the vehicle. If, of course, the engine is being removed for major overhaul, then the opportunity should always be taken to check the clutch for wear and replace worn components as necessary. However, the relatively low cost of the clutch components compared to the time and labor involved in gaining access to them warrants their replacement any time the engine or transaxle is removed, unless they are new or in near-perfect condition. The following procedures assume that the engine will stay in place.

2   Remove the transaxle from the vehicle (see Chapter 7A). Support the engine while the transaxle is out. Preferably, an engine hoist or support fixture should be used to support it from above. However, if a jack is used underneath the engine, make sure a piece of wood is used between the jack and oil pan to spread the load.

3   The release lever and release bearing can remain attached to the transaxle; however, you should inspect them (see Section 7) while the transaxle is removed.

4   Carefully inspect the flywheel and pressure plate for indexing marks. The marks are usually an X, an O or a white letter. If they cannot be found, scribe marks yourself so the pressure plate and the flywheel will be in the same alignment during installation (see illustration). Of course, this won't be necessary if you're planning on replacing the pressure plate with a new one.

5   Slowly loosen the pressure plate-to-flywheel bolts. Work in a criss-cross pattern and loosen each bolt a little at a time until all spring pressure is relieved.

6   Hold the pressure plate securely and completely remove the bolts, followed by the pressure plate and clutch disc (see illustration).

**6.4  Mark the relationship of the pressure plate to the flywheel (in case you're going to re-use the same pressure plate)**

**6.6 When removing the pressure plate, be careful not to let the clutch disc fall out**

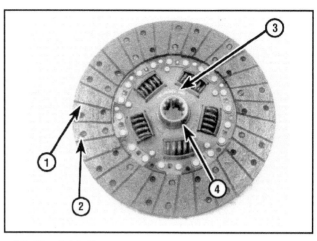

**6.9 The clutch disc**

*1* **Lining** - *this will wear down in use*
*2* **Rivets** - *these secure the lining and will damage the flywheel or pressure plate if allowed to contact the surfaces*
*3* **Markings** - *"Flywheel side" or something similar*
*4* **Hub** - *Be sure this is installed facing the proper direction (see the text)*

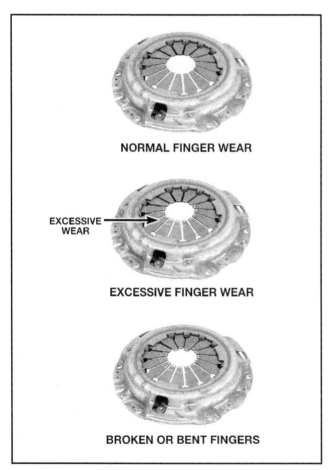

**NORMAL FINGER WEAR**

EXCESSIVE
WEAR

**EXCESSIVE FINGER WEAR**

**BROKEN OR BENT FINGERS**

**6.11a Replace the pressure plate if excessive wear is noted**

**6.11b Examine the pressure plate friction surface for score marks, cracks and evidence of overheating**

## INSPECTION

▶ **Refer to illustrations 6.9, 6.11a and 6.11b**

7   Ordinarily, when a problem occurs in the clutch, it can be attributed to wear of the clutch driven plate assembly (clutch disc). However, all components should be inspected at this time.

8   Inspect the flywheel for cracks, heat checking, score marks and other damage. If the imperfections are slight, a machine shop can resurface it to make it flat and smooth. Refer to Chapter 2A for the flywheel removal procedure.

9   Inspect the lining on the clutch disc. There should be at least 1/16-inch of lining above the rivet heads. Check for loose rivets, distortion, cracks, broken springs and other obvious damage (see illustration). As mentioned above, ordinarily the clutch disc is replaced as a matter of course, so if in doubt about the condition, replace it with a new one.

10   The release bearing should be replaced along with the clutch disc (see Section 7).

11   Check the machined surface and the diaphragm spring fingers of the pressure plate (see illustrations). If the surface is grooved or otherwise damaged, replace the pressure plate assembly. Also check for obvious damage, distortion, cracking, etc. Light glazing can be removed with emery cloth or sandpaper. If a new pressure plate is indicated, new or factory rebuilt units are available.

## INSTALLATION

♦ **Refer to illustration 6.14**

12 Position the clutch disc and pressure plate with the clutch held in place with an alignment tool. Make sure the disc is installed properly. If your vehicle is equipped with a one-piece flywheel, the spring cage must face the pressure plate. If the vehicle is equipped with a two-piece flywheel, the shorter hub end must face the pressure plate. You may find the word "Getriebeseite" stamped on the hub of the disc; if so, this side must face the pressure plate. Also, if the vehicle is equipped with a two-piece flywheel, the white mark on the flywheel must line-up with the white mark on the pressure plate.

13 Tighten the pressure plate-to-flywheel bolts only finger tight, working around the pressure plate.

14 Center the clutch disc by ensuring the alignment tool is through the splined hub and into the recess in the crankshaft (see illustration). Wiggle the tool up, down or side-to-side as needed to bottom the tool. Tighten the pressure plate-to-flywheel bolts a little at a time, working in a criss-cross pattern to prevent distortion of the cover. After all of the bolts are snug, tighten them to the torque listed in this Chapter's Specifications. Remove the alignment tool.

15 Using moly-base grease, lubricate the inner surface of the release bearing and the face of the bearing where it contacts the fingers of the pressure plate diaphragm spring. Also place grease on the release lever contact areas and the transaxle input shaft.

### ✳✳ CAUTION:

**Don't use too much grease.**

16 Install the clutch release bearing (see Section 7).

17 Install the transaxle and all components removed previously, tightening all fasteners to the proper torque specifications.

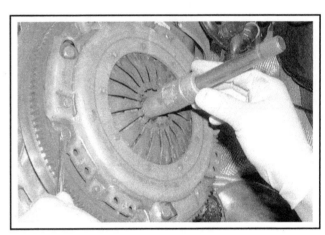

**6.14 Center the clutch disc with a clutch alignment tool, then tighten the pressure plate bolts a little at a time, in a criss-cross pattern, to the torque listed in this Chapter's Specifications**

---

## 7 Clutch release bearing and lever - removal, inspection and installation

♦ **Refer to illustrations 7.2, 7.3, 7.5, 7.6, 7.7a, 7.7b and 7.7c**

### ✳✳ WARNING:

**Dust produced by clutch wear and deposited on clutch components is hazardous to your health. DO NOT blow it out with compressed air and DO NOT inhale it. DO NOT use gasoline or petroleum-based solvents to remove the dust. Brake system cleaner should be used to flush it into a drain pan. After the clutch components are wiped clean with a rag, dispose of the contaminated rags and cleaner in a labeled, covered container.**

## REMOVAL

1 Remove the transaxle as described in Chapter 7A.

2 Use a screwdriver to pry the release lever from the ballstud inside the transaxle bellhousing. If this proves difficult, push the spring clip from the pivot end of the release lever by pushing it through the hole (see illustration). This will release the pivot end of the lever from the ballstud. Now withdraw the lever together with the release bearing from the guide sleeve.

3 Use a screwdriver to depress the plastic tabs and separate the bearing from the lever (see illustration).

4 Remove the plastic pivot from the ballstud. The release lever locates on the plastic pivot.

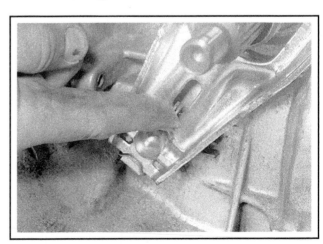

**7.2 Push the spring clip to release the lever from the ballstud**

**7.3 Pry inward on the plastic tabs retaining the bearing to the lever, then separate the bearing from the lever**

**7.5 To check the bearing, hold it by the outer race and rotate the inner race while applying pressure; if the bearing doesn't turn smoothly or if it's noisy, replace it**

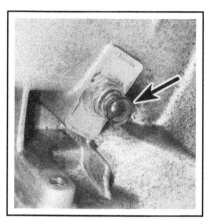

**7.6 Apply copper grease to the release lever ballstud**

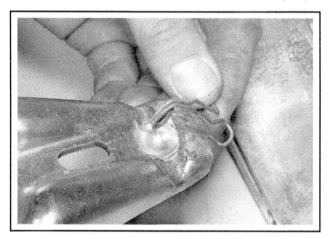

**7.7a Locate the spring over the end of the release lever . . .**

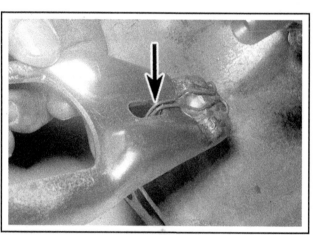

**7.7b . . . and press the spring into the hole . . .**

## INSPECTION

5   Hold the bearing by the outer race and rotate the inner race while applying pressure (see illustration). If the bearing doesn't turn smoothly, or if it's noisy, replace the bearing with a new one. Wipe the bearing with a clean rag and inspect it for damage, wear and cracks. Don't immerse the bearing in solvent; it's sealed for life and to do so would ruin it. Also check the release lever for cracks and bends.

## INSTALLATION

6   Begin installation by lubricating the ballstud and plastic pivot with a little copper grease (see illustration). Smear a little grease on the release bearing surface which contacts the diaphragm spring fingers and the release lever, and also on the guide sleeve.
7   Install the spring onto the release lever, making sure the plastic pivot is in place on the ballstud. Position the lever and bearing and press the release lever onto the ballstud until the spring holds it in position (see illustrations).
8   Install the transaxle (see Chapter 7A).

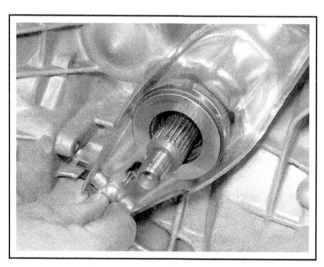

**7.7c . . . then press the release lever onto the ballstud until the spring clip holds it in position**

## 8    Clutch pedal and over-center spring - removal and installation

▶ **Refer to illustration 8.5**

1    Remove the panel from below the steering column (refer to Chapter 11 if necessary).
2    Disconnect the wiring and pull out the switch above the clutch pedal.
3    Release the clip securing the clutch master cylinder pushrod pin from the pedal by twisting it upwards then pulling it from the pedal (see illustration 3.7). Pull up the pedal to disconnect it from the pushrod.
4    Pry the pedal retaining clip from the groove in the left end of the pivot shaft.
5    Remove the clip from the pedal pivot shaft (see illustration), then push the shaft to the right until the clutch pedal can be removed from the bracket.
6    Pry the over-center spring from the slots in the pedal bracket.
7    Installation is the reverse of removal.

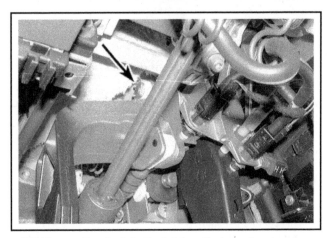

**8.5  Remove the pivot shaft retaining clip (arrow) and push the shaft in to free the pedal**

## 9    Driveaxles - general information and inspection

1    Power is transmitted from the transaxle to the front wheels through a pair of driveaxles. On all-wheel drive models power is also transmitted to the rear wheels through two driveaxles. The inner end of each driveaxle is bolted to a drive flange protruding from the differential; the outer end of each driveaxle has a stub shaft that is splined to the hub and bearing assembly and locked in place with a large bolt.
2    The inner ends of the driveaxles are equipped with sliding constant velocity (CV) joints, which are capable of both angular and axial motion. Each inner CV joint assembly consists of a either a triple rotor-type bearing or a ball-and-cage type bearing and a housing in which the joint is free to slide in-and-out as the driveaxle moves up-and-down with the wheel.
3    The outer ends of the driveaxles are equipped with "ball-and-cage" type CV joints, which are capable of angular but not axial movement. Each outer CV joint consists of six ball bearings running between an inner race and an outer cage.

4    The boots should be inspected periodically for damage and leaking lubricant. Torn CV joint boots must be replaced immediately or the joints will be damaged. If either boot of a driveaxle is damaged, that driveaxle must be removed in order to replace the boot (see Section 10).
5    Should a boot be damaged, the CV joint can be disassembled and cleaned, but if any parts are damaged, the entire driveaxle assembly may have to be replaced as a unit - check with your local auto parts store regarding the availability of replacement parts and CV joints (see Section 11).
6    The most common symptom of worn or damaged CV joints, besides lubricant leaks, is a clicking noise in turns, a clunk when accelerating after coasting and vibration at highway speeds. To check for wear in the CV joints and driveaxle shafts, grasp each axle (one at a time) and rotate it in both directions while holding the CV joint housings, feeling for play indicating worn splines or sloppy CV joints. Also check the driveaxle shafts for cracks, dents and distortion.

## 10   Driveaxle - removal and installation

✳✳ **WARNING:**

The manufacturer recommends replacing the driveaxle/hub bolt with a new one whenever it is removed.

FRONT

**Removal**

▶ **Refer to illustration 10.3**

1    Remove the wheel trim/hub cap (as applicable) and loosen the

**10.3 Remove the bolts retaining the inner CV joint to the drive flange**

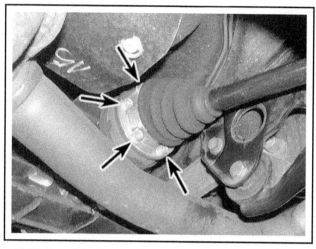

**10.22 Rear driveaxle inner CV joint bolts**

driveaxle/hub bolt with the vehicle resting on its wheels. Also loosen the wheel bolts.

2 Raise the front of the vehicle and support it securely on jackstands. Block the rear wheels to prevent the vehicle from rolling off the stands. Remove the wheel.

3 Unscrew the bolts securing the inner CV joint to the transaxle drive flange and remove the retaining plates from underneath the bolts (see illustration). Support the driveaxle by suspending it with wire or string - do not allow it to hang under its weight, or the joint may be damaged.

4 Pull the ABS wheel speed sensor partially out of its mounting hole in the steering knuckle (see Chapter 9). This will prevent it from becoming damaged as the driveaxle is withdrawn from the hub.

5 Remove the driveaxle/hub bolt.

6 Detach the upper control arms from the steering knuckle (see Chapter 10).

7 Carefully swing the steering knuckle outwards and withdraw the driveaxle outer constant velocity joint from the hub. If the splines of the outer joint are stuck in the hub, tap the joint out of the hub using a brass drift. If this fails to free it from the hub, the joint will have to be pressed out using a puller which is bolted to the hub.

8 Maneuver the driveaxle out from underneath the vehicle and recover the gasket from the end of the inner constant velocity joint, if present. Discard the gasket - a new one should be used on installation.

9 Don't allow the vehicle to rest on its wheels with one (or both) driveaxle(s) removed, as damage to the wheel bearing(s) may result. If moving the vehicle is unavoidable, temporarily insert the outer end of the driveaxle(s) in the hub(s) and tighten the driveaxle/hub bolt(s); in this case, the inner end(s) of the driveaxle(s) must be supported, for example by suspending with string from the vehicle underbody. Do not allow the driveaxle to hang down, as the joint may be damaged.

## Installation

10 Ensure that the transaxle drive flange and inner joint mating surfaces are clean and dry. Install a new gasket to the joint by peeling off its backing foil and sticking it in position.

11 Ensure that the outer joint and hub splines and threads are clean, then lubricate the splines with a light coat of multi-purpose grease.

12 Maneuver the driveaxle into position and engage the outer joint with the hub. Install a new bolt and tighten it to draw the joint fully into

position. Don't tighten it completely yet (wait until the wheel is installed and the vehicle has been lowered to the ground).

13 Connect the control arms to the steering knuckle, tightening the pinch bolt nut to the torque listed in the Chapter 10 Specifications.

14 Align the driveaxle inner joint with the transaxle flange and install the retaining bolts and plates. Tighten the bolts to the torque listed in this Chapter's Specifications.

15 Ensure that the outer joint is drawn fully into position, then install the wheel and lower the vehicle to the ground.

16 The remainder of installation is the reverse of removal, with the following points:

  a) *Tighten the driveaxle/hub bolt to the torque and angle of rotation listed in this Chapter's Specifications.*

  b) *Once the driveaxle/hub bolt is correctly tightened, tighten the wheel lug bolts to the torque listed in the Chapter 1 Specifications and install the wheel trim/hub cap.*

## REAR (ALL-WHEEL DRIVE MODELS)

### Removal

▶ **Refer to illustration 10.22**

17 Remove the wheel trim/hub cap (as applicable) and loosen the driveaxle/hub bolt with the vehicle resting on its wheels. Also loosen the wheel bolts.

18 Raise the rear of the vehicle and support it securely on jackstands. Block the front wheels to prevent the vehicle from rolling off the stands. Remove the wheel.

19 If you're working on an Audi A4 and removing the left-side drive-axle, you may have to remove the rear portion of the exhaust system for access (see Chapter 4).

20 Partially remove the ABS wheel speed sensor to prevent damaging it when the driveaxle is removed (see Chapter 9).

21 If you're working on a VW Passat, unbolt the cover plate from below the driveaxle.

22 Unscrew the bolts securing the inner CV joint to the differential drive flange and remove the retaining plates from underneath the bolts (see illustration). Support the driveaxle by suspending it with wire or string - do not allow it to hang under its weight, or the joint may be damaged.

23 Remove the driveaxle/hub bolt.

24 If you're working on an Audi A4, detach the stabilizer bar link and the upper control arm from the knuckle (see Chapter 10).

25 If you're working on an A4, carefully swing the rear knuckle outwards and withdraw the driveaxle outer constant velocity joint from the hub. If you're working on a Passat, simply lower the inner end of the driveaxle and slide the outer end of the driveaxle out of the hub. On either model, if the splines of the outer joint are stuck in the hub, tap the joint out of the hub using a brass drift. If this fails to free it from the hub, the joint will have to be pressed out using a puller which is bolted to the hub.

26 Maneuver the driveaxle out from underneath the vehicle and recover the gasket from the end of the inner CV joint, if present. Discard the gasket - a new one should be used on installation.

27 Don't allow the vehicle to rest on its wheels with one (or both) driveaxle(s) removed, as damage to the wheel bearing(s) may result. If moving the vehicle is unavoidable, temporarily insert the outer end of the driveaxle(s) in the hub(s) and tighten the driveaxle/hub bolt(s); in this case, the inner end(s) of the driveaxle(s) must be supported, for example by suspending with string from the vehicle underbody. Do not allow the driveaxle to hang down, as the joint may be damaged.

## Installation

28 Ensure that the differential drive flange and inner joint mating surfaces are clean and dry. Install a new gasket to the joint by peeling off its backing foil and sticking it in position.

29 Ensure that the outer joint and hub splines and threads are clean, then lubricate the splines with a light coat of multi-purpose grease.

30 Maneuver the driveaxle into position and engage the outer joint with the hub. Install a new bolt and tighten it to draw the joint fully into position. Don't tighten it completely yet (wait until the wheel is installed and the vehicle has been lowered to the ground).

31 If you're working on an A4, connect the upper control arm and the stabilizer bar link to the steering knuckle, tightening the fasteners to the torque listed in the Chapter 10 Specifications.

### ✳✳ CAUTION:

**Before tightening the fasteners, raise the rear knuckle with a floor jack to simulate normal ride height.**

32 Align the driveaxle inner joint with the differential flange and install the retaining bolts and plates. Tighten the bolts to the torque listed in this Chapter's Specifications.

33 Install the rear portion of the exhaust system, if removed.

34 Ensure that the outer joint is drawn fully into position, then install the wheel and lower the vehicle to the ground.

35 The remainder of installation is the reverse of removal, with the following points:

   a) Tighten the driveaxle/hub bolt to the torque and angle of rotation listed in this Chapter's Specifications.

   b) Once the driveaxle/hub bolt is correctly tightened, tighten the wheel lug bolts to the torque listed in the Chapter 1 Specifications and install the wheel trim/hub cap.

## 11  Driveaxle boot - replacement

1  Remove the driveaxle from the vehicle as described in Section 10.

## OUTER CV JOINT (ALL MODELS)

▶ **Refer to illustrations 11.4, 11.5a, 11.5b, 11.7, 11.8, 11.9, 11.10, 11.17, 11.18, 11.20a, 11.20b and 11.22**

2  Secure the driveaxle in a vise equipped with soft jaws, then loosen the two outer joint boot retaining clamps. If necessary, the clamps can be cut off.

3  Slide the boot down the shaft to expose the constant velocity (CV) joint and wipe off as much grease as possible.

4  Using a hammer and a brass punch, tap the joint off the end of the driveaxle (see illustration).

### ✳✳ CAUTION:

**Place the punch on the inner race of the CV joint only.**

5  Remove the circlip from the driveaxle groove, then slide off the thrust washer and dished washer, noting which way they are installed (see illustration).

6  Slide the boot off the driveaxle and discard it.

7  Clean the outer CV joint assembly to remove as much grease as possible. Mark the relative position of the bearing cage, inner race and housing (see illustration).

8  Mount the outer CV joint in a vise equipped with soft jaws. Push down on one side of the cage and remove the ball bearing from the opposite side. Repeat this procedure until all of the balls are removed (see illustration). If the joint is tight, tap on the inner race (not the cage) with a hammer and brass punch.

9  Remove the cage and inner race assembly from the housing by tilting it vertically and aligning two opposing cage windows in the area between the ball grooves (see illustration).

10  Turn the inner race 90-degrees to the cage and align one of the spherical lands with a cage window. Raise the land into the window and swivel the inner race out of the cage (see illustration).

11  Clean all of the parts with solvent and dry them off.

12  Inspect the housing, splines, balls and races for damage, corrosion, wear and cracks. Check the inner race, for wear and scoring in the races. If any of the components are not serviceable, the entire CV joint assembly must be replaced with a new one. If the joint is in satisfactory condition, obtain a boot replacement kit; kits usually contain a new boot and retaining clamps, a constant velocity joint snap-ring and the correct type of grease. If grease isn't included in the kit, be sure to obtain some

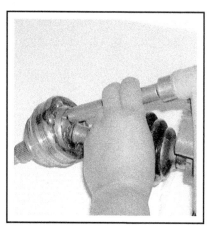

**11.4 Drive the outer CV joint off the shaft with a hammer and a brass punch**

**11.5a Remove the thrust washer . . .**

**11.5b . . . and the dished washer**

**11.7 Mark the relationship of the bearing cage, inner race and housing**

**11.8 If necessary, pry the balls out with a screwdriver**

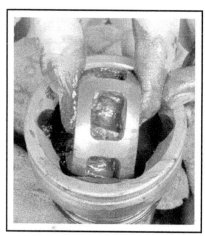

**11.9 Tilt the inner race and cage 90-degrees, then align the windows in the cage with the lands and rotate the inner race up and out of the outer race**

CV joint grease.

13 Coat all of the CV joint components with CV joint grease before beginning reassembly.

14 Install the inner race in the cage and align the marks made in Step 7.

15 Install the inner race and cage assembly into the CV joint housing, aligning the marks on the inner race and cage assembly with the mark on the housing.

16 Install the balls into the holes, one at a time, until they are all in place.

17 Apply CV joint grease through the hole in the inner race, then force a wooden dowel down through the hole (see illustration). This will force the grease into the joint. Continue this procedure until the joint is completely packed. Joints with an outer diameter of 88 mm (3.5-inches) will require 100 grams (3.5 ounces) of grease; joints with an outer diameter of 98 mm (3.8-inches) will require 120 grams (4.2 ounces) of grease. Pack the joint with as much grease as you can, then place the remainder of the grease in the boot.

18 Place the axleshaft in the vise. Clean the end of the axleshaft,

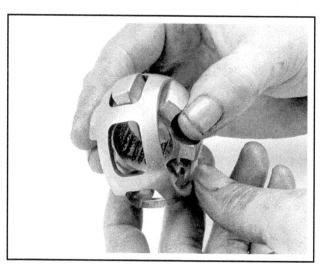

**11.10 Align the inner race with the cage windows and rotate the inner race out of the cage**

**11.17 Apply grease through the splined hole, then insert a wooden dowel into the hole and push down - the dowel will force the grease into the joint**

**11.18 Before sliding the boot onto the shaft, it's a good idea to wrap the splines of the shaft with electrical tape to prevent damaging the boot**

**11.20a Install a new circlip in the driveaxle groove . . .**

**11.20b . . . then tap the joint into place**

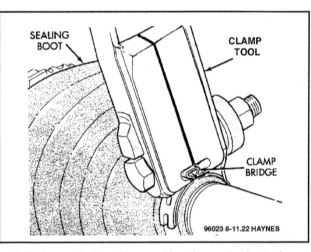

**11.22 Secure the boot clamps with a clamp crimping tool like this, available at most auto parts stores**

then slide the new clamp and boot into place.

➡**Note: It's a good idea to wrap the axleshaft splines with electrical tape to prevent damage to the boot (see illustration). Apply the remainder of the grease from the kit into the CV joint boot.**

19 Remove the protective tape from the driveaxle splines. Slide on the dished washer, convex side first, followed by the thrust washer (see illustrations 11.5b and 11.5a).

20 Install a new circlip in the groove on the driveaxle (see illustration), then tap the joint onto the driveaxle until the circlip engages with the groove in the inner race (see illustration). Make sure the joint is securely retained by the circlip.

21 Ease the boot over the joint, making sure that the boot lips are correctly located on both the driveaxle and CV joint. Lift the outer sealing lip of the boot to equalize air pressure within the boot.

22 Install the large retaining clamp on the boot. Pull the clamp as tight as possible and locate the hooks on the clamp in their slots. Tighten the clamp by crimping the raised area with a special boot clamp tool (see illustration). Due to the relatively hard composition of the boots, this type of tool is required to apply adequate crimping force on

the clamps. Secure the small retaining clamp using the same procedure.

23 Make sure the constant velocity joint moves freely in all directions, then install the driveaxle as described in Section 10.

## INNER CV JOINT

### Triple-rotor type joint

▶ **Refer to illustrations 11.25a, 11.25b, 11.26, 11.27, 11.28, 11.36 and 11.38**

24 Remove the boot clamps and discard them, then pull the boot back on the shaft (or it can be cut off).

25 Mount the driveaxle in a vise equipped with soft jaws, then pry the cover off the inner end of the joint and remove the O-ring from the groove (see illustrations). Discard the cover and O-ring - it isn't necessary to install a new cover or O-ring, as a square-section O-ring will take its place.

26 Mark the relationship of the housing, triple-rotor spider and the end of the axleshaft (see illustration), then remove the axleshaft from the vise and slide it down the shaft.

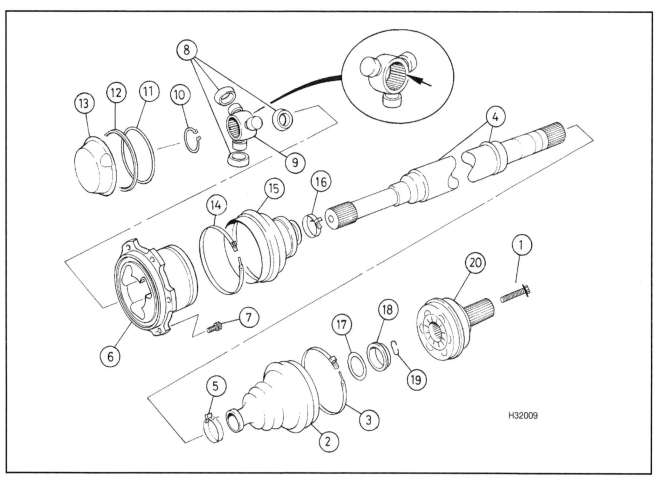

**11.25a  Exploded view of a driveaxle with a triple-rotor inner joint**

| | | | | | |
|---|---|---|---|---|---|
| 1 | Driveaxle/hub bolt | 8 | Rollers | 15 | Inner CV joint boot |
| 2 | Outer CV joint boot | 9 | Triple rotor spider | 16 | Clamp |
| 3 | Clamp | 10 | Snap-ring | 17 | Dished washer |
| 4 | Axleshaft | 11 | O-ring (discard) | 18 | Thrust washer |
| 5 | Clamp | 12 | Square-section O-ring (replacement) | 19 | Circlip |
| 6 | Inner CV joint housing | 13 | Cover (discard) | 20 | Outer CV joint |
| 7 | Bolt | 14 | Clamp | | |

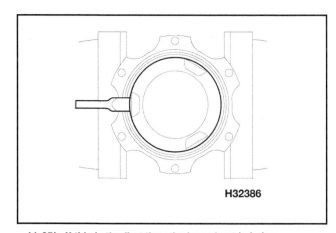

**11.25b  If this is the first time the inner boot is being replaced, you'll have to pry off this cover and remove the O-ring from the groove underneath**

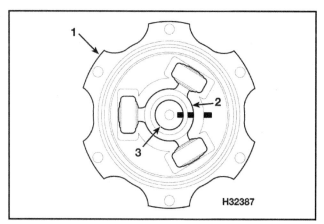

**11.26  Mark the relationship of the inner joint housing, spider and axleshaft**

| | | | |
|---|---|---|---|
| 1 | Inner joint housing | 3 | Axleshaft |
| 2 | Triple-rotor spider | | |

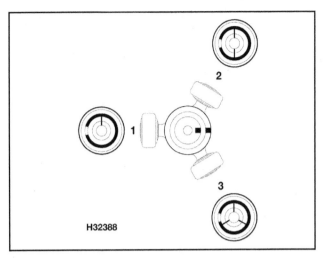

**11.27 Use a felt-tip pen to mark the position of each roller to its post on the spider**

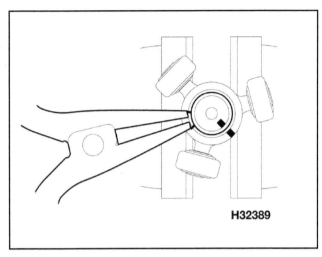

**11.28 Remove the snap-ring from the end of the axleshaft, then remove the spider**

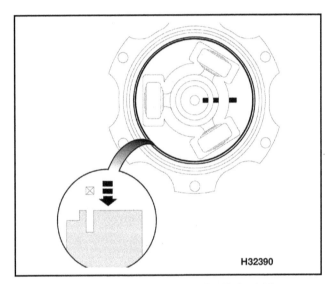

**11.36 When installing the square-section O-ring in the groove in the housing, make sure it seats properly and doesn't get twisted**

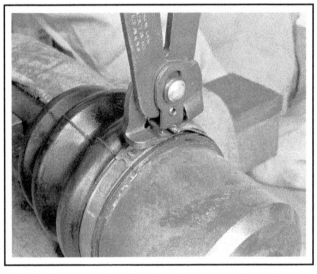

**11.38 Clamp crimping pliers like these are available at most auto parts stores**

27 Mark the relationship of the rollers to the spider (see illustration).

28 Remove the snap-ring from the end of the axleshaft with a pair of snap-ring pliers, then slide the spider off the shaft (see illustration).

➡**Note: If the spider won't slide off or tap off easily, it will be necessary to push it off with a hydraulic press.**

29 Remove the housing from the shaft, then clean all of the components with solvent. Inspect all components for pitting and other signs of wear (shiny, polished spots are normal and won't affect operation). If any signs of wear are found, replace the entire joint.

30 Slide the joint housing onto the shaft, then place the shaft back in the vise.

31 Install the small clamp and the boot onto the shaft. It's a good idea to wrap the splines of the shaft with electrical tape to prevent damage to the boot (see illustration 11.18).

32 Install the spider on the shaft, aligning the marks made in Step 26.

➡**Note: The chamfered side of the spider splines face the opposite side of the shaft. If necessary, use a deep socket or a piece**

of pipe to drive the spider onto the shaft until it contacts its stop. Install a new snap-ring, making sure it seats completely in its groove.

33 Lubricate the posts of the spider with CV joint grease, then install the rollers onto their respective posts. Now coat the outside of the rollers with CV joint grease.

34 Release the shaft from the vise, then slide the housing up onto the triple-rotor spider, aligning the marks made in Step 26.

35 Clamp the housing in the vise, allowing the shaft to hang straight down.

36 Install the square-section O-ring (included in the kit) into the groove in the housing (see illustration).

37 If you're working on a VW Passat, fill the joint with approximately 3.2 ounces (90 grams) of CV joint grease. If you're working on an Audi A4, measure the diameter of the joint housing; 88 mm (3.5-inch) diameter joints require 100 grams (3.5 ounces) of grease; 98 mm (3.9-inch) diameter joints require 120 grams (4.2 ounces) of grease. Regardless

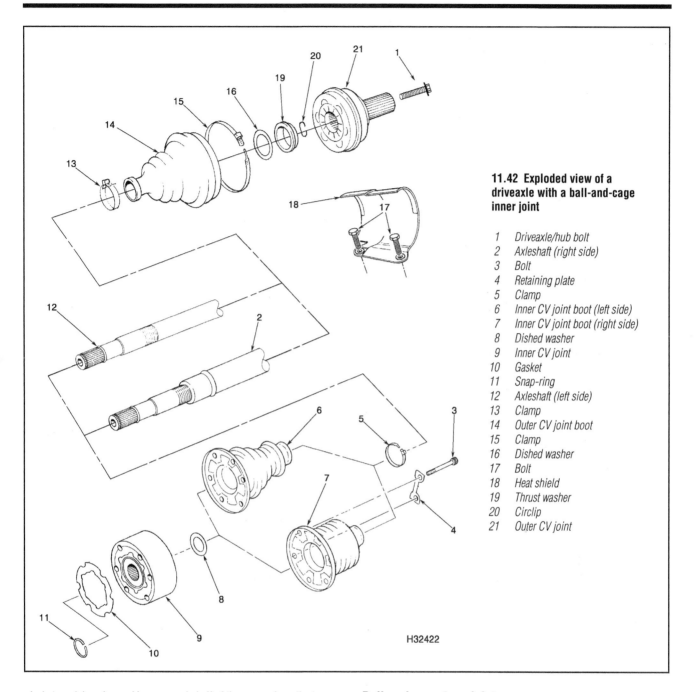

**11.42 Exploded view of a driveaxle with a ball-and-cage inner joint**

1  Driveaxle/hub bolt
2  Axleshaft (right side)
3  Bolt
4  Retaining plate
5  Clamp
6  Inner CV joint boot (left side)
7  Inner CV joint boot (right side)
8  Dished washer
9  Inner CV joint
10  Gasket
11  Snap-ring
12  Axleshaft (left side)
13  Clamp
14  Outer CV joint boot
15  Clamp
16  Dished washer
17  Bolt
18  Heat shield
19  Thrust washer
20  Circlip
21  Outer CV joint

H32422

of what model you're working on, apply half of the grease from the top, then remove the housing from the vise and place the same amount of grease into the other side of the joint.

38  Install the boot onto the housing, making sure it isn't stretched or twisted, then place the clamps in position and tighten them with a pair of clamp crimping pliers (see illustration).

39  Make sure the constant velocity joint moves freely in all directions, then install the driveaxle as described in Section 10.

## ※ CAUTION:

**When handling the driveaxle, be careful not to allow the housing to become pushed back onto the shaft - if this happens, the triple rotor spider and rollers may protrude from the housing and fall apart. Also, some of the grease will be forced from the housing.**

### Ball-and-cage type joint

▶ **Refer to illustrations 11.42, 11.45, 11.52, 11.53a and 11.53b**

40  Remove the boot clamp and discard it.

41  Mount the driveaxle in a vise equipped with soft jaws. Using a hammer and a punch, knock the boot cap off the inner CV joint.

42  Using a pair of snap-ring pliers, remove the snap-ring from its groove in the end of the driveaxle (see illustration).

43  Pull the inner joint off the end of the axleshaft. If it is stuck, use a hammer and a brass punch to drive it off the shaft; apply force to the inner race of the joint only. If it still won't come off, it'll be necessary to push it off with a hydraulic press. Remove the dished washer from the shaft, then pull off the boot.

44  Wipe the grease off the joint and mark the relationship of the inner race, cage and housing.

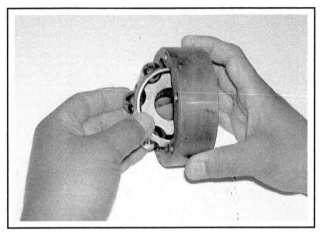

**11.45 Turn the cage and inner race 90-degrees and rotate it out of the housing**

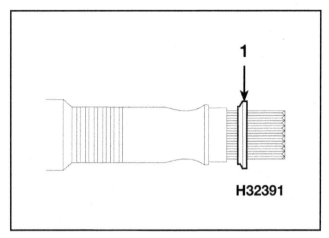

**11.52 The concave side of the dished washer (1) must face the end of the axleshaft (ball-and-cage inner CV joint)**

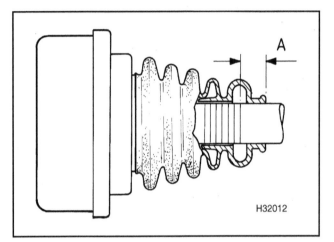

**11.53a The inner end of the left inner CV joint boot must be positioned 11/16-inch (17 mm) past the inner groove on the axleshaft (dimension "A")**

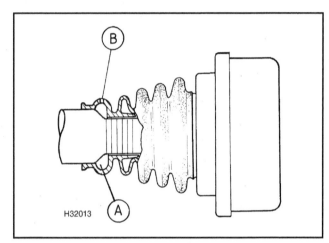

**11.53b The inner end of the right inner CV joint boot must be positioned on the axleshaft like this, so the edge of the vent chamber (A) seats on the tube; the vent hole (B) must not be obstructed**

45  Rotate the cage and inner race 90-degrees and remove it from the housing (see illustration).

46  Remove each ball bearing from the cage, keeping track of their positions so they can be reinstalled in the same spot.

47  Turn the inner race 90-degrees in the cage, align one of the grooves with the edge of the cage and rotate the inner race out.

48  Clean all of the components and inspect for worn or damaged splines, race grooves, ball bearings and cage. Shiny spots are normal and won't affect operation. Replace the joint with a new one if any of the components show signs of wear.

49  Coat the components of the joint with CV joint grease, then assemble the inner race and cage, aligning the marks made in Step 44.

50  Press the ball bearings into their openings, then insert the inner race, cage and balls into the housing. The chamfered side of the splines must face the larger diameter side of the housing. When the components are rotated into place, the wide-spaced grooves of the inner race must be lined up with the wide-spaced grooves in the housing.

51  Install the new boot and clamp on the axleshaft. It's a good idea to wrap the splines of the axleshaft with electrical tape to prevent damage to the boot (see illustration 11.18).

52  Remove the tape and install the dished washer on the axleshaft

with the concave side facing the end of the shaft (see illustration).

53  If you're replacing the boot on the left-side driveaxle, adjust the inner end of the boot so that it is 11/16-inch past the innermost groove on the axleshaft (see illustration). If you're replacing the boot on the right-side driveaxle, seat the inner end of the boot on the larger diameter portion of the axleshaft (see illustration).

54  Place the inner joint assembly on the axleshaft and install a new snap-ring. Make sure the snap-ring seats in its groove completely.

55  Pack the CV joint with CV joint grease. VW Passats with 89 mm diameter joints require 90 grams (3.2 ounces) of grease; models with 88 mm joints require 100 grams (3.5 ounces). Audi A4s with 100 mm diameter joints require 110 grams (3.8 ounces) of grease; models with 108 mm (4.3 inch) diameter joints take 120 grams (4.2 ounces) of grease. Place 1/3 of the grease in the joint and the other 2/3 on the inner side of the joint and in the boot.

56  Seat the cap of the boot on the joint housing, aligning the bolt holes. Make sure the boot is not twisted or deformed in any way.

57  Clean the surface of the joint housing, then stick a new gasket onto the housing.

58  Make sure the constant velocity joint moves freely in all directions, then install the driveaxle as described in Section 10.

## 12 Driveshaft (all-wheel drive models) - removal and installation

◆ Refer to illustrations 12.2 and 12.3

### ✳✳ CAUTION:

**The driveshaft must not be bent at an angle of more than 25-degrees. The manufacturer recommends the use of a holding fixture that secures the two sections of the driveshaft so they are held straight. A similar fixture can be constructed from a 2x4 cut to the proper length, and a block of wood of the correct height at each end. The fixture can then be strapped to the driveshaft before removal or installation. If such a fixture is not used, be sure to support the driveshaft in a level plane either by the use of a floor jack or with the help of an assistant.**

➡Note: At the time of writing, replacement parts for the driveshaft were not available. If the CV joints, U-joint or center support bearing are worn-out, the entire driveshaft must be replaced as an assembly. Before replacing the driveshaft with a new one, however, it would be a good idea to check with a driveline specialist to see if they can carry out repairs to the shaft.

1   Raise the vehicle and support it securely on jackstands.
2   Mark the relationship of the constant velocity (CV) joints to their corresponding flanges at each end of the driveshaft (see illustration).

3   Support each end of the driveshaft (see the **Caution** above), then remove the bolts securing the CV joints to their corresponding companion flanges (see illustration).
4   Remove the nuts securing the center support bearing bracket to the floorpan. Retrieve any shims that may be present between the bracket and the floorpan, noting their installed positions. Carefully lower the driveshaft and remove it out from under the vehicle.
5   Installation is the reverse of removal, noting the following points:
   a)  *Remove all traces of old gasket material from the ends of the CV joints, then install new gaskets.*
   b)  *Use **new** driveshaft CV joint bolts.*
   c)  *If the heat shield above the center support bearing fell out of place, be sure to reinstall it.*
   d)  *Be sure to install any shims that were present between the center support bearing bracket and the floorpan in their original locations.*
   e)  *Before tightening the center support bearing nuts, slide the bearing to the rear as far as it will go and mark the position of the bracket to the floorpan. Next, slide the bearing to the front as far as it will go and mark the position of the bracket to the floorpan. Finally, position the bracket midway between the two marks to centralize the support bearing.*
   f)  *Tighten all fasteners to the torque values listed in this Chapter's Specifications.*

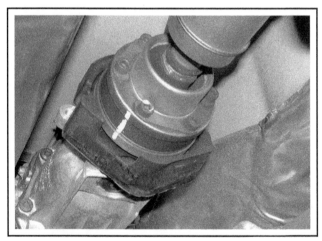

**12.2  Mark the relationship of the driveshaft CV joints to their respective flanges**

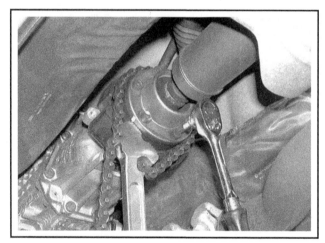

**12.3  Using a chain wrench to prevent the driveshaft from turning as the bolts are removed**

## 13 Rear differential oil seals - replacement

### DRIVE FLANGE (SIDE GEAR SHAFT) SEALS

1   Raise the rear of the vehicle and support it securely on jackstands. Block the front wheels to prevent the vehicle from rolling off the stands.
2   Unbolt the driveaxle from the drive flange (see Section 10). Position the driveaxle out of the way and support it with a piece of wire.

3   Place a drain pan under the differential. Thread two of the drive-axle securing bolts into the drive flange and brace the flange with a large screwdriver to prevent it from turning, then remove the bolt from the center of the flange. Remove the drive flange.
4   Pry the oil seal out of the differential bore with a seal removal tool or a large screwdriver.
5   Pack the open side of the new seal with multi-purpose grease, then drive it into its bore using a seal installation tool or a large socket

with an outside diameter slightly smaller than that of the seal. Make sure the seal enters the bore squarely, and is driven in until it is completely seated.

6   If a groove is worn in the drive flange where it contacts the seal, the drive flange must be replaced.

7   Install a new circlip on the splined area of the drive flange.

8   Lubricate the seal contact area of the flange with multi-purpose grease. Install the drive flange and bolt, tightening the bolt to the torque listed in this Chapter's Specifications. If the drive flange will not go on by hand, slight tapping with a hammer may be required to seat it.

9   Reconnect the driveaxle to the drive flange, tightening the bolts to the torque listed in this Chapter's Specifications.

10   Check the differential lubricant, adding as necessary to bring it to the appropriate level (see Chapter 1).

## PINION SEAL

### VW Passat

▶ **Refer to illustration 13.14**

11   Raise the rear of the vehicle and support it securely on jackstands. Block the front wheels to prevent the vehicle from rolling off the stands.

12   Unbolt the driveshaft from the pinion flange. Support the driveshaft from the underbody with a piece of wire or rope. Don't let it hang, as this will damage the U-joint.

13   Support the differential with a floor jack. Remove the bolts from the differential front mount and lower the front of the differential for access to the pinion flange.

14   Measure the distance from the flange mating surface to the end of the pinion shaft and also to the pinion nut (see illustration). Write these measurements down, as they'll be used during reassembly.

15   Thread two of the driveshaft CV joint bolts into the pinion flange and brace a screwdriver across them to prevent the shaft from turning, then unscrew the pinion nut.

16   Place a drip pan underneath the differential. Using a puller, draw the flange off the pinion shaft.

17   Using a slide hammer and a hook-type seal removal adapter, pull the seal out of the differential housing.

18   Remove the O-ring from the pinion shaft and install a new one.

19   Lubricate the outer edge and lips of the new seal with multi-purpose grease. Also pack the area between the seal lips with grease. Drive the new seal squarely into the bore with a seal driver or a piece of pipe with an outside diameter slightly less than that of the seal.

20   Install the pinion flange and nut. Tighten the nut until the distance from the end of the pinion shaft to the surface of the flange, and the distance from the nut to the surface of the flange, are the same as recorded in Step 14 (within 0.020-inch [0.5 mm]).

21   Stake the collar of the nut to the shaft.

22   Clean off all gasket material from the driveshaft CV joint and attach a new one to it. Connect the driveshaft CV joint to the pinion flange, tightening the bolts to the torque listed in this Chapter's Specifications.

23   Check the differential lubricant and add as necessary to bring it to the appropriate level (see Chapter 1).

### Audi A4

24   Raise the rear of the vehicle and support it securely on jackstands. Block the front wheels to prevent the vehicle from rolling off the stands.

25   Unbolt the driveshaft from the pinion flange. Support the driveshaft from the underbody with a piece of wire or rope. Don't let it hang, as this will damage the U-joint.

26   Support the differential with a floor jack. Remove the bolts from the differential front crossmember and lower the front of the differential.

27   Remove the bolts securing the torque tube to the differential housing, then detach the torque tube.

28   Mount the torque tube in a vise equipped with soft jaws. Remove the sleeve from the shaft with a three-jaw puller.

29   Unbolt and remove the crossmember from the torque tube, then remove the lock-ring from the cover at the front of the torque tube. Thread an M8 bolt into the cover and pull the cover out.

30   Working through the hole, expand the lock-ring on the shaft and slide it back out of its groove.

31   Using a slide hammer puller, remove the shaft from the torque tube.

32   Pry the seal out of the torque tube.

33   Lubricate the outer edge and lips of the new seal with multi-purpose grease. Also pack the area between the seal lips with grease. Drive the new seal squarely into the bore with a seal driver or a piece of pipe with an outside diameter slightly less than that of the seal.

34   Install the shaft in the torque tube, holding the lock-ring expanded with a pair of snap-ring pliers. Once the shaft is in place, seat the lock-ring in its groove.

35   Press the sleeve onto the shaft as far as it will go.

36   Place the cap in its bore and install the lock-ring.

37   Install the crossmember to the differential housing and tighten the bolts to the torque listed in this Chapter's Specifications.

38   Install a new O-ring on the torque tube and connect the torque tube to the differential housing. Tighten the bolts to the torque listed in this Chapter's Specifications.

39   Raise the front of the differential and install the crossmember-to-body bolts. Tighten the bolts to the torque listed in this Chapter's Specifications.

40   Clean off all gasket material from the driveshaft CV joint and attach a new one to it. Connect the driveshaft CV joint to the differential flange, tightening the bolts to the torque listed in this Chapter's Specifications.

41   Check the differential lubricant and add as necessary to bring it to the appropriate level (see Chapter 1).

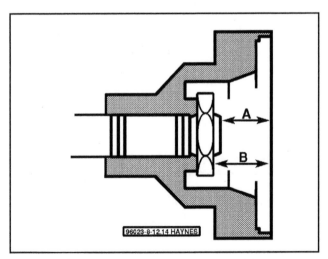

**13.14 Measure the distance from the surface of the flange to the end of the pinion shaft (A) and to the pinion nut (B)**

## 14 Rear differential assembly - removal and installation

1　Raise the rear of the vehicle and support it securely on jackstands. Block the front wheels to prevent the vehicle from rolling off the stands.

2　If you're working on a Passat, remove the cover plate from below the driveaxles.

3　If you're working on an A4, remove the rear portion of the exhaust system (see Chapter 4).

4　Unscrew the bolts securing the inner CV joints to the drive flanges and remove the retaining plates from underneath the bolts. Support the driveaxles by suspending them with wire or string - do not allow them to hang under their weight, or the joints may be damaged.

5　Unbolt the driveshaft from the differential pinion flange. Support the driveshaft from the underbody with a piece of wire or rope. Don't let it hang, as this will damage the U-joint.

6　If you're working on an A4, remove the heat shield from above the driveshaft.

7　Support the differential with a jack. Preferably, a transaxle jack with safety chains should be used; these can be obtained at most equipment rental yards. If one is not available a floor jack can be used, but you'll have to be extra careful not to let the assembly topple off the jack.

8　Unbolt the front and rear mounting brackets from the differential assembly (see illustrations). On the A4, unbolt the front crossmember from the torque tube portion of the differential housing and the body (see illustration).

9　With the help of an assistant, carefully lower the differential assembly and remove it from under the vehicle.

10　Installation is the reverse of the removal procedure. Tighten all fasteners to the torque values listed in this Chapter's Specifications. Check the lubricant level in the rear differential and add, as necessary, to bring it to the appropriate level.

**14.8a  Rear differential rear mounting bolts - VW Passat**

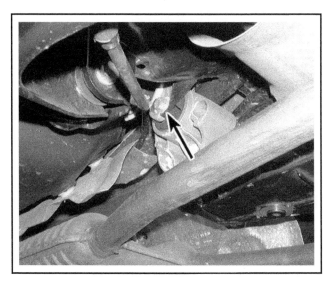

**14.8b  Rear differential front mounting bolt - VW Passat**

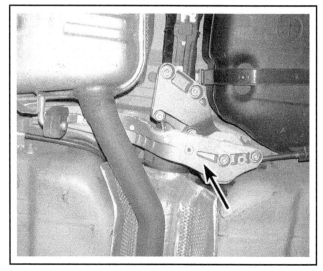

**14.8c  If you're working on an A4, unbolt this crossmember (arrow) from the front of the differential housing (torque tube) and from the body**

| Torque specifications | Ft-lbs (unless otherwise indicated) | Nm |
|---|---|---|
| Clutch master cylinder retaining bolts | 15 | 20 |
| Clutch release cylinder retaining bolts | 18 | 25 |
| Clutch pressure plate bolts | | |
|     VW Passat | 18 | 25 |
|     Audi A4 | 16 | 22 |
| Flywheel bolts | See Chapter 2A or 2B | |
| Inner CV joint-to-drive flange bolts | | |
|     Front | | |
|         Ball-and-cage type | | |
|             M8 | 30 | 40 |
|             M10 | 59 | 80 |
|         Triple rotor type | 59 | 80 |
|     Rear | 30 | 40 |
| Driveaxle/hub bolt* | | |
|     M14 bolt | | |
|         Step 1 | 85 | 115 |
|         Step 2 | Tighten an additional 180-degrees (1/2-turn) | |
|     M16 bolt | | |
|         Step 1 | 140 | 190 |
|         Step 2 | Tighten an additional 180-degrees (1/2-turn) | |
| Driveshaft CV joint bolts* | 41 | 55 |
| Driveshaft center support bearing bracket nuts | 18 | 25 |
| Rear differential crossmember-to-body bolts (A4) | | |
|     M8 bolts | 17 | 23 |
|     M10 bolts | 30 | 40 |
| Torque tube-to-differential housing bolts (A4) | 26 | 35 |
| Rear differential mounting bolts (A4) | | |
|     Front crossmember-to-housing | 30 | 40 |
|     Front crossmember-to-body | | |
|         M8 bolts | 17 | 23 |
|         M10 bolts | 30 | 40 |
|     Front crossmember/exhaust | | |
|         bracket-to-housing | 17 | 23 |
|     Rear crossmember-to-housing | 41 | 55 |
|     Rear crossmember-to-body | 37 | 50 |
| Rear differential mounting bolts (Passat) | | |
|     Front mount-to-housing | 30 | 40 |
|     Rear mount-to-housing | 41 | 55 |
| Side gear shaft to-final drive bolt | 18 | 25 |

*Replace with new bolt(s)

**Section**

**Reference to other Chapters**

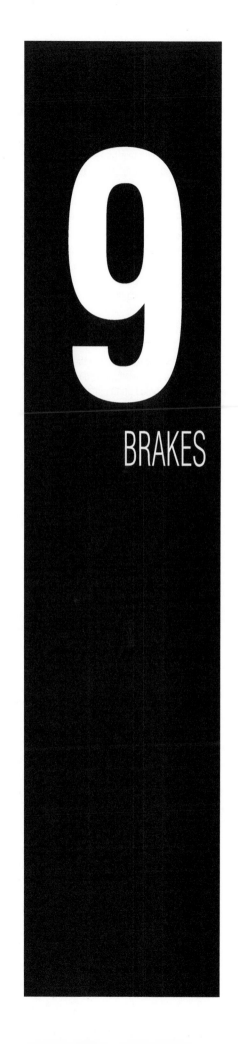

9

BRAKES

## 1  General information

The vehicles covered by this manual are equipped with hydraulically operated front and rear brake systems. The front and rear brakes are disc type and are both self adjusting.

### HYDRAULIC SYSTEM

The hydraulic system consists of two separate circuits. The master cylinder has separate reservoirs for the two circuits, and, in the event of a leak or failure in one hydraulic circuit, the other circuit will remain operative. All models are equipped with an Anti-lock Brake System (ABS) as standard equipment.

### POWER BRAKE BOOSTER

The power brake booster, utilizing engine manifold vacuum and atmospheric pressure to provide assistance to the hydraulically operated brakes, is mounted on the firewall in the engine compartment.

### PARKING BRAKE

The parking brake operates the rear brakes only, through cable actuation. It's activated by a lever mounted in the center console.

### SERVICE

After completing any operation involving disassembly of any part of the brake system, always test drive the vehicle to check for proper braking performance before resuming normal driving. When testing the brakes, perform the tests on a clean, dry, flat surface. Conditions other than these can lead to inaccurate test results.

Test the brakes at various speeds with both light and heavy pedal pressure. The vehicle should stop evenly without pulling to one side or the other.

Tires, vehicle load and wheel alignment are factors which also affect braking performance.

## 2  Anti-lock Brake System (ABS) - general information, component removal and installation

### GENERAL INFORMATION

**♦ Refer to illustration 2.1**

**➡Note: On models equipped with traction control, the ABS unit is a dual function unit, controlling both the anti-lock braking system (ABS) and the electronic differential locking (EDL) system functions.**

ABS is available as an option on the models covered in this manual. The system comprises a hydraulic unit (which contains the hydraulic solenoid valves and accumulators) (see illustration), and four wheel sensors (one installed on each wheel), the ABS control module and the brake light switch. The purpose of the system is to prevent the wheel(s) locking during heavy braking. This is achieved by automatic release of the brake on the relevant wheel, followed by re-application of the brake.

The solenoids are controlled by the control unit, which itself receives signals from the four wheel sensors (one installed on each hub), which monitor the speed of rotation of each wheel. By comparing these signals, the ECU can determine the speed at which the vehicle is traveling. It can then use this speed to determine when a wheel is decelerating at an abnormal rate, compared to the speed of the vehicle, and therefore predicts when a wheel is about to lock. During normal operation, the system functions in the same way as a non-ABS braking system.

If the control unit senses that a wheel is about to lock, it operates the relevant solenoid valve in the modulator block, which then isolates the brake caliper on the wheel which is about to lock from the master cylinder, effectively sealing-in the hydraulic pressure.

If the speed of rotation of the wheel continues to decrease at an abnormal rate, the control unit switches on the electrically driven return

pump, which pumps the brake fluid back into the master cylinder, releasing pressure on the brake caliper so that the brake is released. Once the speed of rotation of the wheel returns to an acceptable rate, the pump stops; the solenoid valve opens, allowing the master cylinder hydraulic pressure to return to the caliper, which then re-applies the brake. This cycle can be carried out at up to 10 times a second.

The action of the solenoid valves and return pump creates pulses in the hydraulic circuit. When the ABS system is functioning, these pulses can be felt through the brake pedal.

The operation of the ABS system is entirely dependent on electrical signals. To prevent the system responding to any inaccurate signals, a

**2.1 The ABS hydraulic unit/control module is mounted just in front of the left-side shock tower**

built-in safety circuit monitors all signals received by the control unit. If an inaccurate signal or low battery voltage is detected, the ABS system is automatically shut down, and the warning light on the instrument panel is illuminated, to inform the driver that the ABS system is not operational. Normal braking should still be available, however.

If a fault does develop in the ABS system, the vehicle must be taken to a VW dealer service department or other qualified repair shop for fault diagnosis and repair.

## COMPONENT REMOVAL AND INSTALLATION

### Hydraulic unit

1   Removal and installation of the hydraulic unit should be entrusted to a VW dealer. Great care has to be taken not to allow any fluid to escape from the unit as the lines are disconnected. If the fluid is allowed to escape, air can enter the unit, causing air locks which cause the hydraulic unit to malfunction.

### Control module

2   The ABS control module is mounted to the side of the hydraulic unit and can only be removed after the hydraulic unit has been removed. New control modules are not coded, but must be coded prior to vehicle operation. This requires a VW scan tool, and even with the proper scan tool the control unit can only be coded after a "dealership code" has been entered into the tool. For this reason, any work involving the ABS control module must be left to a dealer service department or other properly equipped repair facility.

## WHEEL SPEED SENSOR

### Front

▶ **Refer to illustration 2.6**

### Removal

3   Make sure the ignition key is turned to the Off position.
4   Loosen the front wheel bolts. Apply the parking brake, then raise the front of the vehicle and support it securely on jackstands. Remove the wheel.
5   Trace the wiring back from the sensor, detaching it from all the clips and ties while noting its correct routing, then disconnect the electrical connector.
6   Carefully pull the sensor out from the steering knuckle and remove it from the vehicle. With the sensor removed, slide the rubber seal and clamping sleeve out from the hole in the steering knuckle (see illustration).

### Installation

7   Ensure that the mating faces of the sensor, clamping ring and steering knuckle are clean and dry, then lubricate clamping sleeve and wheel sensor surfaces with a film of high-temperature brake grease.
8   Press the clamping sleeve fully into the steering knuckle then insert the wheel sensor together with the rubber seal. Ensure the sensor wiring is correctly positioned then push the sensor firmly into position until it is seated.
9   Ensure the sensor is securely retained then work along the sensor wiring, making sure it is correctly routed, securing it in position with all the relevant clips and ties. Reconnect the electrical connector.
10  Install the wheel and wheel bolts, tightening them securely. Lower the vehicle and tighten the wheel bolts to the torque listed in the Chapter 1 Specifications.

### Rear

▶ **Refer to illustrations 2.15a, 2.15b and 2.15c**

### Removal

11  Make sure the ignition key is turned to the Off position.
12  Remove the rear seat cushion (see Chapter 11) and locate the rear wheel sensor ABS wiring connectors. Unplug the electrical connector and free the wiring from its retaining clips.
13  Loosen the rear wheel lug nuts. Chock the front wheels, then raise the rear of the vehicle and support it securely on jackstands. Remove the wheel.
14  Working underneath the vehicle, trace the wiring back from the sensor, releasing it from its clips. Remove any protective covering, then unseat grommet from the body and pull the wiring through so that it is free to be removed with the sensor.
15  Carefully pull the sensor out from axle assembly or rear knuckle and remove it from the vehicle. With the sensor removed, slide the clamping sleeve out of position (see illustrations).

➡**Note: The rear wheel speed sensor on all-wheel drive models is removed from the rear knuckle just like the front wheel speed sensor is removed from the steering knuckle.**

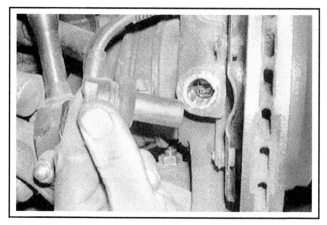

2.6  Removing an ABS front wheel speed sensor

2.15a  On 2WD A4 models the wheel speed sensor is mounted in the axle flange, where the stub axle bolts to (stub axle removed for clarity); pull the sensor out . . .

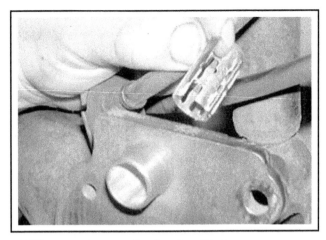

2.15b . . . and remove the clamping sleeve

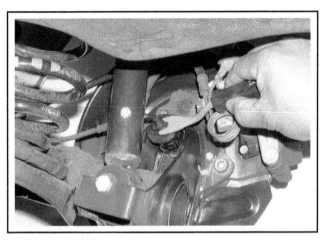

2.15c On 2WD Passat models the wheel speed sensor is mounted in the center of the hub

### Installation

16 Ensure that the mating faces of the sensor, clamping ring and axle are clean and dry then lubricate the clamping sleeve and wheel sensor surfaces with a little high-temperature brake grease.

17 Press the clamping sleeve fully into the axle or rear knuckle then insert the wheel sensor. Make sure the sensor wiring is correctly positioned, then push the sensor firmly into position until it is fully seated.

18 Ensure the sensor is securely retained then work along the sensor wiring, making sure it is correctly routed, securing it in position with all the relevant clips and ties. Reinstall the wiring protective cover to the axle, tighten its retaining screws securely, then feed the wiring connector up through the body and seat the wiring grommet correctly in position.

19 Install the wheel and wheel bolts, tightening them securely. Lower the vehicle to the ground and tighten the wheel bolts to the torque listed in the Chapter 1 Specifications.

20 Reconnect the electrical connector then reinstall the seat cushion.

### RELUCTOR RINGS

21 The reluctor rings are an integral part of the wheel hubs. Examine the rings for damage such as chipped or missing teeth. If replacement is necessary, the complete hub assembly must be disassembled and the bearings replaced as described in Chapter 10.

---

**3    Disc brake pads (front) - replacement**

▶ **Refer to illustration 3.4**

**❊❊ WARNING:**

Disc brake pads must be replaced on both front wheels at the same time - never replace the pads on only one wheel. Also, the dust created by the brake system is harmful to your health. Never blow it out with compressed air and don't inhale any of it. An approved filtering mask should be worn when working on the brakes. Do not, under any circumstances, use petroleum-based solvents to clean brake parts. Use brake system cleaner only!

**❊❊ CAUTION:**

Don't depress the brake pedal with the caliper removed.

1    Remove the cap from the brake fluid reservoir. Remove about

two-thirds of the fluid from the reservoir, then reinstall the cap.

**❊❊ WARNING:**

Brake fluid is poisonous - never siphon it by mouth. Use a suction gun or old poultry baster. If a baster is used, never again use it for the preparation of food.

**❊❊ CAUTION:**

Brake fluid will damage paint. If any fluid is spilled, wash it off immediately with plenty of clean, cold water.

2    Loosen the front wheel bolts, raise the front of the vehicle and support it securely on jackstands. Block the wheels at the opposite end.

3    Remove the wheels. Work on one brake assembly at a time, using the assembled brake for reference if necessary.

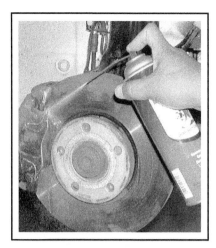

**3.4 Before disassembling the brake, wash it thoroughly with brake system cleaner and allow it to dry - position a drain pan under the brake to catch the residue - DO NOT use compressed air to blow off the brake dust!**

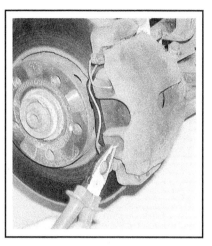

**3.5 Unclip the pad retaining spring and remove it from the caliper**

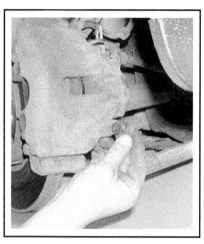

**3.6 Remove the end caps from the guide bushings to gain access to the caliper guide pins**

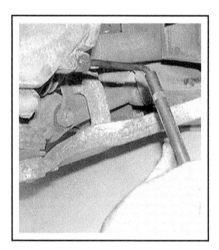

**3.7a Loosen . . .**

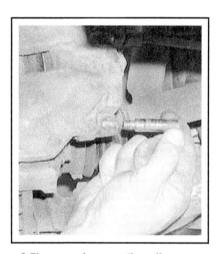

**3.7b . . . and remove the caliper guide pins . . .**

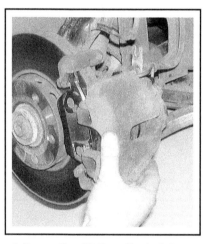

**3.7c . . . then lift the caliper off the mounting bracket**

4   Before removing anything, thoroughly clean the caliper and disc with brake system cleaner (see illustration).

## MODELS WITH TEVES/ATE CALIPERS

▶ **Refer to illustrations 3.5, 3.6, 3.7a, 3.7b, 3.7c, 3.7d, 3.8a and 3.8b**

5   Carefully unclip the pad retaining spring and remove it from the brake caliper (see illustration).

6   Remove the end caps from the guide bushing to gain access to the caliper guide pins (see illustration).

7   Unscrew the caliper guide pins, then lift the caliper away from the mounting bracket (see illustrations). Tie the caliper to the suspension strut using a suitable piece of wire; do not allow it to hang unsupported from the flexible brake hose (see illustration).

**3.7d Hang the caliper from the coil spring with a piece of wire - don't let it hang by the hose**

**3.8a Unclip the inner pad from the caliper piston . . .**

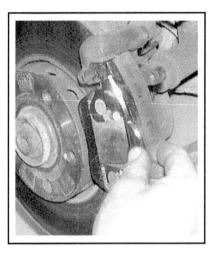

**3.8b . . . and remove the outer pad from the mounting bracket**

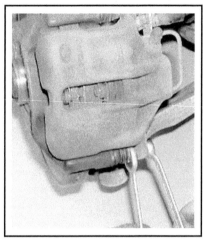

**3.9 Hold the guide pins with an open-end wrench while loosening the caliper mounting bolts**

**3.11a Remove the inner . . .**

**3.11b . . . and outer pads from the caliper mounting bracket**

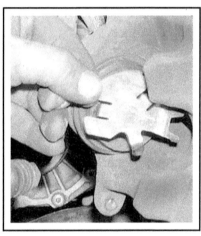

**3.11c Make sure the circular heat shield plate remains clipped into the end face of the piston**

8 Unclip the inner pad from the caliper piston and remove the outer pad from the mounting bracket (see illustrations). Proceed to Step 12.

## MODELS WITH LUCAS CALIPERS

▶ **Refer to illustration 3.9, 3.11a, 3.11b and 3.11c**

➡**Note: The caliper mounting bolts must be replaced with new ones whenever they are removed.**

9 Unscrew the caliper upper and lower mounting bolts while preventing the guide pins from turning with an open-end wrench (see illustration).

➡**Note: New bolts must be used on installation.**

10 Remove the caliper from the mounting bracket and hang it from the coil spring with a length of wire (see illustration 3.7d).

11 Remove the inner and outer pads from the caliper mounting bracket. Ensure that the circular heatshield plate remains clipped into the end face of the piston (see illustrations).

## ALL MODELS

▶ **Refer to illustration 3.14**

12 Inspect the brake disc carefully as outlined in Section 6. If machining is necessary, follow the information in that Section to remove the disc, at which time the pads can be removed as well.

13 Prior to fitting the pads, check that the guide pins are free to slide easily in the caliper body bushings, and are a reasonably tight fit. On Lucas calipers, pull the guide pins out and inspect them for signs of wear. If they're OK, lubricate them with high-temperature brake grease and reinstall them. Use brake system cleaner to clean the caliper and piston. Inspect the dust seal around the piston for damage, and the piston for evidence of fluid leaks, corrosion or damage. If any of these components require attention, the caliper should be replaced.

14 To make room for the new pads, the caliper piston must be pushed back into the cylinder. Either use a piston retraction tool or a C-clamp. Connect a brake bleeding kit to the caliper bleed screw. Open the bleed screw as the piston is retracted; the surplus brake fluid will

**3.14 Use a C-clamp or a piston retraction tool to push the piston back into the caliper. Connecting a bleeder bottle and opening the bleeder screw while doing this will prevent contaminated fluid from being pushed back into the hydraulic system**

then be collected in the bleed kit container (see illustration). Tighten the bleed screw as soon as the piston is retracted to prevent air from being drawn into the system.

➡**Note: The ABS unit contains hydraulic components that are very sensitive to impurities in the brake fluid. Even the smallest particles can cause the system to fail through blockage. The pad retraction method described here prevents any debris in the brake fluid expelled from the caliper from being passed back to the ABS hydraulic unit.**

### Models with Teves/Ate calipers

15  Clip the inner pad into the caliper piston and fit the outer pad to the mounting bracket, ensuring its friction material is against the brake disc. Note that the outer pad has an arrow stamped onto its outer lower edge; this should point in the normal direction of rotation of the brake disc. Remove the adhesive foil backing from the outer pad.

16  Lubricate the caliper guide pins with a light coat of high-temper-

ature brake grease. Maneuver the caliper into position, then install the caliper guide pins and tighten them to the torque listed in this Chapter's Specifications.

17  Install the end caps to the caliper guide bushings.

18  Install the pad retaining spring, ensuring its ends are correctly located in the caliper body holes. Press the inner edge of the spring into position so that its ends are firmly in contact with the surface of the brake pad.

### Models with Lucas calipers

19  Install the new pads in the caliper mounting bracket.

20  Fit the caliper over the brake pads, ensuring that the butterfly springs on the outer edge of the brake pads bear against the inner surface of the caliper body without jamming in the caliper inspection aperture.

➡**Note: New caliper mounting bolts must be used.**

21  Install the new caliper mounting bolts and tighten them to the torque listed in this Chapter's Specifications, holding the guide pins with an open-end wrench, just like during removal.

### All models

22  Depress the brake pedal repeatedly, until the pads are pressed into contact with the brake disc, and a firm brake pedal feel is obtained.

23  Repeat the above procedure on the remaining front brake caliper.

24  Install the wheels and wheel bolts, then lower the vehicle to the ground and tighten the bolts to the torque listed in the Chapter 1 Specifications.

25  Once again, firmly depress the brake pedal a few times to bring the pads into contact with the disc.

26  Check and, if necessary, top-up the brake fluid level as described in Chapter 1.

27  Test the operation of the brakes carefully before placing the vehicle into normal service.

❊❊ **WARNING:**

**New pads will not give full braking efficiency until they have bedded in. Be prepared for this, and avoid hard braking as much as possible for the first hundred miles or so after pad replacement.**

## 4   Disc brake pads (rear) - replacement

▶ **Refer to illustrations 4.4, 4.5a, 4.5b and 4.7**

❊❊ **WARNING:**

**Disc brake pads must be replaced on both rear wheels at the same time - never replace the pads on only one wheel. Also, the dust created by the brake system is harmful to your health. Never blow it out with compressed air and don't inhale any of it. An approved filtering mask should be worn when working on the brakes. Do not, under any circumstances, use petroleum-based solvents to clean brake parts. Use brake system cleaner only!**

❊❊ **CAUTION:**

**Don't depress the brake pedal with the caliper removed.**

➡**Note: The caliper guide pin bolts must be replaced with new ones whenever they are unscrewed.**

1  Loosen the rear wheel bolts. Chock the front wheels, then raise the rear of the vehicle and support it securely on jackstands. Remove the rear wheels.

2  Referring to Section 11, release the parking brake lever then back off the parking brake cable adjusters to obtain maximum freeplay in the

4.4 Remove the caliper from the mounting bracket and support it with a length of wire . . .

4.5a . . . then remove the outer pad . . .

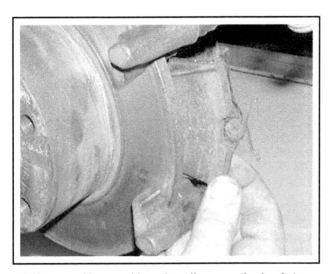

4.5b . . . and inner pad from the caliper mounting bracket

4.7 Using a retraction tool to push the piston back into the caliper

cables and ensure both caliper parking brake levers are against their stops.

3  Unscrew the caliper mounting bolts, counterholding the guide pins with an open-end wrench to prevent them from rotating (see illustration 3.9). Discard the mounting bolts - new bolts must be used on installation.

4  Lift the caliper away from the brake pads (see illustration), and tie it to the suspension strut using a length of wire. Do not allow the caliper to hang unsupported on the flexible brake hose.

5  Remove the brake pads from the caliper mounting bracket (see illustrations).

6  Prior to installing the new pads, check that the guide pins are free to slide easily in the caliper bracket, and check that the rubber guide pin boots are undamaged. Pull the guide pins out and inspect them for signs of wear. If they're OK, lubricate them with high-temperature brake grease and reinstall them. Clean the caliper and piston with brake system cleaner. Inspect the dust seal around the piston for damage, and the piston for evidence of fluid leaks, corrosion or damage. If necessary, replace the caliper.

7  To make room for the new brake pads, it will be necessary to retract the piston fully into the caliper bore by rotating it in a clockwise direction using a retraction tool, or a pair of needle-nose pliers (see illustration). The excess brake fluid should be ejected via the caliper bleed screw (see Section 3, Step 14).

8  Peel the protective sheet from the pad backing plates, then install the pads in the mounting bracket, ensuring that each pad's friction material is facing the brake disc.

9  Slide the caliper back into position over the pads, making sure the pad anti-rattle springs are correctly positioned against the inner surface of the caliper body and are not jammed in the inspection aperture.

➡Note: New mounting bolts must be used when the caliper is installed.

10  Press the caliper into position, then install the new mounting bolts, tightening them to the torque listed in this Chapter's Specifications while preventing the guide pins from turning with an open-end wrench.

11 Repeat the above procedure on the remaining rear brake caliper.

12 Depress the brake pedal repeatedly to force the pads into firm contact with the discs. Once normal pedal feel has returned, check that the discs rotate freely.

13 Adjust the parking brake cables (see Section 11).

14 Install the wheels and wheel bolts, then lower the vehicle to the ground and tighten the wheel bolts to the torque listed in the Chapter 1 Specifications.

15 Check and, if necessary, top-up the brake fluid level as described in Chapter 1.

16 Test the operation of the brakes carefully before placing the vehicle into normal service.

### ✳✳ WARNING:

**New pads will not give full braking efficiency until they have bedded in. Be prepared for this, and avoid hard braking as much as possible for the first hundred miles or so after pad replacement.**

---

## 5  Brake caliper - removal and installation

▶ **Refer to illustrations 5.4a and 5.4b**

### ✳✳ WARNING:

**The dust created by the brake system is harmful to your health. Never blow it out with compressed air and don't inhale any of it. An approved filtering mask should be worn when working on the brakes. Do not, under any circumstances, use petroleum-based solvents to clean brake parts. Use brake system cleaner only!**

## REMOVAL

1  Loosen the front or rear wheel bolts, raise the front or rear of the vehicle and support it securely on jackstands. Block the wheels at the opposite end. Remove the front or rear wheel.

2  If you are removing a rear caliper, unbolt the cable bracket from the caliper and disengage the cable from the toggle lever.

3  If you're removing a front caliper, unscrew the brake line fitting from the caliper, using a flare-nut wrench to prevent rounding-off the corners of the fitting. Unbolt the brake line bracket from the caliper and pull the line away, then plug the fitting to prevent fluid loss and the entry of contaminants.

4  If you're removing a rear caliper, remove the inlet fitting bolt and discard the old sealing washers (see illustration). Disconnect the brake hose from the caliper. Plug the brake hose to keep contaminants out of the brake system and to prevent losing any more brake fluid than is necessary (see illustration).

➡**Note: If you're removing the caliper for access to other components, don't disconnect the hose. Suspend the caliper with a piece of wire to prevent damaging the brake hose.**

5  Remove the caliper guide pins (Teves/Ate calipers) or the mounting bolts (Lucas calipers) and detach the caliper (see Section 3 [front] or 4 [rear]).

## INSTALLATION

6  Installation is the reverse of removal. If you're installing a rear caliper, don't forget to use new sealing washers on each side of the brake hose inlet fitting, and tighten the fitting bolt securely. Be sure to tighten the mounting fasteners to the torque listed in this Chapter's Specifications.

7  Bleed the brake system (see Section 9). Make sure there are no leaks from the hose connections. Pump the brake pedal several times before driving the vehicle, and test the brakes carefully before returning the vehicle to normal service.

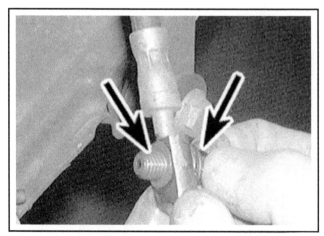

**5.4a  There is a sealing washer on either side of the rear brake hose inlet fitting; be sure to replace these with new ones when reconnecting the hose**

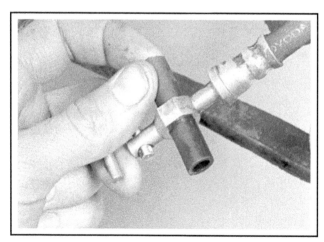

**5.4b  The rear brake hose can be plugged using a snug-fitting piece of tubing**

## 6 Brake disc - inspection, removal and installation

### INSPECTION

▶ **Refer to illustrations 6.3, 6.4a, 6.4b, 6.5a and 6.5b**

1   Loosen the wheel bolts, raise the vehicle and support it securely on jackstands. Remove the wheel and reinstall the wheel bolts, tightening them securely to ensure an accurate runout check.

➡**Note: If the bolts don't contact the disc when screwed on all the way, install washers under them.**

2   Remove the brake caliper (see Section 5). It isn't necessary to disconnect the brake hose. After removing the caliper bolts, suspend the caliper out of the way with a piece of wire (see illustration 3.7d).

3   Visually inspect the disc surface for score marks and other damage. Light scratches and shallow grooves are normal after use and may not always be detrimental to brake operation, but deep scoring requires disc removal and refinishing by an automotive machine shop. Be sure to check both sides of the disc (see illustration). If pulsating has been noticed during application of the brakes, suspect disc runout.

4   To check disc runout, place a dial indicator at a point about 1/2-inch from the outer edge of the disc (see illustration). Set the indicator to zero and turn the disc. The indicator reading should not exceed the specified allowable runout limit. If it does, the disc should be refinished by an automotive machine shop.

➡**Note: When replacing the brake pads, it's a good idea to resurface the discs regardless of the dial indicator reading, as this will impart a smooth finish and ensure a perfectly flat surface, eliminating any brake pedal pulsation or other undesirable symptoms related to questionable discs. At the very least, if you elect not to have the discs resurfaced, remove the glaze from the surface with emery cloth or sandpaper, using a swirling motion (see illustration).**

5   It's absolutely critical that the disc not be machined to a thickness under the specified minimum allowable thickness. The minimum (or discard) thickness is cast into the disc (see illustration). The disc thickness can be checked with a micrometer (see illustration).

**6.3  The brake pads on this vehicle were obviously neglected, as they wore down to the rivets and cut deep grooves into the disc - wear this severe means the disc must be replaced**

**6.4a  Use a dial indicator to check disc runout - if the reading exceeds the specified allowable runout limit, the disc will have to be machined or replaced**

**6.4b  Using a swirling motion, remove the glaze from the disc surface with sandpaper or emery cloth**

**6.5a  The minimum thickness is cast into the outside of the hub area of the disc**

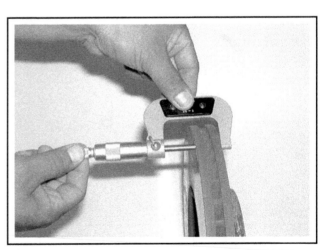

**6.5b  Use a micrometer to measure disc thickness at several points**

## REMOVAL

▶ **Refer to illustrations 6.6a and 6.6b**

6   Remove the two caliper mounting bracket bolts and detach the mounting bracket (see illustrations).

7   Remove the bolts which you installed to hold the disc in place, then slide the disc off the hub.

## INSTALLATION

8   Place the disc in position on the hub flange, aligning the bolt holes.

9   Install the caliper mounting bracket and tighten the bolts to the torque listed in this Chapter's Specifications.

10 Install the caliper, tightening the guide pins (Teves/Ate) or mounting bolts (Lucas) to the torque listed in this Chapter's Specifications.

11 Install the wheel, lower the vehicle and tighten the wheel bolts to the torque listed in the Chapter 1 Specifications. Depress the brake pedal a few times to bring the brake pads into contact with the disc. Bleeding won't be necessary unless the brake hose was disconnected from the caliper. Check the operation of the brakes carefully before driving the vehicle.

**6.6a  Remove the caliper mounting bracket bolts and detach the mounting bracket - this is a front caliper mounting bracket . . .**

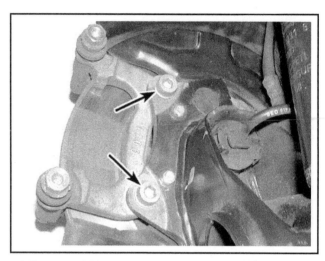

**6.6b  . . . and this is a rear caliper mounting bracket**

## 7   Master cylinder - removal and installation

### ✳✳ WARNING:

**The manufacturer recommends replacing the master cylinder mounting nuts with new ones whenever they are removed.**

➡**Note:** If the master cylinder has been determined to be faulty, it must be replaced with a new one. The only serviceable items are the seals for the fluid reservoir.

## REMOVAL

▶ **Refer to illustrations 7.2 and 7.5**

1   Remove as much fluid as possible from the reservoir with a syringe, suction gun or poultry baster.

### ✳✳ WARNING:

**Brake fluid is poisonous - never siphon it by mouth. If a baster is used, never again use it for the preparation of food.**

### ✳✳ CAUTION:

**Brake fluid will damage paint. If any fluid is spilled, wash it off immediately with plenty of clean, cold water.**

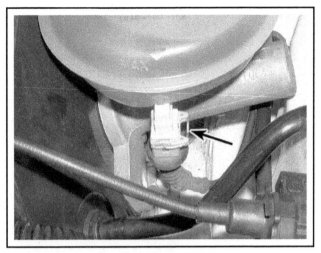

**7.2 Push in on the wire retainer and unplug the electrical connector from the brake fluid level sensor**

**7.5 Master cylinder mounting details**

*1   Hydraulic fittings*          *2   Mounting nuts*

booster (see illustration 7.5), then remove the master cylinder from the cowl area.

➡**Note: On some models it is necessary to first remove the master cylinder reservoir for access to the nuts (see Step 18).**

Again, be careful not to spill the fluid as this is done. Remove the old O-ring on the end of the master cylinder and discard it.

➡**Note: The Torx-head bolts that secure the power brake booster are threaded into the center of the same studs that the master cylinder mounting nuts thread onto. Don't loosen these bolts if you're only removing the master cylinder.**

## INSTALLATION

♦ **Refer to illustrations 7.9, 7.14 and 7.17**

8   Bench bleed the new master cylinder before installing it. Mount the master cylinder in a vise, with the jaws of the vise clamping on the mounting flange.

9   Attach a pair of master cylinder bleeder tubes to the outlet ports of the master cylinder (see illustration).

10  Fill the reservoir with brake fluid of the recommended type (see Chapter 1).

11  Slowly push the pistons into the master cylinder (a large Phillips screwdriver can be used for this) - air will be expelled from the pressure chambers and into the reservoir. Because the tubes are submerged in fluid, air can't be drawn back into the master cylinder when you release the pistons.

12  Repeat the procedure until no more air bubbles are present.

13  Remove the bleed tubes, one at a time, and install plugs in the open ports to prevent fluid leakage and air from entering. Install the reservoir cap.

14  Install a new O-ring onto the end of the master cylinder (see illustration). Guide the master cylinder into place on the power brake booster and tighten the attaching nuts only finger tight at this time.

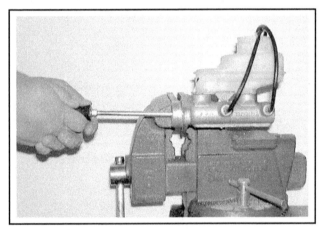

**7.9 The best way to bleed air from the master cylinder before installing it on the vehicle is with a pair of bleeder tubes that direct brake fluid into the reservoir during bleeding**

2   Unplug the electrical connector for the fluid level warning switch (see illustration).

3   Place rags under the fittings and prepare caps or plastic bags to cover the ends of the lines once they're disconnected.

## ✳✳ CAUTION:

**Brake fluid will damage paint. Cover all body parts and be careful not to spill fluid during this procedure.**

4   If the vehicle is equipped with a manual transaxle, detach the fluid feed line from the brake fluid reservoir. Plug the line to prevent contamination and fluid leakage.

5   Loosen the fittings at the ends of the brake lines where they enter the master cylinder (see illustration). To prevent rounding off the flats, use a flare-nut wrench, which wraps around the fitting hex.

6   Pull the brake lines away from the master cylinder and plug the ends to prevent contamination.

7   Remove the nuts attaching the master cylinder to the power

## ✳✳ WARNING:

**Make sure the booster pushrod positively engages the pocket in the master cylinder piston.**

15 Thread the brake line fittings into the master cylinder. Since the master cylinder is still a bit loose, it can be moved slightly so the fittings thread in easily. Don't strip the threads as the fittings are tightened.

16 Tighten the mounting nuts to the torque listed in this Chapter's Specifications. Tighten the brake line fittings securely.

17 Fill the master cylinder reservoir with fluid, then bleed the lines at the master cylinder, followed by bleeding the remainder of the brake system (see Section 9). To bleed the lines at the master cylinder, have an assistant depress the brake pedal and hold it down. Loosen the fitting to allow air and fluid to escape (see illustration). Tighten the fitting, then allow your assistant to return the pedal to its rest position. Repeat this procedure on both fittings until the fluid is free of air bubbles, then bleed the rest of the system. Check the operation of the brake system carefully before driving the vehicle.

### ✳✳ WARNING:

**If you do not have a firm brake pedal at the end of the bleeding procedure, or the ABS light on the instrument panel does not go out, or you have any doubts as to the effectiveness of the brake system, DO NOT drive the vehicle. Have it towed to a dealer service department or other qualified repair shop for diagnosis.**

## RESERVOIR/GROMMET REPLACEMENT

➡Note: The brake fluid reservoir can be replaced separately from the master cylinder body if it becomes damaged. If there is leakage between the reservoir and the master cylinder body, the grommets in the master cylinder body can be replaced.

18 Remove as much fluid as possible from the reservoir with a suction gun, large syringe or a poultry baster.

### ✳✳ WARNING:

**If a poultry baster is used, never again use it for the preparation of food.**

19 Place rags under the master cylinder to absorb any fluid that may spill out once the reservoir is detached from the master cylinder.

### ✳✳ CAUTION:

**Brake fluid will damage paint. Cover all body parts and be careful not to spill fluid during this procedure.**

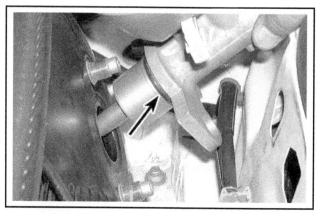

**7.14  Be sure to install a new O-ring on the end of the master cylinder**

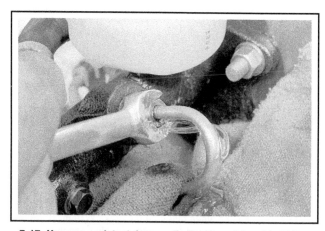

**7.17  Have an assistant depress the brake pedal and hold it down, then loosen the fitting nut, allowing air and fluid to escape; repeat this procedure on both fittings until the fluid is clear of air bubbles (typical)**

20 Pull the reservoir straight out of the master cylinder body. If necessary, carefully pry it up.

21 If you are simply replacing the grommets, carefully pry the old grommets out of the master cylinder body and install new ones.

22 Lubricate the grommets with clean brake fluid, then press the reservoir into place on the master cylinder body.

23 Refill the reservoir with the recommended brake fluid (see Chapter 1) and check for leaks.

24 Bleed the master cylinder (see illustration 7.17).

---

### 8   Brake hoses and lines - inspection and replacement

## INSPECTION

1   About every six months, with the vehicle raised and supported securely on jackstands, the rubber hoses which connect the steel brake lines with the front and rear brake assemblies should be inspected for cracks, chafing of the outer cover, leaks, blisters and other damage. These are important and vulnerable parts of the brake system and inspection should be complete. A light and mirror will be helpful for a thorough check. If a hose exhibits any of the above conditions, replace it with a new one.

## REPLACEMENT

2   If any pipe or hose is to be replaced, minimize fluid loss by first removing the master cylinder reservoir cap, then tightening it down onto a sheet of polyethylene to obtain an airtight seal. Alternatively, flexible hoses can be sealed, if required, using a proprietary brake hose clamp; metal brake line fittings can be plugged (if care is taken not to allow dirt into the system) or capped immediately after they are disconnected. Place a rag under any fitting that is to be disconnected, to catch any spilled fluid.

3   If a flexible hose is to be disconnected, unscrew the brake line fitting before removing the spring clip which secures the hose to its mounting bracket.

4   To unscrew the fitting nuts, use a flare nut wrench; these are available from most auto parts stores. This will prevent rounding-off the corners of the fittings. Always clean the fitting and surrounding area before disconnecting it. If disconnecting a component with more than one fitting, make a careful note of the connections before disturbing any of them.

5   When replacing metal brake lines, be sure to use the correct parts. Don't use copper tubing for any brake system components. Purchase steel brake lines from a dealer or auto parts store.

6   Prefabricated brake line, with the tube ends already flared and fittings installed, is available at auto parts stores and dealer parts departments. These lines must be bent to the proper shapes using a tubing bender.

7   When installing the new line, make sure it's securely supported in the brackets and has plenty of clearance between moving or hot components.

8   Do not overtighten the fitting nuts. It is not necessary to exercise brute force to obtain a sound joint.

9   Ensure that the pipes and hoses are correctly routed, with no kinks, and that they are secured in the clips or brackets provided. After fitting, remove the polyethylene sheet from the reservoir, and bleed the hydraulic system as described in Section 9. Wash off any spilled fluid with water, and check carefully for fluid leaks.

## 9   Brake hydraulic system - bleeding

♦ **Refer to illustration 9.8**

### ✳✳ WARNING:

**Wear eye protection when bleeding the brake system. If the fluid comes in contact with your eyes, immediately rinse them with water and seek medical attention.**

➡**Note: Bleeding the hydraulic system is necessary to remove any air that manages to find its way into the system when it's been opened during removal and installation of a hose, line, caliper or master cylinder.**

1   You'll probably have to bleed the system at all four brakes if air has entered it due to low fluid level, or if the brake lines have been disconnected at the master cylinder.

2   If a brake line was disconnected only at a wheel, then only that caliper or wheel cylinder must be bled.

3   If a brake line is disconnected at a fitting located between the master cylinder and any of the brakes, that part of the system served by the disconnected line must be bled.

4   Remove any residual vacuum from the brake power booster by applying the brake several times with the engine off.

5   Remove the master cylinder reservoir cap and fill the reservoir with brake fluid. Reinstall the cap.

➡**Note: Check the fluid level often during the bleeding operation and add fluid as necessary to prevent the fluid level from falling low enough to allow air bubbles into the master cylinder.**

6   Have an assistant on hand, as well as a supply of new brake fluid, a clear plastic container partially filled with clean brake fluid, a length of clear tubing to fit over the bleeder valve and a wrench to open and close the bleeder valve.

7   Beginning at the right rear wheel, loosen the bleeder valve slightly, then tighten it to a point where it's snug but can still be loosened quickly and easily.

8   Place one end of the tubing over the bleeder valve and submerge the other end in brake fluid in the container (see illustration).

9   Have the assistant slowly depress the brake pedal, then hold the pedal down firmly.

10   While the pedal is held down, open the bleeder valve just enough to allow a flow of fluid to leave the valve. Watch for air bubbles to exit the submerged end of the tube. When the fluid flow slows, close the valve and have your assistant release the pedal.

11   Repeat Steps 9 and 10 until no more air is seen leaving the tube, then tighten the bleeder valve and proceed to the left rear wheel, the right front wheel and the left front wheel, in that order, and perform the same procedure. Be sure to check the fluid in the master cylinder reservoir frequently.

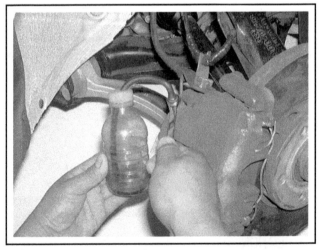

**9.8   When bleeding the brakes, a hose is connected to the bleeder valve at the caliper or wheel cylinder and then submerged in brake fluid - air will be seen as bubbles in the tube and container (all air must be expelled before moving to the next wheel)**

12 Never use old brake fluid. It contains moisture which can cause the fluid to boil, rendering the brake system inoperative.

13 Refill the master cylinder with fluid at the end of the operation.

14 Check the operation of the brakes. The pedal should feel solid when depressed, with no sponginess. If necessary, repeat the entire process.

---

**❋❋ WARNING:**

Do not operate the vehicle if you're in doubt about the effectiveness of the brake system, or if the ABS light on the instrument panel does not go out. Have it towed to a dealer service department or other qualified repair shop for diagnosis.

---

## 10  Power brake booster - check, removal and installation

♦ **Refer to illustrations 10.7a, 10.7b, 10.7c, 10.8 and 10.10**

## CHECK

### Operating check

1  Depress the brake pedal several times with the engine off and make sure there's no change in the pedal reserve distance.

2  Depress the pedal and start the engine. If the pedal goes down slightly, operation is normal. If the pedal does not go down, check the vacuum hose for a leak. If the hose is good, check the intake manifold vacuum with a vacuum gauge (see Chapter 2C).

### Airtightness check

3  Start the engine and turn it off after one or two minutes. Depress the brake pedal slowly several times. If the pedal depresses less each time, the booster is airtight.

4  Depress the brake pedal while the engine is running, then stop the engine with the pedal depressed. If there's no change in the pedal

reserve travel after holding the pedal for 30 seconds, the booster is airtight.

## REMOVAL

➡**Note: A special tool (VW T 10006, or equivalent) is required to disconnect the power brake booster pushrod from the brake pedal. Check on the availability of this tool before proceeding. If this tool is not available, one can be constructed from readily available materials.**

5  Remove the master cylinder as described in Section 7.

6  Carefully pull the vacuum hose fitting out from the grommet in the booster.

7  From inside the vehicle, remove the brake light switch as described in Section 13. Reach up behind the brake pedal, carefully expand the lugs of the pushrod retaining clips and detach the pushrod ball from the pedal. A special tool will be required to do this (see illustrations).

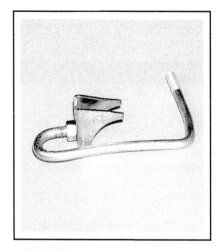

**10.7a  This tool used for detaching the brake pedal from the booster pushrod was fabricated from a modified exhaust clamp, a length of threaded rod and two nuts**

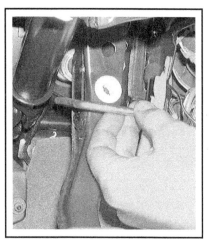

**10.7b  Insert the tool behind the brake pedal and engage it with the pushrod retaining lugs**

**10.7c  Rear view of the brake pedal (pedal removed for clarity) showing the plastic retaining lugs that retain the booster pushrod**

8   Return to the engine compartment, unscrew the two Torx bolts from the booster (see illustration) and maneuver the booster unit out of position, noting the gasket which is installed on the rear of the unit.

## INSTALLATION

9   Check the vacuum hose sealing grommet for signs of damage or deterioration and replace it if necessary.

10  Measure the distance from the booster mounting surface (without the gasket) to the end of the pushrod ball head (see illustration); the distance should be 6.26 +/- 0.020 inches (159 +/- 0.5 mm). If the distance is not correct, adjust the length of the pushrod as necessary.

11  Install a new gasket on the rear of the booster unit, then reposi-tion the unit on the firewall. Install the mounting bolts and tighten them to the torque listed in this Chapter's Specifications.

12  Connect the booster pushrod with the brake pedal, making sure the pedal is securely clipped onto the pushrod ball.

13  Install the brake light switch as described in Section 13.

14  Install the master cylinder as described in Section 7 of this Chapter.

15  Carefully ease the vacuum hose back into position in the booster, taking care not to displace the sealing grommet.

16  Bleed the brake hydraulic system (see Section 9).

17  On completion, start the engine and check for a vacuum leak at the vacuum hose-to-booster grommet. Check the operation of the brake system in an isolated area before returning the vehicle to normal service.

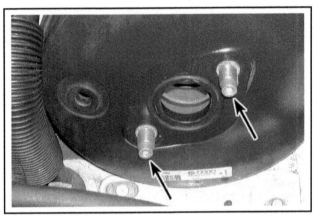

**10.8  Unscrew these two booster mounting Torx bolts, then detach the booster from the firewall**

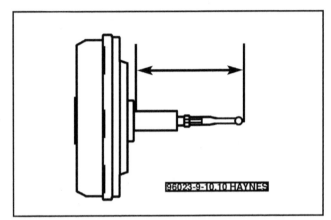

**10.10  The distance from the booster mounting surface (without the gasket) to the end of the pushrod should be 6.26 inches (159 mm)**

## 11  Parking brake - adjustment

▶ **Refer to illustrations 11.4, 11.5a, 11.5b, 11.6a and 11.6b**

➡**Note: The parking brake does not require routine adjustments. It should only be necessary to adjust the parking brake when rear brake components (pads, calipers, discs, parking brake cables) are replaced.**

1   Depress the brake pedal firmly a few times.

2   Chock the front wheels, then raise the rear of the vehicle and support it securely on jackstands. Fully release the parking brake lever.

3   Remove the ashtray from the rear section of the center console.

4   Pass a screwdriver through the hole in the parking brake lever mounting bracket and lock the cable compensator pulley in position, to prevent it from moving (see illustration).

5   Working underneath the vehicle, locate the parking brake adjuster collars, which are located above the exhaust pipe. Access can be gained by removing the exhaust system heat shields (see illustrations).

6   Working on the first adjustment collar, remove the locking ring, then rotate the collar counterclockwise until it reaches its stop. Prevent the rear section of the parking brake cable from turning by using an open-end wrench on the hex nut as you do this (see illustrations).

7   Grasp the parking brake cable on either side of the adjustment collar and push them together firmly, so that the collar takes up all the slack between the two sections of the cable.

8   Rotate the adjustment collar clockwise, until the slot for the locking ring is just visible, then insert the locking ring into its slot.

9   Repeat the Steps 6 through 8 on the other adjustment collar.

10  Pull the two sections of the adjustment collar apart on either cable until the slack in both cables has been taken up. Ensure that the brake caliper parking brake levers remain seated on their stops as you do this.

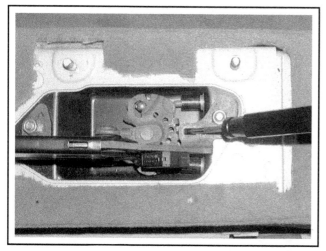

**11.4 Lock the parking brake cable compensator by passing a screwdriver through the hole in the lever mounting bracket (console removed for clarity)**

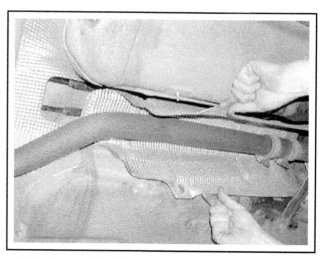

**11.5a Remove the exhaust system heat shields . . .**

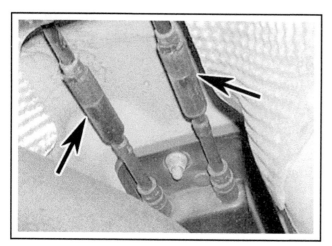

**11.5b . . . to gain access to the parking brake cable adjusters**

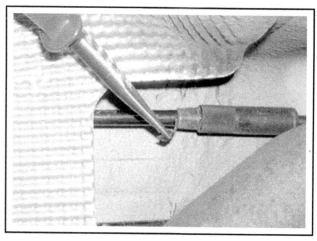

**11.6a Remove the locking ring . . .**

11 Working inside the vehicle, remove the screwdriver to free the cable compensator, then apply and release the parking brake lever firmly at least three times.

12 Return to the underside of the vehicle and turn the adjustment collars on both cables until a gap of approximately 3/64-inch (1 mm), but not greater than 1/16-inch (1.5 mm) can seen between the caliper levers and their stops.

13 Check on both parking brake cables that the adjustment collars have not been turned to the extent that the colored rubber O-rings inside the collars have been exposed. If this is the case, the cables may be stretched beyond the limit of adjustment, or the rear brake pads may be excessively worn.

14 Check the operation of the parking brake and repeat the adjustment procedure as necessary.

15 Once the parking brake is correctly adjusted (both brakes securely lock the wheels with the lever applied and spin freely when the lever is released), lower the vehicle.

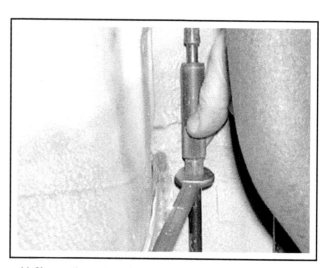

**11.6b . . . then rotate the collar counterclockwise until it reaches its stop; hold the rear section of the cable with a wrench to prevent it from turning**

## 12 Parking brake cables - replacement

▶ **Refer to illustrations 12.4a, 12.4b and 12.8**

1   The parking brake cables are linked to the lever by a compensator plate. Each cable can be removed individually as follows.

2   Raise the vehicle and support it securely on jackstands. Unscrew the nuts and remove the exhaust system heat shields to gain access to the front end of the parking brake cable (see illustration 11.5a).

3   Remove the locking ring from the first parking brake cable adjustment collar and rotate the collar counterclockwise as far as possible (see Section 11). Grasp the front and rear sections of the parking brake cable either side of the adjustment collar and push them firmly together to obtain maximum free play. Repeat this operation on the other parking brake cable.

4   Working at the first rear brake caliper, remove the metal clip and disconnect the parking brake cable casing from its mounting lug. Release the inner cable from its lever and detach the cable from the caliper (see illustrations). Repeat this operation at the other caliper.

5   Work back along the length of each parking brake cable, noting its correct routing, and free it from all the relevant retaining clips and brackets.

6   Unscrew the nuts and remove the exhaust system rear heat shield to gain access to the parking brake cable clamp plate. Release the spring clips and free the cables from the floorpan.

7   Carefully pry the parking brake cable grommets from the underside of the parking brake lever housing.

8   Working inside the vehicle, unhook the parking brake cables from the compensator at the rear of the parking brake lever, by inserting a screwdriver through the hole provided in the parking brake lever mounting bracket (see illustration).

9   Pull the parking brake cables through the floorpan from the underside of the car.

10   Installation is the reverse of the removal procedure, ensuring that the cable is correctly routed and retained by all the necessary clips and ties. In particular, make sure that the cable grommets seat correctly in the underside of the parking brake lever housing. On completion, adjust the operation of the parking brake as described in Section 11.

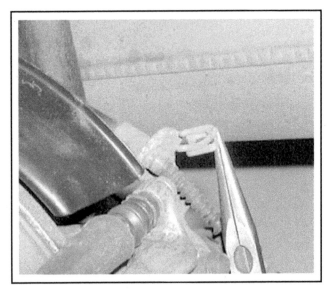

**12.4a  Remove the metal clip and disconnect the cable casing from its mounting lug**

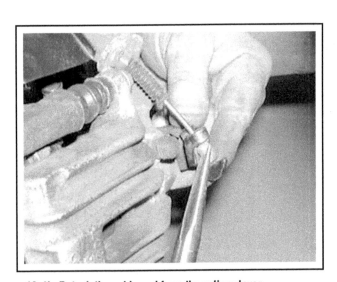

**12.4b  Detach the cable end from the caliper lever**

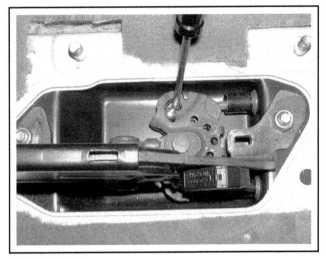

**12.8  Unhook the parking brake cables from the compensator by inserting a screwdriver through the hole, as shown**

## 13 Brake light switch - check, adjustment and replacement

### CHECK

♦ **Refer to illustration 13.1**

1   The brake light switch is located on the brake pedal mounting bracket (see illustration). The switch activates the brake lights at the rear of the vehicle when the pedal is depressed. To gain access to the switch, remove the left-side under-dash panel and the knee bolster (see Chapter 11).

2   If the brake lights are inoperative, check the fuse first (see Chapter 12).

3   If the fuse is good, check for voltage to the switch on the feed wire. If no voltage is present, repair the wire between the switch and the fuse box.

4   If voltage is present, depress the brake pedal and check for voltage at the output wire terminal. If no voltage is present, replace the switch.

5   If voltage is present, check for power on the brake light wires at the tail light housings (with the brake pedal depressed). If voltage is not present, repair the circuit between the switch and the brake lights.

6   If voltage is present, check for a bad ground; using a jumper wire connected to a good ground, probe the ground wire terminal at the tail light connector. If the brake lights go on, repair the ground circuit (follow the ground wire from the tail light housing).

7   Keep in mind that the brake light bulbs could be burned out, but the likelihood of all the bulbs being burned out is very slim.

### ADJUSTMENT

8   With the brake pedal at rest, thread the switch in or out, as necessary, to achieve a clearance of approximately 1/64-inch (0.4 mm) between the end of the switch body and the plunger.

➡**Note: On some models the switch is secured by a locknut that must first be loosened. On other models the switch threads into a pressed-metal nut that remains in the bracket.**

### REPLACEMENT

♦ **Refer to illustration 13.11**

9   Remove left-side under-dash panel and the knee bolster (see Chapter 11).

10   Unplug the electrical connector from the switch.

11   Unscrew the switch from the bracket. On some models the switch is secured by a locknut that must first be loosened. On other models the switch threads into a pressed-metal nut that remains in the bracket. On the latter type, hold the pressed-metal nut with a pair of needle-nose pliers to prevent it from turning while the switch is unscrewed (see illustration).

**13.1  The brake light switch is mounted near the top of the brake pedal bracket - it's the upper switch (the other one is for the cruise control system)**

**13.11  If the brake light switch is threaded into a pressed-metal nut, hold the nut with a pair of pliers while removing or installing the switch**

12  If the switch was equipped with a locknut, remove it from the old switch and thread it onto the new one.

13  To install the new switch, reverse the removal procedure, then adjust it as described in Step 8.

14  Reconnect the electrical connector and check the operation of the brake lights.

15  Install the knee bolster and the under-dash panel.

## Specifications

### General

| | |
|---|---|
| Brake fluid type | See Chapter 1 |

### Disc brakes

| | |
|---|---|
| Minimum pad thickness | See Chapter 1 |
| Brake disc minimum thickness | Cast into disc |
| Maximum disc runout (front and rear) | 0.004 inch (0.1 mm) |
| Maximum disc thickness variation (parallelism) | |
| Front | 0.0004 inch (0.01 mm) |
| Rear | 0.0008 inch (0.02 mm) |

| Torque specifications | Ft-lbs (unless otherwise indicated) | Nm |
|---|---|---|

➡**Note: One foot-pound (ft-lb) of torque is equivalent to 12 inch-pounds (in-lbs) of torque. Torque values below approximately 15 ft-lbs are expressed in inch-pounds, since most foot-pound torque wrenches are not accurate at these smaller values.**

| | | |
|---|---|---|
| Brake booster mounting bolts | 18 | 25 |
| Brake caliper | | |
| Front | | |
| Caliper guide pins (Teves/Ate) | 18 | 25 |
| Caliper mounting bolts (Lucas) | 22 | 30 |
| Caliper mounting bracket bolts | 92 | 125 |
| Brake line bracket-to-caliper bolt | 89 in-lbs | 10 |
| Rear | | |
| Caliper mounting bolts | | |
| A4 | 26 | 35 |
| Passat | 22 | 30 |
| Caliper mounting bracket bolts | 70 | 95 |
| Master cylinder-to-brake booster retaining nuts | 36 | 49 |

**Section**

**Reference to other Chapters**

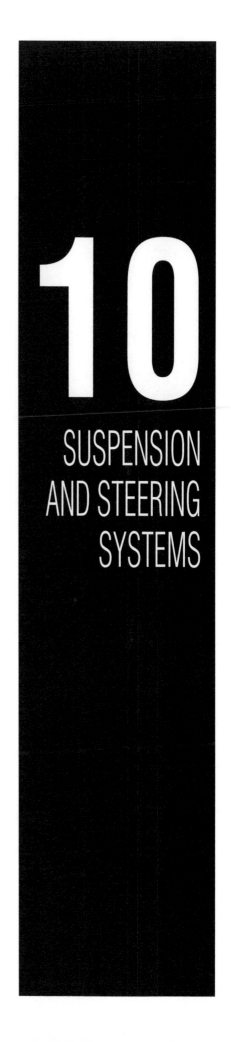

**10**

SUSPENSION
AND STEERING
SYSTEMS

**1.1 Front suspension components**

| | | | | | |
|---|---|---|---|---|---|
| 1 | Stabilizer bar | 4 | Steering knuckle | 7 | Stabilizer bar link |
| 2 | Stabilizer bar clamp | 5 | Shock absorber lower mount | 8 | Rear lower control arm |
| 3 | Front upper control arm (rear arm not visible) | 6 | Front lower control arm | 9 | Subframe |

## 1  General information and precautions

▶ **Refer to illustrations 1.1, 1.2, 1.3a and 1.3b**

### FRONT SUSPENSION

The front suspension is made up of four transverse control arms (two upper and two lower) in an unequal-length, double-wishbone configuration, a steering knuckle/hub assembly, coil-over shock absorbers and a stabilizer bar (see illustration).

### REAR SUSPENSION

The rear suspension on front-wheel drive models consists of a torsion axle with shock absorbers and coil springs (see illustration). A stabilizer bar is incorporated into the rear axle assembly (it's integral with the axle beam and isn't removable).

The rear suspension on all-wheel drive models incorporates upper and lower control arms, knuckle and hub assemblies, track rods connected to the knuckles, coil-over shock absorber assemblies and a stabilizer bar (see illustration).

**1.2  Rear suspension components - front-wheel drive VW Passat**

*1  Rear axle*         *2  Shock absorber*         *3  Coil spring*

## STEERING

The steering column is connected to the steering gear by a universal joint.

The steering gear is mounted on the front subframe, and is connected by two tie-rods, with balljoints at their outer ends, to the steering arms projecting rearwards from the steering knuckles.

Power-assisted steering is standard equipment. The hydraulic steering system is powered by a belt-driven pump, which is driven off the crankshaft pulley.

## PRECAUTIONS

Frequently, when working on the suspension or steering system components, you may come across fasteners which seem impossible to loosen. These fasteners on the underside of the vehicle are continually subjected to water, road grime, mud, etc., and can become rusted or "frozen," making them extremely difficult to remove. In order to unscrew these stubborn fasteners without damaging them (or other components), be sure to use lots of penetrating oil and allow it to soak in for a while. Using a wire brush to clean exposed threads will also ease removal of

**1.3a  Rear suspension components - all-wheel drive Audi A4**

| | | | | | |
|---|---|---|---|---|---|
| 1 | Lower control arm | 4 | Track rod | 6 | Stabilizer bar clamp |
| 2 | Rear knuckle and hub assembly | 5 | Upper control arm | 7 | Stabilizer bar |
| 3 | Shock absorber/coil spring assembly | | | | |

the nut or bolt and prevent damage to the threads. Sometimes a sharp blow with a hammer and punch will break the bond between a nut and bolt threads, but care must be taken to prevent the punch from slipping off the fastener and ruining the threads. Heating the stuck fastener and surrounding area with a torch sometimes helps too, but isn't recommended because of the obvious dangers associated with fire. Long breaker bars and extension, or "cheater," pipes will increase leverage, but never use an extension pipe on a ratchet - the ratcheting mechanism could be damaged. Sometimes tightening the nut or bolt first will help to break it loose. Fasteners that require drastic measures to remove should always be replaced with new ones.

Since most of the procedures dealt with in this Chapter involve jacking up the vehicle sund working underneath it, a good pair of jackstands will be needed. A hydraulic floor jack is the preferred type of jack to lift

the vehicle, and it can also be used to support certain components during various operations.

❋❋ **WARNING:**

**Never, under any circumstances, rely on a jack to support the vehicle while working on it. Whenever any of the suspension or steering fasteners are loosened or removed they must be inspected and, if necessary, replaced with new ones of the same part number or of original equipment quality and design. Torque specifications must be followed for proper reassembly and component retention. Never attempt to heat or straighten any suspension or steering components. Instead, replace any bent or damaged part with a new one.**

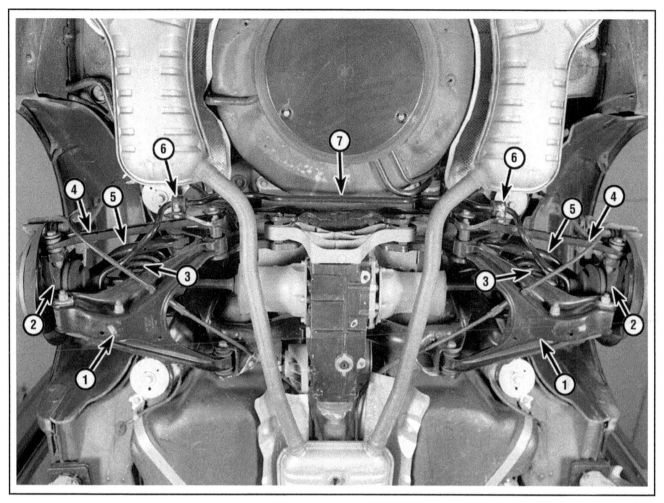

**1.3b  Rear suspension components - all-wheel drive VW Passat Wagon**

| | | |
|---|---|---|
| 1 | Lower control arm | 4 | Track rod | 6 | Stabilizer bar clamp |
| 2 | Rear knuckle and hub assembly | 5 | Upper control arm | 7 | Stabilizer bar |
| 3 | Shock absorber/coil spring assembly | | | | |

## 2   Shock absorber/coil spring assembly (front) - removal, inspection and installation

### REMOVAL

▶ **Refer to illustrations 2.4, 2.8 and 2.9**

1   Remove the wheel trim/hub cap (as applicable) and loosen the wheel bolts by half a turn with the vehicle resting on its wheels.

2   Chock the rear wheels of the car, firmly apply the parking brake, then raise the front of the vehicle and support it securely on jackstands. Remove the wheel.

3   On vehicles equipped with automatic leveling headlamps, disconnect the vehicle level sensor linkage from the front lower control arm.

4   The shock absorber upper mounting nuts are accessed through two holes, located in the cowl area to the rear of the engine compartment, and plugged with rubber covers. Pry out the covers and unscrew the upper mounting nuts using a socket and long extension (see illustration).

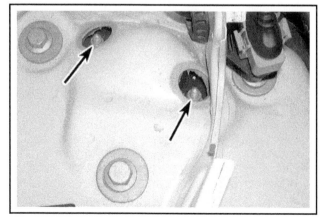

**2.4  The shock absorber/coil spring assembly upper mounting nuts are accessed from the cowl area, and are concealed by rubber plugs**

**2.8 Shock absorber-to-lower control arm bolt/nut**

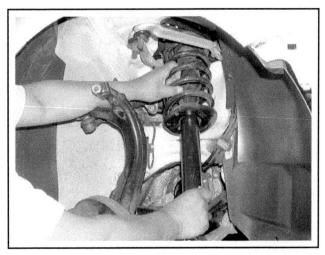

**2.9 Guide the shock absorber/coil spring assembly down and out from between the upper control arms**

5  Release the ABS wheel speed sensor harness from its mounting clips and position it away from the shock absorber (see Chapter 9).

6  Working in the wheelwell, unscrew the securing nut and extract the clamp bolt from the top of the steering knuckle (see Section 4). Separate the front and rear upper control arm balljoints from the top of the knuckle, but do not force the slots apart with a screwdriver or chisel in an attempt to free the balljoint studs. Take care to avoid damaging the balljoint rubber boots.

7  Remove the nut, then separate the rear lower control arm from the steering knuckle (see Section 7). This allows the shock absorber lower mounting bolt to be withdrawn from the front lower control arm.

8  Unscrew the nut and remove the shock absorber lower mounting bolt from the control arm (see illustration).

9  Disengage the shock absorber mounting studs from the mounting bracket and withdraw the unit from the wheelwell (see illustration).

## INSTALLATION

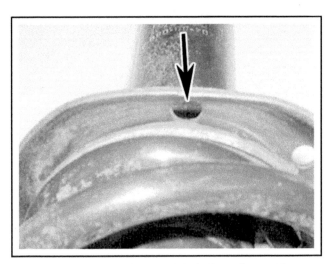

**2.10 Ensure that the alignment hole in the spring seat faces towards the center of the vehicle**

▶ **Refer to illustration 2.10**

10  Guide the shock absorber/coil spring unit into position and engage the upper mounting studs with the mounting bracket. Ensure that the alignment hole in the coil spring lower seat faces inwards towards the vehicle (see illustration).

11  Connect the shock absorber to the lower control arm, install a new securing nut but hand tighten it only at this stage.

12  Reconnect the rear lower control arm to the steering knuckle as described in Section 7. Install a new self-locking nut on the balljoint stud and tighten it to the torque listed in this Chapter's Specifications.

13  Reconnect the upper control arms to the steering knuckle as described in Section 7. Install the clamp bolt and a new self-locking nut, then tighten the nut to the torque listed in this Chapter's Specifications. Press down on both control arms as you tighten the nut, to ensure that the balljoints are properly seated in the steering knuckle.

14  Install new nuts to the upper mounting studs and tighten them

to the torque listed in this Chapter's Specifications. Reinstall the rubber covers.

15  Reconnect the ABS wheel speed sensor harness to its retaining clips.

16  On vehicles equipped with automatic leveling headlamps, fasten the clip to reconnect the vehicle level sensor connecting rod to the front lower control arm.

17  Using a floor jack positioned under the outer ends of the lower control arms, raise the suspension to simulate normal ride height. Now tighten the shock absorber lower mounting nut to the torque listed in this Chapter's Specifications.

18  Install the wheel and wheel bolts, tightening the bolts as securely as possible with the wheel off the ground. Lower the vehicle and tighten the wheel bolts to the torque listed in the Chapter 1 Specifications.

## 3  Shock absorber or coil spring (front) - replacement

◆ **Refer to illustrations 3.3, 3.4, 3.5a, 3.5b, 3.5c, 3.5d, 3.6a, 3.6b, 3.7, 3.13, 3.14a, 3.14b and 3.14c**

### ✲✲ WARNING:

**Always replace the shock absorbers or coil springs in pairs - never replace just one of them.**

1  If the shock absorbers or coil springs exhibit the telltale signs of wear (leaking fluid, loss of damping capability, chipped, sagging or cracked coil springs) explore all options before beginning any work. The shock absorber itself is not serviceable and must be replaced if a problem develops. However, shock absorber/coil spring assemblies complete with springs may be available on an exchange basis, which eliminates much time and work. Whichever route you choose to take, check on the cost and availability of parts before disassembling your vehicle.

### ✲✲ WARNING:

**Disassembling a coil-over type shock absorber is potentially dangerous and utmost attention must be directed to the job, or serious injury may result. Use only a high-quality spring compressor and carefully follow the manufacturer's instructions furnished with the tool. After removing the coil spring from the shock absorber, set it aside in a safe, isolated area.**

2  Remove the shock absorber and spring assembly following the procedure described in the previous Section. Mount the assembly in a vise. Line the vise jaws with wood or rags to prevent damage to the unit and don't tighten the vise excessively.

3  Following the tool manufacturer's instructions, install the spring compressor (which can be obtained at most auto parts stores or equipment yards on a daily rental basis) on the spring and compress it sufficiently to relieve all pressure from the upper spring seat (see illustration). This can be verified by wiggling the spring.

4  Loosen the shock absorber damper shaft nut while preventing the damper shaft from turning with an Allen wrench or hex bit (see illustration). This can achieved using either an offset box-end wrench, or a socket with a center hole large enough to allow the Allen key to pass

through and a hex fitting at the top (such as a spark plug socket); the socket can then be turned with an open-end wrench.

5  Remove the nut, then lift off the mounting plate followed by the washer, the upper spring seat and upper spring support (see illustrations).

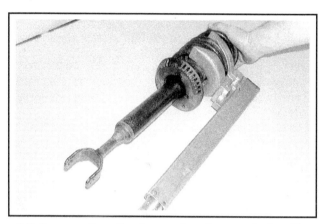

**3.3  Compress the coil spring evenly and progressively until tension is relieved from the spring seats**

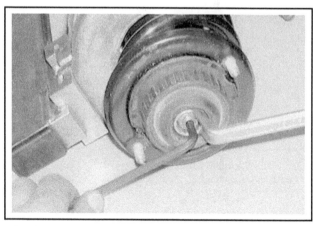

**3.4  Loosen the damper shaft nut while preventing the shaft from turning with an Allen wrench**

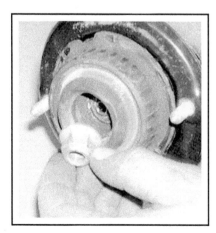

**3.5a  Remove the nut . . .**

**3.5b  . . . then lift off the mounting plate . . .**

**3.5c  . . . followed by the washer . . .**

6   Remove the dust boot, rubber bump stop and protective cap from the shock absorber piston, then lift off the compressed coil spring (see illustrations).

## ✳✳ WARNING:

**Keep the ends of the spring away from your body.**

7   Remove the lower spring support (see illustration), then if required, loosen the lower spring seat  from the shock absorber body by tapping it lightly with a soft faced mallet.

8   With the shock absorber now completely disassembled, examine all the components for wear, damage or deformation, and check the bearing for smoothness of operation. Replace any of the components as necessary.

9   Examine the shock for signs of fluid leakage (a slight amount of seepage is normal). Check the shock damper shaft for signs of pitting along its entire length, and check the body for signs of damage. While holding it in an upright position, test the operation of the shock by moving the damper shaft through a full stroke, and then through several short strokes. In both cases, the resistance felt should be smooth and continuous. If the resistance is jerky, or uneven, or if there is any visible sign of wear or damage to the shock, replacement is necessary.

10   If any doubt exists about the condition of the coil spring, carefully remove the spring compressor and check the spring for distortion and signs of cracking. Replace the spring if it is damaged or distorted, or if there is any doubt as to its condition.

11   Inspect all other components for signs of damage or deterioration, and replace any that are suspect.

12   Slide the bump stop and boot onto the damper shaft.

13   Install the coil spring onto the shock absorber, making sure its end is correctly located against the spring seat stop (see illustration).

14   Install the upper spring support and spring seat, washer and upper mounting to the top of the shock, so that the spring seat alignment hole is positioned at 11-degrees to the longitudinal axis of the shock absorber lower mounting bolt. If the assembly has been carried out correctly, the upper end of the coil spring should be positioned

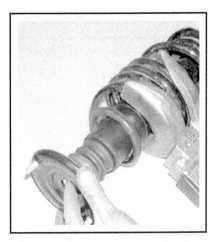

**3.5d . . . the upper spring seat and upper spring support**

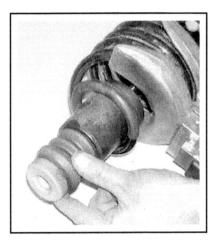

**3.6a  Remove the dust boot, rubber bump stop and protective cap from the shock absorber piston . . .**

**3.6b . . . then carefully lift off the compressed coil spring**

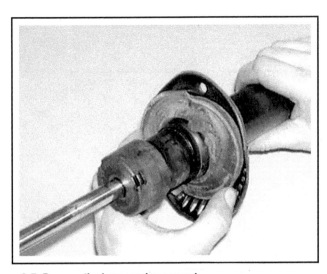

**3.7  Remove the lower spring support**

**3.13  Ensure that the end of the coil spring bears against the stop (arrow) on the spring support**

against the stop on the underside of the upper spring seat (see illustrations).

15 Install a new damper shaft nut, then prevent the damper shaft from turning using the method employed during removal. Tighten the nut to the torque listed in this Chapter's Specifications. Slowly loosen the spring compressor and remove it, making sure the ends of the spring are still properly situated.

16 Install the shock absorber assembly (see Section 2).

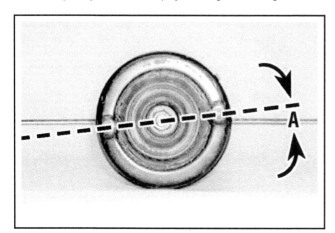

**3.14a  Install the upper spring seat so the alignment hole is positioned at an angle (A) of 11-degrees to the longitudinal axis of the shock absorber lower mounting bolt (shown by a length of welding rod)**

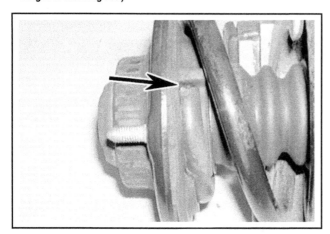

**3.14c  The upper end of the coil spring should be positioned against the stop (arrow) on the underside of the upper spring seat**

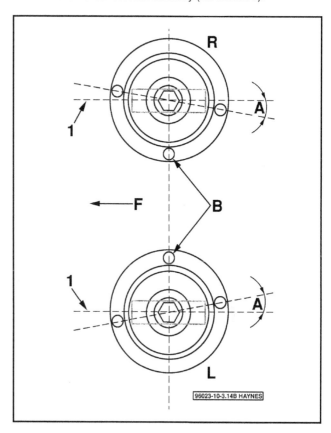

96023-10-3.14B HAYNES

**3.14b  Note the difference in angle offset between the right and left-side shock absorber assemblies**

A    Angle = 11± 2º
B    Lower spring seat alignment holes
F    Direction of travel
L    Left-side shock absorber
R    Right-side shock absorber
1    Axis of shock lower mounting bolt

## 4    Steering knuckle and hub - removal and installation

### ✳✳ WARNING:

**The manufacturer recommends replacing the driveaxle/hub bolt, control arm pivot bolts and nuts and all other self-locking nuts with new ones whenever they are removed.**

### REMOVAL

1    Remove the wheel cover and loosen the driveaxle/hub bolt with the vehicle resting on its wheels. Also loosen the wheel bolts.

2    Chock the rear wheels of the car, firmly apply the parking brake, then raise the front of the car and support it securely on jackstands.

Remove the front wheel.

3    Remove the driveaxle/hub bolt.

4    Remove the ABS wheel speed sensor as described in Chapter 9.

5    Remove the brake caliper (don't disconnect the hose) and brake disc (see Chapter 9). Using a piece of wire or string, tie the caliper to the coil spring - don't let the caliper hang by the hose.

6    Detach the tie-rod end from the steering knuckle (see Section 23).

7    Detach the lower balljoints from the steering knuckle (see Section 7).

8    Detach the upper control arms from the steering knuckle (see Section 7).

9    Carefully pull the hub assembly outwards while pushing the driveaxle from the hub. If necessary, tap the CV joint out of the hub using a soft-faced hammer. If this fails to free it from the hub, the joint will have to be pressed out using a puller.

### ✷✷ CAUTION:

**Be careful not to overextend the inner CV joint.**

### INSTALLATION

10 Lubricate the splines of the driveaxle with multi-purpose grease.

11 Maneuver the knuckle/hub assembly into position and engage it with the driveaxle stub shaft. Install a new driveaxle/hub bolt, but don't attempt to tighten it yet.

12 Connect the upper control arms to the knuckle. Install the pinch bolt and a new nut, then tighten the nut to the torque listed in this Chapter's Specifications.

13 Connect the lower control arms to the knuckle. Install new nuts and tighten them to the torque listed in this Chapter's Specifications.

14 Engage the tie-rod end with the steering knuckle, install the retaining bolt, clamp bolt and a new clamp bolt nut. Tighten these fasteners to the torque values listed in this Chapter's Specifications.

15 Install the brake disc and caliper, tightening the caliper mounting bracket bolts (if equipped) and caliper guide pins to the torque values listed in the Chapter 9 Specifications.

16 Install the ABS wheel speed sensor as described in Chapter 9.

17 Install the wheel and lower the vehicle to the ground.

18 Tighten the driveaxle/hub bolt to the torque listed in the Chapter 8 Specifications, then tighten the wheel bolts to the torque listed in the Chapter 1 Specifications.

## 5    Hub and wheel bearing assembly (front) - removal, bearing replacement and installation

Due to the special tools and expertise required to press the hub and bearing from the steering knuckle, this job should be left to a professional mechanic. However, the steering knuckle and hub may be removed and the assembly taken to an automotive machine shop or other qualified repair facility. See Section 4 for the steering knuckle and hub removal procedure.

## 6    Stabilizer bar and bushings (front) - removal and installation

♦ **Refer to illustrations 6.3 and 6.4**

### ✷✷ WARNING:

**The manufacturer recommends replacing the stabilizer bar link nuts and bushing clamp nuts with new ones whenever they are removed.**

1    Raise the front of the vehicle and support it securely on jackstands.

2    Remove the under-vehicle splash shield.

3    Remove the nuts and detach the stabilizer bar links from the sta-

bilizer bar (see illustration).

➡**Note: Early models use balljoint-type stabilizer bar links; you may have to remove the link-to-control arm nut and fully detach the link in order to detach it from the stabilizer bar.**

4    Remove the nuts and detach the stabilizer bar bushing clamps from the subframe (see illustration).

5    Remove the stabilizer bar. If necessary, unbolt the stabilizer bar links from the control arms.

6    Inspect the clamp bushings and the link bushings. If they're cracked, hardened or deteriorated in any way, replace them.

7    Installation is the reverse of the removal procedure. Be sure to tighten the fasteners to the torque values listed in this Chapter's Specifications.

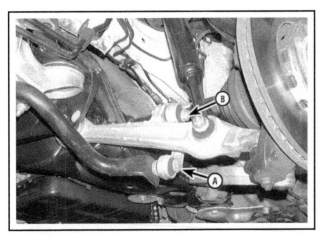

**6.3 Remove the nut and bolt (A) securing the stabilizer bar link to the bar - remove bolt B if you're removing the control arm**

**6.4 Stabilizer bar bushing clamp nuts**

**7   Control arms (front) - removal, bushing replacement and installation**

❋❋ **WARNING:**

**The manufacturer recommends replacing the control arm pivot bolts and nuts, the lower balljoint nuts, the upper control arm balljoint clamp bolt nut and the shock absorber lower mounting nut with new ones whenever they are removed.**

## FRONT AND REAR UPPER ARMS

▸ **Refer to illustrations 7.4a, 7.4b, 7.4c, 7.7, 7.8, 7.10 and 7.12**

➡**Note: Later models equipped with sports or heavy duty suspension have a buffer plate fitted above the upper rear control arm. In addition, the upper rear control arm on all models from 1997 onwards have an integral buffer stop. If the rear upper control arm is replaced, the later revised version should be used regardless of whether the vehicle is equipped with a buffer plate.**

### Removal

1   Loosen the wheel bolts. Chock the rear wheels, firmly apply the parking brake, then raise the front of the vehicle and support it securely on jackstands. Remove the wheel.

2   On vehicles equipped with automatic leveling headlamps, release the clip and disconnect the vehicle level sensor connecting rod from the front lower transverse arm.

3   Detach the ABS wheel speed sensor wiring from its retaining clips.

4   Remove the securing nut and extract the clamp bolt from the top of the steering knuckle. Separate the front and rear upper control arm balljoints from the top of the steering knuckle, but do not force the slots apart with a screwdriver or chisel in an attempt to free the balljoint studs (see illustrations). Take care to avoid damaging the balljoint rubber boots.

5   Remove the securing nut, then separate the rear lower control arm from the base of the steering knuckle, with the aid of a balljoint splitter (avoid damaging the rubber boot) (see illustration 7.23). This allows the shock absorber lower mounting bolt to be withdrawn from the front lower control arm.

6   Unscrew the nut and remove the shock absorber lower mounting bolt from the front lower control arm (see Section 2).

7   Two of the upper control arm mounting bracket bolts are located

in the plenum, to the rear of the engine compartment; the third bolt is located just ahead of the plenum wall (see illustration). Unscrew the three bolts and remove the mounting bracket, the shock absorber/coil spring and both upper control arms as an assembly. There may be plastic washers fitted to the underside of the bolts; these are factory assembly components which do not need to be reinstalled once removed.

**7.4a   Remove the nut . . .**

**7.4b   . . . and pull the clamp bolt out of the top of the steering knuckle . . .**

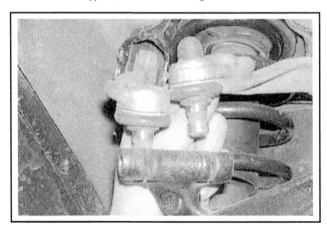

**7.4c   . . . then separate the balljoints from the top of the steering knuckle**

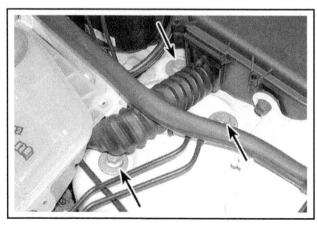

**7.7   Location of the upper control arm bracket bolts**

➡Note: Make a careful note of the positions of any shims installed underneath the heads of the control arm mounting bracket bolts; they must be reinstalled in the same positions to preserve the front suspension alignment settings.

8   Mount the lower end of the shock absorber in a bench vise, then loosen and remove the nut and bolt securing the appropriate upper control arm to the mounting bracket (see illustration).

### Bushing replacement

9   Thoroughly clean the arm, removing all traces of dirt, thread locking compound and undercoating if necessary, then check carefully for cracks, distortion or any other signs of wear or damage, paying particular attention to the inner pivot bushing and balljoint. The balljoint is an integral part of the arm and cannot be replaced separately. If the arm or balljoint are damaged then the complete assembly must be replaced.

10   Replacement of the inner pivot bushing will require the use of a hydraulic press and several spacers and is therefore best entrusted to an automotive machine shop with access to the necessary equipment. If such equipment is available, press out the old bushing and install the new one using a spacer which bears only on the bushing outer

edge. Ensure the bushing is correctly positioned so that the cavities are aligned with the center axis of the arm (see illustration).

### Installation

11   Place the control arm on the mounting bracket, insert a new securing bolt and screw on a new securing nut.

12   Position the transverse arm such that the vertical distance between the front edge of the mounting bracket and the arm is 1-27/32 inches (47 mm) (see illustration). Hold the arm in this position and tighten the securing nut to the specified stage 1 and 2 torque settings. This ensures that the rubber bushing is not stressed when the vehicle is lowered onto its wheels.

13   Attach the shock absorber, control arms and mounting bracket to the wheel housing as an assembly. Insert the mounting bracket securing bolts, together with the shims (where fitted), using the notes made during removal to ensure that the shims are correctly positioned. Tighten the bolts to the torque listed in this Chapter's Specifications.

14   Connect the lower end of the shock absorber to the lower control arm, install a new securing nut but hand tighten it only at this stage.

15   Bolt the rear lower control arm to the steering knuckle, then install a new securing nut and tighten it to the torque listed in this Chapter's Specifications.

16   Reconnect the upper control arm balljoints to the top of the steering knuckle. Install the clamp bolt and a new self locking nut, tightening it to the torque listed in this Chapter's Specifications. Press down on both control arms as you tighten the nut, to ensure that the balljoints are properly seated in the steering knuckle.

17   Connect the ABS wheel speed sensor wiring to its retaining clips.

18   On vehicles equipped with automatic leveling headlamps, fasten the clip to reconnect the vehicle level sensor connecting rod to the front lower control arm.

19   Using a floor jack, raise the outer ends of the lower control arms to simulate normal ride height, then tighten the shock absorber lower mounting nut to the torque listed in this Chapter's Specifications.

20   Install the wheel and tighten the bolts as securely as possible, then lower the vehicle to the ground and tighten the bolts to the torque listed in the Chapter 1 Specifications.

21   On completion, have the front wheel alignment checked and, if necessary, adjusted.

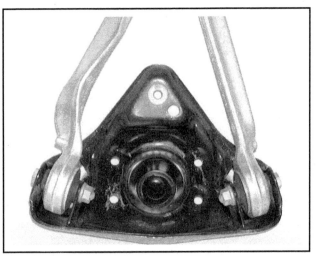

**7.8  Upper control arm pivot bolts/nuts (shock absorber removed for clarity)**

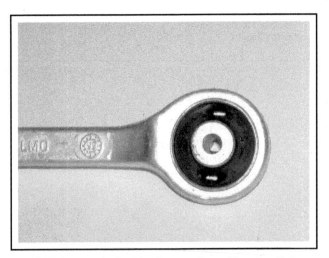

**7.10  Make sure the bushing is correctly positioned so that the cavities are aligned with the center axis of the arm**

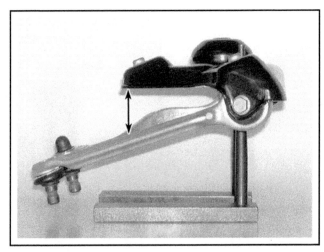

**7.12  Position the control arm so that the distance between the front edge of the mounting bracket and the arm is 47 mm, then tighten the mounting bolt nuts**

## REAR LOWER ARM

▶ **Refer to illustration 7.23**

### Removal

22  Loosen the wheel bolts. Chock the rear wheels, firmly apply the parking brake, then raise the front of the vehicle and support it securely on jackstands. Remove the wheel.

23  Remove the balljoint nut, then separate the control arm from the base of the steering knuckle, with the aid of a ball joint separator (see illustration).

24  Remove the nut from the bolt at the inner end of the control arm. To allow the bolt to be withdrawn, the corner of the subframe must be lowered slightly. To do this, support the subframe with a floor jack, unscrew and remove the two support plate bolts, then loosen and withdraw the subframe securing bolt. Note that the bolt is threaded through the inner of the two sets of subframe bolt holes.

25  Lower the subframe slightly, withdraw the control arm inner pivot bolt then remove the arm from the vehicle.

### Bushing replacement

26  Thoroughly clean the arm, removing all traces of dirt, thread locking compound and undercoating if necessary, then check carefully for cracks, distortion or any other signs of wear or damage, paying particular attention to the inner pivot bushing and balljoint. Note that the inner bushing has a hydraulic action; fluid leakage indicates that the bushing has been damaged and must be replaced. The balljoint is an integral part of the lower arm and cannot be replaced separately.

27  Replacement of the inner pivot bushing will require the use of a hydraulic press and several spacers and is therefore best entrusted to an automotive machine shop. If such equipment is available, press out the old bushing and install the new one using a spacer which bears only on the bushing outer edge. Ensure the bushing is correctly positioned so that the cavities are aligned with the center axis of the arm (see illustration 7.10).

### Installation

28  Installation is the reverse of removal, noting the following points:
  a)  *Use new control arm and subframe securing nuts and bolts.*
  b)  *Ensure that the control arm inboard securing bolt passes through the inner of the two sets of subframe bolt holes.*
  c)  *Raise the outer ends of the control arms with a floor jack to simulate normal ride height before tightening the inner pivot bolt/nut.*
  d)  *Tighten all fasteners to the torque values listed in this Chapter's Specifications.*
  e)  *Lower the vehicle and tighten the wheel bolts to the torque listed in the Chapter 1 Specifications.*
  f)  *Have the front end alignment checked and, if necessary, adjusted.*

## FRONT LOWER ARM

### Removal

29  Loosen the wheel bolts. Chock the rear wheels, firmly apply the parking brake, then raise the front of the vehicle and support it securely on jackstands. Remove the wheel.

30  On vehicles equipped with automatic leveling headlamps, release the clip and disconnect the vehicle level sensor connecting rod from the front lower control arm.

31  Remove the nut, then separate the front lower control arm from the steering knuckle, with the aid of a ball joint separator (see illustration 7.23).

32  Unscrew the nut and remove the shock absorber lower mounting bolt from the front control arm.

33  Remove the nut and detach the stabilizer bar link from the control arm (see Section 6).

34  Unscrew the nut and withdraw the control arm inner pivot bolt, then remove the control arm from the vehicle. Note that the bolt is threaded through the inner of the two sets of subframe bolt holes.

### Bushing replacement

35  Thoroughly clean the arm, removing all traces of dirt, thread locking compound and undercoating if necessary, then check carefully for cracks, distortion or any other signs of wear or damage, paying particular attention to the inner and shock absorber pivot bushings and balljoint. The balljoint is an integral part of the lower arm and cannot be replaced separately.

36  Replacement of the inner and shock absorber mount bushings will require the use of a hydraulic press and several spacers and is therefore best entrusted to an automotive machine shop. If such equipment is available, press out the old bushing and install the new one using a spacer which bears only on the bushing outer edge. Ensure the bushing is correctly positioned so that the cavities are aligned with the center axis of the arm (see illustration 7.10).

### Installation

37  Installation is the reverse of the removal procedure, noting the following points:
  a)  *Use new control arm and shock absorber nuts and bolts.*
  b)  *Ensure that the control arm inboard securing bolt passes through the inner of the two sets of subframe bolt holes.*
  c)  *Raise the outer ends of the control arms with a floor jack to simulate normal ride height before tightening the inner pivot bolt/nut.*
  d)  *Tighten all fasteners to the torque values listed in this Chapter's Specifications.*
  e)  *Lower the vehicle and tighten the wheel bolts to the torque listed in the Chapter 1 Specifications.*
  f)  *Have the front end alignment checked and, if necessary, adjusted.*

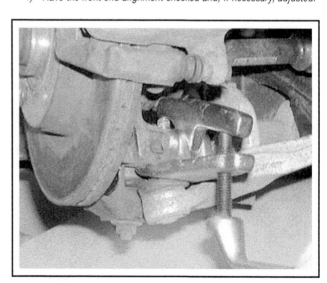

**7.23  This type of balljoint separator won't damage the balljoint boot**

## 8   Balljoints - check and replacement

### CHECK

1   Inspect the control arm balljoints for looseness anytime either of them is separated from the steering knuckle. See if you can turn the ballstud in its socket with your fingers. If the balljoint is loose, or if the ballstud can be turned, replace the balljoint. You can also check the balljoints with the suspension assembled as follows.

#### Upper balljoint

2   Raise the front of the vehicle and support it securely on jackstands placed under the frame rails. Place a floor jack under the lower control arm and raise it slightly.

3   Attempt to move the control arm up and down; a prybar may be helpful. If any play is felt, replace the upper control arm and balljoint as an assembly (the balljoint is not replaceable separately).

4   Also try to move the steering knuckle in-and-out. If any play is felt, replace the upper control arm/balljoint assembly.

5   Check the balljoint boot for cracks and tears. If any are present, replace the upper control arm/balljoint assembly.

#### Lower balljoint

6   Raise the front of the vehicle and support it securely on jackstands placed under the frame rails.

7   Place a floor jack under the front lower control arm and raise it slightly. Attempt to move the steering knuckle up and down; a large prybar underneath the tire, or a prybar placed between the end of the control arm and the steering knuckle will be helpful. If any play is felt, replace the control arm and balljoint as an assembly (the balljoint is not replaceable separately).

8   Also try to move the steering knuckle in-and-out. If any play is felt, replace the control arm/balljoint assembly.

9   Check the balljoint boot for cracks and tears. If any are present, replace the upper control arm/balljoint assembly.

### REPLACEMENT

10  As stated previously, the balljoint is an integral part of the control arm and are not available separately. The entire control arm must be replaced.

## 9   Shock absorber (rear, 2WD Passat) - removal and installation

♦ **Refer to illustrations 9.3 and 9.4**

#### ❊❊ WARNING 1:

he manufacturer recommends replacing the shock absorber lower mounting bolt and nut and the damper rod nut with new ones whenever they are removed.

#### ❊❊ WARNING 2:

Always replace the shock absorbers in pairs - never replace just one of them, as handling peculiarities may result.

1   Loosen the rear wheel bolts. Chock the front wheels to keep the vehicle from rolling, then raise the rear of the vehicle and support it securely on jackstands. Remove the rear wheel.

2   Support the rear axle with a floor jack placed under the rear of the trailing arm.

#### ❊❊ WARNING:

The jack must remain in this position until the shock absorber is reinstalled.

3   Remove the shock absorber lower mounting bolt/nut (see illustration).

4   Remove the shock absorber upper mounting bolts and remove the shock absorber (see illustration).

5   Unscrew the damper rod nut. It'll probably be necessary to hold the damper rod with another wrench to prevent it from turning when loosening the nut.

**9.3 Shock absorber lower mounting bolt/nut - 2WD Passat**

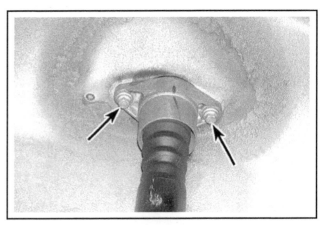

**9.4 Shock absorber upper mounting bolts**

6   Remove the upper mount, bump stop, boot and cap from the damper rod.

7   Install the cap, boot, bump stop and upper mount on the new shock absorber. Install a new nut on the damper shaft and tighten it to the torque listed in this Chapter's Specifications.

8   Guide the shock absorber into position and install the upper mounting bolts and a new lower mounting bolt and nut. Don't tighten the lower mounting bolt/nut yet.

9   Tighten the upper mounting bolts to the torque listed in this Chapter's Specifications.

10  Raise the rear axle to simulate normal ride height, then tighten the lower mounting bolt/nut to the torque listed in this Chapter's Specifications.

11  Repeat the procedure to replace the other rear shock absorber.

12  Install the wheels and lower the vehicle. Tighten the wheel bolts to the torque listed in the Chapter 1 Specifications.

## 10  Shock absorber/coil spring assembly (rear, 2WD A4) - removal and installation

▶ **Refer to illustrations 10.3, 10.5a, 10.5b and 10.7**

### ✳✳ WARNING 1:

**The manufacturer recommends replacing the shock absorber lower mounting bolt and nut and the gasket on the upper mount with new ones whenever they are removed.**

### ✳✳ WARNING 2:

**Always replace the shock absorbers or coil springs in pairs - never replace just one of them, as handling peculiarities may result.**

1   Loosen the rear wheel bolts. Chock the front wheels to keep the vehicle from rolling, then raise the rear of the vehicle and support it securely on jackstands. Remove the rear wheel.

2   Place a floor jack under the rear of the trailing arm and raise it just enough to take the weight of the axle off the shock absorber lower mount.

3   Remove the shock absorber lower mounting bolt/nut (see illustration).

4   Remove the trim panel by the rear seat back for access to the shock absorber upper mount (see Chapter 11).

5   Remove the shock absorber upper mounting bolts, rotate the assembly to disengage the upper mounting lugs and remove the shock absorber/coil spring assembly (see illustrations).

6   If you are replacing either the shock absorber or the coil spring, refer to Section 14.

7   To install the shock absorber, place a new gasket on the upper

mount, guide the assembly into position and rotate it to engage the upper mounting lugs (see illustration). Install the upper mounting bolts and a new lower mounting bolt and nut. Don't tighten the lower mounting bolt/nut yet.

8   Tighten the upper mounting bolts to the torque listed in this Chapter's Specifications.

9   Raise the rear axle to simulate normal ride height, then tighten the lower mounting bolt/nut to the torque listed in this Chapter's Specifications.

10  Install the wheel and lower the vehicle. Tighten the wheel bolts to the torque listed in the Chapter 1 Specifications.

11  Reinstall the trim panel.

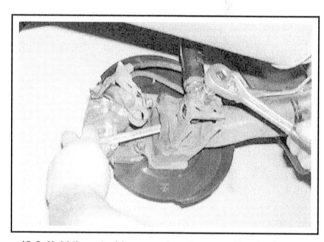

**10.3 Hold the nut with a wrench and remove the shock absorber lower mounting bolt**

**10.5a Remove the shock absorber upper mounting bolts (2WD A4) . . .**

**10.5b . . . then rotate the assembly to disengage the upper mounting lugs and remove the shock from the wheelwell**

**10.7 Rotate the shock absorber assembly until the upper mounting lugs can be felt to engage**

## 11 Shock absorber/coil spring assembly (rear, all-wheel drive Passat) - removal and installation

◆ **Refer to illustrations 11.5 and 11.6**

### ✳✳ WARNING 1:

**The manufacturer recommends replacing the shock absorber lower mounting bolt and nut, the upper control arm-to-steering knuckle bolt and nut and the stabilizer bar clamp nuts and link nuts with new ones whenever they are removed.**

### ✳✳ WARNING 2:

**Always replace the shock absorbers or coil springs in pairs - never replace just one of them, as handling peculiarities may result.**

1   Loosen the rear wheel bolts. Chock the front wheels to keep the vehicle from rolling, then raise the rear of the vehicle and support it securely on jackstands. Remove the rear wheel.

2   Remove the brake caliper (see Chapter 9). Support the caliper with a length of wire or string - don't let it hang by the hose.

3   Detach the stabilizer bar link from the stabilizer bar, then remove the nuts securing the stabilizer bar clamp (see Section 19).

4   Remove the upper control arm-to-rear knuckle bolt (see Section 20).

5   Remove the shock absorber lower mounting bolt (see illustration).

6   Remove the shock absorber upper mounting bolts (see illustration). Pull down on the rear knuckle and guide the shock absorber/coil spring assembly out.

7   If you are replacing either the shock absorber or the coil spring, refer to Section 14.

8   Installation is the reverse of the removal procedure, noting the following points:

a) *Be sure to use new hardware as indicated in the* **Warning** *at the beginning of this Section.*

b) *Don't tighten any of the suspension fasteners until the rear suspension has been raised to simulate normal ride height (this will prevent bushing distortion). Tighten all suspension fasteners to the torque values listed in this Chapter's Specifications.*

c) *Install new brake caliper mounting bolts and tighten them to the torque values listed in the Chapter 9 Specifications.*

d) *Install the wheel and tighten the bolts securely, then lower the vehicle and tighten the bolts to the torque listed in the Chapter 1 Specifications.*

**11.5  Shock absorber lower mounting bolt/nut (all-wheel drive Passat)**

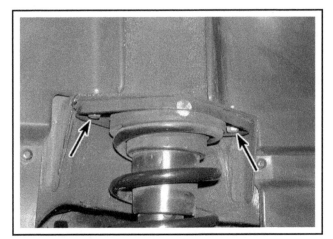

**11.6  Shock absorber upper mounting bolts (all-wheel drive Passat)**

## 12 Shock absorber/coil spring assembly (rear, all-wheel drive A4) - removal and installation

◆ **Refer to illustration 12.4**

### ✳✳ WARNING 1:

**The manufacturer recommends replacing the shock absorber lower mounting bolt and nut, the upper control arm pivot bolt and nut, the upper control arm-to-rear knuckle bolt and nut with new ones whenever they are removed.**

### ✳✳ WARNING 2:

**Always replace the shock absorbers or coil springs in pairs - never replace just one of them, as handling peculiarities may result.**

➡**Note: The upper control arm is removed along with the shock absorber/coil spring assembly.**

1   Loosen the rear wheel bolts. Chock the front wheels to keep the vehicle from rolling, then raise the rear of the vehicle and support it securely on jackstands. Remove the rear wheel.

2   Remove the shock absorber lower mounting bolt.

3   Remove the upper control arm-to-rear knuckle nut and bolt.

4   Remove the shock absorber/coil spring mounting bracket-to-body bolts (see illustration). Pull down on the rear knuckle and guide the shock absorber/coil spring/upper control arm assembly out from the wheelwell.

5   If you are replacing either the shock absorber or the coil spring, refer to Section 14.

6   Installation is the reverse of removal, noting the following points:

a) Be sure to use new hardware as indicated in **Warning 1** at the beginning of this Section.

b) Tighten the mounting bracket bolts to the torque listed in this Chapter's Specifications.

c) Before tightening the shock absorber lower mounting nut/bolt, the upper control arm pivot bolt/nut or the upper control arm-to-rear knuckle bolt/nut, raise the rear knuckle with a floor jack to simulate normal ride height (this will prevent bushing distortion). Tighten all suspension fasteners to the torque values listed in this Chapter's Specifications.

d) Install the wheel and tighten the bolts securely, then lower the vehicle and tighten the bolts to the torque listed in the Chapter 1 Specifications.

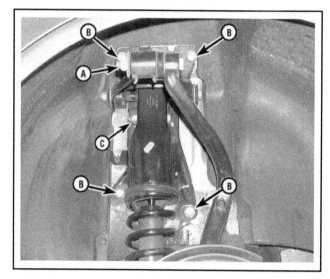

**12.4 Shock absorber/coil spring upper mounting details (all-wheel drive A4)**

A   Upper control arm-to-rear knuckle bolt/nut
B   Mounting bracket-to-body bolts
C   Shock absorber-to-mounting bracket bolt/nut (coil spring must be compressed before removing this bolt)

## 13  Coil spring (rear, 2WD Passat) - removal and installation

♦ **Refer to illustration 13.7**

### ✳✳ WARNING 1:

**Always replace the coil springs in pairs - never replace just one of them.**

### ✳✳ WARNING 2:

**The manufacturer recommends replacing the shock absorber upper mounting bolts with new ones whenever they are removed.**

1   Loosen the wheel bolts. Chock the front wheels to prevent the vehicle from rolling, then raise the rear of the vehicle and support it securely on jackstands placed under the rocker panel flanges. Remove the wheel.

➡**Note: It is not absolutely necessary to remove the wheels, but doing so greatly improves access to the springs.**

2   Support the rear axle with a floor jack placed under the rear of the trailing arm. Raise the jack slightly to take the spring pressure off the shock absorber upper mount.

3   Remove the shock absorber upper mounting bolts (see illustration 9.4).

4   Slowly lower the floor jack; at this point the spring will not be fully extended, as the shock absorber on the other side is limiting the downward travel of the axle.

5   Place the floor jack under the other trailing arm, remove the shock absorber upper mounting bolts, then slowly lower the jack until the coil springs are fully extended.

6   Remove the spring, upper insulator and lower bushing. Check the spring for cracks and chips, replacing the springs as a set if any defects are found. Also check the upper and lower insulators for damage and deterioration, replacing them as necessary.

7   Installation is the reverse of the removal procedure, but make sure the coil springs are positioned properly (see illustration).

8   Tighten the shock absorber upper mounting bolts to the torque listed in this Chapter's Specifications.

9   Lower the vehicle and tighten the wheel bolts to the torque listed in the Chapter 1 Specifications.

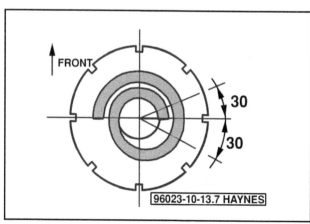

**13.7 Make sure the ends of the coil spring are positioned in their spring seats like this when installed**

## 14 Shock absorber or coil spring (rear, A4 or all-wheel drive Passat models) - replacement

### ✷✷ WARNING 1:

**The manufacturer recommends replacing the damper rod nut with a new one whenever it is removed.**

### ✷✷ WARNING 2:

**Always replace the shock absorbers or coil springs in pairs - never replace just one of them.**

➡Note: The following procedure describes the shock absorber or coil spring replacement on a 2WD A4. While slightly different, the procedure is similar for all-wheel drive A4s and Passats. On these models differences will be apparent as the unit is disassembled; be sure to lay out all parts in order and return them to their original positions, and ignore the steps which do not apply.

1   If the shock absorbers or coil springs exhibit the telltale signs of wear (leaking fluid, loss of damping capability, chipped, sagging or cracked coil springs) explore all options before beginning any work. The shock absorber itself is not serviceable and must be replaced if a problem develops. However, shock absorber/coil spring assemblies

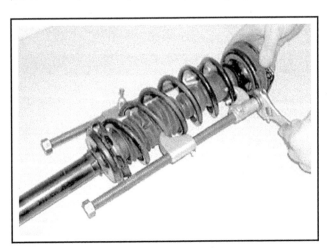

**14.2  Compress the coil spring until all tension is relieved from the spring seats**

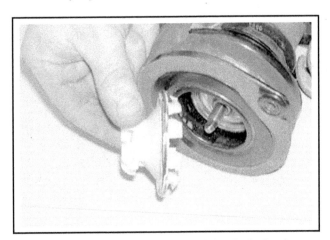

**14.3  Pry the plastic cap from the top of the shock absorber, together with its O-ring seal (2WD A4 models)**

complete with springs may be available on an exchange basis, which eliminates much time and work. Whichever route you choose to take, check on the cost and availability of parts before disassembling your vehicle.

### ✷✷ WARNING:

**Disassembling a coil-over type shock absorber is potentially dangerous and utmost attention must be directed to the job, or serious injury may result. Use only a high-quality spring compressor and carefully follow the manufacturer's instructions furnished with the tool. After removing the coil spring from the shock absorber, set it aside in a safe, isolated area.**

## DISASSEMBLY

◗ **Refer to illustrations 14.2, 14.3, 14.4, 14.5a, 14.5b, 14.5c, 14.5d, 14.6a, 14.6b, 14.6c, 14.6d, 14.6e and 14.6f**

2   Remove the shock absorber and spring assembly following the procedure described in the appropriate Section. Mount the assembly in a vise. Line the vise jaws with wood or rags to prevent damage to the unit and don't tighten the vise excessively.

➡Note: Make alignment marks between the upper mount, coil spring and shock absorber body, as these will be helpful during reassembly.

Following the tool manufacturer's instructions, install the spring compressor (which can be obtained at most auto parts stores or equipment yards on a daily rental basis) on the spring and compress it sufficiently to relieve all pressure from the upper spring seat (see illustration). This can be verified by wiggling the spring.

3   Pry the plastic cap from the top of the unit (2WD A4 models only) (see illustration).

4   If you're working on a 2WD A4 or an all-wheel drive Passat, loosen the shock absorber damper shaft nut while preventing the damper shaft from turning with another wrench (see illustration). If you're working on an all-wheel drive A4, remove the shock absorber-to-mounting bracket bolt. Detach the mounting bracket and upper control arm from the shock absorber.

**14.4  Loosen the damper rod nut while immobilizing the damper rod with another wrench**

➡Note: On all-wheel drive A4 models, note that the axis of the upper mounting eye and the axis of the lower mounting eye differ by 21-degrees (+/- 2-degrees). When reassembling the unit, make sure the angle of the mounting eyes are properly set.

5   Remove the nut then lift off the dished washer, bearing, upper spring plate together with its gasket and upper spring seat, as appli-

cable (see illustrations).

6   Slide off the spacer (if equipped), bearing ring, lower washer and O-ring. Withdraw the bump stop and its lower support ring and plastic bellows followed by the protective cap. Remove the coil spring and compressors, then extract the packing ring and the lower spring seat from the shock absorber body (see illustrations).

14.5a  Remove the nut . . .

14.5b  . . . then lift off the dished washer . . .

14.5c  . . . bearing . . .

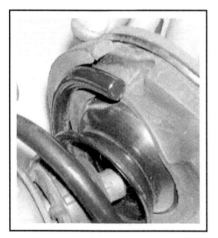

14.5d  . . . and upper spring plate and seat

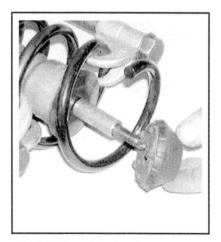

14.6a  Slide off the bearing ring . . .

14.6b  . . . spacer . . .

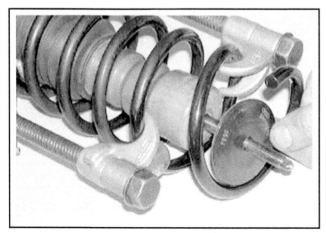

14.6c  . . . lower washer . . .

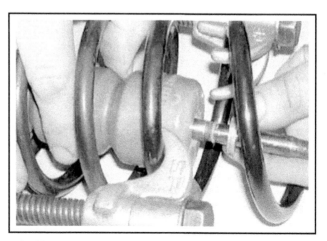

14.6d  . . . and O-ring

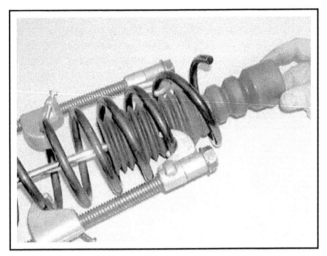

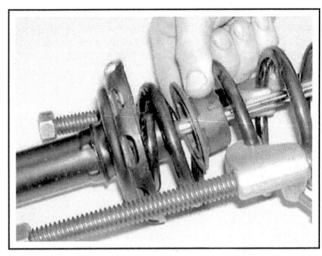

**14.6e** Withdraw the bump stop and its lower support ring and plastic bellows . . .

**14.6f** . . . followed by the protective cap

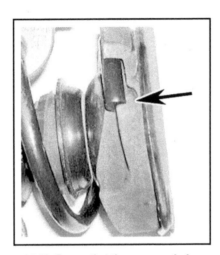

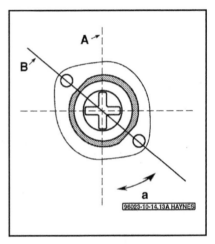

**14.11** Ensure that the lower end of the coil spring abuts the lower spring seat stop (arrow)

**14.12** Ensure that the upper end of the coil spring abuts the upper spring seat stop (arrow)

**14.13a** The axis (B) of the upper mounting bolt holes must be positioned at an angle (a) of 45-degrees to the axis (A) of the lower mounting bolt eye (2WD A4 models)

7   Examine the shock absorber for signs of fluid leakage. Check the piston for signs of pitting along its entire length, and check the shock body for signs of damage. While holding it in an upright position, test the operation of the shock absorber by moving the piston through a full stroke, and then through a series of short strokes. In both cases, the resistance felt should be smooth and continuous. If the resistance is jerky, or uneven, or if there is any visible sign of wear or damage to the shock absorber, replacement will be necessary.

### ✳✳ WARNING:

**Shock absorbers must be replaced in pairs. Different versions of shock absorbers are fitted to different models - make sure you have the correct version for your vehicle.**

8   Inspect all other components for signs of damage or deterioration, and replace any that are suspect.

## REASSEMBLY

▶ **Refer to illustrations 14.11, 14.12, 14.13a, 14.13b and 14.14**

9   Fit the spring lower seat and packing ring to the shock absorber ensuring that their end stops are positioned at 90-degrees to the lower mounting bolt axis.

10  Slide the protective cap over the damper rod then install the plastic bellows, followed by the lower support ring, bump stop, O-ring, lower washer, bearing ring and spacer.

11  Place the coil spring on the shock absorber. Make sure that the spring end abuts the spring seat stop (see illustration).

12  Lubricate the upper spring seat with talcum powder then install it to the upper end of the coil spring ensuring its stop (on some models indicated by the marking 'Federanfang') is correctly located against the spring end (see illustration).

13  Install the upper spring seat to the spring plate, using the alignment marks made on dismantling to ensure it is correctly positioned.

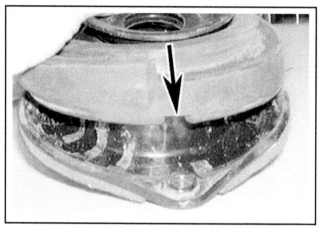

**14.13b  The nuts welded to the underside of the spring plate flange must engage with the corresponding recesses (arrow) in the top of the spring seat (2WD A4 models)**

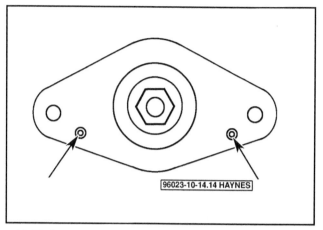

**14.14  On all-wheel drive Passats, the pins on the upper spring seat must face away from the center of the vehicle when the unit is installed, and must be parallel with the lower mounting eye of the shock absorber (2WD A4 models)**

On 2WD A4 models the upper mounting bolt holes must be positioned at 45-degrees to the axis of the shock lower mounting bolt eye. Note also that the nuts welded to the underside of the spring plate flange must engage with the corresponding recesses in the top of the spring seat (see illustrations).

14  Fit the spring seat/plate assembly to the damper rod, then install the bearing followed by the dished washer (2WD A4 models). Install a new gasket to the upper spring plate. On all-wheel drive Passat models the pins on the upper mount must be parallel with the axis of the shock absorber lower mounting eye, and must be positioned so that they are on the outboard side of the vehicle when the unit is installed (see illustration).

15  On 2WD A4s and all-wheel drive Passats, fully extend the damper rod and screw on the new nut. Hold the rod with a wrench to prevent it from rotating and tighten the nut to the torque listed in this Chapter's Specifications.

16  On all-wheel drive A4 models, assemble the mounting bracket to the top of the shock absorber and compressed coil spring, then install a new upper mounting bolt and nut. Tighten the nut to the torque listed in this Chapter's Specifications.

17  With the main components now assembled, check that the ends of the coil spring abut the stops on the spring seats.

18  On 2WD A4 models, press the cap, together with its O-ring, into position on the upper spring plate. Ensure that the x-shaped lugs on the cap are aligned perpendicularly to the axis of the shock absorber lower mounting bolt (see illustration 14.13a).

19  Ensure all components are correctly positioned then carefully release the spring compressor and remove it.

20  Install the shock absorber/coil spring assembly as described in the appropriate Section.

## 15  Hub and wheel bearing assembly (rear) - removal, servicing and installation

1  Loosen the rear wheel bolts, raise the rear of the vehicle and support it securely on jackstands. Remove the wheel.

2  Remove the brake caliper (don't detach the hose), mounting bracket and disc, or the brake drum (see Chapter 9). On models with rear disc brakes, hang the caliper with a piece of wire - don't let it hang by the brake hose.

### 2WD MODELS

### Audi A4 (serviceable bearings)

#### Removal

▶ **Refer to illustrations 15.3, 15.4a, 15.4b, 15.5a, 15.5b, 15.5c and 15.6**

3  Using a hammer and a large chisel, carefully tap and pry the cap out of the center of the brake disc (see illustration). Replace the cap with a new one if it is distorted during removal.

**15.3  Using a hammer and a chisel, carefully tap and pry the cap out of the center of the hub**

**15.4a Pull the cotter pin from the hub nut . . .**

**15.4b . . . and remove the nut lock**

**15.5a Remove the rear hub nut . . .**

**15.5b . . . then slide off the toothed washer . . .**

**15.5c . . . and remove the outer bearing from the center of the disc**

**15.6 The disc can now be slid off the stub axle**

4   Extract the cotter pin from the hub nut and remove the nut lock. Discard the cotter pin; a new one must be used on reassembly (see illustrations).

5   Remove the rear hub nut, then slide off the washer and remove the outer bearing (see illustrations).

6   The disc can now be slid off the stub axle (see illustration).

**15.8 The races can be driven out with a hammer and large punch**

### Servicing, installation and adjustment

▸ **Refer to illustration 15.8, 15.10a, 15.10b, 15.11, 15.13a, 15.13b and 15.13c**

7   Pry the grease seal from the hub. Take care not to damage the ABS wheel speed sensor rotor. Remove the inner bearing from the hub.

8   The races can be driven out using a suitable soft metal drift, but take care not to damage the hub or ABS speed sensor rotor (see illustration).

➡**Note: Only remove the races if you are replacing the bearings. Furthermore, don't install new bearings without installing new races.**

9   Clean the bearing race locations in the hub. Use a bushing or bearing driver to drive or press the new races into position in the hub each side. Ensure that they are squarely and fully inserted. If using the old bearings, be sure to keep the original bearings and races together. Never interchange new bearings with old races, or vice-versa.

10  Lubricate the bearings with grease (see illustration). Insert the inner bearing into position (see illustrations).

11  Pack the area between the seal lip and casing with wheel bearing grease. Carefully drive the new seal into position, taking care not to damage the ABS speed sensor rotor (see illustrations). Lubricate the seal lip with grease, too.

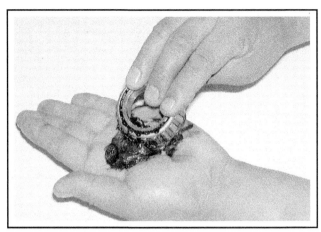

**15.10a If a bearing packing tool is not available, work grease into the bearing rollers by pressing it against the palm of your hand (from the larger diameter side of the bearing)**

**15.10b Install the packed bearing into the inner race**

**15.11 Carefully drive the new grease seal into place**

**15.13a Adjust the nut until the position is found where it is just possible to move the toothed washer from side-to-side using a screwdriver**

**15.13b Install the nut lock and secure it in position with a new cotter pin**

12 Apply a little more grease to the center of the hub, then install it on the stub axle, followed by the outer bearing, thrust washer and hub retaining nut.

13 Rotate the hub while tightening the hub nut slowly, up to approximately 15 ft-lbs to ensure that the bearings are correctly seated. Back the nut off until it is just loose, then tighten it hand-tight. Tighten or loosen the nut, as necessary, until the washer can just be slid to one side when pushed with a screwdriver (see illustration). Install the nut lock and insert a new cotter pin (see illustrations).

14 Beware of overtightening the hub nut, as this will cause premature bearing wear. If you are adjusting a bearing which has been in service for some time, when play has been noted, do not overtighten to compensate for wear - this is potentially dangerous as it could cause the wheel bearing to seize in use.

15 Smear a liberal amount of grease into the dust cap, then carefully drive the cap into position.

16 Install the brake caliper (disc brake models) (see Chapter 9).

17 Install the wheel and bolts. Lower the vehicle and tighten the bolts to the torque listed in the Chapter 1 Specifications.

18 If a new bearing has been installed, it is advisable to check for play after a few hundred miles. Re-adjust the bearing if necessary.

**15.13c Trim the ends of the cotter pin and bend them over to lock the pin in position**

### VW Passat (sealed bearings)

♦ Refer to illustration 15.19

→Note: **The hub and wheel bearing on these models is sealed-for-life and is not serviceable. If problems develop, the unit must be replaced as an assembly.**

19 Working through the holes in the hub flange, unscrew the hub-to-rear axle bolts (see illustration).

20 Remove the hub and bearing assembly from the axle. If it is stuck, tap on it from side-to-side to free it.

21 Installation is the reverse of removal, noting the following points:

a) *Make sure the hub-to-axle mating surfaces are clean before installing the hub and bearing assembly.*

b) *Tighten the mounting bolts to the torque listed in this Chapter's Specifications.*

c) *Install the caliper (see Chapter 9), tightening the mounting bolts to the torque listed in the Chapter 9 Specifications.*

d) *Install the wheel and wheel bolts. Lower the vehicle and tighten the bolts to the torque listed in the Chapter 1 Specifications.*

## ALL-WHEEL DRIVE MODELS

22 Due to the special tools and expertise required to press the hub

**15.19 On 2WD Passat models the rear hub and bearing assembly is retained to the axle with five bolts**

and bearing from the rear knuckle, this job should be left to a professional mechanic. However, the rear knuckle and hub may be removed and the assembly taken to an automotive machine shop or other qualified repair facility. See Section 17 for the rear knuckle and hub removal procedure.

## 16  Stub axle (rear, 2WD A4 models) - removal and installation

**Refer to illustrations 16.6a and 16.6b**

1  Loosen the wheel bolts. Chock the front wheels, then raise the rear of the vehicle and support it securely on jackstands. Remove the rear wheel.

2  Remove the hub and bearing assembly (see Section 11). Be sure to hang the caliper with a piece of wire - don't let it hang by the brake hose.

3  Inspect the stub axle surface for signs of damage such as scoring, and replace if necessary. Do not attempt to straighten the stub axle.

4  Also check the threads of the stub axle; if they are damaged in any way, replace the stub axle. Don't attempt to "clean up" the threads with a thread file or a die.

5  Remove the rear wheel speed sensor (see Chapter 9).

6  Remove the stub axle mounting bolts (see illustration). Remove the stub axle along with the disc splash shield (see illustration).

7  Installation is the reverse of removal, with the following points:

a) *Make sure the stub axle mating surfaces and the mounting bolt holes on the axle beam are clean.*

b) *Tighten the stub axle mounting bolts to the torque listed in this Chapter's Specifications.*

c) *Install the hub and adjust the wheel bearings as described in Section 15.*

d) *Tighten the caliper mounting bracket bolts and the caliper bolts to the torque listed in the Chapter 9 Specifications.*

**16.6a  Remove the stub axle mounting bolts . . .**

**16.6b  . . . and detach the stub axle and disc shield**

## 17  Knuckle and hub assembly (rear, all-wheel drive models) - removal and installation

▶ **Refer to illustrations 17.4, 17.5 and 17.6**

❊❊ **WARNING:**

The manufacturer recommends replacing the following fasteners on VW Passats: the driveaxle/hub bolt, the upper control arm-to-rear knuckle bolt and nut, the lower control arm-to-rear knuckle nut, and the track bar-to-rear knuckle nut. On Audi A4s, the manufacturer recommends replacing the following fasteners: the driveaxle/hub bolt, the upper control arm-to-rear knuckle bolt, washer and nut, the lower control arm-to-rear knuckle nut, and the track bar-to-rear knuckle nut.

1   Loosen the wheel bolts and the rear driveaxle/hub bolt, raise the rear of the vehicle and support it securely on jackstands. Remove the wheel.

2   Remove the rear brake caliper, mount and brake disc (see Chapter 9). Support the caliper with a length of wire or string - don't let it hang by the hose.

3   Remove the ABS rear wheel speed sensor (see Chapter 9).

4   Mark the position of the camber adjusting cam to the lower control arm (see illustration). Unscrew the nut from the other end of the bolt, then remove the eccentric washer and pull the bolt out.

5   Unscrew the nut and remove the track rod-to-rear knuckle bolt (see illustration).

6   Unscrew the nut and remove the upper control arm-to-rear knuckle bolt (see illustration). Remove the knuckle and hub assembly.

7   Installation is the reverse of removal, noting the following points:

   a) *Be sure to use new hardware as indicated in the* **Warning** *at the beginning of this Section.*

   b) *Line up the matchmarks (made in Step 4) on the camber adjusting cam and the lower control arm.*

   c) *Don't tighten any of the suspension fasteners until the rear suspension has been raised to simulate normal ride height (this will prevent bushing distortion). Tighten all suspension fasteners to the torque values listed in this Chapter's Specifications.*

   d) *Install the brake disc, caliper mounting bracket and caliper (see Chapter 9). Tighten the brake fasteners to the torque values listed in the Chapter 9 Specifications.*

   e) *Install the wheel and tighten the bolts securely, then lower the vehicle and tighten the bolts to the torque listed in the Chapter 1 Specifications.*

**17.4  Mark the relationship of the camber adjusting cam to the lower control arm (this will enable you to preserve the camber setting when reassembling)**

**17.5  Track rod-to-rear knuckle bolt**

**17.6  Upper control arm-to-rear knuckle bolt (Passat shown, A4 slightly different)**

## 18  Rear axle assembly (2WD models) - removal and installation

▶ **Refer to illustration 18.10**

❊❊ **WARNING:**

The manufacturer recommends replacing all rear axle mounting fasteners with new ones whenever they are loosened or removed.

### REMOVAL

1   Loosen the rear wheel bolts. Chock the front wheels, then raise the rear of the vehicle and support it securely on jackstands. Remove both rear wheels. Also remove the cover from the front of the axle beam.

2   Loosen the parking brake cable adjuster nut, then detach the parking brake cables from the calipers (see Chapter 9).

3   Remove the brake calipers, mounting brackets and discs (see Chapter 9).

4   Remove the ABS rear wheel speed sensors and unclip the harnesses from the axle.

5   If the axle is to be replaced with a new one, remove the hub and bearing assemblies (Passat models, see Section 15) or the stub axles (A4 models, see Section 16).

6   Detach the parking brake cables from the retaining clips along the

axle trailing arms.

7   Unscrew the brake line fittings from the brake hoses on each side of the rear axle (near the pivot bushings). Use a flare-nut wrench, if available, to prevent rounding-off the fittings (see Chapter 9). Plug the hoses to minimize fluid loss and prevent the entry of dirt into the hydraulic system. If the rear axle is to be replaced with a new one, detach the brake lines from the clips on the axle.

8   If you're working on a Passat, remove the coil springs (see Section 13). If you're working on an A4, unbolt the lower ends of the shock absorbers from the axle (see Section 10).

9   Make a final check that all necessary components have been disconnected and positioned so they will not hinder the removal procedure, then position a floor jack beneath the center of the rear axle assembly. Raise the jack until it is just supporting the weight of the axle.

10  Mark the position of the left-side mounting bracket to the body, then loosen the mounting bracket bolts a few turns (see illustration).

11  Remove the nuts from the axle pivot bolts on each side of the vehicle, then remove the bolts (see illustration 18.13). Slowly lower the jack.

12  Inspect the bushings in the mounting brackets for signs of damage or deterioration. If replacement is necessary, they can be replaced with a drawbolt-type bushing replacement tool (and suitable adapters). If you don't have access to the necessary tools, take the brackets to an automotive machine shop to have the old bushings removed and the new ones installed.

13  Don't unbolt the rear axle mounting brackets from the floor pan unless they are damaged (bent, cracked, or if the pivot bolt holes are worn). If replacement is necessary, mark the position of the bracket to the floor pan and install the new bracket in the same position (this will preserve rear wheel alignment).

## INSTALLATION

14  Installation of the rear axle is the reverse of the removal procedure, noting the following points:

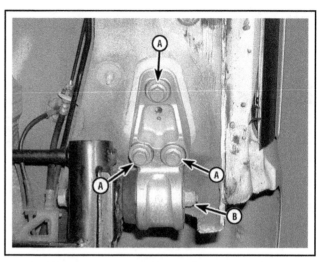

**18.10  Mark the position of the left-side mounting bracket, then loosen the bolts (A); B is the pivot bolt nut**

a)  Replace all loosened or removed suspension fasteners with new ones.

b)  Position the left-side mounting bracket in its original position.

c)  Ensure that the brake lines, parking brake cables and wiring (as applicable) are correctly routed, and retained by all the necessary retaining clips.

d)  Don't tighten the rear axle pivot bolts or the shock absorber lower mounting bolts until the weight of the vehicle is on its wheels (or until the rear suspension has been raised to simulate normal ride height). This will prevent the bushings from "winding-up," which could eventually damage them.

e)  Tighten all fasteners to the proper torque Specifications.

f)  Bleed the brake system (see Chapter 9).

g)  Have the rear wheel alignment checked and, if necessary, adjusted.

## 19  Stabilizer bar and bushings (rear) - removal and installation

### 2WD MODELS

1   The rear stabilizer bar on 2WD models runs along the length of the axle beam. It is an integral part of the axle assembly, and cannot be removed. If the stabilizer bar is damaged, which is very unlikely, the complete axle assembly must be replaced.

### ALL-WHEEL DRIVE MODELS

▶ **Refer to illustrations 19.2a, 19.2b  and 19.3**

### ※ WARNING:

**The manufacturer recommends replacing all self-locking nuts whenever they are removed.**

1   Loosen the rear wheel bolts, raise the rear of the vehicle and support it securely on jackstands. Block the front wheels to keep the vehicle from rolling off the stands. Remove the rear wheels.

2   Remove the nuts from the stabilizer bar links, then separate the links from the bar (see illustrations).

3   Remove the stabilizer bar clamp nuts and bolts (see illustration) and remove the stabilizer bar.

4   Inspect the stabilizer bar bushings and link bushings for cracks, tears and other signs of deterioration. Replace as necessary. Also check the ballstuds at the ends of the links for looseness, replacing the links if necessary.

5   Installation is the reverse of removal. Be sure to tighten all fasteners to the torque values listed in this Chapter's Specifications.

6   Tighten the wheel bolts to the torque listed in the Chapter 1 Specifications.

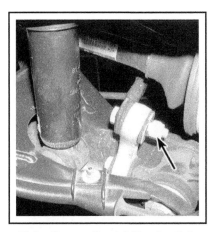

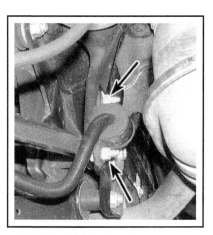

**19.2a** Remove the stabilizer bar link nuts (and bolts, on some models) and detach the links from the bar - this is a Passat . . .

**19.2b** . . . and this is an A4

**19.3** The stabilizer bar clamps are secured by two bolts and nuts

## 20  Rear suspension arms (all-wheel drive models) - removal and installation

1   Loosen the rear wheel bolts, raise the rear of the vehicle and support it securely on jackstands. Block the front wheels to keep the vehicle from rolling off the stands. Remove the rear wheel.

### UPPER CONTROL ARM

#### Audi A4

> **✳✳ WARNING:**
>
> The manufacturer recommends replacing the upper control arm pivot bolt and nut, and the upper control arm-to-rear knuckle bolt, nut and washer with new ones whenever they are removed.

2   Refer to Section 12 and remove the shock absorber/coil spring and mounting bracket assembly.

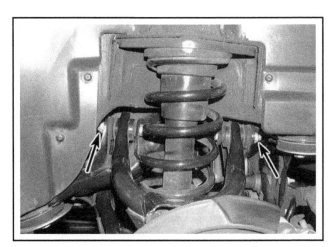

**20.7** Upper control arm pivot bolt/nuts

3   Remove the nut from the pivot bolt, slide out the bolt and detach the control arm from the mounting bracket.

4   Inspect the bushings for damage and wear. The bushings can be replaced, but a press and special adapters are required. If the bushings need to be replaced, take the arm to an automotive machine shop.

5   Installation is the reverse of removal, noting the following points:

  a)  *Use new fasteners as indicated in the* **Warning** *preceding Step 2.*

  b)  *Don't tighten the control arm pivot bolt or the control arm-to-rear knuckle bolt until the suspension has been raised to simulate normal ride height.*

  c)  *Tighten the wheel bolts to the torque listed in the Chapter 1 Specifications.*

### VW Passat

▶ Refer to illustration 20.7

> **✳✳ WARNING:**
>
> The manufacturer recommends replacing the upper control arm pivot bolts and nuts, and the upper control arm-to-rear knuckle bolt and nut with new ones whenever they are removed.

6   Remove the control arm-to-rear knuckle nut, washer and bolt.

7   Remove the nuts from the control arm inner pivot bolts, pull out the bolts and remove the control arm (see illustration).

8   Inspect the bushings for damage and wear. The bushings can be replaced, but a press and special adapters are required. If the bushings need to be replaced, take the arm to an automotive machine shop.

9   Installation is the reverse of removal, noting the following points:

  a)  *Use new fasteners as indicated in the* **Warning** *preceding Step 6.*

  b)  *Don't tighten the control arm pivot bolt/nuts or the control arm-to-rear knuckle bolt/nut until the suspension has been raised to simulate normal ride height.*

  c)  *Tighten the wheel bolts to the torque listed in the Chapter 1 Specifications.*

## LOWER CONTROL ARM

♦ **Refer to illustration 20.14**

**✳✳ WARNING:**

**The manufacturer recommends replacing the lower control arm pivot bolts and nuts, the shock absorber lower mounting bolt and nut, and the lower control arm-to-rear knuckle nut with new ones whenever they are removed.**

10 Remove the bolts and detach the ABS wheel speed sensor harness from the lower arm. Also detach the parking brake cable from the lower arm.

11 If you're working on an Audi A4, disconnect the flexible brake hoses from the rigid line on the lower control arm (see Chapter 9). Plug the hose on the vehicle side to prevent excessive fluid leakage and the entry of contaminants.

12 Unbolt the lower end of the shock absorber from the lower control arm (see Section 11 or 12).

13 Mark the relationship of the camber adjusting cam to the lower control arm (see illustration 17.4). Remove the nut and eccentric washer from the bolt, then remove the bolt.

14 Remove the lower control arm pivot bolts and detach the arm from the vehicle (see illustration).

15 Inspect the bushings for damage and wear. The bushings can be replaced, but special tools are required. If the bushings need to be replaced, take the arm to an automotive machine shop.

16 Installation is the reverse of removal, noting the following points:
- a) *Use new fasteners as indicated in the* **Warning** *preceding Step 10.*
- b) *Bleed the brake system as described in Chapter 9 (A4 models).*
- c) *Align the marks on the camber adjusting cam and the lower arm.*
- d) *Don't tighten the control arm pivot bolt/nuts, the control arm-to-rear knuckle bolt/nut or the shock absorber lower mounting bolt/nut until the suspension has been raised to simulate normal ride height.*
- e) *Tighten the wheel bolts to the torque listed in the Chapter 1 Specifications.*
- f) *Have the rear wheel alignment checked and, if necessary, adjusted.*

## TRACK ROD

♦ **Refer to illustration 20.18**

**✳✳ WARNING:**

**The manufacturer recommends replacing the track rod-to-rear knuckle nut and the track rod-to-subframe nut with new ones whenever they are removed.**

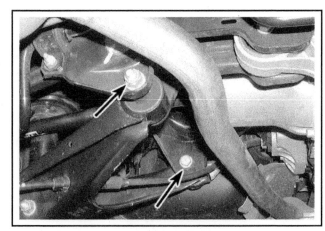

**20.14 Lower control arm pivot bolts/nuts**

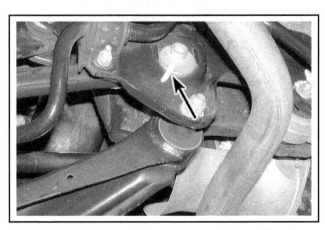

**20.18 Before removing the track rod, mark the position of the toe adjuster cam to the subframe**

17 Remove the track rod-to-rear knuckle nut/bolt (see illustration 17.5).

18 Mark the relationship of the toe adjuster cam to the subframe (see illustration). Remove the nut and eccentric bolt, then detach the rod from the vehicle.

19 Inspect the bushings for damage and wear. The bushings are not replaceable separately; if they are damaged, the track rod must be replaced

20 Installation is the reverse of removal, noting the following points:
- a) *Use new fasteners as indicated in the* **Warning** *preceding Step 17.*
- b) *Align the marks on the toe adjuster cam and the subframe.*
- c) *Don't tighten the track rod-to-subframe bolt/nut or the track rod-to-rear knuckle bolt/nut until the suspension has been raised to simulate normal ride height.*
- d) *Tighten the wheel bolts to the torque listed in the Chapter 1 Specifications.*
- e) *Have the rear wheel alignment checked and, if necessary, adjusted.*

## 21 Steering wheel - removal and installation

**✳✳ WARNING 1:**

**These models are equipped with airbags. Always disable the airbag system before working in the vicinity of any airbag system component to avoid the possibility of accidental deployment of the airbag(s), which could cause personal injury (see Chapter 12).**

**✳✳ WARNING 2:**

**Do not use a memory saving device to preserve the ECM's memory when working on or near airbag system components.**

## ✳✳ CAUTION 1:

These models are equipped with an anti-theft radio. Before performing a procedure that requires disconnecting the battery, make sure you have the activation code.

## ✳✳ CAUTION 2:

Disconnecting the battery can cause driveability problems that require a scan tool to rectify. Additionally, disconnecting the battery may cause one or more warning lights on the instrument panel to illuminate, which will also require the use of a scan tool to turn off. Most scan tools available to the public do not have the capability to perform either of these tasks, which will necessitate taking the vehicle to a dealer service department or other properly equipped repair facility after service work has been performed.

## REMOVAL

▶ **Refer to illustrations 21.3a, 21.3b, 21.4, 21.7a, 21.7b and 21.9**

1   Park the vehicle with the wheels pointing straight ahead. Disconnect the cable from the negative terminal of the battery.
2   Refer to Chapter 12 and disable the airbag system.
3   If you're working on a 1996 or 1997 A4, remove the two Torx screws to detach the airbag module from the steering wheel. On all other models, turn the steering wheel 90-degrees to gain access to the hole in the backside (the side facing the dash) of the steering wheel. Insert a screwdriver into the hole for the spring clip that retains the airbag module and push the spring aside to release the pin (see illustrations). Now turn the steering wheel 180-degrees in the other direction and do the same thing to release the other pin.

➡**Note: This step can be very difficult and frustrating. Take your time and be careful not to tear the rubber steering wheel trim when prying.**

4   Disconnect the airbag module electrical connector (see illustration).

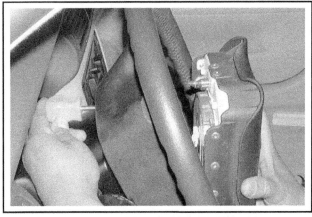

**21.3a  On all models except 1996 and 1997 A4s, insert a screwdriver into the hole in the steering wheel and pry up, which will release the clip securing one side of the airbag module (repeat this on the other side)**

5   Set the module aside in a safe, isolated area, with the airbag side of the module facing UP.

## ✳✳ WARNING:

When carrying the airbag module, keep the driver's (trim) side of it away from your body and, when you set it down, make sure the driver's side is facing up.

6   Center the steering wheel.
7   Remove the steering wheel bolt. Passats with three-spoke wheels have a standard hex bolt which can be reused. Passats with four-spoke wheels could have either a hex-head bolt or a 12-point internal spline drive bolt. The spline drive bolt can be used up to five times. Audi A4 models also use this type of spline-drive bolt, but the manufacturer recommends installing a new one each time the steering wheel Is removed. On models with this type of bolt, use a 12 mm 12-point spline-drive bit to remove the steering wheel retaining bolt (see illustration). On all

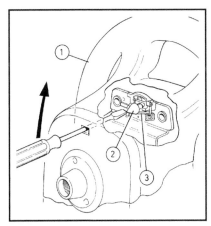

**21.3b  Here's a ghost view showing the screwdriver prying the spring clip away from the airbag module retaining post**

| 1 | Steering wheel | 2 | Pin |
| 3 | Clip |

**21.4  Unplug the airbag module electrical connector (arrow)**

**21.7a  To unscrew the steering wheel bolt on models with a spline drive bolt, a 12 mm, 12-point spline-drive bit is required**

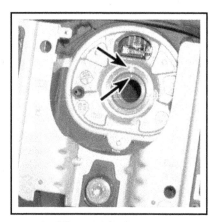

**21.7b After removing the bolt, check for alignment marks on the steering wheel and steering shaft - if there aren't any, make your own**

**21.9 Airbag clockspring retaining clips (not all are visible in this photo)**

**21.10 After centering the clockspring, make sure the yellow flag appears in the window**

models, mark the position of the steering wheel to the shaft if marks don't already exist or don't line up (see illustration).

8 Remove the steering wheel from the shaft. If it is tight, tap it up near the center using the palm of your hand, or twist it from side-to-side while pulling upwards.

### ❋❋ CAUTION:

Don't hammer on the shaft to remove the wheel.

9 Lift the steering wheel from the shaft.

### ❋❋ WARNING:

Don't allow the steering shaft to turn with the steering wheel removed. If the shaft turns, the airbag clockspring will become uncentered, which may cause the wire inside to break when the vehicle is returned to service.

If it is necessary to remove the clockspring, disengage the retaining clips and detach it from the steering column, then disconnect the electrical connector (see illustration).

## INSTALLATION

♦ **Refer to illustrations 21.10 and 21.14**

10 Before installing the steering wheel, make sure the airbag clockspring is centered (see illustration).

11 If the airbag system clockspring is not centered, remove the steering column covers (see Chapter 11). Release the locking tabs (see illustration 21.9), unplug the electrical connector and lift the clockspring off the steering column. Unplug the electrical connector.

12 To center the clockspring, depress the spring-loaded plunger and turn the hub in either direction until it stops (don't apply too much force). Now, rotate the hub in the other direction, counting the number of turns it takes to reach the opposite stop. Divide that number by two, then turn the hub back that many turns, approximately, until the yellow flag appears in the window (see illustration 21.10). Install the clockspring, making sure the locking tabs engage securely. Install the

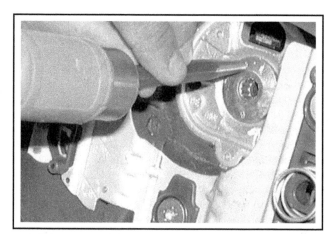

**21.14 On Passat models with a spline-drive type bolt, make a punch mark on the steering wheel bolt after installing it (it can be used up to five times)**

steering column covers.

13 Install the wheel on the steering shaft, aligning the marks.

14 Install the steering wheel bolt and tighten it to the torque listed in this Chapter's Specifications.

### ❋❋ WARNING:

If you're working on an A4, replace the spline-drive bolt with a new one.

➡Note: On Passat models the spline-drive bolt can be re-used up to five times. Using a center punch, place a mark on the head of the bolt after you tighten it, to keep track of how many times the bolt has been tightened (see illustration).

15 Plug in the electrical connector for the airbag and horn into the clockspring. Attach the ground leads to the airbag securing plate.

16 Position the airbag module on the steering wheel. On 1996 and 1997 Audi A4 models, install the Torx screws and tighten them to the torque listed in this Chapter's Specifications. On other models, push it in until the pins on the module engage with the spring clips.

17 Refer to Chapter 12 for the procedure to enable the airbag system.

## 22  Steering column - removal and installation

▶ Refer to illustrations 22.8, 22.9 and 22.10

### ※※ WARNING 1:

These models are equipped with airbags. Always disable the airbag system before working in the vicinity of any airbag system component to avoid the possibility of accidental deployment of the airbag(s), which could cause personal injury (see Chapter 12).

### ※※ WARNING 2:

Do not use a memory saving device to preserve the ECM's memory when working on or near airbag system components.

### ※※ WARNING 3:

The manufacturer recommends replacing the universal joint-to-steering gear pinch bolt nut with a new one whenever it is removed.

### ※※ CAUTION 1:

These models are equipped with an anti-theft radio. Before performing a procedure that requires disconnecting the battery, make sure you have the activation code.

### ※※ CAUTION 2:

Disconnecting the battery can cause driveability problems that require a scan tool to rectify. Additionally, disconnecting the battery may cause one or more warning lights on the instrument panel to illuminate, which will also require the use of a scan tool to turn off. Most scan tools available to the public do not have the capability to perform either of these tasks, which will necessitate taking the vehicle to a dealer service department or other properly equipped repair facility after service work has been performed.

### REMOVAL

1   Park the vehicle with the wheels in the straight-ahead position. Disconnect the cable from the negative terminal of the battery. Disable the airbag system (see Chapter 12).

2   Remove the steering wheel (see Section 21).

3   Remove the lower instrument panel trim (under the steering column) (see Chapter 11).

4   Remove the steering column covers (see Chapter 11).

5   Remove the airbag system clockspring (see Section 21).

6   Remove the multi-function switch (see Chapter 12).

7   On models equipped with an automatic transmission, detach the shift interlock cable from the ignition switch (see Chapter 7B).

8   Secure the lower section of the steering column to the upper section by passing a length of wire through the hole in the lower section of the column and through the spring on the upper section (see illustration).

9   Mark the relationship of the steering shaft lower universal joint to the steering gear input shaft, then remove the nut from the pinch bolt securing the universal joint (see illustration). Turn the Torx-head bolt clockwise a little to relieve tension on the bolt, then remove the bolt and separate the universal joint from the steering gear input shaft.

10  Remove the steering column mounting bolts (see illustration), then guide the column out from the instrument panel.

### INSTALLATION

11  Guide the column into position, connecting the U-joint with the steering gear input shaft.

12  Install the mounting bolts, but don't tighten them yet.

13  Remove the wire installed in Step 8. If a new steering column was installed, remove the transport protection rod and clip from the column.

14  Turn the U-joint bolt counterclockwise to tension it (it's an eccentric bolt), then install the nut. Tighten the nut to the torque listed in this Chapter's Specifications.

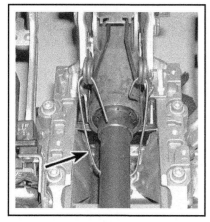

**22.8  Secure the lower portion of the shaft to the upper portion by passing a length of wire through the hole and spring**

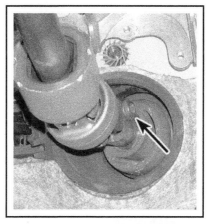

**22.9  Here's the pinch bolt securing the steering shaft U-joint to the steering gear; remove the nut from the other side, then turn this bolt a little and pull it out**

**22.10  Steering column mounting bolts**

15 The remainder of installation is the reverse of the removal procedure, noting the following points:

a) *When installing the steering wheel, be sure the airbag clockspring is centered, and tighten the steering wheel bolt to the torque listed in this Chapter's Specifications (see Section 21 for special instructions regarding the spline-drive bolt, if equipped).*

b) *Position the column as necessary to even-up the gaps between the trim panels, then tighten the column mounting bolts to the torque listed in this Chapter's Specifications.*

## 23 Tie-rod ends - removal and installation

▶ **Refer to illustrations 23.2 and 23.3**

### ❋❋ WARNING:

**The manufacturer recommends replacing the tie-rod end-to-steering knuckle pinch bolt nut with a new one whenever it is removed.**

## REMOVAL

1 Loosen the wheel bolts, raise the front of the vehicle and support it securely on jackstands. Apply the parking brake and block the rear wheels to keep the vehicle from rolling off the jackstands. Remove the wheel.

2 Break loose the tie-rod end jam nut (see illustration). Don't back the nut off; once it has just been loosened, it will serve as the point to which the tie-rod end will be threaded. If you are replacing the tie-rod end, count the number of threads exposed between the nut and the tie-rod end.

3 Remove the tie-rod to steering knuckle bolt, then remove the pinch bolt/nut (see illustration). Push the tie-rod end out of the steering knuckle.

4 Unscrew the tie-rod end from the tie-rod.

## INSTALLATION

5 If the jam nut was removed, thread it onto the tie-rod end until the number of threads counted in Step 2 are visible. Thread the tie-rod end into the tie-rod until it contacts the jam nut, then connect the tie-rod end to the steering knuckle. Install the tie-rod end-to-steering knuckle bolt and tighten it to the torque listed in this Chapter's Specifications.

6 Install the pinch bolt and a new nut, tightening the nut to the torque listed in this Chapter's Specifications.

7 Tighten the jam nut securely and install the wheel. Lower the vehicle and tighten the wheel bolts to the torque listed in the Chapter 1 Specifications.

8 Have the front end alignment checked and, if necessary, adjusted.

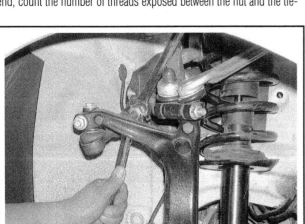

**23.2  Hold the tie-rod still with a wrench, then break loose the jam nut**

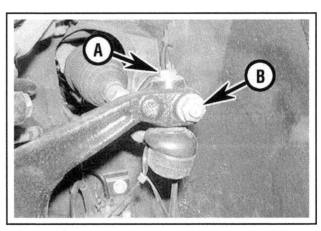

**23.3  Remove the tie-rod end-to-steering knuckle bolt (A), followed by the nut and pinch bolt (B), then detach the tie-rod end from the knuckle**

## 24 Steering gear boots - replacement

▶ **Refer to illustration 24.3**

### ❋❋ WARNING:

**The manufacturer recommends replacing the tie-rod end-to-steering knuckle pinch bolt nut with a new one whenever it is removed.**

1 Loosen the wheel bolts, raise the front of the vehicle and support it securely on jackstands. Remove the wheels.

2 Remove the tie-rod end from the tie-rod (see Section 23).

3 Remove the inner and outer boot clamps and discard them (see illustration). The clamps can be pried apart at the crimped area, or cut off with a pair of cutting pliers.

4 Remove the boot.

5 Install a new clamp on the inner end of the boot.

6   Slide the boot into place, making sure each end of the boot seats in its groove.

7   Make sure the boot isn't twisted, then tighten the inner clamp with a pair of clamp crimping pliers.

8   Install and tighten the outer clamp.

9   Install the tie-rod end (see Section 23).

10  Have the front end alignment checked and, if necessary, adjusted.

**24.3  Pry open or cut off the steering gear boot clamps**

## 25  Steering gear - removal and installation

### ✳✳ WARNING 1:

**These models are equipped with airbags. Always disable the airbag system before working in the vicinity of the any airbag system component to avoid the possibility of accidental deployment of the airbag, which could cause personal injury (see Chapter 12). Also, don't allow the steering wheel to turn after the steering gear has been removed. To prevent this, pass the seat belt through the steering wheel and plug it into its latch.**

### ✳✳ WARNING 2:

**The manufacturer recommends replacing the tie-rod end clamp bolt nuts and the steering shaft universal joint-to-steering gear pinch bolt nut with new ones whenever they are removed.**

### ✳✳ CAUTION 1:

**These models are equipped with an anti-theft radio. Before performing a procedure that requires disconnecting the battery, make sure you have the activation code.**

### ✳✳ CAUTION 2:

**Disconnecting the battery can cause driveability problems that require a scan tool to rectify. Additionally, disconnecting the battery may cause one or more warning lights on the instrument panel to illuminate, which will also require the use of a scan tool to turn off. Most scan tools available to the public do not have the capability to perform either of these tasks, which will necessitate taking the vehicle to a dealer service department or other properly equipped repair facility after service work has been performed.**

## REMOVAL

▶ **Refer to illustrations 25.11, 25.13a, 25.13b and 25.13c**

1   Park the vehicle with the front wheels pointing straight ahead. Apply the parking brake and chock the rear wheels, then loosen the front wheel bolts. Raise the front of the vehicle and support it securely on jackstands. Remove both front wheels.

2   Remove the battery from the engine compartment (see Chapter 5).

3   Turn the steering wheel to the center position then remove the ignition key to engage the steering lock.

### ✳✳ CAUTION:

**Ensure that the steering column remains in the straight-ahead position throughout the remainder of this procedure, or the airbag contact unit may become misaligned, leading to the failure of the airbag system.**

4   Remove the driver's side under-dash panel.

5   Secure the lower section of the steering column to the upper section as described in Section 22.

### ✳✳ CAUTION:

**Do not allow the upper and lower sections of the steering column to become separated while the steering column is detached from the steering gear, as this can cause the internal components to become detached and misaligned.**

6   Unscrew the nut from the pinch bolt securing the universal joint at the base of the steering column to the steering gear. Rotate the eccentric bolt to free it up and withdraw it from the joint.

7   Pull the steering column universal joint off the steering gear input shaft and move it to one side. Pull the input shaft cover from the firewall and into the vehicle.

8   Siphon the fluid from the power steering fluid reservoir. If a suitable implement is not readily available to siphon the fluid from the system, it can be drained into a container when the hydraulic lines are detached from the steering gear.

9   To minimize fluid leakage, apply hose clamps to the fluid lines leading to and from the steering gear. Take care to avoid causing damage to the hoses by pinching.

10  Unscrew the pressed-metal nut, pry out the clips and remove the section of the inner wheel well liner that shrouds the point where the tie-rod end enters the engine compartment.

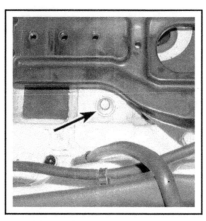

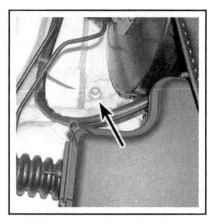

**25.11  Unscrew the banjo bolts from the pressure and return line fittings at the steering gear, then plug the fittings to prevent leakage and the entry of contaminants; the use of a ratchet (arrow), long extension and swivel socket is the easiest way to access the banjo bolts**

**25.13a  Remove this steering gear mounting bolt from the right side of the cowl area . . .**

**25.13b  . . . and this one from the left side**

11  Unscrew the fittings and disconnect the fluid supply and return lines from the steering gear (see illustration). Clean the connections before they are detached. Drain any fluid remaining in the system into a container for disposal. Tie the lines back away from the work area, and seal off their ends to prevent further leakage and the possible ingress of dirt.

12  Refer to Section 23 and unbolt the tie-rod ends from the steering knuckles.

13  Unscrew the steering gear mounting bolts (see illustrations).

14  Check that all connections are free and clear of the steering gear, then unclip the plastic collar from the pinion housing and withdraw the steering gear from the vehicle through the left-side wheel well.

15  If the steering gear is known to be damaged or worn beyond an acceptable level, it may have to be replaced. However, it is possible to have the steering gear overhauled - consult a dealer service department or other qualified repair shop for further advice.

## INSTALLATION

▶ **Refer to illustration 25.16**

16  Before the steering gear can be installed, it must be centered as follows. Remove the socket head bolt from the tapped inspection hole at the side of the input shaft housing. Move the right-side tie-rod by hand until the alignment hole - drilled into the surface of the steering rack - is visible through the inspection hole. Obtain a bolt of the same thread as that removed from the inspection hole and file the end of it to a conical point. Thread the bolt into the inspection hole and turn it until the pointed end engages with the drilled alignment hole in the steering rack; check that the rack is immobilized by trying to move the right hand tie-rod end. The steering gear is now locked in the center position (see illustration).

17  Maneuver the steering gear into position in the engine compartment. Insert the three securing bolts, then tighten the one accessible from underneath to the torque listed in this Chapter's Specifications.

**25.13c  The remaining steering gear mounting bolt is difficult to access - you can reach it from below with a socket and long extension, inserted just to the rear of the left-side driveaxle**

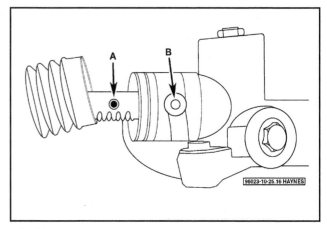

**25.16  Steering rack centering details - align the hole in the rack (A) with the inspection hole (B) and thread the specially fabricated bolt into the hole**

Now tighten the two upper mounting bolts to the torque listed in this Chapter's Specifications. On completion, remove the locking bolt from the alignment hole, then reinstall the original plug to seal the steering gear.

18 The remainder of the installation procedure is a reversal of removal, noting the following points:

a) *Connect the pressure and return line fittings to the steering gear, using new sealing washers. Tighten the banjo bolts to the torque listed in this Chapter's Specifications.*

b) *Refer to Section 23 to reconnect the tie-rod ends.*

c) *Top up the fluid level as described in Chapter 1 and bleed the system as described in Section 27.*

d) *Tighten the wheel bolts to the torque listed in the Chapter 1 Specifications.*

e) *Finally, have the wheel alignment checked and, if necessary, adjusted.*

## 26  Power steering pump - removal and installation

### ✳✳ CAUTION 1:

**These models are equipped with an anti-theft radio. Before performing a procedure that requires disconnecting the battery, make sure you have the activation code.**

### ✳✳ CAUTION 2:

**Disconnecting the battery can cause driveability problems that require a scan tool to rectify. Additionally, disconnecting the battery may cause one or more warning lights on the instrument panel to illuminate, which will also require the use of a scan tool to turn off. Most scan tools available to the public do not have the capability to perform either of these tasks, which will necessitate taking the vehicle to a dealer service department or other properly equipped repair facility after service work has been performed. See Chapter 5, Section 1 for the use of an auxiliary voltage input device ("memory saver") before disconnecting the battery and for other precautions related to battery disconnection.**

1  Disconnect the cable from the negative terminal of the battery.
2  Remove the ribbed drivebelt (see Chapter 1).

## FOUR-CYLINDER MODELS

3  Raise the front of the vehicle and support it securely on jackstands.
4  Remove the under-vehicle splash shield.
5  Remove the front bumper, then move the radiator support panel to the service position (see Chapter 11).
6  Remove the water pump pulley and drivebelt (see Chapter 3).
7  Using brake hose clamps, clamp both the supply and return hoses near the power steering fluid reservoir. This will minimize fluid loss during subsequent operations.
8  Wipe clean the area around the power steering pressure and return line fittings.
9  Unscrew the banjo bolt and disconnect the pressure line from the pump; be prepared for fluid spillage, and position a container beneath the pipe while unscrewing the fitting bolt. Disconnect the line and recover the sealing washers; discard the washers (new ones must be used when reassembling). Plug the end of the line and the steering pump orifice, to minimize fluid leakage and to keep dirt out of the hydraulic system.
10  Loosen the clamp and disconnect the fluid supply hose from the rear of the power steering pump. Plug the end of the hose and cover the pump fluid port to prevent contamination.
11  Remove the pump mounting bolts and detach the pump from its bracket.

12  If the power steering pump is faulty it must be replaced. The pump is a sealed unit and cannot be overhauled.
13  If a new pump is to be installed, it must be primed with fluid first, to ensure adequate lubrication during its initial stages of operation. Failure to do this could cause noisy operation and may lead to early pump failure. To prime the pump, pour the specified grade of hydraulic fluid into the fluid supply port on the pump, and simultaneously rotate the pump pulley. When the fluid exits from the pressure port, it is primed and ready for use.
14  Maneuver the pump into position, then install its mounting bolts and tighten them to the torque listed in this Chapter's Specifications.
15  Using new sealing washers, connect the pressure line to the pump. Ensure the line is correctly routed, then tighten the banjo bolt to the torque listed in this Chapter's Specifications.
16  Reconnect the supply hose to the pump and secure it in position with the retaining clip. Remove the hose clamps used to minimize fluid loss.
17  Install the water pump pulley and drivebelt (see Chapter 1).
18  Install the ribbed drivebelt (see Chapter 1).
19  Install the radiator support panel and front bumper (Chapter 11).
20  Install the under-vehicle splash shield. Reconnect the negative battery cable.
21  Top up the hydraulic system (see Chapter 1), then bleed the system as described in Section 27.

## V6 MODELS

▶ **Refer to illustrations 26.25 and 26.27**

22  Remove the top engine cover (see Chapter 2B).
23  Mark the positions of the spark plug wires, then remove the ignition coil assembly (see Chapter 5).
24  Remove the bolts and detach the pulley from the pump. A strap wrench or chain wrench can be used to prevent the pulley from turning as the bolts are loosened.

### ✳✳ CAUTION:

**Wrap the pulley with a leather belt or an old drivebelt to prevent damage to the pulley.**

25  Unscrew the banjo bolt and disconnect pressure line from the pump (see illustration); be prepared for fluid spillage. Disconnect the line and recover the sealing washers; discard the washers (new ones must be used when reassembling). Plug the end of the line and the steering pump orifice, to minimize fluid leakage and to keep dirt out of the hydraulic system.

**26.25 Power steering pump and bracket mounting details - DOHC V6 engine**

| | |
|---|---|
| 1  *Pressure line banjo bolt* | 3  *Mounting bolts* |
| 2  *Fluid supply hose* | |

**26.27 There's another power steering pump mounting bracket bolt near the upper intake manifold**

26  Loosen the clamp and disconnect the fluid supply hose from the pump (see illustration 26.25). Plug the end of the hose and cover the pump fluid port to prevent contamination.

27  Remove the pump bracket mounting bolts and detach the pump and bracket from the engine (see accompanying illustration and illustration 26.25). The pump can now be separated from the mounting bracket by removing the fasteners.

28  If the power steering pump is faulty it must be replaced. The pump is a sealed unit and cannot be overhauled.

29  If a new pump is to be installed, it must be primed with fluid first, to ensure adequate lubrication during its initial stages of operation. Failure to do this could cause noisy operation and may lead to early pump failure. To prime the pump, pour the specified grade of hydraulic fluid into the fluid supply port on the pump, and simultaneously rotate the pump pulley. When the fluid exits from the pressure port, it is primed and ready for use.

30  Maneuver the pump and bracket into position, then install its mounting bolts and tighten them to the torque listed in this Chapter's Specifications.

31  Using new sealing washers, connect the pressure line to the pump. Ensure the line is correctly routed, then tighten the banjo bolt to the torque listed in this Chapter's Specifications.

32  Reconnect the supply hose to the pump and secure it in position with the retaining clip. Remove the hose clamps used to minimize fluid loss.

33  Install the pulley, tightening the bolts to the torque listed in this Chapter's Specifications.

34  Install the ignition coil assembly (see Chapter 5) and the engine top cover.

35  Reconnect the negative battery cable, top up the hydraulic system (see Chapter 1), then bleed the system as described in Section 27.

## 27  Power steering system - bleeding

1  Following any operation in which the power steering fluid lines have been disconnected, the power steering system must be bled to remove all air and obtain proper steering performance.

2  With the front wheels in the straight ahead position, check the power steering fluid level and, if low, add fluid until it is between the Cold marks on the reservoir.

3  Raise the front of the vehicle and support it securely on jackstands.

4  Turn the steering wheel back-and-forth repeatedly, lightly hitting the stops.

### ✳✳ CAUTION:

**Don't hold the steering wheel in the full-right or full-left position, as this could damage the pump.**

5  Start the engine and allow it to run at fast idle. Recheck the fluid level and add more if necessary to reach the Cold marks.

6  Bleed the system by turning the wheels from side to side, just barely contacting the stops. This will work the air out of the system. Keep the reservoir full of fluid as this is done.

7  When the air is worked out of the system and the fluid level stabilizes, return the wheels to the straight ahead position and leave the vehicle running for several more minutes before shutting it off. Lower the vehicle.

8  Road test the vehicle to be sure the steering system is functioning normally and noise-free.

9  Recheck the fluid level to be sure it is between the Hot marks on the reservoir while the engine is at normal operating temperature. Add fluid if necessary (see Chapter 1).

## 28  Wheels and tires - general information

▶ **Refer to illustration 28.1**

1  All vehicles covered by this manual are equipped with metric-sized steel belted radial tires (see illustration). Use of other size or

type of tires may affect the ride and handling of the vehicle. Don't mix different types of tires, such as radials and bias belted, on the same vehicle as handling may be seriously affected. It's recommended that tires be replaced in pairs on the same axle, but if only one tire is being replaced, be sure it's the same size, structure and tread design as the other.

2   Because tire pressure has a substantial effect on handling and wear, the pressure on all tires should be checked at least once a month or before any extended trips (see Chapter 1).

3   Wheels must be replaced if they are bent, dented, leak air, have elongated bolt holes, are heavily rusted, out of vertical symmetry or if the lug bolts won't stay tight. Wheel repairs that use welding or peening are not recommended.

4   Tire and wheel balance is important in the overall handling, braking and performance of the vehicle. Unbalanced wheels can adversely affect handling and ride characteristics as well as tire life. Whenever a tire is installed on a wheel, the tire and wheel should be balanced by a shop with the proper equipment.

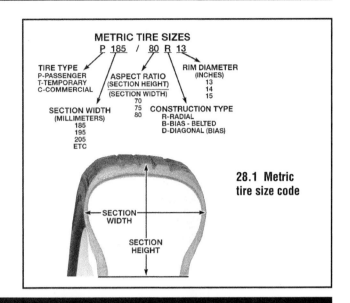

**28.1  Metric tire size code**

## 29  Wheel alignment - general information

♦ **Refer to illustration 29.1**

A wheel alignment refers to the adjustments made to the wheels so they are in proper angular relationship to the suspension and the ground. Wheels that are out of proper alignment not only affect vehicle control, but also increase tire wear.

The front end angles normally measured are camber, caster and toe-in (see illustration). Camber and caster are preset at the factory on the vehicle covered by this manual, but are usually checked to see if any suspension components are worn or damaged; at the front end, toe-in is the only adjustable angle on these vehicles.

No angles are adjustable on the rear wheels of 2WD models. On 2WD models the rear toe-in should be checked to make sure it is equal on each side, but each wheel is not adjustable individually (the entire axle can be shifted if the toe angles are not equal. Camber is usually measured to check for a bent axle.

Camber and toe-in are adjustable on the rear wheels of all-wheel drive models. Camber is adjusted by turning the lower control arm-to-rear knuckle bolt, which is equipped with an eccentric washer on either end. Toe-in is adjusted by turning the track rod-to-subframe bolt, which is also equipped with eccentric washers.

Getting the proper wheel alignment is a very exacting process, one in which complicated and expensive machines are necessary to perform the job properly. Because of this, you should have a technician with the proper equipment perform these tasks. We will, however, use this space to give you a basic idea of what is involved with a wheel alignment so you can better understand the process and deal intelligently with the shop that does the work.

Toe-in is the turning in of the wheels. The purpose of a toe specification is to ensure parallel rolling of the wheels. In a vehicle with zero toe-in, the distance between the front edges of the wheels will be the same as the distance between the rear edges of the wheels. The actual amount of toe-in is normally only a fraction of an inch. On the front end, toe-in is controlled by the tie-rod end position on the tie-rod. Incorrect toe-in will cause the tires to wear improperly by making them scrub against the road surface.

Camber is the tilting of the wheels from vertical when viewed from one end of the vehicle. When the wheels tilt out at the top, the camber is said to be positive (+). When the wheels tilt in at the top the camber is negative (-). The amount of tilt is measured in degrees from vertical

and this measurement is called the camber angle. This angle affects the amount of tire tread which contacts the road and compensates for changes in the suspension geometry when the vehicle is cornering or traveling over an undulating surface.

Caster is the tilting of the front steering axis from the vertical. A tilt toward the rear is positive caster and a tilt toward the front is negative caster.

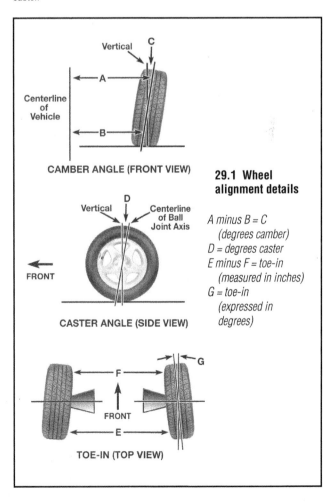

**CAMBER ANGLE (FRONT VIEW)**

**CASTER ANGLE (SIDE VIEW)**

**TOE-IN (TOP VIEW)**

**29.1  Wheel alignment details**

$A$ minus $B = C$ (degrees camber)
$D$ = degrees caster
$E$ minus $F$ = toe-in (measured in inches)
$G$ = toe-in (expressed in degrees)

| Torque specifications | Ft-lbs (unless otherwise indicated) | Nm |
|---|---|---|

➡Note: One foot-pound (ft-lb) of torque is equivalent to 12 inch-pounds (in-lbs) of torque. Torque values below approximately 15 ft-lbs are expressed in inch-pounds, since most foot-pound torque wrenches are not accurate at these smaller values.

### Front suspension

| | Ft-lbs | Nm |
|---|---|---|
| Stabilizer bar | | |
|   Link-to stabilizer bar nut | | |
|     A4 with balljoint-type link | 74 | 100 |
|     A4 with rubber bushing-type link | | |
|       Step 1 | 30 | 40 |
|       Step 2 | Tighten an additional 1/4-turn (90-degrees) | |
|     Passat | 66 | 90 |
|   Link-to-control arm nut | | |
|     A4 | 30 | 40 |
|     Passat | 52 | 70 |
|   Clamp bolts | | |
|     A4 | 18 | 25 |
|     Passat | 22 | 30 |
| Upper balljoint clamp bolt/nut | 30 | 40 |
| Lower balljoint-to-steering knuckle nut | 74 | 100 |
| Upper control arm pivot bolt/nut | | |
|   Step 1 | 30 | 40 |
|   Step 2 | Tighten an additional 1/4-turn (90-degrees) | |
| Upper control arm mounting bracket-to-body bolts | | |
|   A4 | 55 | 75 |
|   Passat | 48 | 65 |
| Lower control arm pivot bolt/nut | | |
|   A4 | | |
|     Front arm | | |
|       Step 1 | 59 | 80 |
|       Step 2 | Tighten an additional 1/4-turn (90-degrees) | |
|     Rear arm | | |
|       Step 1 | 66 | 90 |
|       Step 2 | Tighten an additional 1/4-turn (90-degrees) | |
|   Passat | | |
|     Step 1 | 66 | 90 |
|     Step 2 | Tighten an additional 1/4-turn (90-degrees) | |
| Subframe mounting bolts | | |
|   Step 1 | 81 | 110 |
|   Step 2 | Tighten an additional 1/4-turn (90-degrees) | |
| Shock absorber damper shaft nut | | |
|   A4 | 44 | 60 |
|   Passat | 37 | 50 |
| Shock absorber-to-upper control arm mounting bracket nuts | 15 | 20 |
| Shock absorber-to-lower control arm bolt/nut | 66 | 90 |
| Driveaxle/hub bolt | See Chapter 8 | |

**Torque specifications (continued)**   Ft-lbs (unless otherwise indicated)   Nm

➡Note: One foot-pound (ft-lb) of torque is equivalent to 12 inch-pounds (in-lbs) of torque. Torque values below approximately 15 ft-lbs are expressed in inch-pounds, since most foot-pound torque wrenches are not accurate at these smaller values.

## Rear suspension - 2WD models

| | Ft-lbs | Nm |
|---|---|---|
| Rear axle | | |
|   Pivot bolt/nut | | |
|     A4 | | |
|       Step 1 | 59 | 80 |
|       Step 2 | Tighten an additional 1/4-turn (90-degrees) | |
|     Passat | | |
|       Step 1 | 89 | 120 |
|       Step 2 | Tighten an additional 1/4-turn (90-degrees) | |
|   Mounting bracket retaining bolts | | |
|     A4 | 55 | 75 |
|     Passat | | |
|       Step 1 | 81 | 110 |
|       Step 2 | Tighten an additional 1/4-turn (90-degrees) | |
| Shock absorber-to-body bolts | | |
|   A4 | 18 | 25 |
|   Passat | 33 | 45 |
| Shock absorber-to-rear axle bolt/nut | | |
|   Step 1 | 37 | 50 |
|   Step 2 | Tighten an additional 1/4-turn (90-degrees) | |
| Shock absorber upper mount-to-damper shaft nut | 18 | 25 |
| Stub axle mounting bolts (A4) | 22 | 30 |
| Hub/bearing assembly-to-rear axle bolts (Passat) | 44 | 60 |

## Rear suspension - all-wheel drive models

| | Ft-lbs | Nm |
|---|---|---|
| Upper control arm inner pivot bolts/nuts | | |
|   A4 | | |
|     Step 1 | 37 | 50 |
|     Step 2 | Tighten an additional 1/4-turn (90-degrees) | |
|   Passat | 70 | 95 |
| Upper control arm-to-rear knuckle bolt/nut | | |
|   A4 | | |
|     Step 1 | 37 | 50 |
|     Step 2 | Tighten an additional 1/4-turn (90-degrees) | |
|   Passat | | |
|     Step 1 | 52 | 70 |
|     Step 2 | Tighten an additional 1/4-turn (90-degrees) | |
| Lower control arm inner pivot bolts/nuts | | |
|   Step 1 | 52 | 70 |
|   Step 2 | Tighten an additional 1/4-turn (90-degrees) | |
| Lower control arm-to-rear knuckle bolts/nuts | 70 | 95 |
| Track rod-to-subframe bolt/nut | | |
|   A4 | 66 | 90 |
|   Passat | 70 | 95 |
| Track rod-to-rear knuckle bolt/nut | | |
|   A4 | 37 | 50 |
|   Passat | 70 | 95 |

| **Torque specifications (continued)** | **Ft-lbs (unless otherwise indicated)** | **Nm** |
|---|---|---|

➡️**Note: One foot-pound (ft-lb) of torque is equivalent to 12 inch-pounds (in-lbs) of torque. Torque values below approximately 15 ft-lbs are expressed in inch-pounds, since most foot-pound torque wrenches are not accurate at these smaller values.**

## Rear suspension - all-wheel drive models (continued)

| | | |
|---|---|---|
| Shock absorber/upper control arm mounting bracket-to-body bolts (A4) | 41 | 55 |
| Shock absorber upper mounting bolts (Passat) | 33 | 45 |
| Shock absorber-to-lower control arm bolt/nut | | |
| Step 1 | 52 | 70 |
| Step 2 | Tighten an additional 1/4-turn (90-degrees) | |
| Shock absorber-to-mounting bracket bolt/nut (A4) | | |
| Step 1 | 52 | 70 |
| Step 2 | Tighten an additional 1/4-turn (90-degrees) | |
| Shock absorber damper rod nut (Passat) | 20 | 27 |
| Stabilizer bar link bolts/nuts | 37 | 50 |
| Stabilizer bar clamp bolts/nuts | | |
| A4 | 18 | 25 |
| Passat | 15 | 20 |

## Steering

| | | |
|---|---|---|
| Driver's airbag module retaining screws (1996 and 1997 A4) | 62 in-lbs | 7 |
| Steering column | | |
| Mounting bolts | 18 | 25 |
| Universal joint pinch bolt/nut | 30 | 40 |
| Steering gear | | |
| Mounting bolts | 48 | 65 |
| Pressure line banjo fitting bolt | 30 | 40 |
| Return line banjo fitting bolt | 37 | 50 |
| Tie-rod end-to-steering knuckle | | |
| Vertical bolt | 62 in-lbs | 7 |
| Clamp bolt/nut | 33 | 45 |
| Steering wheel bolt | | |
| Hex-head bolt | 55 | 75 |
| Internal spline-drive bolt | 44 | 60 |
| Power steering pump | | |
| Four-cylinder engine | | |
| Mounting bolts | 15 | 20 |
| Pressure line banjo fitting bolt | 37 | 50 |
| Pulley retaining bolts | 18 | 25 |
| V6 engine | | |
| Pump-to-mounting bracket bolts | 18 | 25 |
| Mounting bracket-to-engine bolts | 18 | 25 |
| Pressure line banjo fitting bolt | 30 | 40 |
| Pulley retaining bolts | 18 | 25 |

## Wheels

| | | |
|---|---|---|
| Wheel bolts | See Chapter 1 | |

**Section**

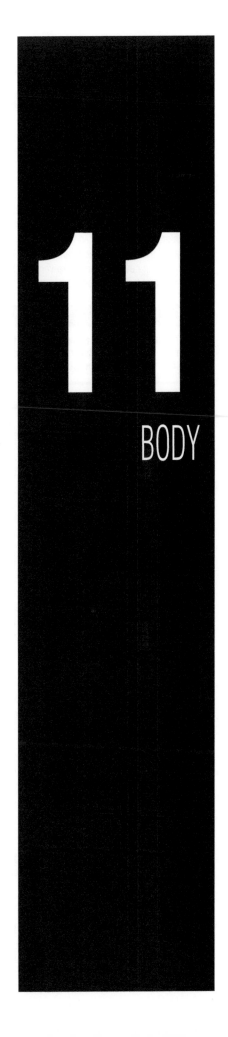

11

BODY

## 1 General information

The Volkswagen Passat and the Audi A4 feature a "unibody" layout, using a floor pan with front and rear frame side rails which support the body components, front and rear suspension systems and other mechanical components. Certain components are particularly vulnerable to accident damage and can be unbolted and repaired or replaced. Among these parts are the bumpers, front fenders, doors, the hood, the trunk lid (sedan models) or the liftgate (wagon models). Only general body maintenance practices and body panel repair procedures within the scope of the do-it-yourselfer are included in this Chapter.

Although all covered models are very similar, some procedures may differ somewhat from one body to another.

### ✳✳ CAUTION 1:

These models are equipped with an anti-theft radio. Before performing a procedure that requires disconnecting the battery, make sure you have the proper activation code.

### ✳✳ CAUTION 2:

Disconnecting the battery can cause severe driveability problems that require a scan tool to rectify. Additionally, disconnecting the battery may cause one or more warning lights on the instrument panel to illuminate, which will also require the use of a scan tool to turn off. Most scan tools available to the public do not have the capacity to perform either of these tasks, which will necessitate taking the vehicle to a dealer service department or other properly equipped repair facility after service work has been performed. See Chapter 5, Section 1 for the use of an auxiliary voltage input ("memory saver") device before disconnecting the battery and for other precautions related to battery disconnection.

## 2 Body - maintenance

1   The condition of your vehicle's body is very important, because the resale value depends a great deal on it. It's much more difficult to repair a neglected or damaged body than it is to repair mechanical components. The hidden areas of the body, such as the wheel wells, the frame and the engine compartment, are equally important, although they don't require as frequent attention as the rest of the body.

2   Once a year, or every 12,000 miles, it's a good idea to have the underside of the body steam cleaned. All traces of dirt and oil will be removed and the area can then be inspected carefully for rust, damaged brake lines, frayed electrical wires, damaged cables and other problems.

3   At the same time, clean the engine and the engine compartment with a steam cleaner or water-soluble degreaser.

4   The wheel wells should be given close attention, since undercoating can peel away and stones and dirt thrown up by the tires can cause the paint to chip and flake, allowing rust to set in. If rust is found, clean down to the bare metal and apply an anti-rust paint.

5   The body should be washed about once a week. Wet the vehicle thoroughly to soften the dirt, then wash it down with a soft sponge and plenty of clean soapy water. If the surplus dirt is not washed off very carefully, it can wear down the paint.

6   Spots of tar or asphalt thrown up from the road should be removed with a cloth soaked in solvent.

7   Once every six months, wax the body and chrome trim. If a chrome cleaner is used to remove rust from any of the vehicle's plated parts, remember that the cleaner also removes part of the chrome, so use it sparingly. After cleaning chrome trim, apply paste wax to preserve it.

## 3 Vinyl trim - maintenance

Don't clean vinyl trim with detergents, caustic soap or petroleum-based cleaners. Plain soap and water works just fine, with a soft brush to clean dirt that may be ingrained. Wash the vinyl as frequently as the rest of the vehicle. After cleaning, application of a high-quality rubber and vinyl protectant will help prevent oxidation and cracks. The protectant can also be applied to weatherstripping, vacuum lines and rubber hoses, which often fail as a result of chemical degradation, and to the tires.

## 4 Upholstery and carpets - maintenance

1   Every three months remove the floormats and clean the interior of the vehicle (more frequently if necessary). Use a stiff whiskbroom to brush the carpeting and loosen dirt and dust, then vacuum the upholstery and carpets thoroughly, especially along seams and crevices.

2   Dirt and stains can be removed from carpeting with basic household or automotive carpet shampoos available in spray cans. Follow the directions and vacuum again, then use a stiff brush to bring back the "nap" of the carpet.

3   Most interiors have cloth or vinyl upholstery, either of which can be cleaned and maintained with a number of material-specific cleaners or shampoos available in auto supply stores. Follow the directions on the product for usage, and always spot-test any upholstery cleaner on an inconspicuous area (bottom edge of a backseat cushion) to ensure that it doesn't cause a color shift in the material.

4   After cleaning, vinyl upholstery should be treated with a protectant.

**→Note: Make sure the protectant container indicates the product can be used on seats - some products may make a seat too slippery.**

### ✳ CAUTION:

**Do not use protectant on vinyl-covered steering wheels.**

5   Leather upholstery requires special care. It should be cleaned regularly with saddlesoap or leather cleaner. Never use alcohol, gasoline, water, nail polish remover or thinner to clean leather upholstery.

6   After cleaning, regularly treat leather upholstery with a leather conditioner, rubbed in with a soft cotton cloth. Never use car wax on leather upholstery.

7   In areas where the interior of the vehicle is subject to bright sunlight, cover leather seating areas of the seats with a sheet if the vehicle is to be left out for any length of time.

### 5   Body repair - minor damage

## PLASTIC BODY PANELS

The following repair procedures are for minor scratches and gouges. Repair of more serious damage should be left to a dealer service department or qualified auto body shop. Below is a list of the equipment and materials necessary to perform the following repair procedures on plastic body panels. Although a specific brand of material may be mentioned, it should be noted that equivalent products from other manufacturers may be used instead.

*Wax, grease and silicone removing solvent*
*Cloth-backed body tape*
*Sanding discs*
*Drill motor with three-inch disc holder*
*Hand sanding block*
*Rubber squeegees*
*Sandpaper*
*Non-porous mixing palette*
*Wood paddle or putty knife*
*Curved-tooth body file*
*Flexible parts repair material*

1   Remove the damaged panel, if necessary or desirable. In most cases, repairs can be carried out with the panel installed.

2   Clean the area(s) to be repaired with a wax, grease and silicone removing solvent applied with a water-dampened cloth.

3   If the damage is structural, that is, if it extends through the panel, clean the backside of the panel area to be repaired as well. Wipe dry.

4   Sand the rear surface about 1-1/2 inches beyond the break.

5   Cut two pieces of fiberglass cloth large enough to overlap the break by about 1-1/2 inches. Cut only to the required length.

6   Mix the adhesive from the repair kit according to the instructions included with the kit, and apply a layer of the mixture approximately 1/8-inch thick on the backside of the panel. Overlap the break by at least 1-1/2 inches.

7   Apply one piece of fiberglass cloth to the adhesive and cover the cloth with additional adhesive. Apply a second piece of fiberglass cloth to the adhesive and immediately cover the cloth with additional adhesive in sufficient quantity to fill the weave.

8   Allow the repair to cure for 20 to 30 minutes at 60-degrees to 80-degrees F.

9   If necessary, trim the excess repair material at the edge.

10  Remove all of the paint film over and around the area(s) to be repaired. The repair material should not overlap the painted surface.

11  With a drill motor and a sanding disc (or a rotary file), cut a "V" along the break line approximately 1/2-inch wide. Remove all dust and loose particles from the repair area.

12  Mix and apply the repair material. Apply a light coat first over the damaged area; then continue applying material until it reaches a level slightly higher than the surrounding finish.

13  Cure the mixture for 20 to 30 minutes at 60-degrees to 80-degrees F.

14  Roughly establish the contour of the area being repaired with a body file. If low areas or pits remain, mix and apply additional adhesive.

15  Block sand the damaged area with sandpaper to establish the actual contour of the surrounding surface.

16  If desired, the repaired area can be temporarily protected with several light coats of primer. Because of the special paints and techniques required for flexible body panels, it is recommended that the vehicle be taken to a paint shop for completion of the body repair.

## STEEL BODY PANELS

**→See photo sequence**

### Repair of minor scratches

17  If the scratch is superficial and does not penetrate to the metal of the body, repair is very simple. Lightly rub the scratched area with a fine rubbing compound to remove loose paint and built up wax. Rinse the area with clean water.

18  Apply touch-up paint to the scratch, using a small brush. Continue to apply thin layers of paint until the surface of the paint in the scratch is level with the surrounding paint. Allow the new paint at least two weeks to harden, then blend it into the surrounding paint by rubbing with a very fine rubbing compound. Finally, apply a coat of wax to the scratch area.

19  If the scratch has penetrated the paint and exposed the metal of the body, causing the metal to rust, a different repair technique is required. Remove all loose rust from the bottom of the scratch with a pocketknife, then apply rust inhibiting paint to prevent the formation of rust in the future. Using a rubber or nylon applicator, coat the scratched area with glaze-type filler. If required, the filler can be mixed with thinner to provide a very thin paste, which is ideal for filling narrow scratches. Before the glaze filler in the scratch hardens, wrap a piece of smooth cotton cloth around the tip of a finger. Dip the cloth in thinner and then quickly wipe it along the surface of the scratch. This will ensure that the surface of the filler is slightly hollow. The scratch can now be painted over as described earlier in this Section.

These photos illustrate a method of repairing simple dents. They are intended to supplement Body repair - minor damage in this Chapter and should not be used as the sole instructions for body repair on these vehicles.

1  If you can't access the backside of the body panel to hammer out the dent, pull it out with a slide-hammer-type dent puller. In the deepest portion of the dent or along the crease line, drill or punch hole(s) at least one inch apart . . .

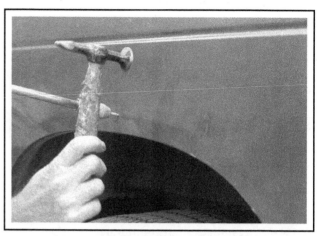

2  . . . then screw the slide-hammer into the hole and operate it. Tap with a hammer near the edge of the dent to help 'pop' the metal back to its original shape. When you're finished, the dent area should be close to its original contour and about 1/8-inch below the surface of the surrounding metal

3  Using coarse-grit sandpaper, remove the paint down to the bare metal. Hand sanding works fine, but the disc sander shown here makes the job faster. Use finer (about 320-grit) sandpaper to feather-edge the paint at least one inch around the dent area

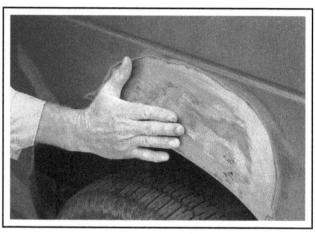

4  When the paint is removed, touch will probably be more helpful than sight for telling if the metal is straight. Hammer down the high spots or raise the low spots as necessary. Clean the repair area with wax/silicone remover

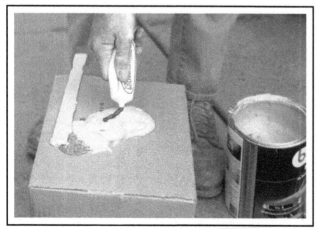

5  Following label instructions, mix up a batch of plastic filler and hardener. The ratio of filler to hardener is critical, and, if you mix it incorrectly, it will either not cure properly or cure too quickly (you won't have time to file and sand it into shape)

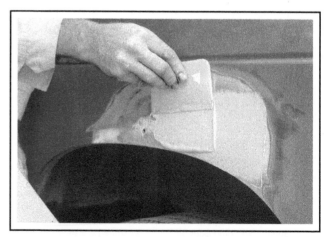

6  Working quickly so the filler doesn't harden, use a plastic applicator to press the body filler firmly into the metal, assuring it bonds completely. Work the filler until it matches the original contour and is slightly above the surrounding metal

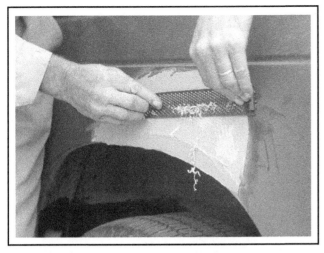

7 Let the filler harden until you can just dent it with your fingernail. Use a body file or Surform tool (shown here) to rough-shape the filler

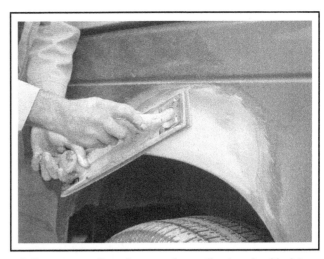

8 Use coarse-grit sandpaper and a sanding board or block to work the filler down until it's smooth and even. Work down to finer grits of sandpaper - always using a board or block - ending up with 360 or 400 grit

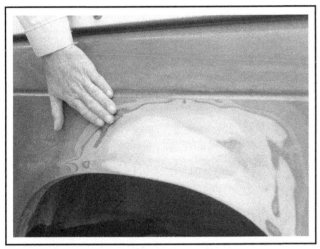

9 You shouldn't be able to feel any ridge at the transition from the filler to the bare metal or from the bare metal to the old paint. As soon as the repair is flat and uniform, remove the dust and mask off the adjacent panels or trim pieces

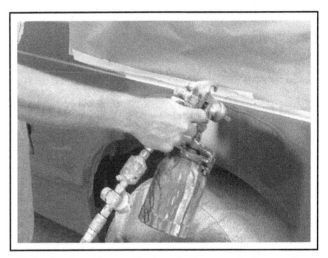

10 Apply several layers of primer to the area. Don't spray the primer on too heavy, so it sags or runs, and make sure each coat is dry before you spray on the next one. A professional-type spray gun is being used here, but aerosol spray primer is available inexpensively from auto parts stores

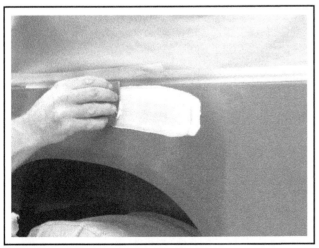

11 The primer will help reveal imperfections or scratches. Fill these with glazing compound. Follow the label instructions and sand it with 360 or 400-grit sandpaper until it's smooth. Repeat the glazing, sanding and respraying until the primer reveals a perfectly smooth surface

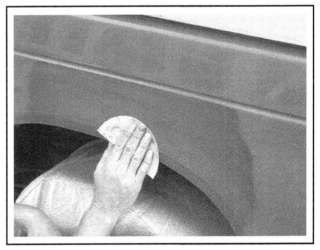

12 Finish sand the primer with very fine sandpaper (400 or 600-grit) to remove the primer overspray. Clean the area with water and allow it to dry. Use a tack rag to remove any dust, then apply the finish coat. Don't attempt to rub out or wax the repair area until the paint has dried completely (at least two weeks)

## Repair of dents

20 When repairing dents, the first job is to pull the dent out until the affected area is as close as possible to its original shape. There is no point in trying to restore the original shape completely as the metal in the damaged area will have stretched on impact and cannot be restored to its original contours. It is better to bring the level of the dent up to a point that is about 1/8-inch below the level of the surrounding metal. In cases where the dent is very shallow, it is not worth trying to pull it out at all.

21 If the backside of the dent is accessible, it can be hammered out gently from behind using a soft-face hammer. While doing this, hold a block of wood firmly against the opposite side of the metal to absorb the hammer blows and prevent the metal from being stretched.

22 If the dent is in a section of the body which has double layers, or some other factor makes it inaccessible from behind, a different technique is required. Drill several small holes through the metal inside the damaged area, particularly in the deeper sections. Screw long, self-tapping screws into the holes just enough for them to get a good grip in the metal. Now pulling on the protruding heads of the screws with locking pliers can pull out the dent.

23 The next stage of repair is the removal of paint from the damaged area and from an inch or so of the surrounding metal. This is easily done with a wire brush or sanding disk in a drill motor, although it can be done just as effectively by hand with sandpaper. To complete the preparation for filling, score the surface of the bare metal with a screwdriver or the tang of a file or drill small holes in the affected area. This will provide a good grip for the filler material. To complete the repair, see the Section on filling and painting.

## Repair of rust holes or gashes

24 Remove all paint from the affected area and from an inch or so of the surrounding metal using a sanding disk or wire brush mounted in a drill motor. If these are not available, a few sheets of sandpaper will do the job just as effectively.

25 With the paint removed, you will be able to determine the severity of the corrosion and decide whether to replace the whole panel, if possible, or repair the affected area. New body panels are not as expensive as most people think and it is often quicker to install a new panel than to repair large areas of rust.

26 Remove all trim pieces from the affected area except those which will act as a guide to the original shape of the damaged body, such as headlight shells, etc. Using metal snips or a hacksaw blade, remove all loose metal and any other metal that is badly affected by rust. Hammer the edges of the hole in to create a slight depression for the filler material.

27 Wire-brush the affected area to remove the powdery rust from the surface of the metal. If the back of the rusted area is accessible, treat it with rust inhibiting paint.

28 Before filling is done, block the hole in some way. This can be done with sheet metal riveted or screwed into place, or by stuffing the hole with wire mesh.

29 Once the hole is blocked off, the affected area can be filled and painted. See the following subsection on filling and painting.

## Filling and painting

30 Many types of body fillers are available, but generally speaking, body repair kits which contain filler paste and a tube of resin hardener are best for this type of repair work. A wide, flexible plastic or nylon applicator will be necessary for imparting a smooth and contoured finish to the surface of the filler material. Mix up a small amount of filler on a clean piece of wood or cardboard (use the hardener sparingly). Follow the manufacturer's instructions on the package, otherwise the filler will set incorrectly.

31 Using the applicator, apply the filler paste to the prepared area. Draw the applicator across the surface of the filler to achieve the desired contour and to level the filler surface. As soon as a contour that approximates the original one is achieved, stop working the paste. If you continue, the paste will begin to stick to the applicator. Continue to add thin layers of paste at 20-minute intervals until the level of the filler is just above the surrounding metal.

32 Once the filler has hardened, the excess can be removed with a body file. From then on, progressively finer grades of sandpaper should be used, starting with a 180-grit paper and finishing with 600-grit wet-or-dry paper. Always wrap the sandpaper around a flat rubber or wooden block, otherwise the surface of the filler will not be completely flat. During the sanding of the filler surface, the wet-or-dry paper should be periodically rinsed in water. This will ensure that a very smooth finish is produced in the final stage.

33 At this point, the repair area should be surrounded by a ring of bare metal, which in turn should be encircled by the finely feathered edge of good paint. Rinse the repair area with clean water until all of the dust produced by the sanding operation is gone.

34 Spray the entire area with a light coat of primer. This will reveal any imperfections in the surface of the filler. Repair the imperfections with fresh filler paste or glaze filler and once more smooth the surface with sandpaper. Repeat this spray-and-repair procedure until you are satisfied that the surface of the filler and the feathered edge of the paint are perfect. Rinse the area with clean water and allow it to dry completely.

35 The repair area is now ready for painting. Spray painting must be carried out in a warm, dry, windless and dust free atmosphere. These conditions can be created if you have access to a large indoor work area, but if you are forced to work in the open, you will have to pick the day very carefully. If you are working indoors, dousing the floor in the work area with water will help settle the dust that would otherwise be in the air. If the repair area is confined to one body panel, mask off the surrounding panels. This will help minimize the effects of a slight mismatch in paint color. Trim pieces such as chrome strips, door handles, etc., will also need to be masked off or removed. Use masking tape and several thickness of newspaper for the masking operations.

36 Before spraying, shake the paint can thoroughly, then spray a test area until the spray painting technique is mastered. Cover the repair area with a thick coat of primer. The thickness should be built up using several thin layers of primer rather than one thick one. Using 600-grit wet-or-dry sandpaper, rub down the surface of the primer until it is very smooth. While doing this, the work area should be thoroughly rinsed with water and the wet-or-dry sandpaper periodically rinsed as well. Allow the primer to dry before spraying additional coats.

37 Spray on the top coat, again building up the thickness by using several thin layers of paint. Begin spraying in the center of the repair area and then, using a circular motion, work out until the whole repair area and about two inches of the surrounding original paint is covered. Remove all masking material 10 to 15 minutes after spraying on the final coat of paint. Allow the new paint at least two weeks to harden, then use a very fine rubbing compound to blend the edges of the new paint into the existing paint. Finally, apply a coat of wax.

## 6    Body repair - major damage

1    Major damage must be repaired by an auto body shop specifi-cally equipped to perform unibody repairs. These shops have the spe-cialized equipment required to do the job properly.

2    If the damage is extensive, the body must be checked for proper alignment or the vehicle's handling characteristics may be adversely affected and other components may wear at an accelerated rate.

3    Due to the fact that all of the major body components (hood, fenders, etc.) are separate and replaceable units, any seriously damaged components should be replaced rather than repaired. Sometimes the components can be found in a wrecking yard that specializes in used vehicle components, often at considerable savings over the cost of new parts.

## 7    Hinges and locks - maintenance

Once every 3000 miles, or every three months, the hinges and latch assemblies on the doors, hood and trunk should be given a few drops of light oil or lock lubricant. The door latch strikers should also be lubricated with a thin coat of grease to reduce wear and ensure free movement. Lubricate the door and trunk locks with spray-on graphite lubricant.

## 8    Windshield and fixed glass - replacement

Replacement of the windshield and fixed glass requires the use of special fast-setting adhesive/caulk materials and some specialized tools and techniques. These operations should be left to a dealer service department or a shop specializing in glass work.

## 9    Radiator support panel - repositioning, removal and installation

1    The radiator support panel is the name given to the section of bodywork that is mounted across the front of the vehicle. A number of major components, including the hood latch mechanism, front bum-per, radiator, condenser (if equipped with air conditioning), automatic transmission fluid cooler and the front headlight/side marker light assemblies are mounted on the radiator support panel. The construc-tion of these vehicles is such that the radiator support panel and its associated components can be removed as an assembly without being extensively dismantled. In addition, the radiator support panel can be moved forward several inches to the 'service position' without having to disconnect the various hoses, pipes and wiring harnesses from the components mounted on it. Once the radiator support panel is placed in the "service position", access to components at the front of the engine are greatly improved. Although, sometimes because of limited access to the components at the front of the engine compartment, it may be easier to have the air conditioning system properly discharged and completely remove the radiator support panel from the vehicle to facilitate an adequate amount of working room.

### PLACING THE RADIATOR SUPPORT PANEL IN THE SERVICE POSITION

▶ **Refer to illustration 9.3, 9.5, 9.6, 9.9a, 9.9b, 9.9c, 9.10, 9.11 and 9.12**

➡**Note: To carry out this procedure, it will be necessary to fabri-cate two service tools, using two 10 inch lengths of threaded rod and a selection of hex nuts.**

2    Raise the front of the vehicle and support it securely on jackstands.

3    Working under the vehicle, remove the lower splash shield below the engine (see illustration).

**9.3  Remove the splash shield from below the engine**

**9.5 Remove the screws and detach the air intake duct**

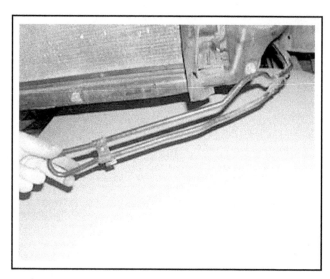

**9.6 Unbolt the power steering oil cooler line from the radiator support panel and position it aside without disconnecting the hoses**

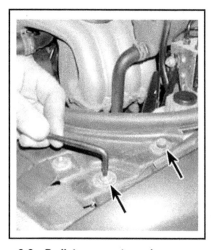

**9.9a Radiator support panel upper mounting bolts (arrows) - (A4 model shown, Passat models have only one bolt on each side securing the top of the panel)**

**9.9b Radiator support panel-to-fender retaining bolt (Audi A4)**

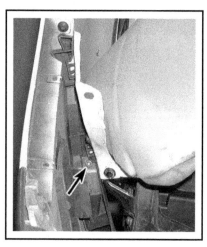

**9.9c Radiator support panel-to-fender retaining bolt (VW Passat) - view here is from below the vehicle with the bumper cover still installed**

4   Refer to Section 12 and remove the front bumper.

5   Remove the securing screw(s) and detach the air intake duct from the radiator support panel and the air cleaner housing (see illustration).

6   Unbolt the power steering oil cooler from the bottom of the radiator (see illustration).

7   Release the wiring from the retaining clips on the left hand side of the radiator.

8   Release the ends of the rubber seal at the top of the radiator support panel.

9   Remove the upper bolts securing the radiator support panel to the fenders (see illustration). Also remove the bolts (one each side, located underneath the front turn signal housing) securing the radiator support panel to the side of each front fender (see illustrations).

10  Remove one of the upper bolts from each of the bumper support brackets. Thread the homemade service tools into the holes vacated by

the bumper bracket bolts, then thread the hex nuts onto the end of the service tools (see illustration).

11  Remove the remainder of the bumper support bracket bolts (see illustration).

12  Carefully draw the radiator support panel away from the front of the engine compartment as far as possible without disconnecting the coolant hoses (see illustration).

➡**Note: On Audi A4 models the radiator support panel can be pulled outward and secured in the service position by aligning the rear bolt holes at the top with the front bolt holes and inserting the securing bolts to retain the lock carrier in position.**

13  The radiator support panel can be refitted by following the repositioning procedure in reverse. Ensure that all bolts are tightened to the correct torque specification, where specified. On completion, have the headlights checked for correct alignment.

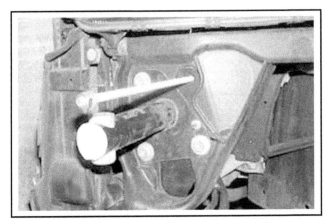

**9.10 Use lengths of threaded rod to support the radiator support panel**

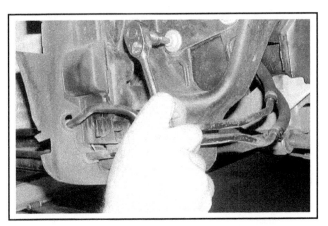

**9.11 Remove the remaining bolts securing the bumper support brackets to the frame**

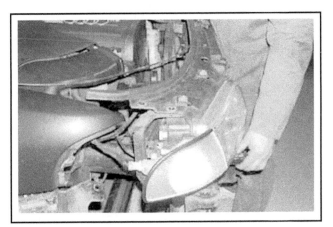

**9.12 Placing the radiator support panel in the "service position"**

**9.19 Disconnecting the wiring harness connectors from the headlights**

## REMOVAL AND INSTALLATION

▶ **Refer to illustration 9.19**

14 Have the air conditioning system discharged by a dealership service department or an automotive air conditioning facility.

### ✳✳ WARNING 1:

**The air conditioning system is under high pressure. DO NOT loosen any fittings or remove any components until after the system has been discharged. Air conditioning refrigerant should be properly discharged into an EPA-approved container at a dealership service department or an automotive air conditioning facility. Always wear eye protection when disconnecting air conditioning system fittings.**

### ✳✳ WARNING 2:

**Wait until the engine is completely cool before beginning this procedure.**

15 Refer to Steps 1 through 12 and place the radiator support panel in the service position.

16 Drain the coolant as described in Chapter 1.

17 Disconnect the coolant hoses from the radiator as described in Chapter 3. On models with air conditioning, also unbolt the refrigerant lines from the condenser, then disconnect the electrical connectors from the A/C high pressure switch referring to Chapter 3.

18 Disconnect the hood release cable from the hood latch mechanism with reference to Section 11.

19 Disconnect wiring harness connectors for the headlights, the side marker lights and if equipped the headlight aiming control motors (see illustration). Also disconnect the wiring harness connectors located at the left hand corner of the engine compartment.

20 On four-cylinder models, disconnect the electrical connector for the auxiliary fan thermo switch located at the lower left edge of the radiator. Also remove the screws securing the intercooler air duct to the radiator support panel and detach the air duct from the vehicle.

21 On models with automatic transaxles, remove the transmission cooler lines from the bottom of the radiator.

22 With the help of an assistant to support the panel, remove the nuts from the ends of the service tools, then withdraw the radiator support panel from the front of the vehicle.

23 Installation is the reverse of removal. Check the operation of the front lights, and the hood lock and safety catch, on completion. Refill and bleed the cooling system as described in Chapter 1, and have the headlights checked for correct alignment.

## 10 Hood - removal, installation and adjustment

➡️**Note: The hood is heavy and somewhat awkward to remove and install - at least two people should perform this procedure.**

### REMOVAL AND INSTALLATION

▶ **Refer to illustration 10.2**

1   Use blankets or pads to cover the cowl area of the body and the fenders. This will protect the body and paint as the hood is lifted off.

2   Scribe alignment marks around the hinge plate to insure proper alignment during installation (a permanent-type felt-tip marker also will work for this) (see illustration).

3   Disconnect the windshield washer hoses from the jets at the rear of the hood, and the underhood light, if applicable.

4   Have an assistant support the weight of the hood and detach the upper end of the hood support strut (see Section 11).

5   Remove the hinge-to-hood bolts (see illustration 10.2) and lift off the hood.

6   Installation is the reverse of removal.

### ADJUSTMENT

▶ **Refer to illustrations 10.10 and 10.11**

7   Fore-and-aft and side-to-side adjustment of the hood is done by moving the hood in relation to the hinge plates after loosening the bolts.

8   Scribe or trace a line around the entire hinge plate so you can judge the amount of movement (see illustration 10.2).

9   Loosen the bolts and move the hood into correct alignment. Move it only a little at a time. Tighten the hinge bolts or nuts and carefully lower the hood to check the alignment.

10  Adjust the hood bumpers on the front edge of the hood, so the hood is flush with the fenders when closed (see illustration).

11  The safety latch assembly can also be adjusted up-and-down and side-to-side after loosening the nuts (see illustration).

12  The hood latch assembly, as well as the hinges, should be periodically lubricated with white lithium-base grease to prevent sticking and wear.

**10.2 Use a marking pen to outline the hinge plate**

**10.10 Adjust the hood height by screwing the hood bumpers in or out**

**10.11 Loosen the bolts (arrows) and move the latch to adjust the hood fit in the closed position**

## 11 Hood latch, release cable and support struts - removal and installation

### LATCH

▶ **Refer to illustrations 11.3 and 11.4**

1   Open the hood.

2   Remove the nuts/bolts and detach the latch assembly (see illustration 10.11).

3   On Passat models, pry off the retaining clip and detach the safety latch handle (see illustration).

4   Detach the cable from the latch assembly, then remove the latch (see illustration).

5   Installation is the reverse of removal.

### CABLE

▶ **Refer to illustration 11.7**

6   Detach the cable case from the bracket and release the cable end from the latch in the engine compartment following steps 1 through 4.

7   In the passenger compartment, remove the driver's side kick panel and remove the screws securing the hood release handle (see illustration).

8   Pull the handle downward and release the cable from the handle.

9   Attach a wire or string to the end of the old cable in the passenger compartment.

**11.3  Pry off the retaining clip and detach the safety latch handle from the latch**

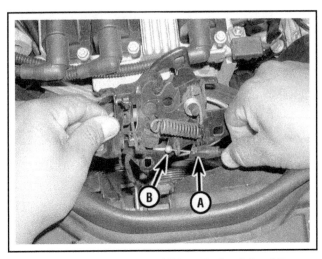

**11.4  Detach the cable casing (A) from the bracket and the cable end (B) from the latch**

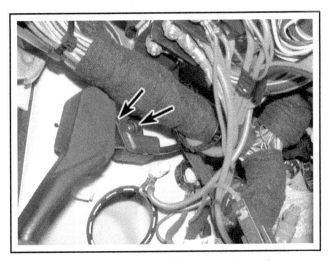

**11.7  Hood release handle mounting screws (arrows)**

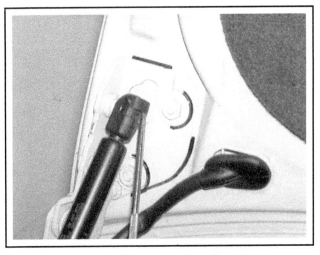

**11.15a  On Passat models, pry the clip outward to release the hood support strut**

10  Working in the engine compartment release the cable grommet from the cowl, then release the clips at the inner edge of the left front fenderwell until the cable is free of the body.

11  Pull the cable through the cowl and into the engine compartment.

12  Connect the string or wire to the new cable and pull it through the cowl into the passenger  compartment.

13  The remainder of installation is the reverse of removal. Be sure to fasten all of the cable retaining clips in their original location and seat the cable grommet in the cowl properly.

## SUPPORT STRUT

▶ **Refer to illustrations 11.15a and 11.15b**

14  Open the hood and support it securely.

15  Use a small screwdriver to detach the retaining clips at both ends of the support strut. Then pry or pull sharply to detach it from the vehicle (see illustration).

16  Installation is the reverse of removal.

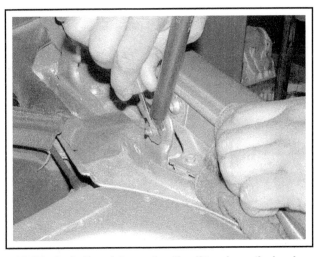

**11.15b  On Audi models, pry the clip off to release the hood support strut**

## 12 Bumpers - removal and installation

1  Front and rear bumpers on all models are composed of a fascia, or bumper (exterior) cover, and a metal structural beam.

## FRONT BUMPER

♦ **Refer to illustration 12.5**

2  Apply the parking brake, raise the vehicle and support it securely on jackstands.
3  Open the hood.
4  Remove the splash shield from below the engine (see illustration 9.3).
5  Working at each front wheelwell, remove the screws securing the edges of the bumper cover (see illustration).

### Audi A4

♦ **Refer to illustrations 12.6, 12.8, 12.9a and 12.9b**

6  Release the retaining tab and remove the lower grilles from both sides of the bumper (see illustration).

7  If equipped, remove the foglights from the bumper, referring to Chapter 12, if necessary. On certain models, the ambient temperature sensor is mounted on the rear surface of the bumper; unplug the wiring from it at the connector.
8  Unscrew and remove the two bumper retaining bolts from below (see illustration).
9  Carefully withdraw the bumper assembly by sliding it squarely away from the front of the vehicle (see illustrations). If headlight wipers are equipped, pull the bumper out to the point where the washer hose can be detached from its bumper-to-crosspanel connection before fully withdrawing the bumper.

➠**Note: The height of the bumper can be altered by adjusting the position of the threaded sleeves, inside the bumper mounting tubes.**

10  If the bumper cover needs to be removed from the metal bumper, simply remove the screws and/or push pins and separate the two components.

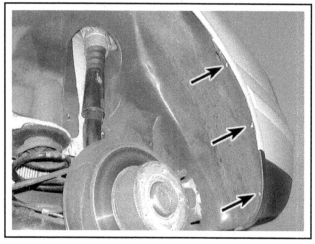

**12.5 Remove the screws (arrows) securing the bumper cover to the front fenderwells**

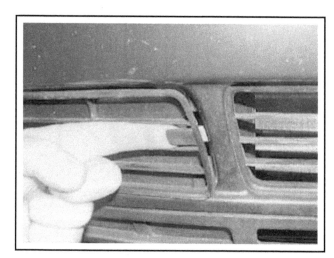

**12.6 On Audi models detach the lower grilles from the bumper**

**12.8 Remove the bolts securing the structural beam to the bumper support brackets**

**12.9a Unclip the trailing edge of the bumper from its retaining bracket . . .**

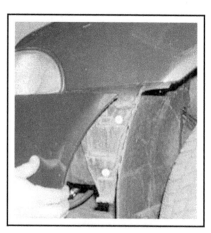

**12.9b . . . then pull the front bumper away from the vehicle**

## Volkswagen Passat

### 1998 through 2001 models

▶ **Refer to illustrations 12.11, 12.13, 12.14 and 12.15**

11 Remove the screws securing the bottom of the bumper cover (see illustration).

12 Remove the hood latch mechanism and release the safety catch handle from the latch (see illustration 11.3).

13 Remove the front turn signal lights (see Chapter 12) and detach the bolt securing the bumper cover in each of the turn signal light openings (see illustration).

14 Remove the screws securing the top of the bumper cover to the radiator support panel and remove the cover from the vehicle (see illustration).

15 Remove the bolts securing the metal bumper to the bumper support brackets and detach bumper (see illustration).

16 Installation is the reverse of the removal procedure.

### 2002 and later models

17 Remove the screws from the top of the grille, then pull it up and out of the vehicle, releasing the clips as you do so

18 Remove the screws from the bottom of the bumper cover.

19 Remove the screws from the top of the bumper cover at the grille opening.

20 Pull back the inner fender liner and remove the screws from each end of the bumper cover.

21 Carefully slide the bumper cover horizontally forward and off of the two alignment guides.

22 Installation is the reverse of removal.

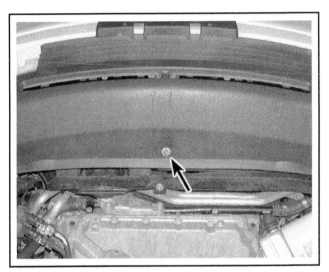

**12.11 On Passat models, remove the screws (arrow) securing the lower half of the bumper cover - the two outer screws are not shown in this photo**

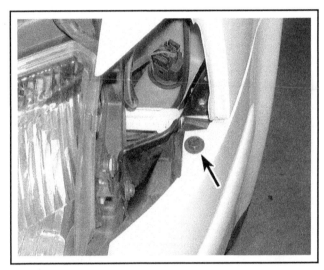

**12.13 Remove the bolts (arrow) located in the turn signal light openings**

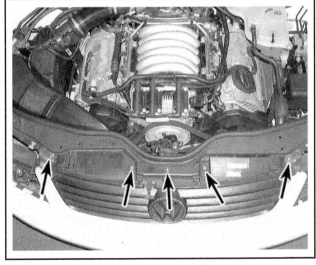

**12.14 Bumper cover upper retaining bolts (arrows) (Passat models)**

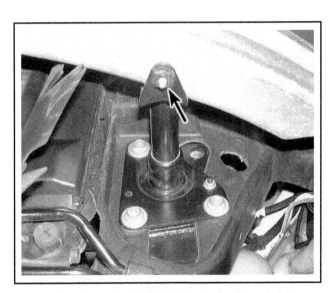

**12.15 Front bumper retaining bolt (arrow) - although this view is taken from below the bumper, the bolts are removed from the top**

## REAR BUMPER

♦ **Refer to illustration 12.23**

23 Working at the rear wheelwells, remove the screws securing the bumper cover to the wheelwell (see illustration).

## Audi A4

♦ **Refer to illustrations 12.24, 12.25a and 12.25b**

24 Working in the trunk area, fold back the floor covering for access, then unscrew and remove the two retaining bolts at each side (see illustration). On Avant models, it will be necessary to remove the floor sill plate and the side trim panels in the rear cargo area in order to peel back the floor carpet and gain access to the bolts.

25 Carefully withdraw the bumper assembly by sliding it squarely away from the rear of the vehicle. If the bumper sticks in position, free the edges one at a time by grasping the lower edge of the bumper just to the rear of the wheelwell and pivoting it upwards and away from the

rear quarter panel, to release it (see illustrations).

26 The bumper support brackets can be removed from the bumper by removing the mounting screws.

27 Installation is the reverse of removal. Loosely fit all retaining bolts and screws before fully tightening them.

## Volkswagen Passat

### Wagons and 1998 through 2001 sedan models

♦ **Refer to illustrations 12.29, 12.31 and 12.32**

28 Raise the rear of the vehicle and support it securely on jackstands.

29 Remove the screws securing the bottom of the bumper cover (see illustration).

30 Open the trunk.

31 Remove the screws securing the bumper cover in the trunk opening and detach the cover from the vehicle (see illustration).

32 Remove the bolts securing the metal to the bumper support

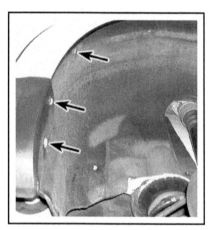

**12.23 Rear bumper cover-to-wheelwell mounting bolts (arrows) - lower bolt not shown in this photo**

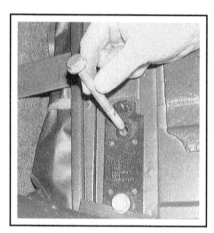

**12.24 On Audi models, remove the bumper support bracket bolts in the trunk**

**12.25a Unclip the trailing edge of the bumper from its retaining bracket . . .**

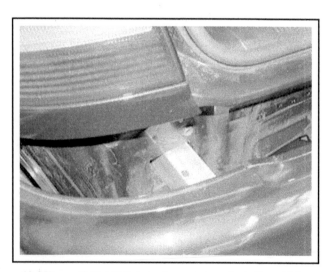

**12.25b . . . then pull the rear bumper away from the vehicle**

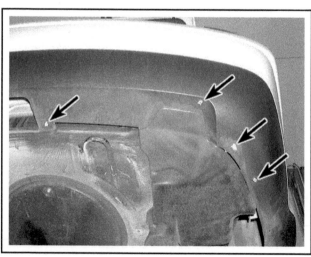

**12.29 Rear bumper cover lower retaining screws (arrows) - Passat models (right side shown)**

brackets and detach the bumper (see illustration).

33 Installation is the reverse of the removal procedure. Loosely fit all retaining bolts and screws before fully tightening them.

**2002 and later sedan models**

34 Remove the seven bolts around the bottom of the bumper cover.

35 Remove the three bolts from each forward end of the bumper cover. It will be necessary to loosen and pull back the inner fender cover to gain access to the bolts.

36 Refer to Chapter 12 and remove the taillight housings.

37 Remove the bolts and push-pin retainers at the top of the bumper cover.

38 Carefully slide the bumper cover horizontally rearward and off of the two alignment guides.

39 Installation is the reverse of removal.

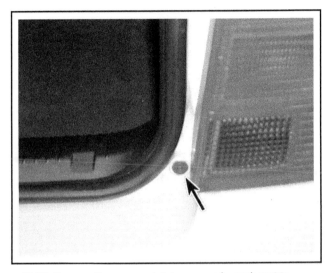

**12.31 Remove the upper retaining screw (arrow) next to each tail light**

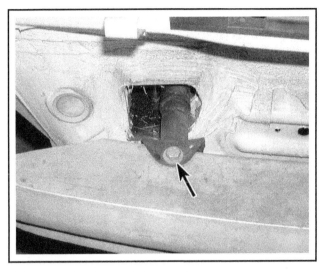

**12.32 Rear bumper retaining bolt (arrow) - Passat models (left side shown)**

## 13 Front fender - removal and installation

▶ **Refer to illustrations 13.3, 13.5, 13.6, 13.8 and 13.9**

1 Loosen the wheel bolts, raise the vehicle, support it securely on jackstands and remove the front wheel.

2 Refer to Section 12 and remove the front bumper.

3 Remove the inner fenderwell from the side on which the fender is to be removed (see illustration).

4 Working at the rear of the front fenderwell, remove the closeout panel between the fender and the door. This will allow access to the retaining bolts securing the fender to the "A" pillar.

➡**Note: Use a heat gun to soften the adhesive on this panel (as well as any other similar components) before trying to pull it off. Use care to avoid overheating the plastic.**

5 Remove the bolt securing the fender to the "A" pillar, then open

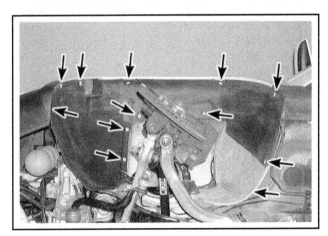

**13.3 Remove the bolts (arrows) and remove the inside fenderwell**

the door and remove the upper fender-to-door pillar bolt in the door-jamb (see illustration).

6 Remove plastic moulding and detach the bolts securing the lower half of the fender to the rocker panel (see illustration).

7 On Audi models remove the head-light/side marker light assembly (see Chap-ter 12).

8 Remove the bolt securing the fender to the front support brace (see illustration).

9 Remove the mounting bolts securing the top of the fender (see

illustration). Then detach the hood release cable from the fender.

10 Detach the fender. It's a good idea to have an assistant support the fender while it's being moved away from the vehicle to prevent damage to the surrounding body panels.

11 Installation is the reverse of the removal procedure. Tighten all nuts, bolts and screws securely.

12 Tighten the wheel bolts to the torque listed in the Chapter 1 Specifications.

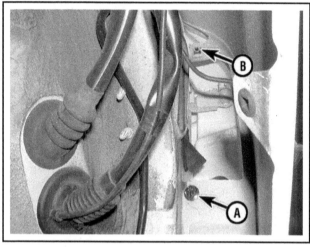

**13.5 Fender-to-"A" pillar bolts - bolt (A) can be accessed through the wheelwell, bolt (B) can be accessed through the door opening**

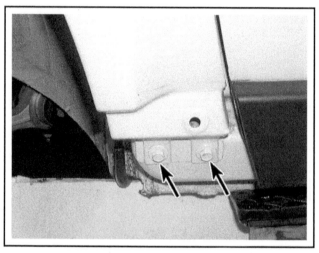

**13.6 The fender-to-rocker panel bolts (arrows) can be accessed after the plastic moulding is removed**

**13.8 Remove the fender support brace bolt (arrow) at the front of the fender - this bolt is removed from below**

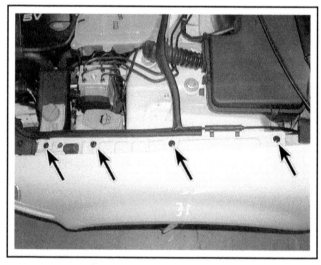

**13.9 Front fender upper retaining bolts (arrows)**

## 14 Cowl cover - removal and installation

▶ **Refer to illustrations 14.2 and 14.3**

1   Open the hood, then refer to Chapter 12 and remove the windshield wiper arms.

2   Remove the hood hinge trim covers and the bolt securing the cowl cover to the pollen filter housing (see illustration).

3   Remove the cowl cover retaining clips (see illustration).

4   Lift off the plastic cowl cover.

5   Installation is the reverse of removal.

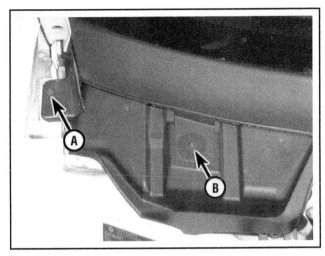

14.2  With the wiper arms removed, remove the hood hinge trim covers (A) and the pollen filter cover bolt (B)

14.3  Cowl cover retaining clips (arrows)

## 15 Door trim panels - removal and installation

### AUDI A4

▶ **Refer to illustrations 15.1a, 15.1b, 15.2, 15.3a, 15.3b, 15.3c, 15.4 and 15.6**

1   Open the door and remove the screws securing the inside pull handle trim. Withdraw the moulding from the door (see illustrations).

2   Remove the screws in the pull handle trim opening (see illustration).

3   On models with manual front windows, insert a flat bladed screwdriver underneath the winder handle knob and lever against the trim to detach it from the winder handle. Undo the securing screw and remove

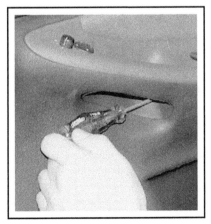

15.1a  Remove the screw . . .

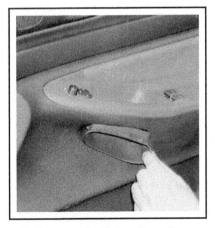

15.1b  . . . and withdraw the pull handle trim cover

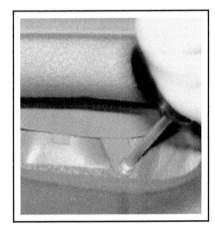

15.2  Remove the two screws exposed by the removal of the pull handle trim cover

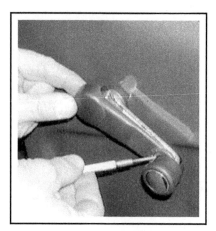

**15.3a On models with manual windows, lever against the trim cover to detach it from the handle**

**15.3b Undo the handle retaining screw and remove the handle from its shaft . . .**

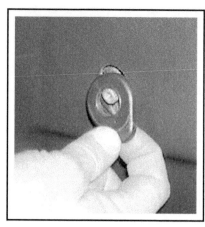

**15.3c . . . together with its spacer**

**15.4 Remove the screws at the front and rear of the upper edge of the door trim panel**

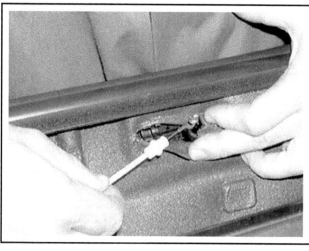

**15.6 Unhook the operating cable from the rear of the interior handle**

the handle from its shaft, together with its spacer (see illustrations).

4   Undo the screw at the front and rear of the upper edge of the door trim panel (see illustrations).

5   Lift the trim panel up squarely, to disengage the mounting hooks on the rear of the panel from the door.

6   Unhook the operating cable from the rear of the interior handle, as it becomes accessible (see illustration).

7   Unplug the wiring from the electric window/mirror/central locking switches (where applicable), and the door speakers. Release the wiring from its retaining clips on the rear of the door panel.

8   Lift the trim panel away from the door, noting the fitted positions of the moulded clips mounted along the sides.

9   If required, the watershield can be removed by carefully passing the wiring harness connectors through it and releasing it from the mounting hooks on the rear of the trim panel.

10  Installation is the reverse of removal. Guide the door locking knob through the hole at the upper edge of the trim panel, and ensure

that the wiring and connections are secure and correctly routed, clear of the window regulator and latch/lock components.

## VOLKSWAGEN PASSAT

▶ **Refer to illustrations 15.11a, 15.11b, 15.11c, 15.11d, 15.12a, 15.12b, 15.13, 15.14 and 15.15**

11  If you're removing the driver's side door trim panel, remove pull handle grip and the power window switch assembly. Then remove the screws securing the center of the panel (see illustrations).

12  If you're removing the passenger's side trim panel or the rear door trim panels, remove the door pull cover and the two door panel screws behind it (see illustrations).

13  Remove the screws retaining the bottom of the door panel to the door (see illustration).

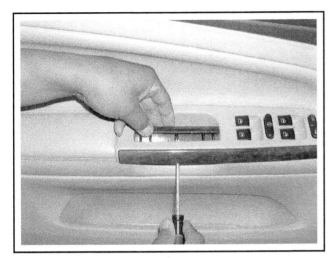

15.11a On the driver's side door trim panel, use a small screwdriver to detach the inside pull handle grip . . .

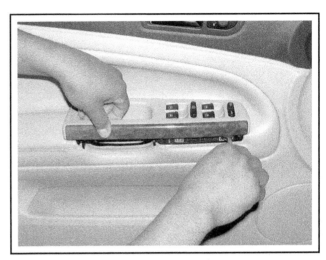

15.11b . . . then pry the switch plate out . . .

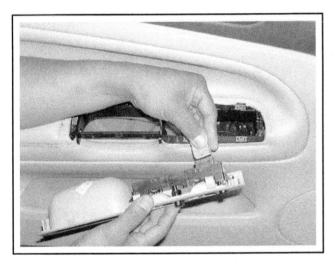

15.11c . . . and disconnect the electrical connectors

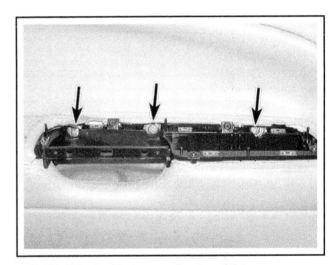

15.11d Driver's side door panel center retaining screws (arrows)

15.12a On the passenger's side and the rear door trim panels pry off the pull handle trim cover . . .

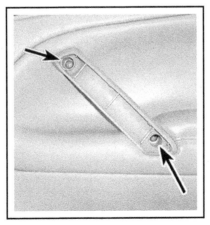

15.12b . . . to access the screws behind it (arrows)

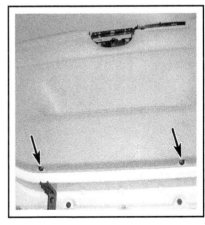

15.13 Remove the screws (arrows) securing the bottom of the door panel

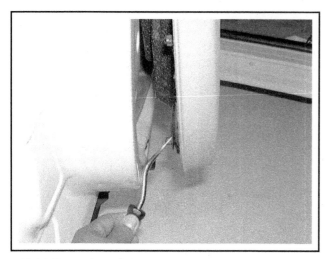

**15.14  Use a trim tool along the front and rear edges of the panel to release the clips**

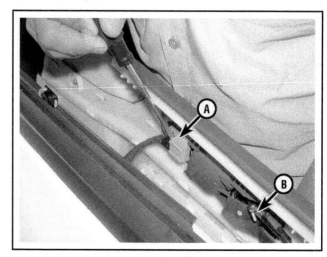

**15.15  Disconnect the electrical connectors (A) and the cable (B) from the rear of the door panel**

14  Using a trim tool, pry along the sides of the door panel to disengage the clips (see illustration).

15  Grasp the trim panel and pull up sharply to detach it from the door. Disconnect the inner door handle lock cable from the handle and unplug any electrical connectors (see illustration).

16  To install the panel, connect the wire harness connectors and door handle cable, then place the panel in position in the door. Press the trim panel down into place until the clips are seated.

17  Install the screws and the screw covers, where used.

## 16  Door - removal, installation and adjustment

▶ **Refer to illustrations 16.2, 16.5, 16.7 and 16.9**

1  Remove the door trim panel (see Section 15).

2  If you're removing the front doors refer to Section 26 and remove the kick panel. Then disconnect the top two electrical connectors from the multi-terminal connector (see illustration).

3  If you're removing the rear doors, detach the trim panel from the door pillar and disconnect the electrical connectors from the multi-terminal connector.

4  Pull the rubber harness boot out of the body and feed the harness connectors out.

5  Pry off the trim cap and remove the screw from the upper hinge (see illustration).

6  Place a jack under the door or have an assistant on hand to support it when the lower hinge bolts are removed.

➡ **Note: If a jack is used, place a rag between it and the door to protect the door's painted surfaces.**

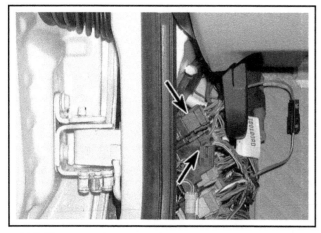

**16.2  Disconnect the electrical connectors (arrows) behind the kick panel and feed the wires out through the body pillar**

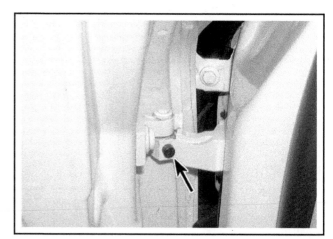

**16.5  Pry off the trim cap and remove the retaining screw from the upper hinge**

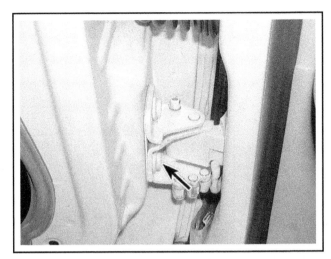

**16.7 Remove the lower hinge bolt (arrow) and lift the door off**

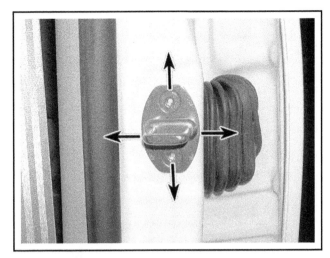

**16.9 The striker can be loosened and moved slightly to achieve secure latch engagement**

7   Scribe a line around the lower hinge with a marking pen, remove the lower hinge bolt (see illustration) and carefully lift off the door.

8   Installation is the reverse of removal, making sure to align the hinge with the marks made during removal before tightening the bolts.

9   Following installation of the door, check the alignment and adjust it if necessary as follows:

a)  *Up-and-down and in-and-out adjustments are made by loosening the hinge-to-door bolts/nuts and moving the door as necessary.*

b)  *Forward-and-backward adjustments are made by loosening the hinge-to-body bolts and moving the door as necessary.*

c)  *The door lock striker can also be adjusted both up-and-down and sideways to provide positive engagement with the lock mechanism. This is done by loosening the mounting screws and moving the striker as necessary (see illustration).*

## 17  Door latch, lock cylinder and outside handle - removal and installation

## LATCH

### Audi A4

▶ **Refer to illustrations 17.4, 17.5, 17.6, 17.7a, 17.7b, 17.8a and 17.8b**

1   Close the window. Refer to Section 15 and remove the door trim panel and the watershield.

2   Refer to Section 19 and remove the window regulator carrier from the door.

➡**Note: On some early models where an anti-theft panel is not fitted above the lock mechanism, it is not necessary to remove the window component carrier to gain access to the lock.**

3   Unclip and remove the anti-theft panel from the lock mechanism (where fitted).

4   Carefully release the locking button, exterior handle and lock cylinder operating rods from the lock mechanism (see illustration). Make a note of the locations of each rod.

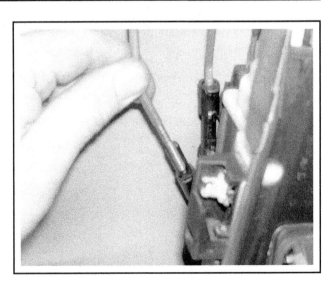

**17.4 Release the lock rods from the lock mechanism**

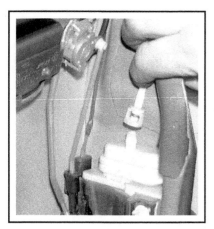

**17.5 Disconnect the vacuum hose from the central locking actuator**

**17.6 Remove the latch retaining screws and withdraw the latch unit from the door**

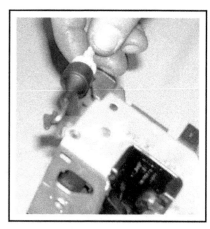

**17.7a Unhook the interior handle cable from the latch . . .**

**17.7b . . . then unplug the switch wiring**

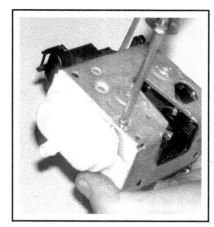

**17.8a To remove the vacuum actuator unit, remove the screw . . .**

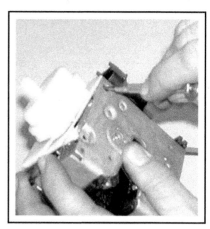

**17.8b . . . and depress its locking tab with a small screwdriver**

5   Disconnect the vacuum hose from the central locking actuator (see illustration).

6   Remove the two retaining screws, and withdraw the lock unit, together with its base plate from the door (see illustrations).

7   Unhook the interior handle operation cable from the lock unit, then unplug the switch wiring (see illustrations).

8   If required the vacuum actuator unit can be removed by removing the securing screw, depressing its locking tab with a small screwdriver and withdrawing the unit (see illustrations).

9   Installation is the reverse of the removal.

### Volkswagen Passat

▶ **Refer to illustration 17.16**

10   Remove the door trim panel (see Section 15). The door latch, radio speaker and window regulator are all attached to a large metal carrier, which must be removed from the door as a unit.

11   Refer to Steps 23 and 24 and remove the lock cylinder, then refer to step 34 and remove the handle.

12   Lower the window glass until the glass track bolts can be loos-

ened (see Section 18).

13   With the glass separated from the regulator, pull the glass up by hand and tape it in the UP position.

14   Refer to Steps 2 and 3 in Section 16 and disconnect the electrical harness for the door, feeding the wires into the door cavity.

15   Refer to Section 19 and remove the window regulator carrier.

16   To remove the latch from the carrier, use a punch to drive out the two pins, then flip the latch over and disconnect the cable and the linkage rod (see illustration).

17   Installation is the reverse of the removal procedure.

## LOCK CYLINDER

### Audi A4

▶ **Refer to illustrations 17.19 and 17.20**

18   Remove the exterior door handle as described in Steps 26 through 31.

19   Where applicable, trace the wiring for the central locking switch

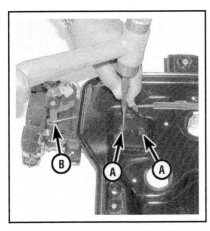

**17.16 Tap out these pins (A) with a punch and hammer, then disconnect the cable (B)**

**17.19 Disconnecting the wiring for the lock cylinder de-icer heating element**

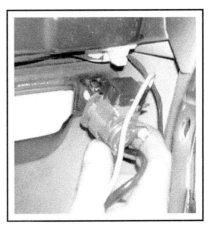

**17.20 Withdraw the lock cylinder housing from the door**

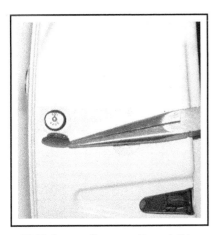

**17.23 Loosen the screw through this hole (arrow) after removing the plug**

**17.28a Rotate the locking element . . .**

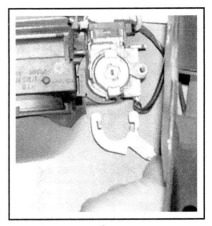

**17.28b . . . and release it from the rear of the handle and lock cylinder**

and/or lock de-icing element, from the rear of the lock cylinder housing to the door wiring harness and unplug at the wiring connectors (see illustration).

20 Withdraw the lock cylinder housing from the door (see illustration).

21 To remove the lock cylinder from its housing, insert the key, then remove the circlip from the rear of the lock cylinder. Slide off the driver plate followed by the spring, turn the lock cylinder through 180 degress then withdraw it from the handle assembly.

22 To fit the new lock cylinder, insert the lock cylinder into the housing and turn it to the left using the key. Refit the spring, driver plate and circlip. To align the driver plate correctly, ensure that the half-tooth on the driver plate engages with the corresponding half-tooth on the microswitch drive pinion.

### Volkswagen Passat

▶ **Refer to illustration 17.23**

23 Remove the plastic plug in the latch end of the door and loosen the setscrew at the lock cylinder (see illustration).

**✳✳ CAUTION:**

**Do not turn the setscrew all the way out, or it will fall into the door.**

24 Pull the lock cylinder out of the door.
25 Installation is the reverse of removal.

## OUTSIDE HANDLE

### Audi A4

▶ **Refer to illustrations 17.28a, 17.28b, 17.29a and 17.29b**

26 Fully close the window. Refer to Section 15 and remove the door trim panel and the watershield.

27 Refer to Section 19 and remove the window component carrier from the door.

28 Rotate the locking element to release it from the rear of the handle and lock cylinder (see illustrations).

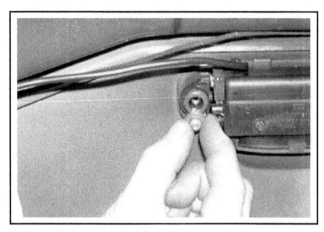

**17.29a Remove the retaining screw . . .**

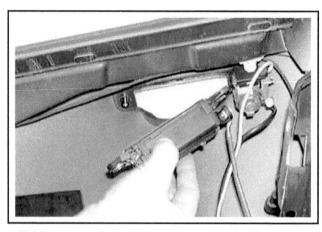

**17.29b . . . and the handle assembly from the door while holding the outside handle trim in place**

29 Remove the retaining screw, then remove the handle assembly from the door (see illustrations), leaving the lock cylinder housing in place.

30 Remove the trim from the outer surface of the door.

31 If required, the locking bar can be removed from the rear of the handle using a small flat bladed screwdriver and detach the trim panel.

32 Installation is the reverse of removal.

## Volkswagen Passat

▶ **Refer to illustration 17.34**

33 Refer to Steps 23 and 24 and remove the lock cylinder.

34 Unclip the cable, swing the key-end of the handle outward from the door, then rearward to remove it from the door (see illustration).

35 Installation is the reverse of removal.

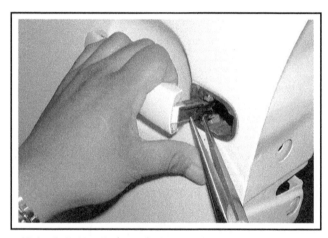

**17.34 Pull this clip (arrow), then remove the door handle**

## 18 Door window glass - removal and installation

## FRONT DOOR GLASS

▶ **Refer to illustration 18.2**

1 Remove the door trim panel (see Section 15).

2 If you're working on the front window glass, lower the glass until the window retaining clamps are accessible through the two holes in the carrier, then use a socket and extension to loosen the bolts retaining the regulator glass channel to the glass (see illustration). Separate the clamps to free the glass.

3 Secure the glass in this position with tape, then lower the window regulator all the way down.

4 Tilt the rear end of the glass up and out of the door until the glass clears the door.

5 Raise the regulator until the glass channel bolts align with the holes in the carrier (see illustration 18.2).

6 Lower the glass into the door front end first, then tilt it until the clamps fit around the glass and the bolts can be tightened.

7 The remainder of the installation is the reverse of the removal procedure.

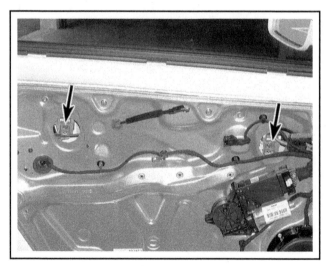

**18.2 Remove the two rubber plugs from the carrier, then loosen the glass retaining bolts (arrows)**

## REAR DOOR GLASS

8 Remove the door trim panel (see Section 15).

### Audi A4

▶ **Refer to illustration 18.11**

9 Refer to Section 19 and remove the window regulator carrier and the rear window glass as an assembly from the door.

10 Lower the window down in the regulator carrier.

11 Remove the glass retaining bolts and remove the rear window from the regulator carrier (see illustration).

### Volkswagen Passat

12 Lower the window in the door.

13 Pry off the outer window trim from the base of the window.

14 Pull out the weatherstripping from the front and rear endposts.

15 Remove the screws securing the front and rear endpost trim to the door.

16 Working through the access hole in the door frame press the glass retaining pin out from the glass retaining plug using a hammer and a punch. Being careful not to damage the outer door panel.

17 Lift the rear window glass out of the door.

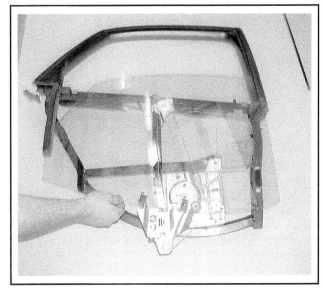

**18.11 Removing the rear window glass on an Audi A4**

## 19 Door window glass regulator carrier - removal and installation

▶ **Refer to illustration 19.3a, 19.3b, 19.3c, 19.3d, 19.3e and 19.3f**

1 Remove the door trim panel (see Section 15). Unplug the wiring from the electrical components attached to the regulator carrier.

2 On Passat models refer to Section 18 and remove the door window glass.

3 Remove the bolts securing the window regulator carrier to the door (see illustrations).

4 On Passat models, the regulator is part of the carrier assembly and can't be replaced separately. On Audi models it will be necessary to a drill out the regulator mounting rivets, then separate the regulator from the carrier.

5 Installation is the reverse of removal.

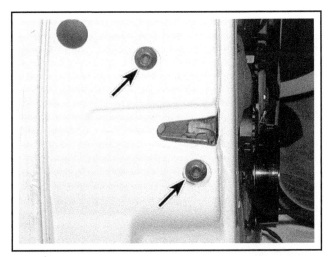

**19.3a On Passat models, remove the two Torx bolts (arrows) securing the latch to the door frame . . .**

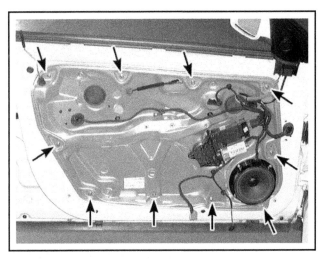

**19.3b . . . and remove the window regulator carrier mounting bolts (arrows)**

19.3c On Audi models remove the two bolts at the leading edge (arrows) . . .

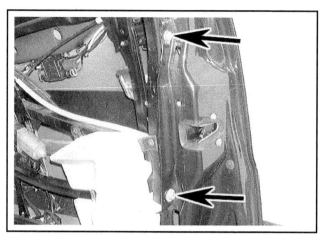

19.3d . . . and two at the trailing edge (arrows) . . .

19.3e . . . then remove the window regulator carrier from the door

19.3f Remove the wedge-shaped adjustment block

## 20 Mirrors - removal and installation

### INTERIOR

▶ Refer to illustrations 20.1 and 20.2

1   Pull the mirror downward at an angle to remove the retaining plate (see illustration). If the vehicle is equipped with an automatic mirror, unclip the cover and detach the wiring harness and mirror connector.

2   To install the mirror, align it 90 degrees from the installed position (vertical with the front and rear of the vehicle) and engage the mirror with the retaining plate, then rotate it clockwise until the mirror locks in place (see illustration).

3   If the vehicle is equipped with an automatic mirror, connect the wiring harness, mirror connector and the mirror trim cover.

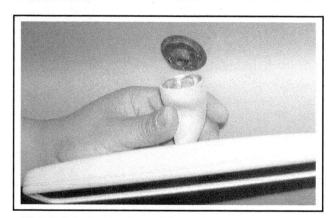

20.1 Pull the inside mirror off its retaining bracket at an angle

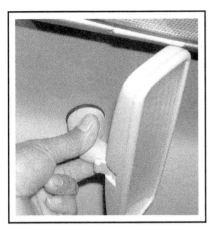

**20.2 Align the mirror vertical in the vehicle and rotate it 90 degrees until it locks into place**

**20.6 Loosen the mirror trim cover mounting screws (arrows)**

**20.7 Outside mirror mounting screw (arrow)**

## EXTERIOR

▶ **Refer to illustrations 20.6 and 20.7**

4    Refer to Section 15 and remove the door trim panel.

5    Disconnect the electrical connector for the mirror, inside the door, then apply tape to the connector so that it can be withdrawn through the mirror harness hole in the outside of the door without snagging.

6    With the door open, remove the two screws securing the mirror trim cover, then remove the trim cover (see illustration).

7    Remove the screw and detach the mirror from the door (see illustration). Work the mirror harness through the hole in the door.

8    Installation is the reverse of removal.

## 21 Trunk lid (sedan models) - removal, installation and adjustment

▶ **Refer to illustrations 21.2a, 21.2b, 21.3a, 21.3b, 21.8 and 21.9**

➡**Note: The trunk lid is heavy and somewhat awkward to remove and install - at least two people should perform this procedure.**

1    Open the trunk lid and cover the edges of the trunk compartment with pads or cloths to protect the painted surfaces when the lid is removed.

2    Remove warning triangle from the trunk lid, then remove the warning triangle holder if equipped. On Audi models it will also be necessary to remove the screws securing the grab handle to the trunk lid (see illustrations).

3    Remove the flock-covered screws, then unclip the trim panel from the trunk lid (see illustrations).

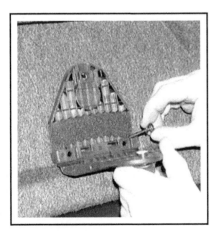

**21.2a Remove the warning triangle holder from the trunk lid if equipped**

**21.2b On Audi models, remove the retaining screws securing the grab handle**

**21.3a Remove the flock-covered screws . . .**

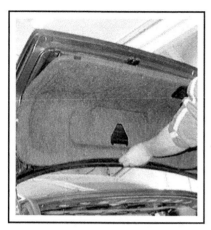

**21.3b . . . then unclip the trim panel from the trunk lid**

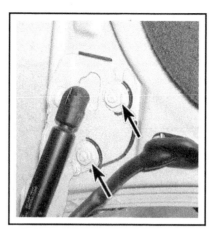

**21.8 Trunk lid retaining nuts (arrows)**

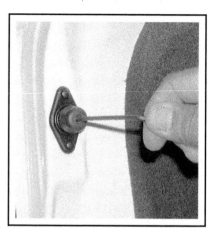

**21.9 Trunk lid adjustment bumper**

4   Unplug the wiring connectors from the lock switch and rear light units, then release the grommet and withdraw the wiring harness from the trunk lid. Unclip the wiring from the guide channel clipped to the side of the hinge.

5   Disconnect the central locking vacuum hose at the line connector and unclip it from the hinge guide channel (if equipped).

6   Mark the relationship between the trunk lid and the hinges by drawing around the outside of each hinge with a marker pen.

7   Refer to Section 22 and remove the trunk lid support struts.

8   With the help of an assistant to support the trunk lid, unscrew and remove the hinge-to-trunk lid retaining nuts (see illustration), and lift the lid clear.

9   Installation is the reverse of removal. Check the trunk lid for correct alignment, and if necessary loosen the hinge bolts to adjust, then retighten them; there should be an even gap of approximately 1/8 inch (3 mm) between the outside edge of the trunk lid and the surrounding bodywork. The height of the trunk lid can be adjusted by loosening the bolts and altering the position of the rubber bumpers located on the rear lower corners of the trunk lid (see illustration).

## 22  Trunk lid latch, lock cylinder and support struts - removal and installation

### LATCH

▶ **Refer to illustrations 22.2, 22.3a and 22.3b**

1   Refer to Section 21 and remove the inner trim panel from the trunk lid.

2   Disconnect the lock operating rod from the latch (see illustration).

3   Mark the fitted position of the latch with a marker pen, then unscrew the retaining nuts and withdraw the lock unit from the trunk lid. As it is withdrawn, unplug the wiring connector from the lock (see illustrations).

4   Installation is the reverse of removal.

➡**Note: When reinstalling the trunk lid latch, align the bolt heads with the marks made during removal.**

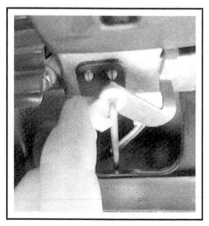

**22.2 Disconnect the actuating rod from the latch**

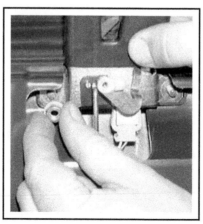

**22.3a Unscrew the retaining nuts and withdraw the latch from the trunk lid**

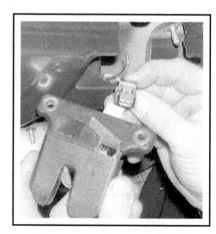

**22.3b As it is withdrawn, unplug the wiring connector from the latch**

## LOCK CYLINDER

**Refer to illustration 22.6**

5   Unplug the central locking vacuum hose from the lock cylinder unit if equipped.

6   Unscrew the retaining nuts and withdraw the lock cylinder unit from the trunk lid (see illustrations).

7   On models with central locking, detach the operating rod from the joint by unclipping the plastic fastener and rotating the joint while pulling the rod from the lever.

8   Disconnect the wiring from the lock cylinder unit as it becomes accessible (see illustration 22.6).

9   If equipped, the vacuum unit can be removed by depressing the locking tab with a screwdriver. The central locking switch unit can be removed in a similar manner.

10   Installation is the reverse of removal.

## SUPPORT STRUTS

11   The trunk lid supports are removed in the same manner as the hood support struts. Refer to Section 11 for this procedure.

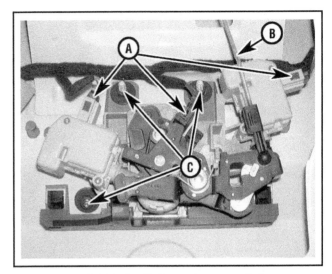

**22.6  The trunk lid lock cylinder can be removed after disconnecting the following components**

A    *Electrical connectors*          C    *Mounting nuts*
B    *Actuating rod*

## 23  Liftgate (wagon models) - removal, installation and adjustment

▶ **Refer to illustrations 23.1, 23.2a, 23.2b and 23.2c**

1   Open the liftgate, then unscrew the trim panel/grab handle securing bolt at the center of the lower edge of the liftgate (see illustration).

2   Remove the screws at the left and right hand sides, then carefully pry the lower section of the trim panel from the liftgate using just enough force to overcome the spring clips. Similarly, unclip the upper section of the trim panel from the liftgate rear window opening (see illustrations).

3   Disconnect the wiring from the liftgate components (lock switch, wiper motor, rear light units at the connectors. Note the routing and attachment locations of the wires.

4   Unplug the central locking vacuum hose from the liftgate at the line connector.

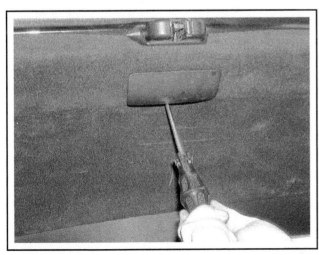

**23.1  Unscrew the trim panel/grab handle securing bolt at the center of the lower edge of the liftgate**

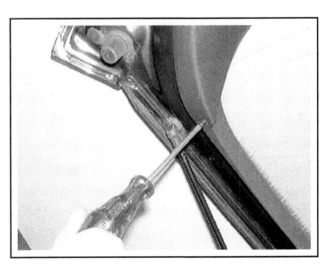

**23.2a  Remove the screws . . .**

**23.2b ... then carefully pry the lower section of the trim panel from the liftgate**

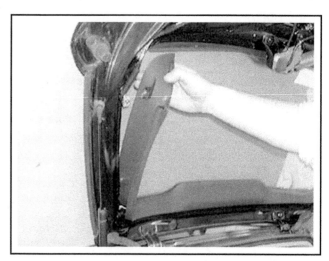

**23.2c Similarly, unclip the upper section of the trim panel from the liftgate rear window opening**

5   Mark the relationship between the liftgate and its hinges using a felt tip pen.

6   with the help of an assistant to help support the liftgate, detach the liftgate strut with reference to Section 24.

7   Unscrew and remove the liftgate-to-hinge securing bolts, and lift the liftgate clear of the vehicle.

8   Installation is the reverse of removal. Check that the liftgate is correctly aligned before fully tightening the liftgate hinge bolts.

9   The closed position of the liftgate can be adjusted by altering the positions of the rubber buffers at the upper and lower edges of the liftgate.

## 24  Liftgate latch, lock cylinder and support struts (wagon models) - removal and installation

The liftgate latch, lock cylinder and support struts are removed in the same manner as the trunk lid latch, lock cylinder and support struts.

Refer to Section 22 for this procedure.

## 25  Center console - removal and installation

### ✳✳ WARNING:

**The models covered by this manual are equipped with Supplemental Restraint Systems (SRS), more commonly known as airbags. Always disable the airbag system before working in the vicinity of any airbag system components to avoid the possibility of accidental deployment of the airbags, which could cause personal injury (see Chapter 12).**

### REAR SECTION

♦ **Refer to illustrations 25.1a, 25.1b, 25.2, 25.3, 25.4 and 25.6**

1   With the handbrake lever applied, remove the securing pin at the base of the parking brake grip, and pull the lever grip free. Withdraw the trim from the lever (see illustrations).

2   Remove the plastic plugs from the lower edges at the front of the console and remove the securing screws beneath (see illustration).

3   Pull the cupholder out at the rear of the console, then remove the ashtray and the retaining nuts in the base of the ashtray recess (see illustration).

4   Place the center armrest in the upright position, then remove the rubber mat and unscrew the nut securing the center of the rear console (see illustration).

5   Where applicable, release the diagnostic connector from its position adjacent to the ashtray.

6   Partially withdraw the console, disconnect the ashtray/cigar lighter wiring, then remove the console, by lifting it from the rear up and over the brake lever (see illustration).

7   Installation is the reverse of the removal. Ensure that the ashtray/cigar lighter wiring is reconnected and that the diagnostic connector is installed in the correct position.

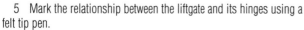

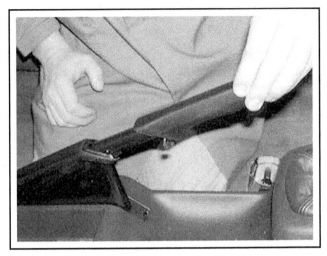

25.1a With the parking brake applied, remove the brake lever grip . . .

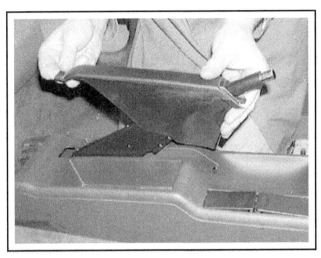

25.1b . . . and the trim at the base of the handle (if equipped)

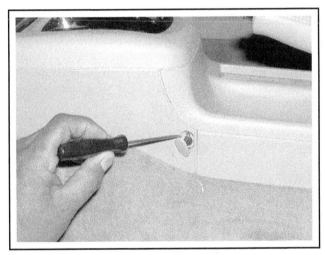

25.2 Pry out the plastic plugs at the front of the rear console and remove the screws beneath

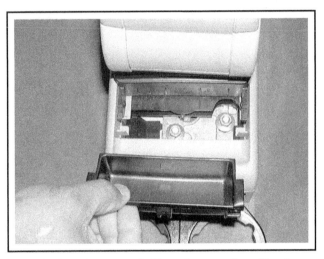

25.3 Pull out the rear cupholder and remove the ashtray to access the retaining nuts at the rear

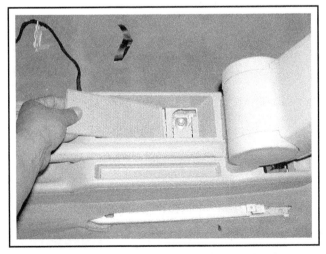

25.4 Place the center armrest in the upright position, then remove the rubber mat and unscrew the nut securing the center of the rear console

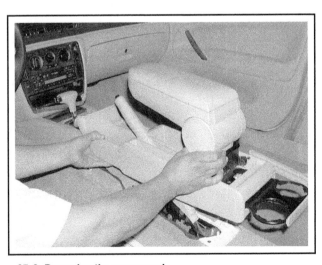

25.6 Removing the rear console

**25.10 On models with manual transaxles, remove the gear shift knob and the boot**

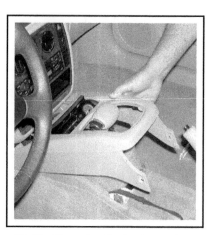

**25.11a Removing the front console on a Volkswagen Passat**

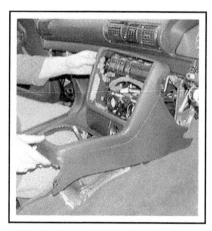

**25.11b Removing the front console on a Audi A4**

## FRONT SECTION

▶ **Refer to illustrations 25.10, 25.11a and 25.11b**

8   Remove the rear section of the center console as described in Steps 1 through 6.

9   On Audi models, remove the radio/cassette unit as described in Chapter 12. Also pull the knobs from the heater controls (see Chapter 3) and remove the four securing screws at the corners of the radio opening and release the console trim panel from its retaining clips. Then remove the plastic plugs from the front lower edges of the console and remove the securing nuts beneath.

10   On models with manual transaxles, unscrew the gear lever knob, then unclip and remove the gear lever boot (see illustration). On models with automatic transaxles, carefully unclip the trim/gear indicator panel, lift it over the gear shift lever and remove it from the console.

11   Partially withdraw the console, disconnect the ashtray/cigar lighter wiring, then remove the console from the vehicle (see illustrations).

12   Installation is a reverse of removal. On Audi models, ensure that the clips at the front lower edges of the console engage with the instrument panel support framework, and that all wiring is correctly routed.

## 26 Dashboard trim panels - removal and installation

✳✳ **WARNING:**

**The models covered by this manual are equipped with Supplemental Restraint Systems (SRS), more commonly known as airbags. Always disable the airbag system before working in the vicinity of any airbag system components to avoid the possibility of accidental deployment of the airbags, which could cause personal injury (see Chapter 12).**

## DASHBOARD END CAPS

▶ **Refer to illustration 26.2**

1   The end caps are held in place by clips.

2   Grasp the cover securely and pull sharply to remove it. If necessary, gently pry with a trim tool (see illustration).

3   Installation is the reverse of removal.

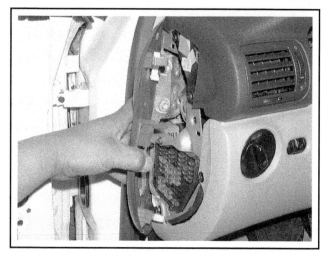

**26.2 Pull outward to detach the clips from the end caps**

## LOWER KICK PANELS

▶ **Refer to illustration 26.4**

4   Pry out the plastic trim plugs and remove the screws securing the kick panel to the door pillar (see illustration).

5   Pull the kick panel away from the door pillar and remove it from the vehicle.

6   Installation is the reverse of removal.

## DRIVER'S KNEE BOLSTER

▶ **Refer to illustration 26.7**

7   Remove the left dashboard end cap to access two side screws (see illustration 26.2). Remove the screws retaining the bottom of the bolster to the instrument panel (see illustration).

8   Remove the driver side kick panel as described in Steps 4 and

5. Pull backward on the top of the bolster to release it from the clips, and disconnect the electrical connectors at the headlight and dimmer switches.

9   If the bolster reinforcement panel needs to be removed to perform a repair procedure, remove the reinforcement panel screws.

10 Installation is the reverse of removal.

## CENTER TRIM PANEL (PASSAT MODELS)

▶ **Refer to illustrations 26.11 and 26.12**

11 To access the upper screws for the center trim panel, pry out the heater control panel bezel (see illustration)

12 Remove the screws at the top of the center trim panel, then the two at the bottom (see illustration).

13 Lift the center trim panel out and disconnect the electrical connectors from the switches.

14 Installation is the reverse of removal.

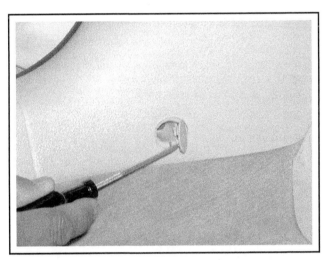

**26.4  Pry out the plastic trim plugs to access the kick panel retaining screws**

**26.7  The bolster is held by screws at the side, bottom (arrows) and clips at the top**

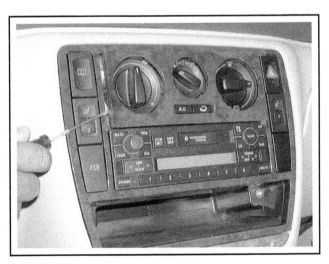

**26.11  Pry off the heater control panel bezel**

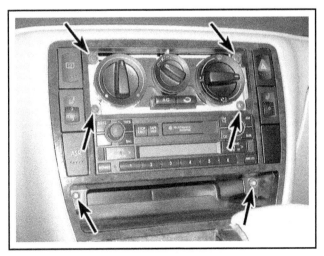

**26.12  Center trim panel retaining screws (arrows) (Passat models)**

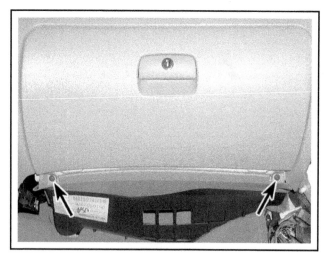

**26.15 Glove box lower retaining screws**

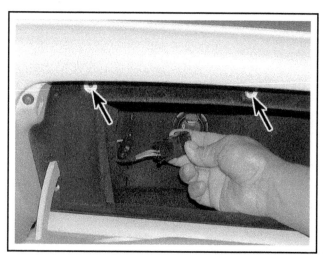

**26.16 Pry the glove box lamp and disconnect the electrical connector (two of the upper retaining screws shown)**

## GLOVE BOX

▶ **Refer to illustrations 26.15 and 26.16**

15 Remove the lower retaining screws (see illustration).

16 Open the glove box, then pry the light out and disconnect the electrical connector (see illustration).

17 Remove the upper and lower retaining screws from the opening and remove the glove box from the instrument panel.

18 Installation is the reverse of removal.

## 27 Steering column covers - removal and installation

▶ **Refer to illustration 27.1**

1 Remove the screws from the lower steering column cover, then remove the upper cover (see illustration).

2 To remove the lower cover, the steering wheel must be removed (see Chapter 10). Pull down on the steering column tilt lever, then remove the lower cover.

3 Installation is the reverse of removal.

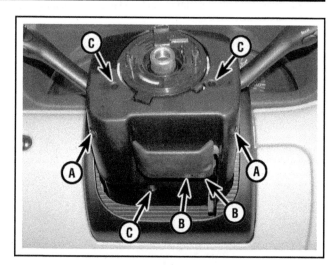

**27.1 Steering column cover mounting details**

A   Upper cover mounting screws
B   Tilt lever handle mounting screws
C   Lower cover mounting screws

## 28  Instrument panel - removal and installation

♦ Refer to illustrations 28.7, 28.8a, 28.8b, 28.8c and 28.9

### ✳✳ WARNING:

The models covered by this manual are equipped with Supplemental Restraint Systems (SRS), more commonly known as airbags. Always disable the airbag system before working in the vicinity of any airbag system components to avoid the possibility of accidental deployment of the airbags, which could cause personal injury (see Chapter 12).

### ✳✳ CAUTION:

This is a difficult procedure for the home mechanic, involving tedious disassembly and the disconnection/reconnection of numerous electrical connectors. If you do attempt this procedure, make sure you take good notes and mark all matching connectors (and their mounting points) to aid reassembly.

1   Disconnect the negative battery cable (see Chapter 5). Read the **Cautions** in Section 1.

2   Remove the center console (see Section 25).

3   Remove the steering column covers (see Section 27).

4   Remove all of the dashboard trim panels (see Section 26).

5   Remove the audio components, the multi-function switch, the instrument cluster and the passenger airbag (see Chapter 12).

6   Remove the heater control assembly (see Chapter 3).

7   Remove the bolts securing the instrument panel on either side of the steering column (see illustration).

8   Remove the bolts securing the center of the instrument panel (see illustrations)

9   Remove the screws securing the ends of the instrument panel (see illustration).

10 Refer to Chapter 10 and remove the steering column lower pinch bolt.

11 Remove any remaining  fasteners securing the instrument panel, and any electrical connectors still attached to the panel. Have an assistant help you pull the panel back and out of the vehicle.

12 Installation is the reverse of removal.

28.7  Remove the screws on the left and right side of the steering column

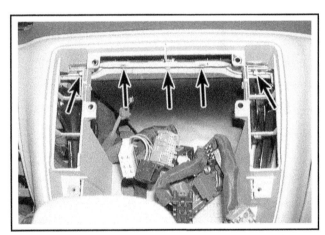

28.8a  Center instrument panel mounting screws (arrows) (Passat model shown, Audi models similar)

28.8b On Passat models there are also lower retaining screws at the center (arrows) . . .

28.8c  . . . and at the side (arrow)

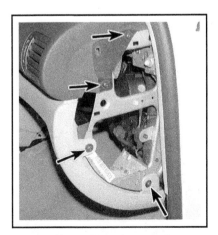

28.9  Remove the end caps and remove the mounting screws (arrows) at the left and right side of the instrument panel

## 29 Seats - removal and installation

**※ WARNING:**

**The models covered by this manual are equipped with Supplemental Restraint Systems (SRS), more commonly known as airbags. Always disable the airbag system before working in the vicinity of any airbag system components to avoid the possibility of accidental deployment of the airbags, which could cause personal injury (see Chapter 12).**

### FRONT

▶ **Refer to illustrations 29.2a and 29.2b**

1   The front seats on some models are equipped with side-impact airbags at the upper outside of the seatback. Refer to Chapter 12 to disable the airbag system before working on front seats.

2   Slide the seat forward and remove the seat track covers, move the seat rearward and remove the front seat track bolts and unplug any electrical connectors attached to the seat (see illustrations).

3   Slide the seat all the way to the back of the tracks and off. Remove the seat from the vehicle.

4   Installation is the reverse of removal.

### REAR

▶ **Refer to illustrations 29.5, 29.6a and 29.6b**

5   Remove the seat bottom cushion by pushing in on the front edge of the cushion in the vicinity of the clips, to disengage part of the seat's wire frame from the clips (see illustration). With the clips unhooked, lift up and remove the seat bottom from the vehicle.

6   Fold the seat back down, then remove the bolt and the retaining bracket. Pull the seat back off the rotating pin and remove it from the vehicle (see illustrations).

7   Installation is the reverse of removal. Tighten the seat back bolts securely.

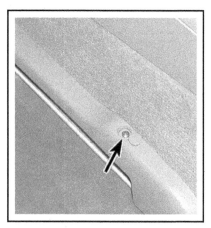

**29.2a  Remove the plastic buttons in the seat track covers, then remove the screws (arrow) and the covers**

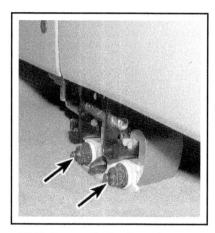

**29.2b Remove the front seat front mounting bolts (arrows), then slide the seat all the way back and off the tracks**

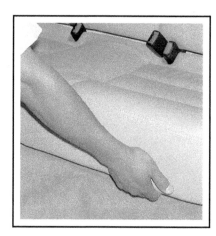

**29.5  Pull up the center tab of the rear seat cushion to disengage the cushion from two clips (arrows)**

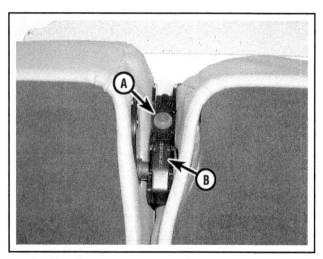

**29.6a  Remove the bolt (A) and the seat back retaining bracket (B)**

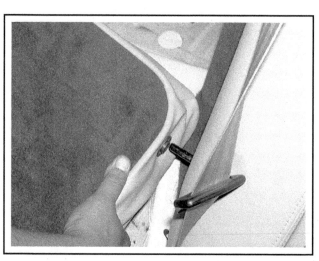

**29.6b  Pull the seat back off the rotating pin and remove it from the vehicle**

## 30 Sunroof - adjustment

▶ **Refer to illustrations 30.3a, 30.3b and 30.4**

1   The position of the glass panel can be adjusted in the following manner.

2   Open the interior sunshade all the way back and operate the sunroof until it is tilted open.

3   There are glass mounting screws on each side of the sunroof opening, inside the vehicle. Remove the plastic covers (see illustrations).

4   Loosen the front screws and lower the sunroof (see illustration). Set the front edge of the glass approximately 0.039-inch (1 mm) below the surface of the roof when closed. Tighten the front screws.

5   Loosen the rear glass adjustment screws and adjust the glass approximately 0.039-inch (1 mm) above the roof at the rear. Tighten the rear screws.

6   When the adjustment is correct, tilt the sunroof open and reinstall the plastic covers over the glass rails.

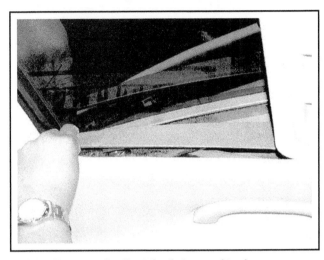

**30.3a  Unsnap and pull out the first sunroof track plastic cover . . .**

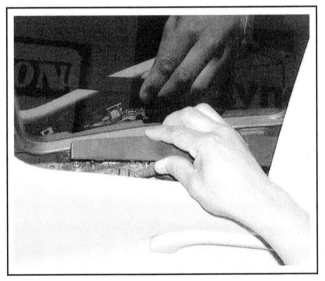

**30.3b  . . . then the second cover to reveal the tracks**

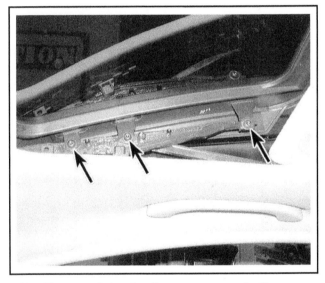

**30.4  The sunroof glass has three screws on each side (arrows) used to adjust the fit of the glass to the roof**

**Notes**

**Section**

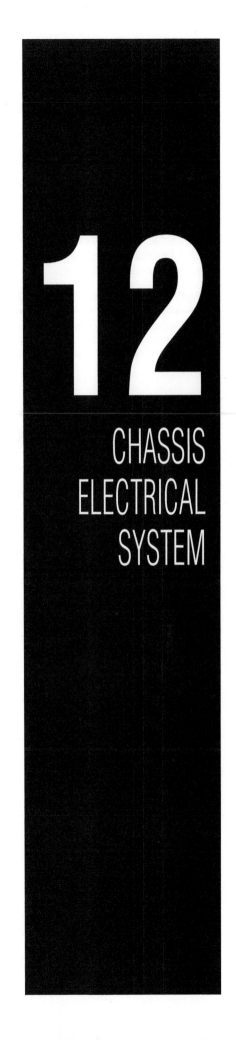

12

CHASSIS
ELECTRICAL
SYSTEM

## 1   General information

The electrical system is a 12-volt, negative ground type. Power for the lights and all electrical accessories is supplied by a lead/acid-type battery that is charged by the alternator.

This Chapter covers repair and service procedures for the various electrical components not associated with the engine. Information on the battery, alternator, ignition system and starter motor can be found in Chapter 5.

It should be noted that when portions of the electrical system are serviced, the negative cable should be disconnected from the battery to prevent electrical shorts and/or fires.

### ✳✳ CAUTION 1:

**Disconnecting the battery can cause driveability problems that require a scan tool to rectify. Additionally, disconnecting the battery may cause one or more warning lights on the instrument panel to illuminate, which will also require the use of a scan tool to turn off. Most scan tools available to the public do not have the capability to perform either of these tasks, which will necessitate taking the vehicle to a dealer service department or other properly equipped repair facility after service work has been performed. See Chapter 5, Section 1 for the use of an auxiliary voltage input device (memory saver) before disconnecting the battery and for other precautions related to battery disconnection.**

### ✳✳ CAUTION 2:

**These models are equipped with an anti-theft radio. Before performing a procedure that requires disconnecting the battery, make sure you have the proper activation code.**

## 2   Electrical troubleshooting - general information

▸ **Refer to illustrations 2.5a, 2.5b, 2.6 and 2.9**

A typical electrical circuit consists of an electrical component, any switches, relays, motors, fuses, fusible links or circuit breakers related to that component and the wiring and connectors that link the component to both the battery and the chassis. To help you pinpoint an electrical circuit problem, wiring diagrams are included at the end of this Chapter.

Before tackling any troublesome electrical circuit, first study the appropriate wiring diagrams to get a complete understanding of what makes up that individual circuit. Trouble spots, for instance, can often be narrowed down by noting if other components related to the circuit are operating properly. If several components or circuits fail at one time, chances are the problem is in a fuse or ground connection, because several circuits are often routed through the same fuse and ground connections.

Electrical problems usually stem from simple causes, such as loose or corroded connections, a blown fuse, a melted fusible link or a failed relay. Visually inspect the condition of all fuses, wires and connections in a problem circuit before troubleshooting the circuit.

If test equipment and instruments are going to be utilized, use the diagrams to plan ahead of time where you will make the necessary connections in order to accurately pinpoint the trouble spot.

The basic tools needed for electrical troubleshooting include a circuit tester or voltmeter (a 12-volt bulb with a set of test leads can also be used), a continuity tester, which includes a bulb, battery and set of test leads, and a jumper wire, preferably with a circuit breaker incorporated, which can be used to bypass electrical components (see illustrations). Before attempting to locate a problem with test instruments, use the wiring diagram(s) to decide where to make the connections.

### VOLTAGE CHECKS

Voltage checks should be performed if a circuit is not functioning properly. Connect one lead of a circuit tester to either the negative battery terminal or a known good ground. Connect the other lead to a

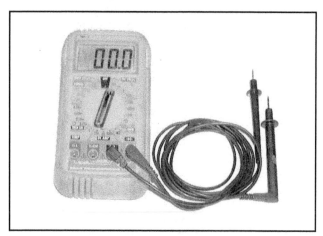

**2.5a  The most useful tool for electrical troubleshooting is a digital multimeter that can check volts, amps, and test continuity**

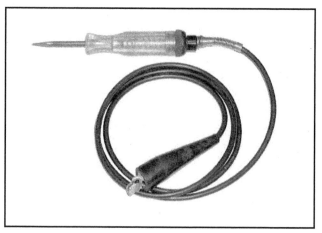

**2.5b  The easiest way to check for the presence of voltage in a circuit is with the use of a simple test light**

connector in the circuit being tested, preferably nearest to the battery or fuse (see illustration). If the bulb of the tester lights, voltage is present, which means that the part of the circuit between the connector and the battery is problem free. Continue checking the rest of the circuit in the same fashion. When you reach a point at which no voltage is present, the problem lies between that point and the last test point with voltage. Most of the time the problem can be traced to a loose connection.

➡**Note: Keep in mind that some circuits receive voltage only when the ignition key is in the Accessory or Run position.**

## FINDING A SHORT

One method of finding shorts in a live circuit is to remove the fuse and connect a test light in place of the fuse terminals (fabricate two jumper wires with small spade terminals, plug the jumper wires into the fuse box and connect the test light). There should be voltage present in the circuit. Move the suspected wiring harness from side-to-side while watching the test light. If the bulb goes off, there is a short to ground somewhere in that area, probably where the insulation has rubbed through.

## GROUND CHECK

Perform a ground test to check whether a component is properly grounded. Disconnect the battery and connect one lead of a continuity tester or multimeter (set to the ohms scale), to a known good ground. Connect the other lead to the wire or ground connection being tested. If the resistance is low (less than 5 ohms), the ground is good. If the bulb on a self-powered test light does not go on, the ground is not good.

## CONTINUITY CHECK

A continuity check is done to determine if there are any breaks in a circuit - if it is passing electricity properly. With the circuit off (no power in the circuit), a self-powered continuity tester or multimeter can be used to check the circuit. Connect the test leads to both ends of the circuit (or to the "power" end and a good ground), and if the test light comes on the circuit is passing current properly (see illustration). If the resistance is low (less than 5 ohms), there is continuity; if the reading is 10,000 ohms or higher, there is a break somewhere in the circuit.

The same procedure can be used to test a switch, by connecting the continuity tester to the switch terminals. With the switch turned On, the test light should come on (or low resistance should be indicated on a meter).

## FINDING AN OPEN CIRCUIT

When diagnosing for possible open circuits, it is often difficult to locate them by sight because the connectors hide oxidation or terminal misalignment. Merely wiggling a connector on a sensor or in the wiring harness may correct the open circuit condition. Remember this when an open circuit is indicated when troubleshooting a circuit. Intermittent problems may also be caused by oxidized or loose connections.

Electrical troubleshooting is simple if you keep in mind that all electrical circuits are basically electricity running from the battery, through the wires, switches, relays, fuses and fusible links to each electrical component (light bulb, motor, etc.) and to ground, from which it is passed back to the battery. Any electrical problem is an interruption in the flow of electricity to and from the battery.

## CONNECTORS

Most electrical connections on these vehicles are made with multiwire plastic connectors. The mating halves of many connectors are secured with locking clips molded into the plastic connector shells. The mating halves of large connectors, such as some of those under the instrument panel, are held together by a bolt through the center of the connector.

To separate a connector with locking clips, use a small screwdriver to pry the clips apart carefully, then separate the connector halves. Pull only on the shell, never pull on the wiring harness as you may damage the individual wires and terminals inside the connectors. Look at the connector closely before trying to separate the halves. Often the locking clips are engaged in a way that is not immediately clear. Additionally, many connectors have more than one set of clips.

Each pair of connector terminals has a male half and a female half. When you look at the end view of a connector in a diagram, be sure to understand whether the view shows the harness side or the component side of the connector. Connector halves are mirror images of each other, and a terminal shown on the right side end-view of one half will be on the left side end view of the other half.

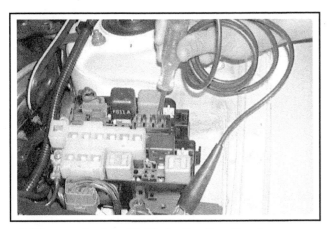

**2.6  In use, a basic test light's lead is clipped to a known good ground, then the pointed probe can test connectors, wires or electrical sockets - if the bulb lights, the circuit being tested has battery voltage**

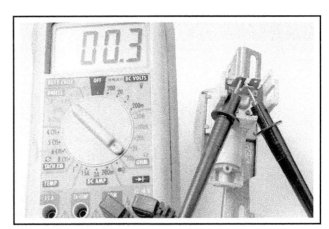

**2.9  With a multimeter set to the ohms scale, resistance can be checked across two terminals - when checking for continuity, a low reading indicates continuity, a high reading or infinity indicates lack of continuity**

## 3  Fuses - general information

▶ **Refer to illustrations 3.1a, 3.1b, 3.3a and 3.3b**

The electrical circuits of the vehicle are protected by a combination of fuses, circuit breakers and fusible links. The interior fuse panel is located at the left end of the instrument panel, while the main fuse/relay panel is located just below the interior fuse panel (see illustrations).

Each of the fuses is designed to protect a specific circuit, and the various circuits are identified on the fuse panel itself.

Several sizes of fuses are employed in the fuse blocks. There are small, medium and large sizes of the same design, all with the same blade terminal design, as well as five "metal" fuses located in the engine compartment fuse panel. The medium and large fuses can be removed with your fingers, but the small fuses require the use of pliers or the small plastic fuse-puller tool found in most fuse boxes. The metal fuses are for heavy loads, and if the metal strip melts due to an overload, it is easily seen, although the battery should be disconnected while replacing this type fuse (see the Cautions in Chapter 5, Section 1). If an electrical com-

ponent fails, always check the fuse first. The best way to check the fuses is with a test light. Check for power at the exposed terminal tips of each fuse (see illustration). If power is present at one side of the fuse but not the other, the fuse is blown. A blown fuse can also be identified by visually inspecting it (see illustration).

Be sure to replace blown fuses with the correct type. Fuses (of the same physical size) of different ratings may be physically interchangeable, but only fuses of the proper rating should be used. Replacing a fuse with one of a higher or lower value than specified is not recommended. Each electrical circuit needs a specific amount of protection. The amperage value of each fuse is molded into the top of the fuse body.

If the replacement fuse immediately fails, don't replace it again until the cause of the problem is isolated and corrected. In most cases, this will be a short circuit in the wiring caused by a broken or deteriorated wire.

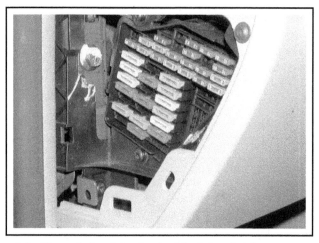

**3.1a  The main fuse/relay box is at the left end of the instrument panel, behind the end cover - the inside of the cover has a legend to identify the fuses and relays**

**3.1b  The under-dash fuse/relay panel is located under the sound insulator at the left end of the instrument panel - this panel contains relays, fuses and circuit breakers if equipped**

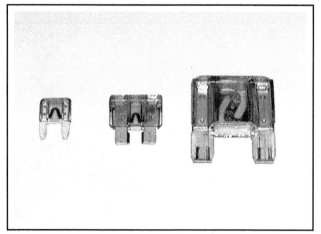

**3.3a  All three of these fuses are of 30-amp rating, yet are different sizes - make sure you get the right amperage and size when purchasing replacement fuses**

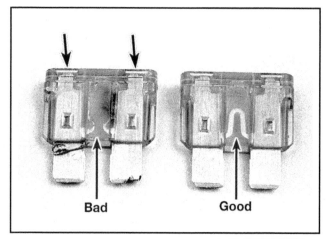

**3.3b  When a fuse blows, the element between the terminals melts - the fuse on the left is blown, the one on the right is good - these fuses can be tested on top (arrows) with a test light without removing the fuse**

## 4  Circuit breakers - general information and check

Circuit breakers protect certain circuits, such as the power windows or heated seats. Depending on the vehicle's accessories, there may be one or two 25-amp circuit breakers, located in the under-dash fuse/relay box under the left end of the dashboard (see illustration 3.1b).

Because the circuit breakers reset automatically, an electrical overload in a circuit-breaker-protected system will cause the circuit to fail momentarily, then come back on. If the circuit does not come back on, check it immediately.

For a basic check, pull the circuit breaker up out of its socket on the fuse panel, but just far enough to probe with a voltmeter. The breaker should still contact the sockets.

With the voltmeter negative lead on a good chassis ground, touch each end prong of the circuit breaker with the positive meter probe. There should be battery voltage at each end. If there is battery voltage only at one end, the circuit breaker must be replaced.

## 5  Relays - general information and testing

### GENERAL INFORMATION

1   Several electrical accessories in the vehicle, such as the fuel injection system, horns, starter, and fog lamps use relays to transmit the electrical signal to the component. Relays use a low-current circuit (the control circuit) to open and close a high-current circuit (the power circuit). If the relay is defective, that component will not operate properly. Most relays are mounted in the under-dash fuse/relay box (see illustration 3.1b). If a faulty relay is suspected, it can be removed and tested using the procedure below or by a dealer service department or a repair shop. Defective relays must be replaced as a unit.

### TESTING

‣ **Refer to illustrations 5.2a and 5.2b**

2   Most of the relays used in these vehicles are of a type often called "ISO" relays, which refers to the International Standards Organization. The terminals of ISO relays are numbered to indicate their usual circuit connections and functions. There are two basic layouts of terminals on the relays used in the covered vehicles (see illustrations).

3   Refer to the wiring diagram for the circuit to determine the proper connections for the relay you're testing. If you can't determine the correct connection from the wiring diagrams, however, you may be able to determine the test connections from the information that follows.

4   Two of the terminals are the relay control circuit and connect to the relay coil. The other relay terminals are the power circuit. When the relay is energized, the coil creates a magnetic field that closes the larger contacts of the power circuit to provide power to the circuit loads.

5   Terminals 85 and 86 are normally the control circuit. If the relay contains a diode, terminal 86 must be connected to battery positive (B+) voltage and terminal 85 to ground. If the relay contains a resistor, terminals 85 and 86 can be connected in either direction with respect to B+ and ground.

6   Terminal 30 is normally connected to the battery voltage (B+) source for the circuit loads. Terminal 87 is connected to the ground side of the circuit, either directly or through a load. If the relay has several alternate terminals for load or ground connections, they usually are numbered 87A, 87B, 87C, and so on.

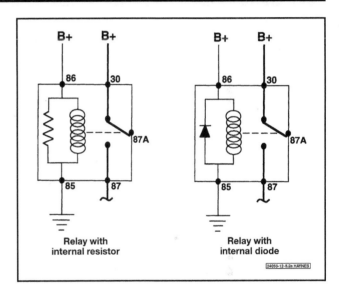

**5.2a  Typical ISO relay designs, terminal numbering and circuit connections**

**5.2b  Most relays are marked on the outside to easily identify the control circuit and power circuits - this one is of the four-terminal type**

7   Use an ohmmeter to check continuity through the relay control coil.

   a) *Connect the meter according to the polarity shown in the illustration for one check; then reverse the ohmmeter leads and check continuity in the other direction.*
   b) *If the relay contains a resistor, resistance will be indicated on the meter, and should be the same value with the ohmmeter in either direction.*
   c) *If the relay contains a diode, resistance should be higher with the ohmmeter in the forward polarity direction than with the meter leads reversed.*
   d) *If the ohmmeter shows infinite resistance in both directions, replace the relay.*

8   Remove the relay from the vehicle and use the ohmmeter to check for continuity between the relay power circuit terminals. There should be no continuity between terminal 30 and 87 with the relay de-energized.

9   Connect a fused jumper wire to terminal 86 and the positive battery terminal. Connect another jumper wire between terminal 85 and ground. When the connections are made, the relay should click.

10  With the jumper wires connected, check for continuity between the power circuit terminals. Now, there should be continuity between terminals 30 and 87.

11  If the relay fails any of the above tests, replace it.

## 6   Turn signal and hazard flashers - check and replacement

1   The flasher function on the covered models is performed by the hazard-warning switch, headlight and parking light assembly located on the switch panel at the left section of the assembly. For testing and replacement of the switch, refer to Section 9 and the wiring diagrams at the end of this Chapter.

2   When the flasher unit is functioning properly, an audible click may be heard during its operation. If the turn signal indicator on one side of the vehicle flashes much more rapidly than normal, a faulty turn signal bulb is indicated.

3   If both turn signals fail to blink, the problem may be due to a blown fuse, a faulty flasher unit switch or a loose or open connection. If a quick check of the fuse box indicates that the turn signal fuse has blown, check the wiring for a short before installing a new fuse.

4   If the flasher switch is not the problem, refer to Section 7 and test the turn signal portion of the multi-function switch.

## 7   Steering column switches - check and replacement

### ✳✳ WARNING:

**The models covered by this manual are equipped with Supplemental Restraint Systems (SRS), more commonly known as airbags. Always disable the airbag system before working in the vicinity of any airbag system components to avoid the possibility of accidental deployment of the airbags, which could cause personal injury (see Section 26).**

1   The multi-function switch is located on the top of the steering column. It incorporates into one switch the turn signal, headlight dimmer (and headlight switch on A4 models), windshield wiper/washer and, if equipped, cruise control functions.

➡**Note: The following checks can be made with simple equipment, but thorough testing of the systems controlled by the multi-function switch can only be made at a dealer or other repair shop with a scan tool.**

## CHECK

▶ **Refer to illustration 7.2**

2   Remove the steering wheel (see Chapter 10), the steering column covers (see illustration) and the knee bolster (see Chapter 11) for access to the steering column switches. Follow the wiring diagrams at the end of this Chapter to determine which terminals supply the left and right turn signal circuits, the cruise control and the headlight dimming systems.

3   Use a test light to check for power at the designated terminals with the switch in each position. Look at the wiring diagrams at the end of this Chapter to see how power and ground are routed. If any portion of the multi-function switch fails the tests, the switch must be replaced as a unit. The 12-pin connector handles the turn signals, headlight dimmer and parking lights; the 4-pin and 8-pin connectors control the wipe/wash functions, and the cruise control connector is the 7-pin connector.

## REPLACEMENT

▶ **Refer to illustration 7.6**

4   Remove the steering wheel (see Chapter 10).

5   Remove the steering column covers (see Chapter 11).

6   Remove the pinch bolt at the rear of the multi-function switch (see illustration).

7   Pull the multi-function switch out from the steering column enough to disconnect the electrical connectors.

8   Installation is the reverse of removal.

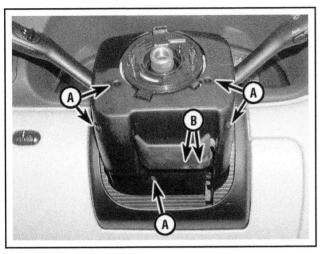

7.2  Remove the column cover screws (A) and the tilt handle screws (B) and separate the covers from the steering column

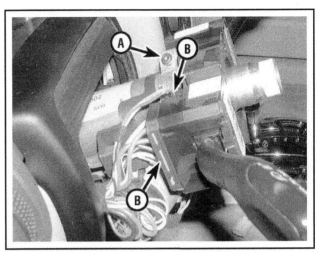

7.6  Remove the pinch bolt (A) at the back of the multi-function switch, slide the multi-function switch forward and then disconnect the harness connector (B)

## 8  Ignition switch and key lock cylinder - check and replacement

▶ **Refer to illustrations 8.1, 8.3, 8.7, 8.8 and 8.10**

### ✳✳ WARNING:

The models covered by this manual are equipped with Supplemental Restraint Systems (SRS), more commonly known as airbags. Always disable the airbag system before working in the vicinity of any airbag system components to avoid the possibility of accidental deployment of the airbags, which could cause personal injury (see Section 26).

### ✳✳ CAUTION 1:

Disconnecting the battery can cause driveability problems that require a scan tool to rectify. Additionally, disconnecting the battery may cause one or more warning lights on the instrument panel to illuminate, which will also require the use of a scan tool to turn off. Most scan tools available to the public do not have the capability to perform either of these tasks, which will necessitate taking the vehicle to a dealer service department or other properly equipped repair facility after service work has been performed. See Chapter 5, Section 1 for the use of an auxiliary voltage input device (memory saver) before disconnecting the battery and for other precautions related to battery disconnection.

### ✳✳ CAUTION 2:

These models are equipped with an anti-theft radio. Before performing a procedure that requires disconnecting the battery, make sure you have the proper activation code.

1    The ignition switch, located under the steering column (switch on the left, key-lock cylinder on the right), is comprised of a cast-metal housing, an ignition lock cylinder and an electrical component, the switch device (see illustration).

### CHECK

2    Access the switch and disconnect the electrical connector from it (see Steps 6 and 7).

8.1  Location of the ignition lock assembly (arrow)

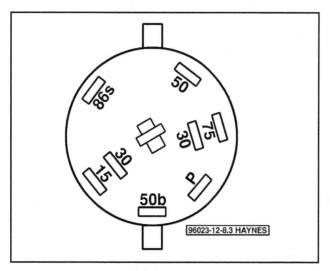

8.3 Terminal designations for testing the ignition switch

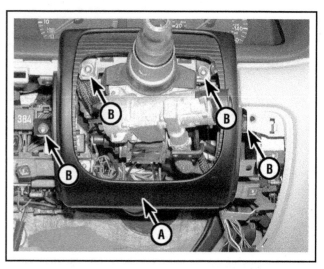

8.6 Remove the mounting bolts (B) from the ignition assembly shroud (A)

8.7 Remove the park lock cable retainer clip (arrow)

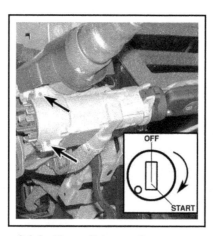

8.8 Ignition switch mounting screws (arrows)

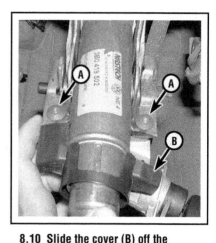

8.10 Slide the cover (B) off the ignition lock assembly to expose the break-off bolts (A)

3  Backprobe the switch with an ohmmeter (see illustration). Power is supplied to the switch through the two red wires (the terminals marked "30" on the back of the switch). Check for continuity between terminal(s) 30 and other terminals in different switch positions. With the switch in Off, there should be no continuity. With the switch in On, there should be continuity at the terminals 75, 86s and 15, and in the Start position, only at terminal 50b.

4  If the switch does not have the correct continuity, replace it.

## REPLACEMENT

5  Disconnect the negative battery cable (see the **Cautions** in Chapter 5, Section 1).

6  Remove the steering column covers (see Chapter 11) and the ignition switch shroud (see illustration).

7  On models with automatic transmissions, the shift linkage must be in Park before removing/installing the ignition switch. Remove the

retainer clip and disconnect the park/lock cable from the ignition switch (see illustration).

8  To remove the switch, disconnect the large electrical connector on the left side, then remove the two switch mounting screws and pull the switch out of the ignition assembly (see illustration).

➡Note: The screw heads may be coated with a special "locking" coating. You'll have to chip this off to remove the screws.

9  To remove the key lock cylinder, place the key in the lock and rotate clockwise to the Start position. Insert the straightened end of a large paper clip into the hole in the cylinder and pull the key and switch from the lock cylinder housing (see illustration 8.8).

10  To remove the ignition assembly, chisel the shear bolts from the steering column bracket (see illustration).

11  When installing the key lock cylinder, align the cylinder as it was (still in On position with the key in), then push the cylinder in until it snaps in position. Remove the paper clip, turn the key to the Lock position and remove the key.

12  The remainder of installation is the reverse of removal.

## 9  Instrument panel switches - check and replacement

### ✳✳ WARNING:

The models covered by this manual are equipped with Supplemental Restraint Systems (SRS), more commonly known as airbags. Always disable the airbag system before working in the vicinity of any airbag system components to avoid the possibility of accidental deployment of the airbags, which could cause personal injury (see Section 26).

### ✳✳ CAUTION 1:

Disconnecting the battery can cause driveability problems that require a scan tool to rectify. Additionally, disconnecting the battery may cause one or more warning lights on the instrument panel to illuminate, which will also require the use of a scan tool to turn off. Most scan tools available to the public do not have the capability to perform either of these tasks, which will necessitate taking the vehicle to a dealer service department or other properly equipped repair facility after service work has been performed. See Chapter 5, Section 1 for the use of an auxiliary voltage input device (memory saver) before disconnecting the battery and for other precautions related to battery disconnection.

### ✳✳ CAUTION 2:

These models are equipped with an anti-theft radio. Before performing a procedure that requires disconnecting the battery, make sure you have the proper activation code.

## HEADLIGHT SWITCH

### Check

1  The headlight switch must be removed  for testing. If you're working on a Passat, see Steps 4 and 5 for removal. If you're working on an A4, remove the multi-function switch as described in Section 7.

2  Using an ohmmeter or self-powered continuity tester, check the switch for proper continuity between the terminals. Refer to the wiring diagrams at the end of this Chapter for the wire colors and circuits. There should be continuity between the power side and the lighting side when the switch is in the indicated position.

3  If the switch fails any of the tests, replace the switch.

### Replacement

4  Turn the headlight switch knob counterclockwise until it stops at the zero position. Push in on the switch and twist it to the right (clockwise) then withdraw it from the instrument panel.

5  Pull the switch out far enough to disconnect the electrical connector.

6  Installation is the reverse of removal.

## DASH LIGHT DIMMER SWITCH

▶ **Refer to illustration 9.7**

7  The dimmer switch is mounted on the dashboard trim panel (see illustration).

8  Remove the trim panel from the dash (see Chapter 11).

9  Apply battery voltage to the input side of the switch (see the wiring diagrams at the end of this Chapter), and use a digital ohmmeter to check for resistance on the output side. If the resistance doesn't vary as the knob is turned, replace the switch.

10  Installation is the reverse of removal.

## ON/OFF SWITCHES

11  Depending on the options of the vehicle, there may be one or more switches on the instrument panel, including seat heaters, hazard flasher and rear window defogger.

12  All of the above-mentioned switches are located in the switch panel at the center of the dashboard.

13  All of the switches need to be removed for testing, and all are removed the same way. Use a screwdriver or trim tool to pry behind the lip of the switches to release them from the switch panel, then pull them out far enough to disconnect the electrical connector.

14  With simple on/off switches, use an ohmmeter or self-powered continuity tester to check the switch for proper continuity between the terminals. Refer to the wiring diagrams at the end of this Chapter for input and output terminals and wire colors. There should be continuity between input and output terminals only when the switch is engaged. If any switch fails the test, replace the switch.

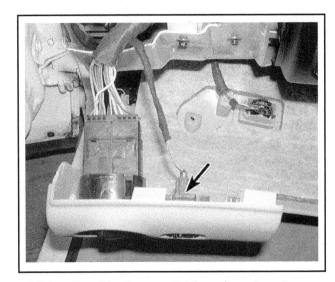

**9.7  Location of the dimmer switch (arrow) on a Passat**

## 10  Instrument cluster - removal and installation

▶ Refer to illustrations 10.4 and 10.5

### ✳✳ WARNING:

The models covered by this manual are equipped with Supplemental Restraint Systems (SRS), more commonly known as airbags. Always disable the airbag system before working in the vicinity of any airbag system components to avoid the possibility of accidental deployment of the airbags, which could cause personal injury (see Section 26).

### ✳✳ CAUTION 1:

Disconnecting the battery can cause driveability problems that require a scan tool to rectify. Additionally, disconnecting the battery may cause one or more warning lights on the instrument panel to illuminate, which will also require the use of a scan tool to turn off. Most scan tools available to the public do not have the capability to perform either of these tasks, which will necessitate taking the vehicle to a dealer service department or other properly equipped repair facility after service work has been performed. See Chapter 5, Section 1 for the use of an auxiliary voltage input device (memory saver) before disconnecting the battery and for other precautions related to battery disconnection.

### ✳✳ CAUTION 2:

These models are equipped with an anti-theft radio. Before performing a procedure that requires disconnecting the battery, make sure you have the proper activation code.

➡Note 1: If the instrument cluster is replaced in these vehicles, the new cluster must be programmed with a factory scan tool. For this reason it is best to have a Volkswagen/Audi dealer perform this job.

➡Note 2: Later models equipped with clusters with blue backlighting do not have replaceable bulbs or other components. They must be replaced as a unit even if only one part is malfunctioning.

1  Disconnect the negative cable from the battery. See **Cautions** in Chapter 5, Section 1.
2  Remove the steering wheel (see Chapter 10) and the steering column covers (see Chapter 11).
3  Remove the ignition assembly shroud (see Section 8).
4  Remove the screws securing the cluster to the instrument panel (see illustration).
5  Pull the cluster forward enough to disengage the two large electrical connectors at the back, then pull the cluster out, tilting the bottom out first (see illustration).
6  Installation is the reverse of removal.

10.4  Remove the instrument cluster mounting screws (arrows)

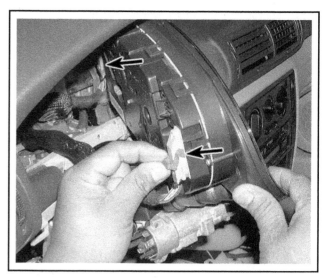

10.5  Pull the cluster out enough to disconnect the electrical connectors

## 11  Radio and speakers - removal and installation

### ✳✳ WARNING:

The models covered by this manual are equipped with Supplemental Restraint Systems (SRS), more commonly known as airbags. Always disable the airbag system before working in the vicinity of any airbag system components to avoid the possibility of accidental deployment of the airbags, which could cause personal injury (see Section 26).

### ✳✳ CAUTION 1:

Disconnecting the battery can cause driveability problems that require a scan tool to rectify. Additionally, disconnecting the battery may cause one or more warning lights on the instrument panel to illuminate, which will also require the use of a scan tool to turn off. Most scan tools available to the public do not have the capability to perform either of these tasks, which will necessitate taking the vehicle to a dealer service department

or other properly equipped repair facility after service work has been performed. See Chapter 5, Section 1 for the use of an auxiliary voltage input device (memory saver) before disconnecting the battery and for other precautions related to battery disconnection.

### ✳✳ CAUTION 2:

These models are equipped with an anti-theft radio. Before performing a procedure that requires disconnecting the battery, make sure you have the proper activation code.

➡Note: The audio system is part of the diagnostic network of the vehicle. Any problems with the radio, antenna or speakers may set a trouble code that can be retrieved with a scan tool.

## RADIO

▶ Refer to illustrations 11.2, 11.3a and 11.3b

1   Turn the ignition switch to the Off position and remove the key. Also turn the radio to Off.

2   Two special tools, available at most auto parts stores, must be used to remove the radio (see illustration). These tools can be fabricated from the pieces of sheetmetal (like feeler gauges) if you can't purchase them.

3   Push the two tools into the slots on either side of the radio, pull the radio out of the instrument panel, disconnect the connectors, then remove it from the vehicle (see illustrations).

4   Installation is the reverse of removal. To disengage the plastic tools from the radio before installation, depress the lugs on the side of the radio.

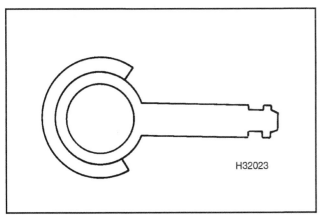

H32023

**11.2  Radio removal tool**

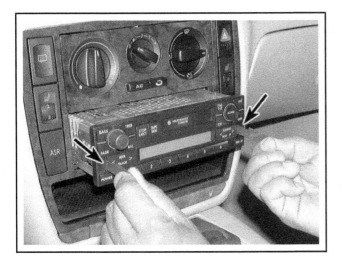

**11.3a  Push the tools (arrows) into the slots and pull the radio out**

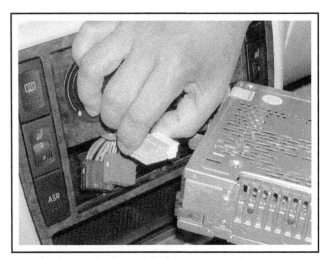

**11.3b  Disconnect the connectors from the back of the radio, then disconnect the ground wire and the antenna cable**

## SPEAKERS

♦ **Refer to illustration 11.7**

5   There are small tweeter speakers in each of the front doors forward of the window trim near the outside mirrors, and in the rear doors inside the door handle trim, and if found defective, the trim/speaker must be replaced as a unit. The remaining bass speakers include one in each door panel, front and rear.

6   To remove the door speakers, remove the door panel first (see Chapter 11).

7   Disconnect the electrical connector at the speaker, then drill out the rivets securing it to the door carrier (see illustration).

8   Installation is the reverse of the removal procedure. Self-tapping screws may be used to replace the speaker mounting rivets.

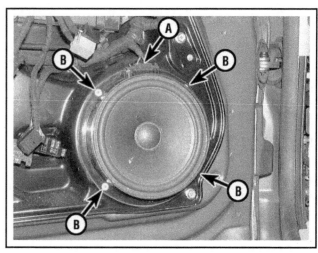

**11.7  To remove a door or rear speaker, disconnect the connector (A), then drill out the rivets (B)**

## 12  Antenna - removal and installation

♦ **Refer to illustration 12.1**

1   Grip the base of the antenna mast firmly and twist to unscrew it from the mounting base (see illustration).

2   If the base or the cable itself must be replaced, the headliner must be removed to access the nut on the bottom of the antenna base, on the interior side. This is a complex job for the home mechanic, and should be left to a dealer or professional auto stereo shop.

3   Installation of the antenna mast is the reverse of the removal procedure.

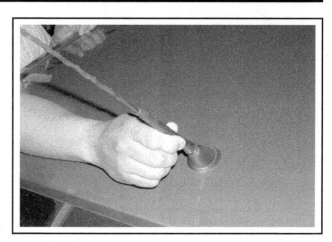

**12.1  The antenna mast can be unscrewed from its base by twisting it by hand**

## 13  Headlight bulb - replacement

♦ **Refer to illustrations 13.3a and 13.3b**

### ❊❊ WARNING:

**Halogen and Xenon bulbs are gas-filled and under pressure and may shatter if the surface is scratched or the bulb is dropped. Wear eye protection and handle the bulbs carefully, grasping only the base whenever possible. Don't touch the surface of the bulb with your fingers because the oil from your skin could cause it to overheat and fail prematurely. If you do touch the bulb surface, clean it with rubbing alcohol.**

### 2001 AND EARLIER MODELS

1   Refer to Section 15 and remove the headlight housing.

2   Disconnect the bulb electrical connectors.

3   Remove the outer (low beam) or inner (high beam) bulbholder from the back of the headlight housing (see illustrations). Remove the old bulb from the holder.

4   Handling the new bulb only with gloves or a clean rag, insert the new bulb in the holder.

5   Installation of the housing is the reverse of the removal procedure.

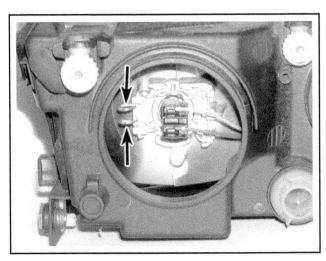

**13.3a Pinch the spring levers (arrows) to release the low beam headlight bulb assembly**

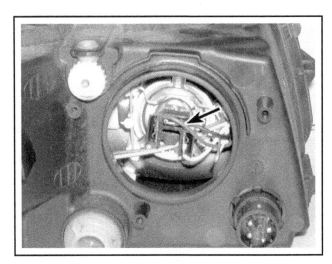

**13.3b Remove the retaining clip (arrow), then pull out the headlight bulb holder (high beam bulb shown)**

## 2002 AND LATER MODELS

6   Remove the plastic cover from the back of the headlight housing.
7   Rotate the bulb holder counterclockwise and pull it out of the housing.

8   Pull the bulb straight out of the socket. Make sure to avoid touching the new bulb with your fingers (see **Warning** in this Section).
9   Installation is the reverse of removal.

## 14  Headlights - adjustment

▶ Refer to illustrations 14.1 and 14.2

### �֍ WARNING:

The headlights must be aimed correctly. If adjusted incorrectly, they could temporarily blind the driver of an oncoming vehicle and cause an accident or seriously reduce your ability to see the road. The headlights should be checked for proper aim every 12 months and any time a new headlight is installed or front-end bodywork is performed. The following procedure is only an interim step to provide temporary adjustment until the headlights can be adjusted by a properly equipped shop.

➡Note: Some models are equipped with a headlight leveling system. This adjustment procedure will not apply. Have the headlights adjusted by a dealer service department or other qualified repair shop.

1   These models are equipped with headlight housings with two adjustment screws, one controlling left-and-right movement and one for up-and-down movement (see illustration).

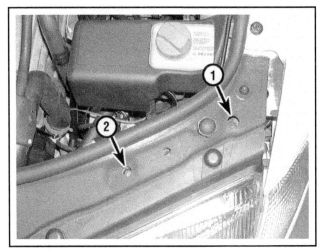

**14.1 Open the hood and adjust the headlight alignment with these screws - (2) is the vertical (height) adjuster, (1) is the horizontal (lateral) adjuster**

2   There are several methods of adjusting the headlights. The simplest method requires an open area with a blank wall and a level floor (see illustration).

3   Position masking tape vertically on the wall in reference to the vehicle centerline and the centerlines of both headlights.

4   Position a horizontal tape line in reference to the centerline of all the headlights.

➡**Note: It may be easier to position the tape on the wall with the vehicle parked only a few inches away.**

5   Adjustment should be made with the vehicle parked 25 feet from the wall, sitting level, the gas tank half-full and no unusually heavy load in the vehicle.

6   Starting with the low beam adjustment, position the high intensity zone so it is two inches below the horizontal line and two inches to the side of the vertical headlight line, away from oncoming traffic. Twist the adjustment screws until the desired level has been achieved.

7   With the high beams on, the high intensity zone should be vertically centered with the exact center just below the horizontal line.

➡**Note: It may not be possible to position the headlight aim exactly for both high and low beams. If a compromise must be made, keep in mind that the low beams are the most used and have the greatest effect on driver safety.**

8   Have the headlights adjusted by a dealer service department at the earliest opportunity.

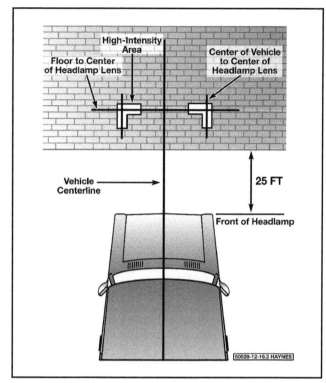

**14.2  Headlight adjustment details**

---

## 15  Headlight housing - replacement

### 2001 AND EARLIER MODELS

▶ **Refer to illustration 15.3**

1   Remove the front turn signal housing (see Section 16).

2   Apply masking tape onto the body areas surrounding the headlight housing to prevent damage.

3   Pull off the covers and remove the headlight housing mounting screws (see illustration).

4   Pull the headlight housing far enough forward and then slightly sideways to clear the front grille trim. Disconnect the electrical connectors, then slide the headlight housing all the way out.

5   To install the housing, slide it back in, aligning the grille tabs on each side.

6   When the housing is in far enough, reach through from inside the engine compartment to reattach the electrical connectors. The remainder of installation is the reverse of the removal procedure. Refer to Section 14 for aiming procedures after the headlight housing is installed.

### 2002 AND LATER MODELS

7   Refer to Chapter 11 and remove the front bumper cover as well as the reinforcing plastic piece.

8   Disconnect the wiring from the headlight.

9   Pry off the cap from the concealed headlight bolt and then remove the bolt.

10  Remove the other two headlight mounting bolts.

11  Slide the headlight housing forward and out of the vehicle.

12  Installation is the reverse of removal.

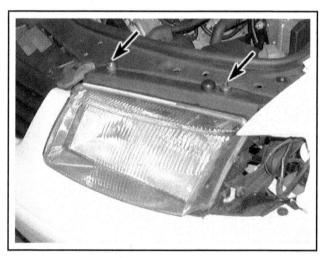

**15.3  Headlight housing mounting screws (arrows)**

## 16  Bulb replacement

### ✷✷ WARNING:

**Bulbs can remain hot for up to twenty minutes after they're turned off. Be sure bulbs are off and cool before you touch them.**

## FRONT TURN SIGNAL/SIDE-MARKER LIGHTS

**♦ Refer to illustrations 16.1, 16.3 and 16.5**

1   Release the securing lock from the turn signal bulb assembly by pulling back toward the engine (see illustration).
2   Pull the turn signal assembly forward, out of the fender. Disconnect the electrical connectors.
3   Each bulb is replaced by twisting the bulbholder out (see illustration).
4   Installation is the reverse of removal.

5   Guide the turn signal assembly into the tabs in the body of the vehicle (see illustration).
6   Install the retainer lock to the backside of the turn signal assembly.

## TAIL/STOP/TURN/BACK-UP LIGHTS

**♦ Refer to illustrations 16.8 and 16.9**

7   On all models, these rear lights are all in one housing, accessible through the trunk area in sedans or in the side storage compartment in wagons.
8   Locate the door cover, remove the three nuts and swing the door open (see illustration). Squeeze the release clips to unlock the bulb assembly.
9   Pull the taillight housing out and disconnect the electrical connectors. Twist the bulbholders counterclockwise to remove them for

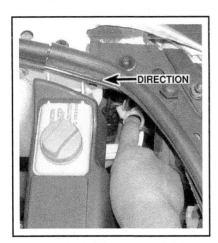

**16.1  Pull back the securing lock to release the turn signal bulb assembly**

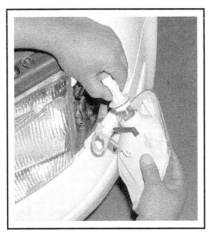

**16.3  Twist and then pull the bulb from the turn signal assembly**

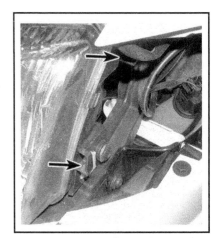

**16.5  Guide the turn signal assembly into the tabs (arrows) on the vehicle**

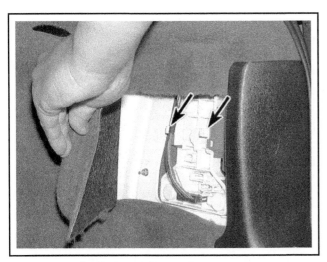

**16.8  Swing the door open and squeeze the release clips (arrows) to unlock the bulb assembly**

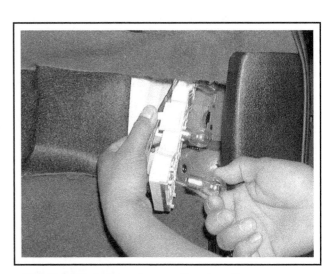

**16.9  Twist the bulbs and remove them from the rear light assembly**

bulb replacement (see illustration).

10 Installation is the reverse of removal. Make sure the washer and spacer are in place before installing the plastic nut on the housing.

## SIDE-MARKER LIGHTS

**▶ Refer to illustrations 16.12 and 16.13**

11 The side marker lights are in the fenders.

12 Use a small screwdriver to depress the tab on the back of the marker light housing and push the light housing to the rear (see illustration).

13 Twist the bulbholder counterclockwise to remove it and replace the bulb (see illustration).

14 Installation is the reverse of the removal procedure. Make sure the hook end of the light housing goes in first, then snap the housing in

place until the tab is engaged.

## CENTER HIGH-MOUNTED STOP LIGHT (CHMSL)

### Sedans

**▶ Refer to illustrations 16.16 and 16.17**

15 Refer to Chapter 11 and remove the inner trim panel from the rear shelf, directly against the rear window.

16 Press the lock retaining springs for the CHMSL (see illustration).

17 Pull the housing from the hatch and disconnect the electrical connector(s) (see illustration). Pull the bulbs straight out of the bulb holders, then replace the bulbs.

### Wagons

18 Refer to Chapter 11 and remove the inner trim panel from the rear hatch.

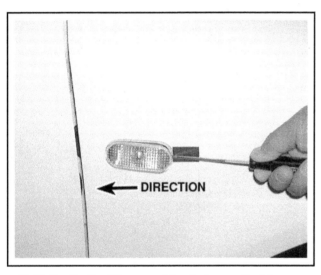

**16.12  Use the tip of the screwdriver to depress the tab on the back of the marker light housing then push the housing to the rear**

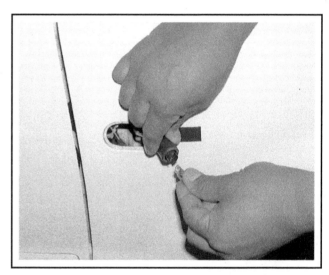

**16.13  Twist the bulb and remove it from the bulbholder**

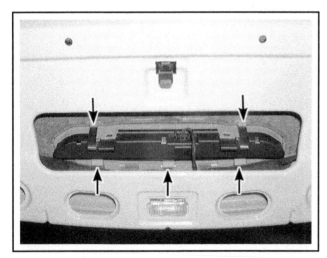

**16.16  Press the lock retaining springs for the CHMSL and withdraw the assembly**

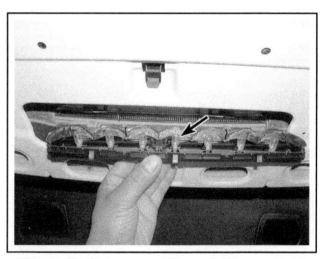

**16.17  Pull the bulb(s) out to remove it from the bulbholder(s)**

19 Press the lock retaining springs for the CHMSL.

20 Pull the housing from the hatch and disconnect the electrical connector(s). Pull the bulbs straight out of the bulb holders, then replace the bulbs.

### All models

21 Installation is the reverse of the removal procedure. Make sure the gasket on the back of the housing is intact, or a water leak could develop.

## LICENSE PLATE BULB

▶ **Refer to illustrations 16.22 and 16.23**

22 Use a small screwdriver and remove the mounting screws of the license plate light, then tilt and pull out the light housing (see illustration).

23 Remove the bulb from the bulbholder (see illustration).

24 Installation is the reverse of removal.

## INSTRUMENT CLUSTER LIGHTS

25 The instrument cluster is illuminated by LED's that are part of the printed circuit board. Thus, there are no user-replaceable bulbs behind the instrument cluster. Remove the instrument cluster and have it replaced at a dealer parts department.

## OVERHEAD LIGHTS

### Reading lights

▶ **Refer to illustrations 16.26a and 16.26b**

26 Remove the dome light lens (see illustration). Remove the bulb (see illustration).

27 Installation is the reverse of removal.

### Vanity lights

▶ **Refer to illustration 16.28**

28 Pry the light out of the headliner carefully with a small screwdriver, then replace the bulb by pulling it straight out (see illustration).

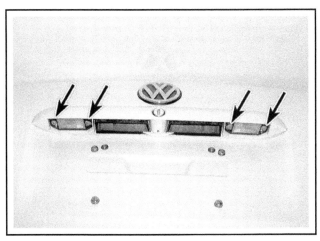

**16.22  Remove the license plate light mounting screws (arrows)**

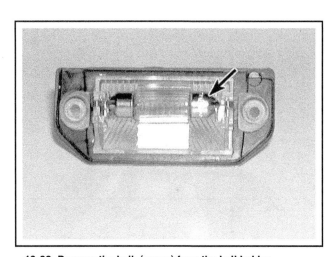

**16.23  Remove the bulb (arrow) from the bulbholder**

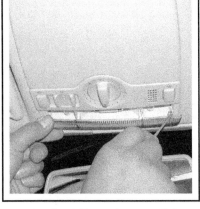

**16.26a  Carefully pry the dome light lens from the overhead light assembly**

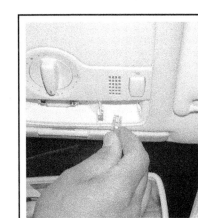

**16.26b  Replace the reading light bulb**

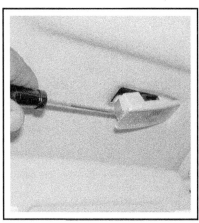

**16.28  Pry the vanity light bulbholder using a small flat-bladed screwdriver**

29 Installation is the reverse of removal.

## Dome lights

▶ **Refer to illustration 16.30**

30 Remove the dome light lens (see illustration 16.26a), then replace the bulb by pulling it straight out (see illustration).

31 Installation is the reverse of removal.

## Glovebox lights

▶ **Refer to illustrations 16.33a and 16.33b**

32 Remove the glovebox door (see Chapter 11).

33 Remove the glovebox liner (see illustration) then replace the bulb by pulling the bulb assembly straight out (see illustration).

34 Installation is the reverse of removal.

## Luggage compartment lights

▶ **Refer to illustration 16.35**

35 Pry the lens from the trunk lid (see illustration), then replace the bulb.

36 Installation is the reverse of removal.

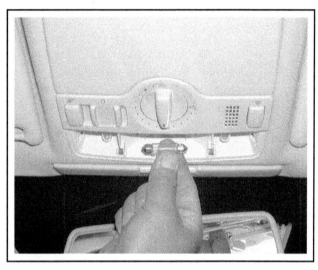

**16.30  Pull the dome light bulb straight down to remove**

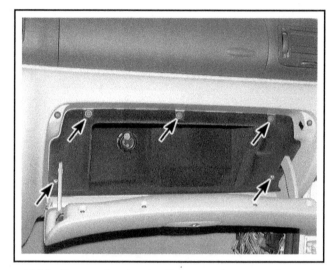

**16.33a  Remove the glovebox liner mounting screws (arrows)**

**16.33b  Pull the bulb assembly out and replace the bulb**

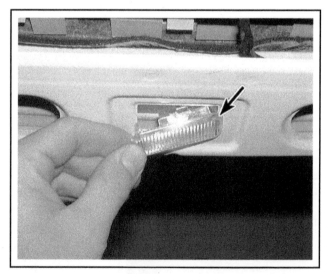

**16.35  Pry the luggage light bulbholder from the trunk lid**

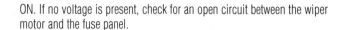

## 17  Wiper motor - check and replacement

### WIPER MOTOR CIRCUIT CHECK

➡**Note: Refer to the wiring diagrams for wire colors and locations in the following checks. When checking for voltage, probe a grounded 12-volt test light to each terminal at a connector until it lights; this verifies voltage (power) at the terminal. If the following checks fail to locate the problem, have the system diagnosed by a dealer service department or other properly equipped repair facility.**

1   If the wipers work slowly, make sure the battery is in good condition and has a strong charge (see Chapter 5). If the battery is in good condition, remove the wiper motor (see below) and operate the wiper arms by hand. Check for binding linkage and pivots. Lubricate or repair the linkage or pivots as necessary. Reinstall the wiper motor. If the wipers still operate slowly, check for loose or corroded connections, especially the ground connection. If all connections look OK, replace the motor.

2   If the wipers fail to operate when activated, check the fuse. If the fuse is OK, connect a jumper wire between the wiper motor's ground terminal and ground, then retest. If the motor works now, repair the ground connection. If the motor still doesn't work, turn the wiper switch to the HI position and check for voltage at the motor.

➡**Note: The cowl cover will have to be removed (see Chapter 11) to access the electrical connector.**

3   If there's voltage at the connector, remove the motor and check it off the vehicle with fused jumper wires from the battery. If the motor now works, check for binding linkage (see Step 1). If the motor still doesn't work, replace it. If there's no voltage to the motor, check for voltage at the wiper control relays. If there's voltage at the wiper control relays and no voltage at the wiper motor, check the switch for continuity (see Section 7). If the switch is OK, the wiper control relay is probably bad. See Section 5 for relay testing.

4   If the interval (delay) function is inoperative, check the continuity of all the wiring between the switch and wiper control module.

5   If the wipers stop at the position they're in when the switch is turned off (fail to park), check for voltage at the park feed wire of the wiper motor connector when the wiper switch is OFF but the ignition is

ON. If no voltage is present, check for an open circuit between the wiper motor and the fuse panel.

### REPLACEMENT

▶ **Refer to illustrations 17.7a, 17.7b, 17.11, 17.13 and 17.16**

6   Turn the ignition switch to the Off position and remove the key.

7   Mark the positions of the wiper arm(s) on the windshield, then remove the wiper arm(s) (see illustrations).

➡**Note: Disconnect the washer hose from the wiper arm.**

#### Front wiper motor

8   Remove the windshield cowl cover (see Chapter 11).

9   Remove the cover from the ECM (see Chapter 6).

10  Remove the ECM housing mounting nuts and push the housing forward for added clearance.

11  Disconnect the wiper motor harness connector and remove the windshield wiper motor/linkage assembly mounting nuts (see illustration).

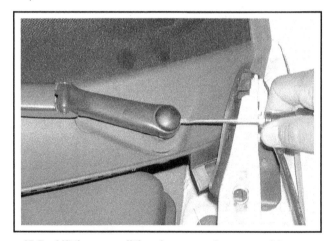

**17.7a  Lift the cover off the wiper arm using a screwdriver**

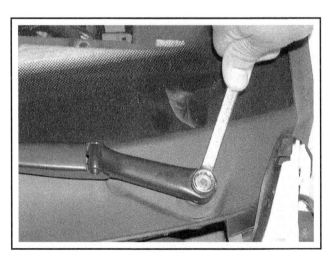

**17.7b  Use a wrench to remove the nut on the wiper shaft, then grasp the wiper arm and use a rocking motion to detach it from the shaft**

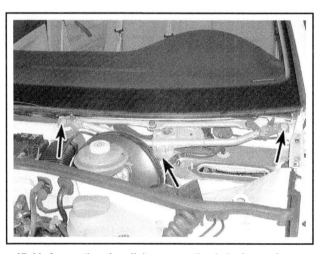

**17.11  Access the wiper linkage mounting bolts (arrows) after the windshield cowl cover has been removed**

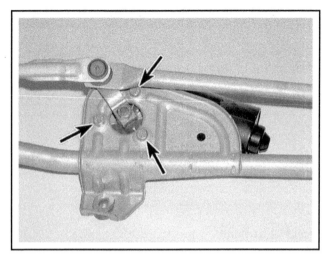

**17.13 Remove the motor arm nut and remove the motor-to-linkage mounting bolts (arrows)**

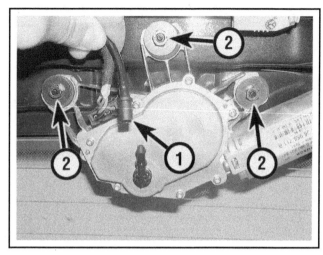

**17.16 Tailgate wiper motor**

| 1 | Washer pipe | 2 | Wiper motor mounting nuts |

12 Lift the windshield wiper motor assembly from the cowl area.
13 Remove the wiper motor mounting nuts and separate the motor from the assembly (see illustration).
14 Installation is the reverse of removal.

**Rear wiper motor**

15 Remove the inner trim panel from the hatch area (see Chapter 11).
16 Disconnect the wiper motor harness connector and remove the windshield wiper motor mounting bolts (see illustration).
17 Lift the windshield wiper motor assembly from the hatch area.
18 Installation is the reverse of removal.

## 18 Horn - check and replacement

### CHECK

1 Remove the cover from the under-dash fuse/relay box (see illustration 3.1b) and check the relay (see Section 5). Check the horn fuse number 40 (25A) in the fuse panel at the left end of the instrument panel.
2 Raise the vehicle and secure it on jackstands.
3 Disconnect the electrical connectors from the horns. There are two horns, located in the front fenderwell, just behind the front bumper cover, accessible from under the vehicle.
4 Have an assistant press the horn button and use a voltmeter to make sure there is battery voltage at the voltage supply wire terminal to the horns. If the relay is good and there's no voltage at the horn, the wire (which leads to the relay) has a fault.
5 Use an ohmmeter to measure the resistance between the wiring connector brown wire and a good ground. There should be less than 5.0 ohms. If not, repair the fault in the ground circuit.
6 If there's voltage at the horn and the wiring circuits are good, the horn is faulty and must be replaced.

### REPLACEMENT

7 Disconnect the electrical connectors, remove the mounting bolts and detach the horns.
8 Installation is the reverse of removal.

## 19 Daytime Running Lights (DRL) - general information

The Daytime Running Lights (DRL) system used on all models illuminates the low-beam headlight bulbs (at reduced power), whenever the ignition is On. The only exception is with the engine running and the parking brake engaged (standard-shift models) or with the shift lever in Park (automatic-transmission models). Once the parking brake is released or the shift lever is moved, the lights will remain on as long as the ignition switch is on.

## 20  Rear window defogger - check and repair

1   The rear window defogger consists of a number of horizontal heating elements baked onto the inside surface of the glass. Power is supplied through a large fuse from the fuse/relay box in the dash area. Refer to the wiring diagrams at the end of Chapter 12. The heater is controlled by the instrument panel switch.

2   Small breaks in the element can be repaired without removing the rear window.

## CHECK

▶ **Refer to illustrations 20.5, 20.6 and 20.8**

3   Turn the ignition switch and defogger switch to the ON position.

4   Using a voltmeter, place the positive probe against the defogger grid positive terminal and the negative probe against the ground terminal. If battery voltage is not indicated, check the fuse, defogger switch, defogger relay and related wiring. If voltage is indicated, but all or part of the defogger doesn't heat, proceed with the following tests.

5   When measuring voltage during the next two tests, wrap a piece of aluminum foil around the tip of the voltmeter positive probe and press the foil against the heating element with your finger (see illustration). Place the negative probe on the defogger grid ground terminal.

6   Check the voltage at the center of each heating element (see illustration). If the voltage is 5 to 6 volts, the element is okay (there is no break). If the voltage is 0 volts, the element is broken between the center of the element and the positive end. If the voltage is 10 to 12 volts the element is broken between the center of the element and the ground side. Check each heating element.

7   If none of the elements are broken, connect the negative probe to a good chassis ground. The voltage reading should stay the same, if it doesn't the ground connection is bad.

8   To find the break, place the voltmeter negative probe against the defogger ground terminal. Place the voltmeter positive probe with the

foil strip against the heating element at the positive side and slide it toward the negative side. The point at which the voltmeter deflects from several volts to zero is the point where the heating element is broken (see illustration).

## REPAIR

▶ **Refer to illustration 20.14**

9   Repair the break in the element using a repair kit specifically for this purpose, such as Dupont paste No. 4817 (or equivalent). The kit includes conductive plastic epoxy.

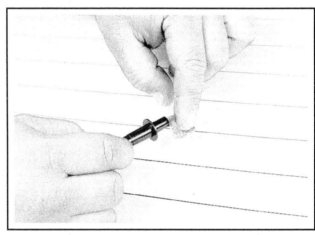

**20.5  When measuring the voltage at the rear window defogger grid, wrap a piece of aluminum foil around the positive probe of the voltmeter and press the foil against the wire with your finger**

**20.6 To determine if a heating element has broken, check the voltage at the center of each element - if the voltage is 6-volts, the element is unbroken**

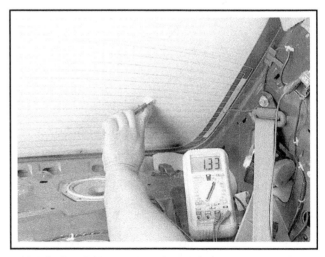

**20.8  To find the break, place the voltmeter negative lead against the defogger ground terminal, place the voltmeter positive lead with the foil strip against the heat wire at the positive terminal end and slide it toward the negative terminal end - the point at which the voltmeter deflects from several volts to zero volts is the point at which the wire is broken**

10 Before repairing a break, turn off the system and allow it to cool for a few minutes.

11 Lightly buff the element area with fine steel wool; then clean it thoroughly with rubbing alcohol.

12 Use masking tape to mask off the area being repaired.

13 Thoroughly mix the epoxy, following the kit instructions.

14 Apply the epoxy material to the slit in the masking tape, overlapping the undamaged area about 3/4-inch on either end (see illustration).

15 Allow the repair to cure for 24 hours before removing the tape and using the system.

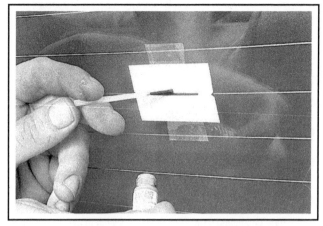

**20.14  To use a defogger repair kit, apply masking to the inside of the window at the damaged area, then brush on the special conductive coating**

## 21  Cruise control system - description and check

1  The cruise control system maintains vehicle speed with a servo motor that is connected to the throttle valve by linkage. The system consists of the servo motor, the cruise control module, the vacuum pump, the clutch vent valve and the brake vent valve and vacuum signal amplifier. Some features of the system require special testers and diagnostic procedures that are beyond the scope of the home mechanic. Listed below are some general procedures that may be used to locate common problems.

2  Check the cruise control system fuses in the main fuse box at the left end of the instrument panel.

3  Test the clutch vent valve and the brake vent valve located under the dash near the clutch pedal and the brake pedal. Depress the brake pedal while you check the stop lamp operation.

4  If the brake lights do not operate properly, correct the problem and retest the cruise control.

5  The cruise control system uses information from the vehicle speed sensor, which is located in the transmission. To locate the vehicle speed sensor, see Chapter 6.

6  Test the cruise control switch mounted in the multi-function lever (see Section 7).

7  Check the vacuum system for leaks or damaged components.

8  Test-drive the vehicle to determine if the cruise control is now working. If it isn't, take it to a dealer service department or an automotive electrical specialist for further diagnosis.

## 22  Power window system - description and check

♦ **Refer to illustrations 22.10a and 22.10b**

1  The power window system operates electric motors, mounted in the doors, which lower and raise the windows. The system consists of the control switches, the motors, regulators, glass mechanisms and associated wiring.

2  The power windows can be lowered and raised from the master control switch by the driver or by the switch located at the passenger window. Each window has a separate motor that is reversible. The position of the control switch determines the polarity and therefore the direction of operation.

3  The circuit is protected by fuses and a circuit breaker. Check the fuses in the fuse panel at the left end of the instrument panel. Each motor is equipped with an internal circuit breaker; this prevents one stuck window from disabling the whole system. Refer to the wiring diagrams at the end of Chapter 12.

4  The power window system will only operate when the ignition switch is ON. In addition, many models have a window lockout switch at the master control switch, when activated, disables the switch at the passenger's window also. Always check these items before troubleshooting a window problem.

5  These procedures are general in nature, so if you can't find the problem using them, take the vehicle to a dealer service department or other properly equipped repair facility.

6  Check the wiring between the switches and fuse panel for continuity. Repair the wiring, if necessary. Refer to the wiring diagrams at the end of Chapter 12.

7  If the passenger window is inoperative from the master control switch, try the other control switch at the passenger window.

8  If the same window works from one switch, but not the other, check the switch for continuity.

9  If the switch tests OK, check for a short or open in the circuit between the affected switch and the window motor.

10  If one window is inoperative from both switches, use a flat-bladed trim tool to pry up and remove the switch panel from the affected door. Check for voltage at the switch (see illustrations) and at the motor (refer to Chapter 11 for door panel removal) while the switch is operated.

11  If voltage is reaching the motor, disconnect the glass from the regulator (see Chapter 11). Move the window up and down by hand while checking for binding and damage. Also check for binding and damage to the regulator. If the regulator is not damaged and the window moves up and down smoothly, replace the motor. If there's binding or damage, lubricate, repair or replace parts, as necessary.

12  If voltage isn't reaching the motor, check the wiring in the circuit for continuity between the switches and motors. You'll need to consult the wiring diagram at the end of this Chapter.

13  Test the windows after you are done to confirm proper repairs.

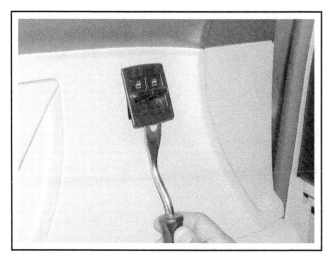

**22.10a  Use a trim tool to pry the door switches from the door panel**

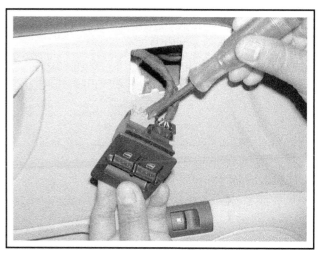

**22.10b  Pry up the driver's door switch assembly from the door panel and backprobe the connector**

## 23  Power door lock and keyless entry system - description and check

▶ **Refer to illustration 23.10**

1  The power door lock system operates the door lock actuators mounted in each door. The system consists of the switches, actuators, relays and associated wiring. Diagnosis can usually be limited to simple checks of the wiring connections and actuators for minor faults that can be easily repaired.

2  Power door lock systems are operated by bi-directional solenoids located in the doors. The lock switches have two operating positions: Lock and Unlock. These switches activate a relay, which in turn connects voltage to the door lock solenoids. Depending on which way the relay is activated, it reverses polarity, allowing the two sides of the circuit to be used alternately as the feed (positive) and ground side.

3  Some models may have keyless entry, electronic control modules and anti-theft systems incorporated into the power locks. If you are unable to locate the trouble using the following general steps, consult your dealer service department.

4  Always check the circuit protection first. Refer to the wiring diagrams at the end of this Chapter.

5  Operate the door lock switches in both directions (Lock and Unlock) with the engine off. Listen for the faint click of the relay operating.

6  If there's no click, check for voltage at the switches. If no voltage is present, check the wiring between the fuse panel and the switches for shorts and opens.

7  If voltage is present but no click is heard, test the switch for continuity. Replace it if there's not continuity in both switch positions. To remove the switch, use a flat-bladed trim tool to pry out the door/ window switch assembly (see Chapter 11).

8  If the switch has continuity but the relay doesn't click, check the wiring between the switch and relay for continuity. Repair the wiring if there's no continuity.

9  If the relay is receiving voltage from the switch but is not sending voltage to the solenoids, check for a bad ground at the relay case. If the relay case is grounding properly, replace the relay. Relay tests are described in Section 5.

10  If only one lock solenoid operates, remove the trim panel from the affected door (see Chapter 11) and check for voltage at the solenoid while the lock switch is operated (see illustration). One of the wires should have voltage in the Lock position; the other should have voltage in the Unlock position.

11  If the inoperative solenoid is receiving voltage, replace the solenoid.

12  If the inoperative solenoid isn't receiving voltage, check for an open or short in the wire between the lock solenoid and the relay.

➡**Note 1: It's common for wires to break in the portion of the harness between the body and door - opening and closing the door fatigues and eventually breaks the wires.**

➡**Note 2: This system is governed by the Central Comfort Control Module. Problems within this module must be diagnosed by a technician with a factory scan tool. If you have eliminated the obvious causes of a problem, have the vehicle checked at a dealership service department or other properly equipped repair shop.**

## KEYLESS ENTRY SYSTEM

13  The keyless entry system consists of a remote control transmitter that sends a coded infrared signal to a receiver, which then operates the door-lock system.

14  Replace the battery when the transmitter doesn't operate the locks at a distance of 10 feet. Normal range should be about 30 feet.

15  Use a small screwdriver to carefully separate the case halves.

16  Replace the two (3-volt) CR2016 lithium batteries.

17  Snap the case halves together.

**23.10  To test a door lock solenoid, check that power is getting to the solenoid connector (arrow) while the switch is operated**

## 24  Electric side view mirrors - description and check

1  The electric rear view mirrors use two motors to move the glass; one for up and down adjustments and one for left-right adjustments.

2  The control switch has a selector portion that sends voltage to the left or right side mirror. With the ignition ACC position and the engine OFF, roll down the windows and operate the mirror control switch through all functions (left-right and up-down) for both the left and right side mirrors.

➡**Note: On some models, the mirrors are also heated electrically, by turning the outside mirror "joystick" control to the "heat" position, which is between the left and right positions.**

3  Listen carefully for the sound of the electric motors running inside the mirrors.

4  If the motors can be heard but the mirror glass doesn't move, there's probably a problem with the drive mechanism inside the mirror. Power mirrors have no user-serviceable parts inside - a defective mirror must be replaced as a unit (see Chapter 11).

5  If the mirrors don't operate and no sound comes from the mirrors, check the Comfort system fuse in the fuse panel at the left end of the instrument panel.

6  If the fuses are OK, remove the switch panel for access to the back of the mirror control switch without disconnecting the wires attached to it. Turn the ignition ON and check for voltage at the switch. There should be voltage at one terminal. If there's no voltage at the

switch, check for an open or short in the wiring between the fuse panel and the switch.

7  If there's voltage at the switch, disconnect it. Check the switch for continuity in all its operating positions. If the switch does not have continuity, replace it.

8  Locate the wire going from the switch to ground. Leaving the switch connected, connect a jumper wire between this wire and ground. If the mirror works normally with this wire in place, repair the faulty ground connection.

9  If the mirror still doesn't work, remove the mirror and check the wires at the mirror for voltage. Check with ignition ON and the mirror selector switch on the appropriate side. Operate the mirror switch in all its positions. There should be voltage at one of the switch-to-mirror wires in each switch position (except the neutral "off" position).

10  If voltage is absent in each switch position, check the wiring between the mirror and control switch for opens and shorts.

11  If there's voltage, remove the mirror and test it off the vehicle with jumper wires. Replace the mirror if it fails this test.

➡**Note: This system is governed by the Central Comfort Control Module. Problems within this module must be diagnosed by a technician with the manufacturer's scan tool. If you have eliminated the obvious causes of a problem, have the vehicle checked at a dealership service department or other properly equipped repair shop.**

## 25  Sunroof switch - check and replacement

1   The sunroof switch is located in a panel in the headliner, above the rear-view mirror. The switch is accessible, but the sunroof module and drive can only be accessed using special tools, which is a job for a dealership service department or other properly equipped repair facility.

2   To check the switch, the panel must be pulled down (see Steps 6 and 7).

3   Check the fuse and relay first. Refer to the wiring diagrams at the end of Chapter 12.

4   Disconnect the electrical connector on the back of the sunroof switch and check for continuity (see illustration 25.7).

5   Check the continuity of the switch using the electrical schematic at the end of Chapter 12. If the switch fails the tests, replace the switch.

### REPLACEMENT

▶ **Refer to illustrations 25.6 and 25.7**

6   Unclip the sunroof switch panel from the headliner panel (see illustration).

7   Remove the two screws and pull down the switch and it's frame (see illustration).

8   The switch can be removed from the frame by disconnecting the electrical connector and releasing the two clips at the back of the switch.

9   In case of sunroof switch or module failure, the sunroof can be operated manually. A crank handle is snapped into clips on the back of the switch panel. Use the crank handle in the hole in the sunroof drive to open or close the sunroof.

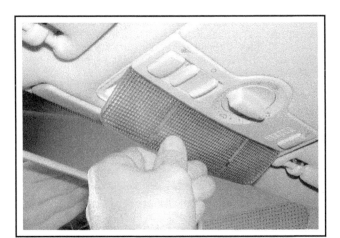

**25.6  Use a trim tool to pry down the switch panel from the headliner**

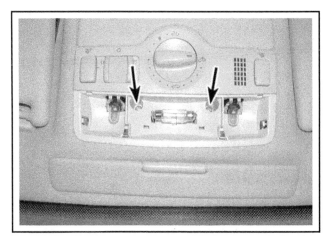

**25.7  Remove the screws (arrows), pull down the switch frame, and disconnect the two clips on the back securing the switch**

## 26  Airbag system - general information

▶ **Refer to illustrations 26.1 and 26.12**

1   These models are equipped with a Supplemental Restraint System (SRS), more commonly known as airbags, designed to protect the driver and the front seat passenger from serious injury in the event of a head-on or side collision. All models have a sensing/diagnostic control unit, located on the floor under the center of the instrument panel (see illustration).

### ❊❊ WARNING:

**If your vehicle is ever involved in a flood, or the interior carpeting is soaked for any reason, disconnect the battery and do not start the vehicle until the airbag system can be checked by your dealer. If the SRS system is subjected to flooding, the airbags could go off upon starting the vehicle, even without an accident taking place.**

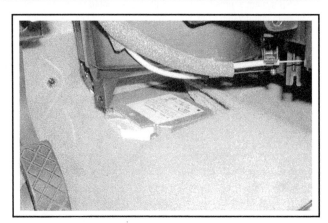

**26.1  The airbag control module is located on the floor ahead of the console - do not tamper with the yellow connectors attached to it**

## AIRBAG MODULES

2   The airbag modules consist of a housing incorporating the cushion (airbag) and inflator unit. The inflator assembly is mounted on the back of the housing over a hole through which gas is expelled, inflating the bag almost instantaneously when an electrical signal is sent from the system. The specially-wound wire on the driver's side that carries this signal to the driver's module is called a spiral cable. The spiral cable is a flat, ribbon-like electrically conductive tape that is wound many times so that it can transmit an electrical signal regardless of steering wheel position. Airbag modules are located in the steering wheel, on the passenger side above the glove box, at the upper side of each front seat (side-impact airbags) and, on some models, head-level airbags located along the roof rails.

## SENSING/DIAGNOSTIC CONTROL UNIT AND SENSORS

3   The sensing/diagnostic control unit contains an on-board microprocessor which monitors the operation of the system, and also contains a crash sensor. It checks this system every time the vehicle is started, causing the "AIRBAG" light to illuminate for five seconds, then go off, if the system is operating properly. If there is a fault in the system, the light may not come on at all, the light will go on and continue, either illuminated steadily or blinking, and the unit will store fault codes indicating the nature of the fault.

## OPERATION

4   For the airbag(s) to deploy, the impact sensor must be activated. When this condition occurs, the circuit to the airbag inflator is closed and the airbag inflates.

## SELF-DIAGNOSIS SYSTEM

5   A self-diagnosis circuit in the SRS unit displays a light on the instrument panel when the ignition switch is turned to the On position. If the system is operating normally, the light should go out after about five seconds. If the light doesn't come on, or doesn't go out after a short time, or if it comes on while you're driving the vehicle, or if it blinks at any time, there's a malfunction in the SRS system. Have it inspected and repaired as soon as possible. Do not attempt to troubleshoot or service the SRS system yourself. Even a small mistake could cause the SRS system to malfunction when you need it.

## SERVICING COMPONENTS NEAR THE SRS SYSTEM

6   Nevertheless, there are times when you need to remove the steering wheel, radio or service other components on or near the dashboard. At these times, you'll be working around components and wire harnesses for the SRS system.

**✳✳ WARNING:**

**Do not use electrical test equipment on airbag system wires; it could cause the airbag(s) to deploy. ALWAYS DISABLE THE SRS SYSTEM BEFORE WORKING NEAR THE SRS SYSTEM COMPONENTS OR RELATED WIRING.**

## DISABLING THE SRS SYSTEM

**✳✳ WARNING 1:**

**Any time you are working in the vicinity of airbag wiring or components, DISABLE THE SRS SYSTEM.**

**✳✳ WARNING 2:**

**An auxiliary voltage input device (memory saver) must not be used when working near airbag system components.**

**✳✳ CAUTION 1:**

**Disconnecting the battery can cause driveability problems that require a scan tool to rectify. Additionally, disconnecting the battery may cause one or more warning lights on the instrument panel to illuminate, which will also require the use of a scan tool to turn off. Most scan tools available to the public do not have the capability to perform either of these tasks, which will necessitate taking the vehicle to a dealer service department or other properly equipped repair facility after service work has been performed. See Chapter 5, Section 1 for other precautions related to battery disconnection.**

**✳✳ CAUTION 2:**

**These models are equipped with an anti-theft radio. Before performing a procedure that requires disconnecting the battery, make sure you have the proper activation code.**

7   To disable the airbag system, perform the following steps:
   a) *Turn the steering wheel to the straight-ahead position and turn the ignition switch to the Lock position, then remove the key.*
   b) *Disconnect the negative battery cable. Refer to the Cautions in Chapter 5, Section 1.*
   c) *Before touching any airbag system component, ground yourself to a metal part of the vehicle to discharge any static electricity built up in your body.*

## ENABLING THE SYSTEM

8   To enable the airbag system, perform the following steps:
   a) *Turn the ignition switch to the On position.*
   b) *Make sure nobody is inside the vehicle.*
   c) *Connect the battery cable.*
   d) *Turn the ignition to the Off position, then with your body out of the path of the airbag, turn the ignition switch to the On position. Confirm that the airbag warning light is functioning properly.*
   e) *Take the vehicle to a dealer service department or other qualified repair facility and have the airbag system checked and the diagnostic light canceled, if it remains lit.*

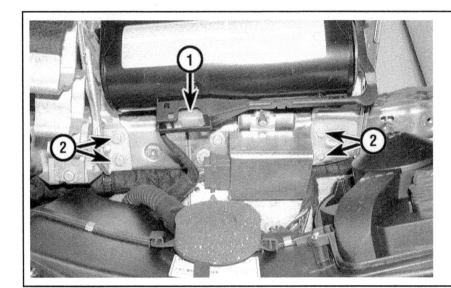

**26.12 Disconnect the connector (1) from the passenger airbag, then remove the airbag mounting bolts (2) for the passenger side airbag assembly**

## REMOVAL AND INSTALLATION

### Driver's airbag

9   Refer to Chapter 10 for removal and installation of the driver's airbag located in the steering wheel.

### Passenger's airbag

10  Disable the airbag system as described earlier in this section.
11  Remove the glovebox and its dash panel (see Chapter 11).
12  From under the instrument panel, disconnect the passenger airbag connector (see illustration).
13  Remove the airbag mounting bolts.
14  Gently pry the airbag module from the brackets on the dash support.

> ✳✳ **WARNING:**
>
> **Whenever handling an airbag module, always carry the airbag module with the trim panel side facing away from your body. Place the airbag module in a safe location with the trim panel side facing up.**

> ✳✳ **CAUTION:**
>
> **The airbag assembly is heavier than it looks; use both hands when removing it from the dash.**

15  Installation is the reverse of the removal procedure.

### Side impact airbags

16  Disable the airbag system as described earlier in this section.
17  Remove the front seats (see Chapter 11).
18  Have the side airbags removed from the seats by a dealer service department or other qualified auto repair facility.
19  Installation is the reverse of the removal procedure.

### Side curtain airbags (2001 and later models)

20  Disable the airbag system as described earlier in this Section.

21  The side curtain airbags are concealed beneath the door pillar trim pieces. Refer to Chapter 11, remove the clips and pry off the upper and side covers of the door pillars.

> ✳✳ **CAUTION:**
>
> **The upper front pillar fasteners are almost always damaged during removal.**

22  Disconnect the airbag wiring.
23  Remove the screws and detach the airbag from the retaining clips.

> ✳✳ **WARNING:**
>
> **Make sure to ground your body using a special wrist strap or a similar device connected to a solid ground on the vehicle's body. Place the airbag module in a safe place with the trim panel side facing up.**

24  Installation is the reverse of removal.

## IMPACT SEAT BELT RETRACTORS

25  All models are equipped with pyrotechnic (explosive) units in the front seat belt retracting mechanisms for both the lap and shoulder belts. During an impact that would trigger the airbag system, the airbag control unit also triggers the seat belt retractors. When the pyrotechnic charges go off, they accelerate the retractors to instantly take up any slack in the seat belt system to more fully prepare the driver and front seat passenger for impact.
26  The airbag system should be disabled any time work is done to or around the seats.

> ✳✳ **CAUTION:**
>
> **Never strike the pillars or floorpan with a hammer or use an impact-driver tool in these areas unless the system is disabled.**

## 27  Wiring diagrams - general information

Since it isn't possible to include all wiring diagrams for every year and model covered by this manual, the following diagrams are those that are typical and most commonly needed.

Prior to troubleshooting any circuits, check the fuses and circuit breakers (if equipped) to make sure they're in good condition. Make sure the battery is properly charged and check the cable connections (see Chapter 1).

When checking a circuit, make sure that all connectors are clean, with no broken or loose terminals. When disconnecting a connector, do not pull on the wires. Pull only on the connector housings themselves.

**Audi A4 1996 to 2001 wiring diagrams**                                    **Diagram 1**

## Key to symbols

| | |
|---|---|
| Bulb | |
| Switch | |
| Multiple contact switch (ganged) | |
| Fuse/fusible link and current rating | F5 30A |
| Resistor | |
| Variable resistor | |
| Connecting wires | |
| Plug and socket contact | |
| Item no. | 2 |
| Pump/motor | M |
| Ground point and location | E22 |
| Gauge/meter | |
| Diode | |
| Wire splice or soldered joint | |
| Solenoid actuator | |
| Light emitting diode (LED) | |
| Wire colour (brown with black tracer) | Br/Sw |
| Screened cable | |

Dashed outline denotes part of a larger item, containing in this case an electronic or solid state device.
**6**    -  unspecified connector pin 6.
**T14/9** -  14 pin connector, pin 9.

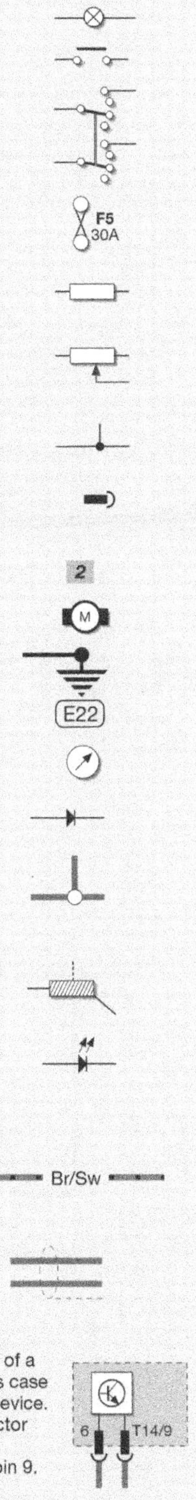

## Ground locations

| | |
|---|---|
| E1 | Ground strap, battery to body |
| E2 | Instrument panel ground connection 1 |
| E3 | Instrument panel ground connection 2 |
| E4 | Ground connection in trunk lid/rear left pillar |
| E5 | Drivers door wiring harness connection 1 |
| E6 | Electric window wiring harness |
| E7 | Lower left 'A' pillar |
| E8 | Drivers door wiring harness connection 2 |
| E9 | Rear wiring harness ground connection |
| E10 | Passengers door wiring harness |
| E11 | In left headlight wiring harness |
| E12 | In right headlight wiring harness |
| E13 | Air conditioning wiring harness |
| E14 | In airbag wiring harness |
| E15 | Behind instrument panel |

## Key to circuits (Audi models)

| | |
|---|---|
| Diagram 1 | Information for wiring diagrams |
| Diagram 2 | Wash/wipe, automatic day/night interior mirror |
| Diagram 3 | Power windows and power sunroof |
| Diagram 4 | Central locking |
| Diagram 5 | Interior lights, instrument lights and cigarette lighter |
| Diagram 6 | Headlights, fog lights and back-up lights |
| Diagram 7 | Brake lights, turn signal lights. tail, side marker and license plate light |
| Diagram 8 | Radio, Rear window defogger, heated washer jets and dual tone horn |
| Diagram 9 | Automatic climate control |
| Diagram 10 | Automatic climate control continued, airbag and ABS |

Refer to VW diagrams 23 through 30 for fuel, emissions and engine electrical diagrams

## Fuse holder

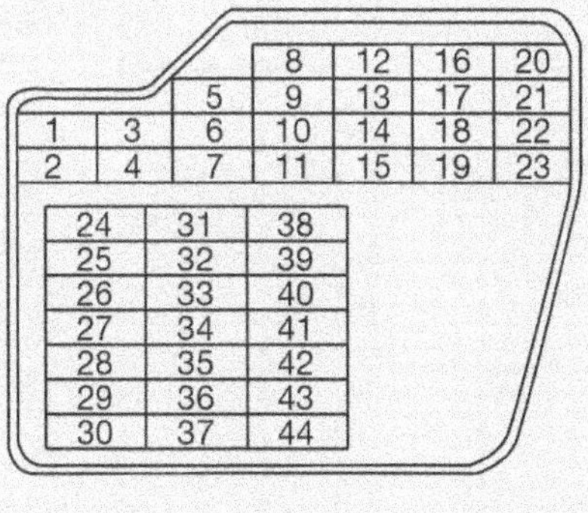

| | | | | 8 | 12 | 16 | 20 |
|---|---|---|---|---|---|---|---|
| | | | 5 | 9 | 13 | 17 | 21 |
| 1 | 3 | 6 | 10 | 14 | 18 | 22 | |
| 2 | 4 | 7 | 11 | 15 | 19 | 23 | |

| | | |
|---|---|---|
| 24 | 31 | 38 |
| 25 | 32 | 39 |
| 26 | 33 | 40 |
| 27 | 34 | 41 |
| 28 | 35 | 42 |
| 29 | 36 | 43 |
| 30 | 37 | 44 |

H32275

**Diagram 1:  Audi Models - Information for wiring diagrams**

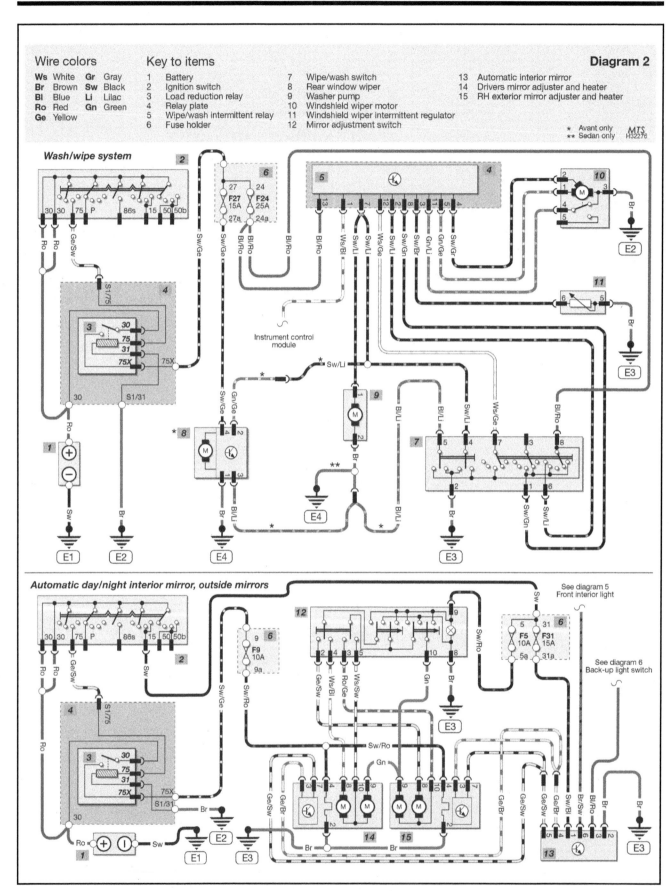

**Diagram 2:** Audi Models - Wash/wipe, automatic day/night interior mirror, outside mirrors

### Wire colors

**Ws** White  **Gr** Gray
**Br** Brown  **Sw** Black
**Bl** Blue  **Li** Lilac
**Ro** Red  **Gn** Green
**Ge** Yellow

### Key to items

1 Battery
2 Ignition switch
4 Relay plate
6 Fuse holder
16 Power window fuse
17 LH front window switch
18 LH front window motor
19 RH front window switch (in LH door)
20 RH front window switch (in RH door)
21 RH front window motor
22 LH rear window switch (in console)
23 LH rear window switch (in LH rear door)
24 LH rear window motor
25 RH rear window switch (in console)
26 RH rear window switch (in RH rear door)
27 RH rear window motor
28 Window lock out switch
29 Sunroof control module/motor
30 Sunroof switch

**Diagram 3**

MTS
H32277

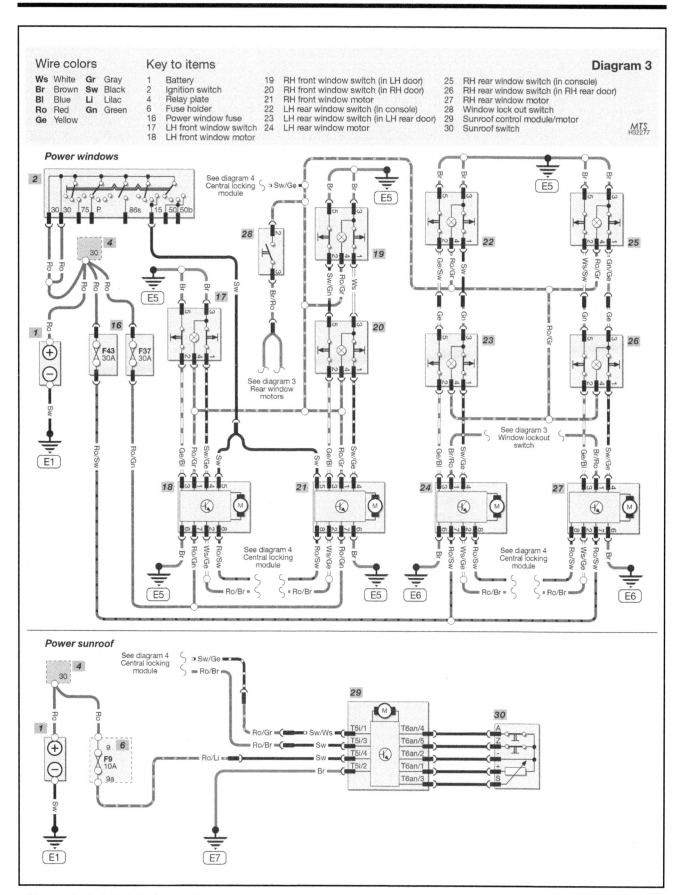

**Diagram 3:  Audi Models - Power windows and power sunroof**

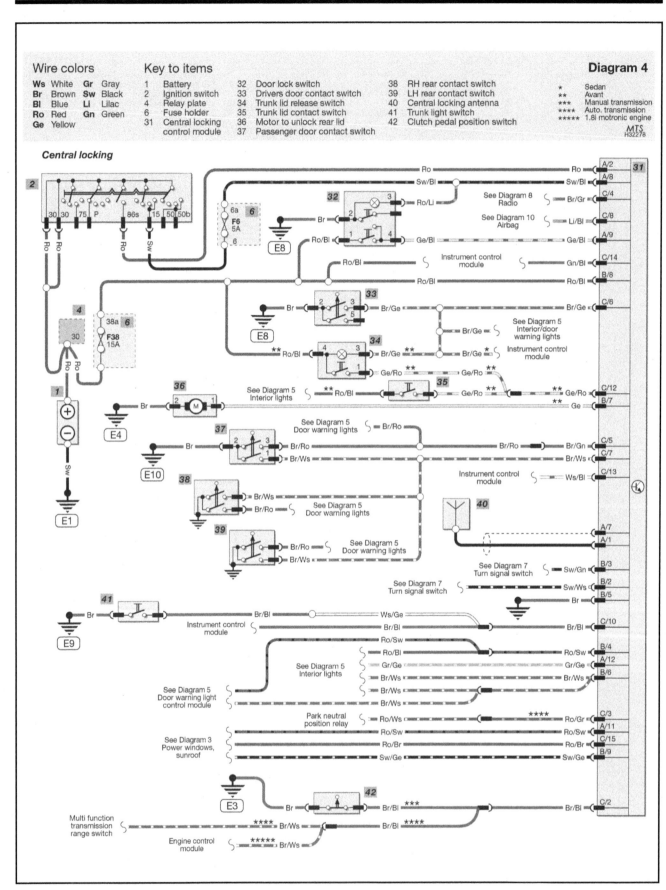

**Wire colors**

| | | | |
|---|---|---|---|
| **Ws** | White | **Gr** | Gray |
| **Br** | Brown | **Sw** | Black |
| **Bl** | Blue | **Li** | Lilac |
| **Ro** | Red | **Gn** | Green |
| **Ge** | Yellow | | |

**Key to items**

| | |
|---|---|
| 1 | Battery |
| 2 | Ignition switch |
| 4 | Relay plate |
| 6 | Fuse holder |
| 31 | Central locking control module |
| 32 | Door lock switch |
| 33 | Drivers door contact switch |
| 34 | Trunk lid release switch |
| 35 | Trunk lid contact switch |
| 36 | Motor to unlock rear lid |
| 37 | Passenger door contact switch |
| 38 | RH rear contact switch |
| 39 | LH rear contact switch |
| 40 | Central locking antenna |
| 41 | Trunk light switch |
| 42 | Clutch pedal position switch |

**Diagram 4**

| | |
|---|---|
| * | Sedan |
| ** | Avant |
| *** | Manual transmission |
| **** | Auto. transmission |
| ***** | 1.8i motronic engine |

*Central locking*

**Diagram 4: Audi Models - Central locking**

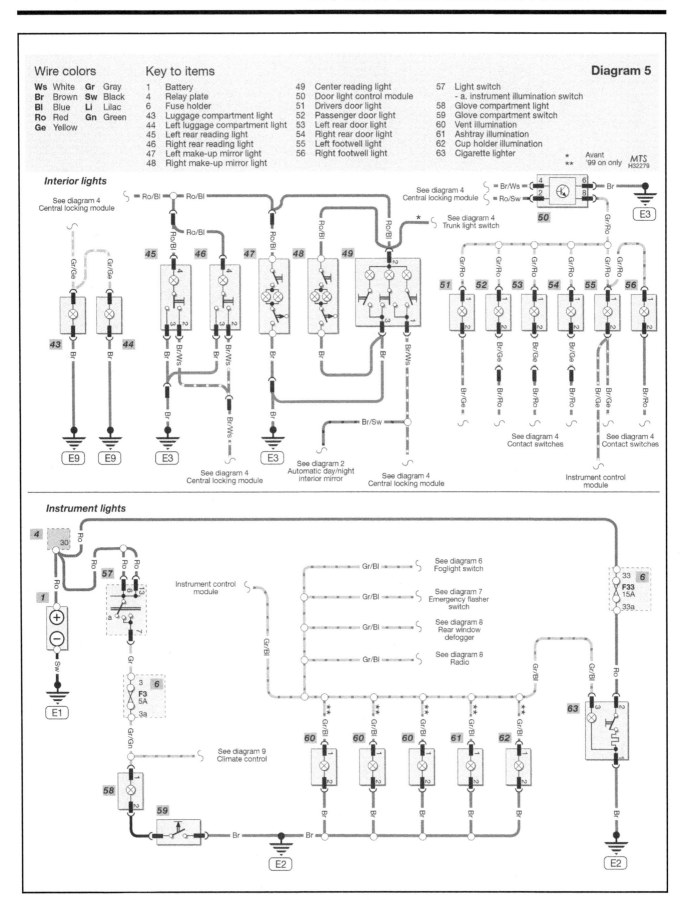

**Diagram 5: Audi Models - Interior lights, instrument lights and cigarette lighter**

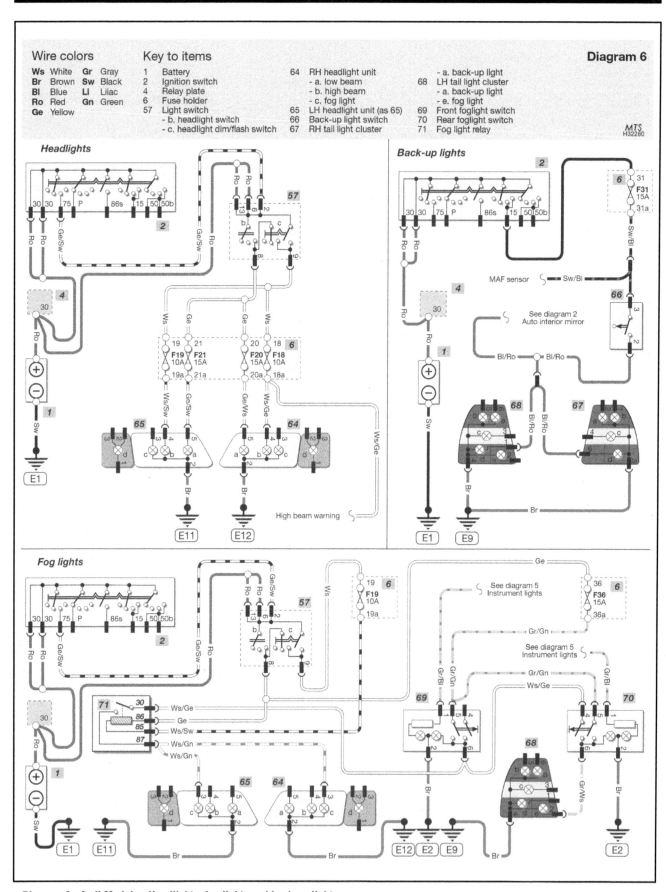

**Diagram 6: Audi Models - Headlights, fog lights and back-up lights**

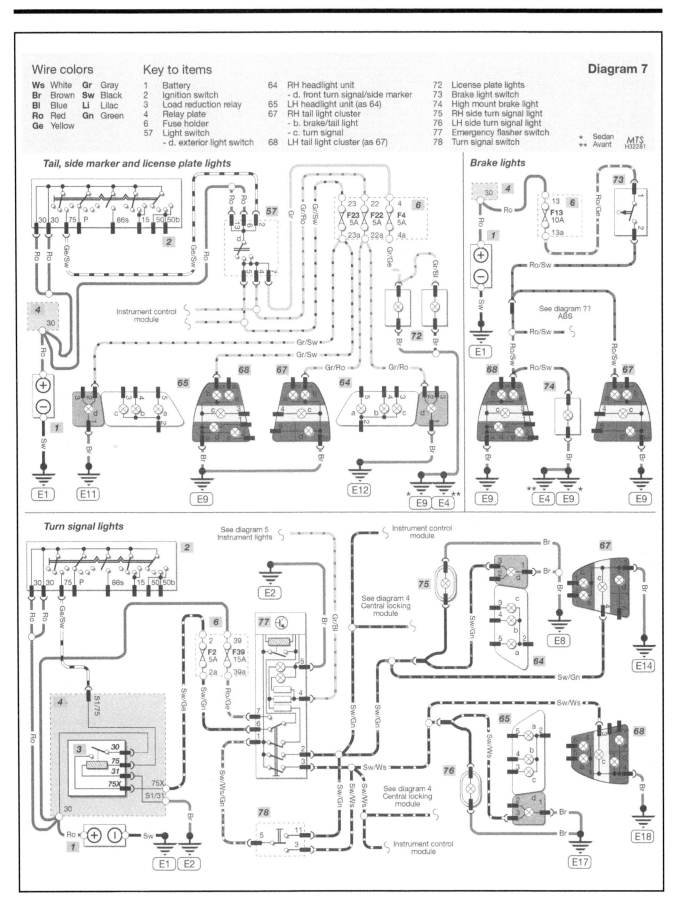

**Diagram 7: Audi Models - Brake lights, turn signal lights, tail, side marker and license plate lights**

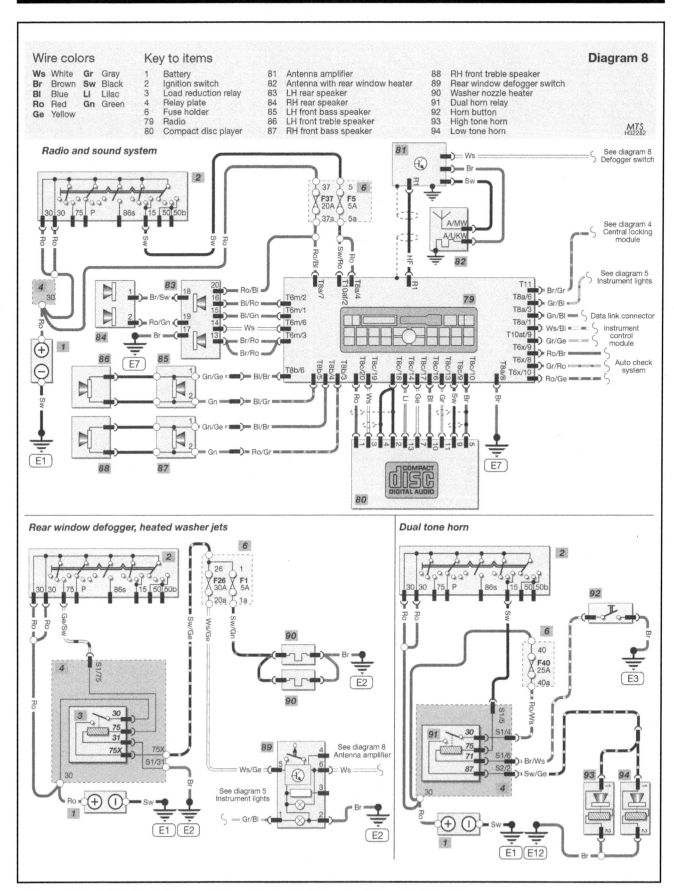

**Diagram 8:** Audi Models - Radio, rear window defogger, heated washer jets and dual tone horn

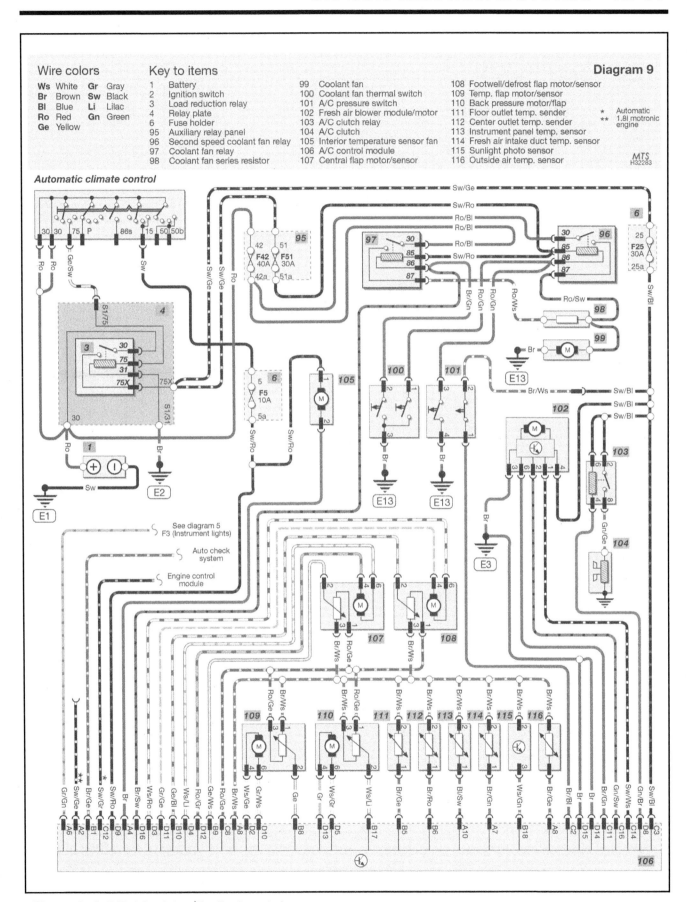

## Wire colors

| | | | |
|---|---|---|---|
| **Ws** | White | **Gr** | Gray |
| **Br** | Brown | **Sw** | Black |
| **Bl** | Blue | **Li** | Lilac |
| **Ro** | Red | **Gn** | Green |
| **Ge** | Yellow | | |

## Key to items

1 Battery
2 Ignition switch
3 Load reduction relay
4 Relay plate
6 Fuse holder
95 Auxiliary relay panel
96 Second speed coolant fan relay
97 Coolant fan relay
98 Coolant fan series resistor
99 Coolant fan
100 Coolant fan thermal switch
101 A/C pressure switch
102 Fresh air blower module/motor
103 A/C clutch relay
104 A/C clutch
105 Interior temperature sensor fan
106 A/C control module
107 Central flap motor/sensor
108 Footwell/defrost flap motor/sensor
109 Temp. flap motor/sensor
110 Back pressure motor/flap
111 Floor outlet temp. sender
112 Center outlet temp. sender
113 Instrument panel temp. sensor
114 Fresh air intake duct temp. sensor
115 Sunlight photo sensor
116 Outside air temp. sensor

**Diagram 9**

\* Automatic
\*\* 1.8l motronic engine

MTS
H32283

*Automatic climate control*

**Diagram 9:  Audi Models - Automatic climate control**

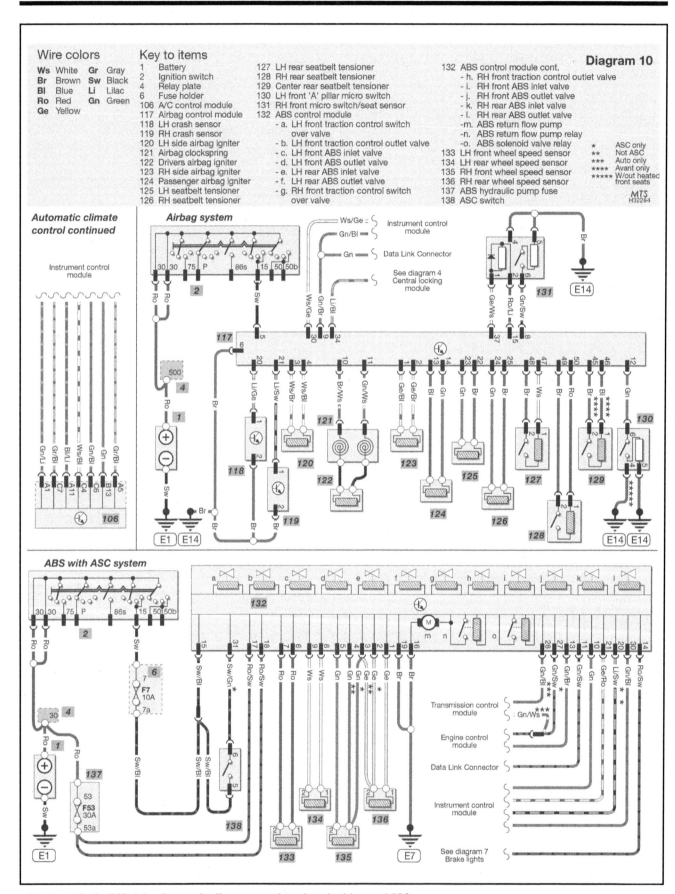

### Wire colors

**Ws** White   **Gr** Gray
**Br** Brown   **Sw** Black
**Bl** Blue   **Li** Lilac
**Ro** Red   **Gn** Green
**Ge** Yellow

### Key to items

1   Battery
2   Ignition switch
4   Relay plate
6   Fuse holder
106   A/C control module
117   Airbag control module
118   LH crash sensor
119   RH crash sensor
120   LH side airbag igniter
121   Airbag clockspring
122   Drivers airbag igniter
123   RH side airbag igniter
124   Passenger airbag igniter
125   LH seatbelt tensioner
126   RH seatbelt tensioner

127   LH rear seatbelt tensioner
128   RH rear seatbelt tensioner
129   Center rear seatbelt tensioner
130   LH front 'A' pillar micro switch
131   RH front micro switch/seat sensor
132   ABS control module
  - a. LH front traction control switch
    over valve
  - b. LH front traction control outlet valve
  - c. LH front ABS inlet valve
  - d. LH front ABS outlet valve
  - e. LH rear ABS inlet valve
  - f. LH rear ABS outlet valve
  - g. RH front traction control switch
    over valve

132   ABS control module cont.
  - h. RH front traction control outlet valve
  - i. RH front ABS inlet valve
  - j. RH front ABS outlet valve
  - k. RH rear ABS inlet valve
  - l. RH rear ABS outlet valve
  - m. ABS return flow pump
  - n. ABS return flow pump relay
  - o. ABS solenoid valve relay
133   LH front wheel speed sensor
134   LH rear wheel speed sensor
135   RH front wheel speed sensor
136   RH rear wheel speed sensor
137   ABS hydraulic pump fuse
138   ASC switch

\*   ASC only
\*\*   Not ASC
\*\*\*   Auto only
\*\*\*\*   Avant only
\*\*\*\*\*   W/out heated
    front seats

**Diagram 10**

MTS
H32284

**Automatic climate control continued**

**Airbag system**

**ABS with ASC system**

**Diagram 10: Audi Models - Automatic climate control continued, airbag and ABS**

**VW Passat 1998 and later wiring diagrams**                                      Diagram 11

## Key to symbols

| | |
|---|---|
| Bulb | |
| Switch | |
| Multiple contact switch (ganged) | |
| Fuse/fusible link and current rating | F5 30A |
| Resistor | |
| Variable resistor | |
| Connecting wires | |
| Plug and socket contact | |
| Item no. | 2 |
| Pump/motor | M |
| Ground point and location | E22 |
| Gauge/meter | |
| Diode | |
| Wire splice or soldered joint | |
| Solenoid actuator | |
| Light emitting diode (LED) | |
| Wire colour (brown with black tracer) | Br/Sw |
| Screened cable | |

Dashed outline denotes part of a larger item, containing in this case an electronic or solid state device.
**6** - unspecified connector pin 6.
**T14/9** - 14 pin connector, pin 9.

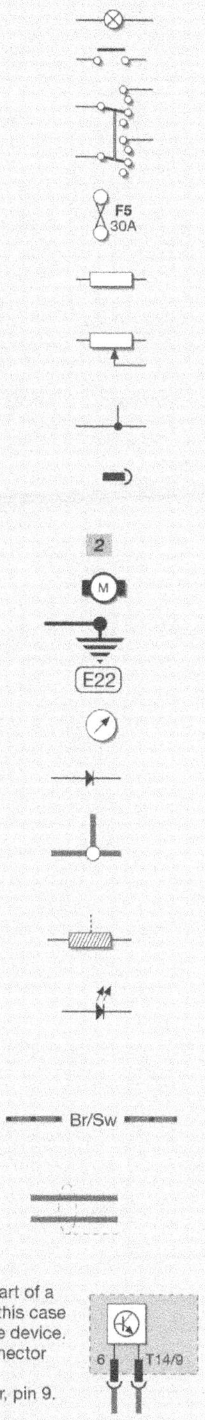

6    T14/9

## Ground locations

| | | | |
|---|---|---|---|
| E1 | Ground strap, battery to body | E16 | On roof bow (wagon only) |
| E2 | Instrument panel ground connection 1 | E17 | In left headlight wiring harness |
| E3 | Instrument panel ground connection 2 | E18 | LH tail light bulb holder |
| E4 | Ground connection in trunk lid/rear left pillar | E19 | Under center console |
| | | E20 | Below rear shelf |
| E5 | Interior light harness ground connection 1 | E21 | In airbag wiring harness |
| E6 | Lower left 'A' pillar | E22 | Ground connection beside fuse/relay panel |
| E7 | Ground connection near steering wheel | E23 | Ground connection on valve cover |
| E8 | In right headlight wiring harness | E24 | Ground connection on hydraulic unit |
| E9 | On rear left pillar | E25 | Engine wiring harness ground connection 1 |
| E10 | Drivers door wiring harness | | |
| E11 | Lower RH 'A' pillar | E26 | Engine wiring harness ground connection 2 |
| E12 | Passengers door wiring harness | | |
| E13 | Rear right pillar | E27 | Rear wiring harness ground connection 4 |
| E14 | Rear wiring harness ground connection 2 | E28 | Ground connection in engine compartment left side |
| E15 | Rear wiring harness ground connection 3 | E29 | Ground connection in engine compartment right side |

## Key to circuits (VW models)

| | |
|---|---|
| Diagram 11 | Information for wiring diagrams |
| Diagram 12 | Wash/wipe, automatic day/night interior mirror and dual tone horn |
| Diagram 13 | Central locking and convenience system - drivers door |
| Diagram 14 | Central locking and convenience system - passenger door |
| Diagram 15 | Central locking and convenience system - rear left and rear right sides |
| Diagram 16 | Central locking and convenience system - central control module |
| Diagram 17 | Interior lights and instrument lights and cigarette lighter |
| Diagram 18 | Headlights, fog lights and back-up lights |
| Diagram 19 | Brake lights, turn signal lights. tail, side marker and license plate light |
| Diagram 20 | Radio, Fresh air blower and rear window defogger |
| Diagram 21 | Typical air conditioning |
| Diagram 22 | Airbag and ABS |
| Diagram 23 | 1.8 Motronic engine AEB includes starting and charging |
| Diagram 24 | 1.8 Motronic engine AEB continued. |
| Diagram 25 | 1.8 Motronic engine ATW includes starting and charging |
| Diagram 26 | 1.8 Motronic engine ATW continued. |
| Diagram 27 | 2.8 Motronic engine AHA includes starting and charging |
| Diagram 28 | 2.8 Motronic engine AHA continued. |
| Diagram 29 | 2.8 Motronic engine ATQ includes starting and charging |
| Diagram 30 | 2.8 Motronic engine ATQ continued. |

## Fuse holder

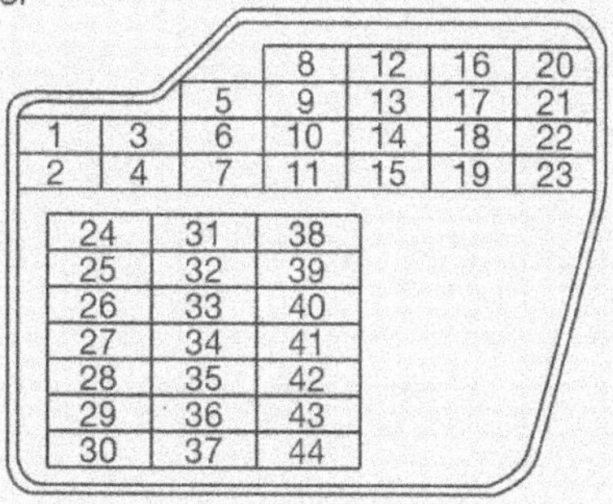

| | | | 8 | 12 | 16 | 20 |
|---|---|---|---|---|---|---|
| | | 5 | 9 | 13 | 17 | 21 |
| 1 | 3 | 6 | 10 | 14 | 18 | 22 |
| 2 | 4 | 7 | 11 | 15 | 19 | 23 |

| | | |
|---|---|---|
| 24 | 31 | 38 |
| 25 | 32 | 39 |
| 26 | 33 | 40 |
| 27 | 34 | 41 |
| 28 | 35 | 42 |
| 29 | 36 | 43 |
| 30 | 37 | 44 |

H32263

**Diagram 11: VW Models - Information for wiring diagrams**

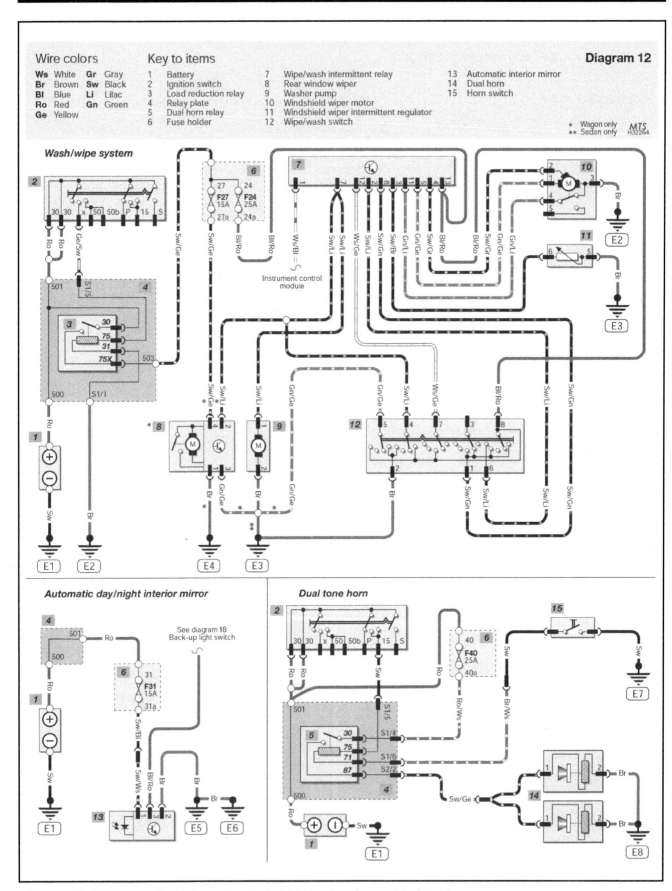

**Diagram 12:** VW Models - Wash/wipe, automatic day/night interior mirror and dual tone horn

**Diagram 13**

## Wire colors

| | | | |
|---|---|---|---|
| **Ws** | White | **Gr** | Gray |
| **Br** | Brown | **Sw** | Black |
| **Bl** | Blue | **Li** | Lilac |
| **Ro** | Red | **Gn** | Green |
| **Ge** | Yellow | | |

## Key to items

| | | | |
|---|---|---|---|
| 1 | Battery | 17 | Power window fuse |
| 2 | Ignition switch | 18 | Mirror adjustment unit |
| 4 | Relay plate | 19 | Power window switch |
| 6 | Fuse holder | 20 | Drivers door control module |
| 16 | Window motor | 21 | Central locking SAFE light |
| | | 22 | Drivers door central locking unit |
| | | 23 | Drivers mirror adjuster and heater |
| | | 24 | Drivers door entry light |

\* Without in door rear
  lid/fuel tank switch only
\*\* With in door rear
  lid/fuel tank switch only

MTS
H32265

*Central locking and convenience system - drivers door*

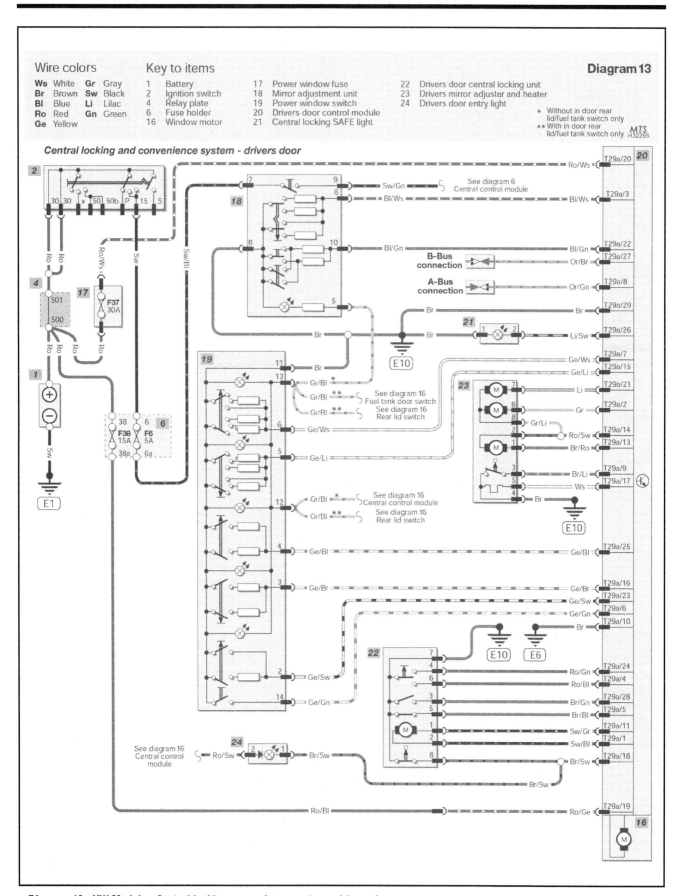

**Diagram 13:  VW Models - Central locking convenience system - drivers door**

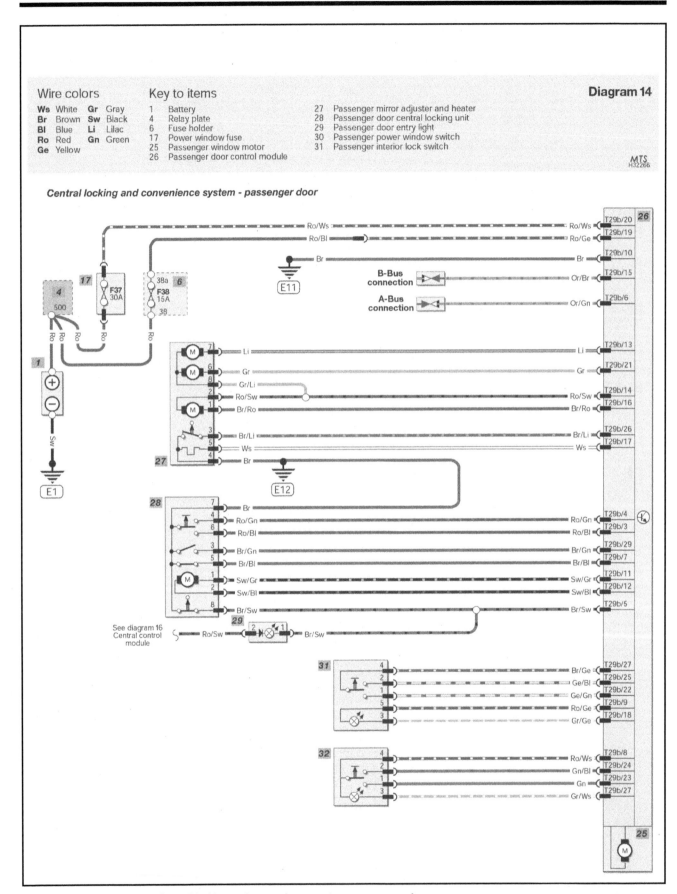

**Wire colors**

| | | | |
|---|---|---|---|
| **Ws** | White | **Gr** | Gray |
| **Br** | Brown | **Sw** | Black |
| **Bl** | Blue | **Li** | Lilac |
| **Ro** | Red | **Gn** | Green |
| **Ge** | Yellow | | |

**Key to items**

1  Battery
4  Relay plate
6  Fuse holder
17  Power window fuse
25  Passenger window motor
26  Passenger door control module

27  Passenger mirror adjuster and heater
28  Passenger door central locking unit
29  Passenger door entry light
30  Passenger power window switch
31  Passenger interior lock switch

**Diagram 14**

MTS
H32266

*Central locking and convenience system - passenger door*

**Diagram 14: VW Models - Central locking and convenience system - passenger door**

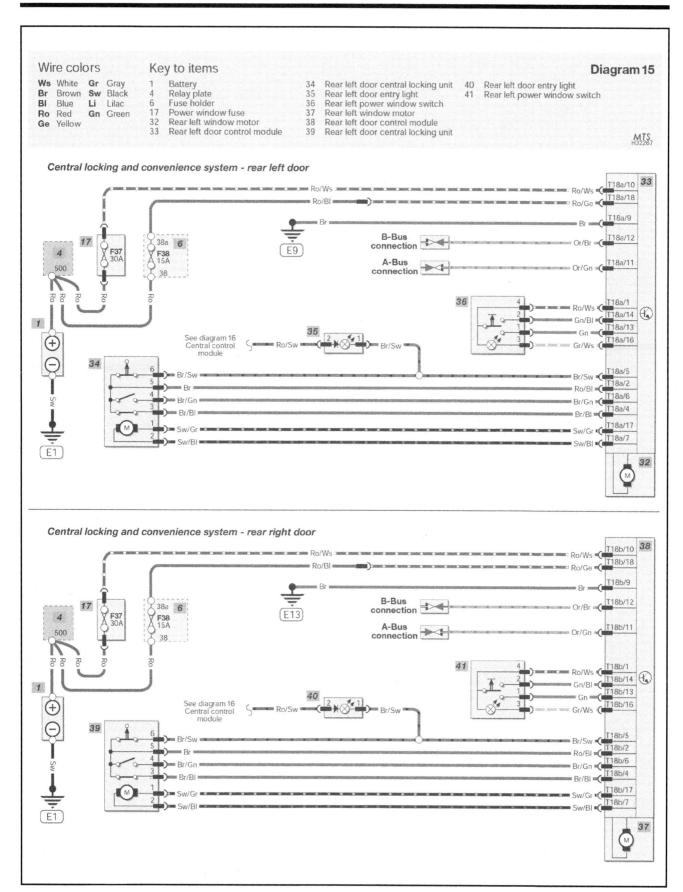

**Wire colors**

| | | | |
|---|---|---|---|
| **Ws** | White | **Gr** | Gray |
| **Br** | Brown | **Sw** | Black |
| **Bl** | Blue | **Li** | Lilac |
| **Ro** | Red | **Gn** | Green |
| **Ge** | Yellow | | |

**Key to items**

| | |
|---|---|
| 1 | Battery |
| 4 | Relay plate |
| 6 | Fuse holder |
| 17 | Power window fuse |
| 32 | Rear left window motor |
| 33 | Rear left door control module |
| 34 | Rear left door central locking unit |
| 35 | Rear left door entry light |
| 36 | Rear left power window switch |
| 37 | Rear left window motor |
| 38 | Rear left door control module |
| 39 | Rear left door central locking unit |
| 40 | Rear left door entry light |
| 41 | Rear left power window switch |

**Diagram 15**

MTS
H32267

*Central locking and convenience system - rear left door*

*Central locking and convenience system - rear right door*

**Diagram 15:  VW Models - Central locking and convenience system - rear left and rear right sides**

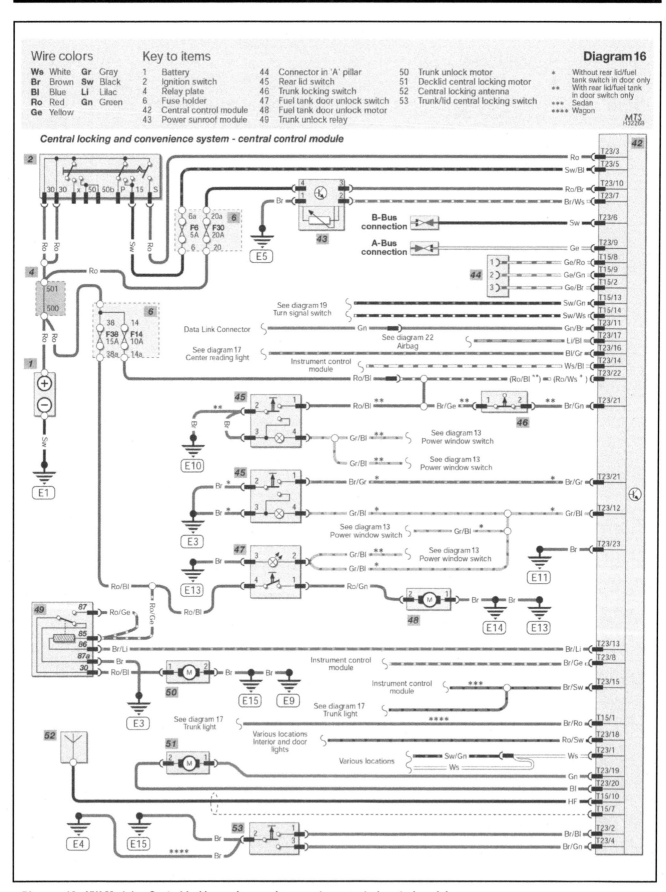

**Diagram 16**

### Wire colors

| | | | |
|---|---|---|---|
| **Ws** | White | **Gr** | Gray |
| **Br** | Brown | **Sw** | Black |
| **Bl** | Blue | **Li** | Lilac |
| **Ro** | Red | **Gn** | Green |
| **Ge** | Yellow | | |

### Key to items

1  Battery
2  Ignition switch
4  Relay plate
6  Fuse holder
42  Central control module
43  Power sunroof module
44  Connector in 'A' pillar
45  Rear lid switch
46  Trunk locking switch
47  Fuel tank door unlock switch
48  Fuel tank door unlock motor
49  Trunk unlock relay
50  Trunk unlock motor
51  Decklid central locking motor
52  Central locking antenna
53  Trunk/lid central locking switch

\*  Without rear lid/fuel tank switch in door only
\*\*  With rear lid/fuel tank switch in door only
\*\*\*  Sedan
\*\*\*\*  Wagon

MTS
H32268

*Central locking and convenience system - central control module*

**Diagram 16:  VW Models - Central locking and convenience system - central control module**

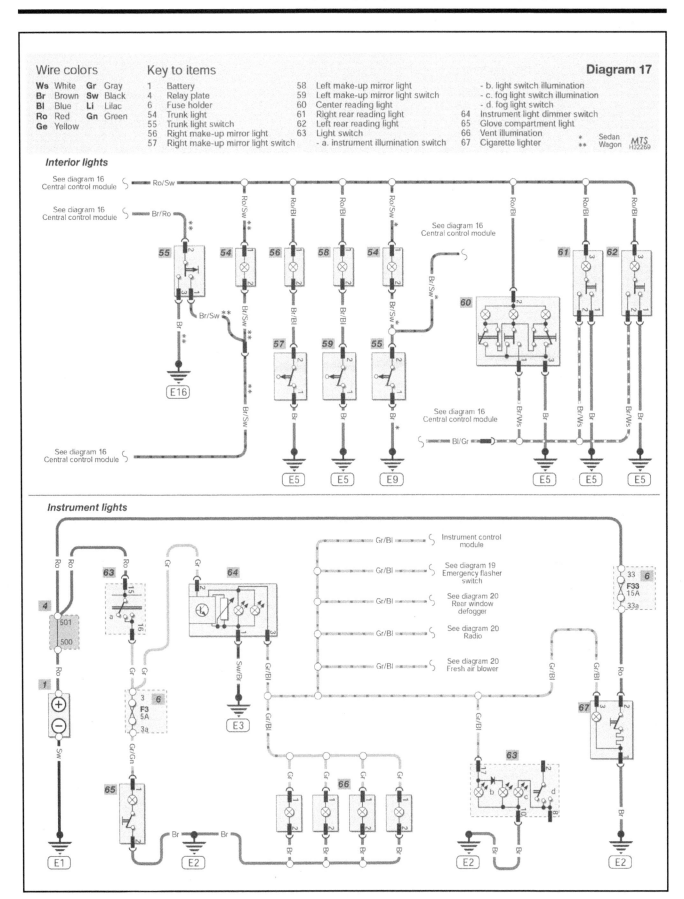

**Diagram 17: VW Models - Interior lights, instrument lights and cigarette lighter**

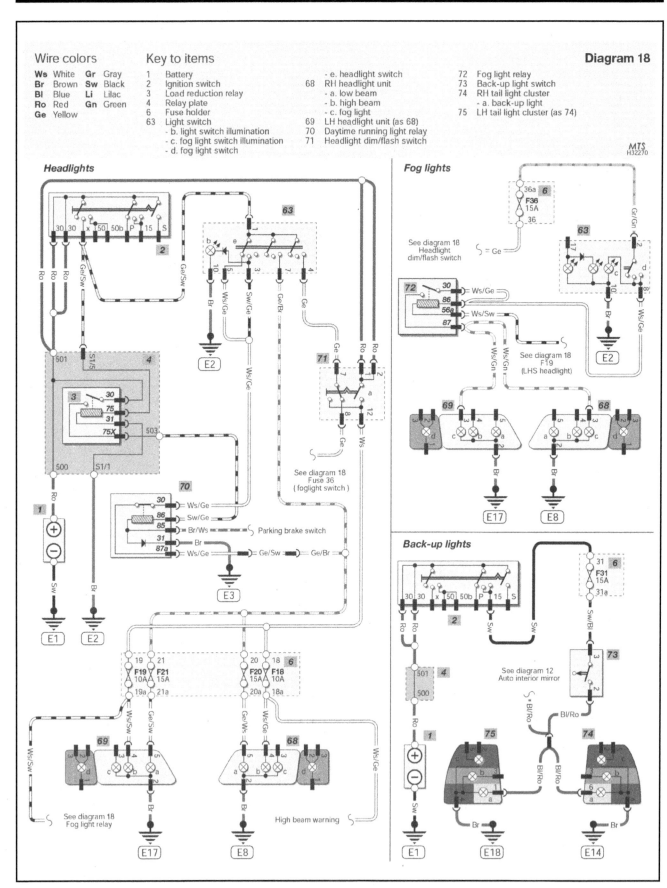

**Diagram 18: VW Models - Headlights, fog lights and back-up lights**

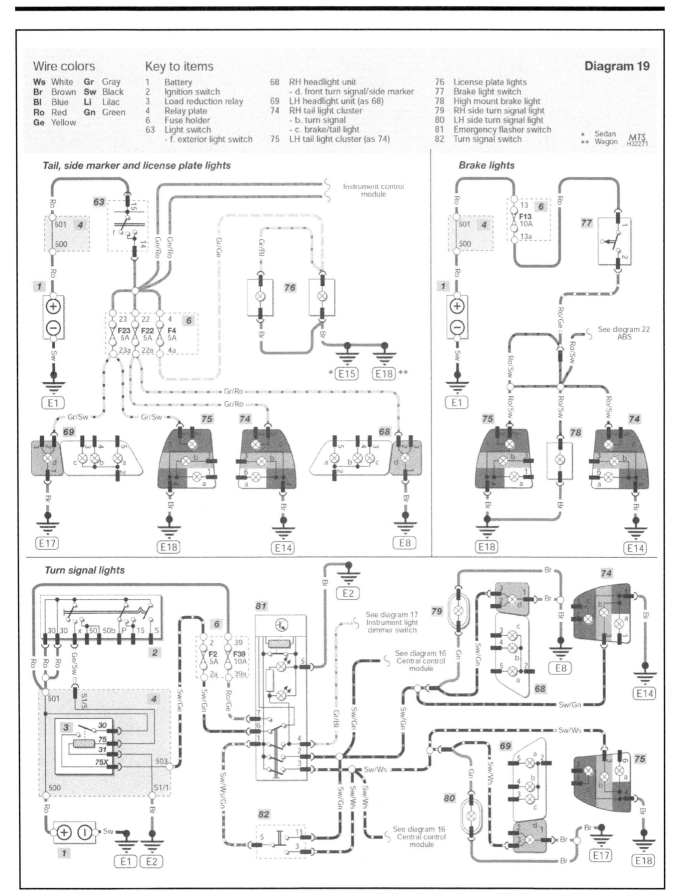

**Diagram 19: VW Models - Brake lights, turn signal lights, tail, side marker and license plate light**

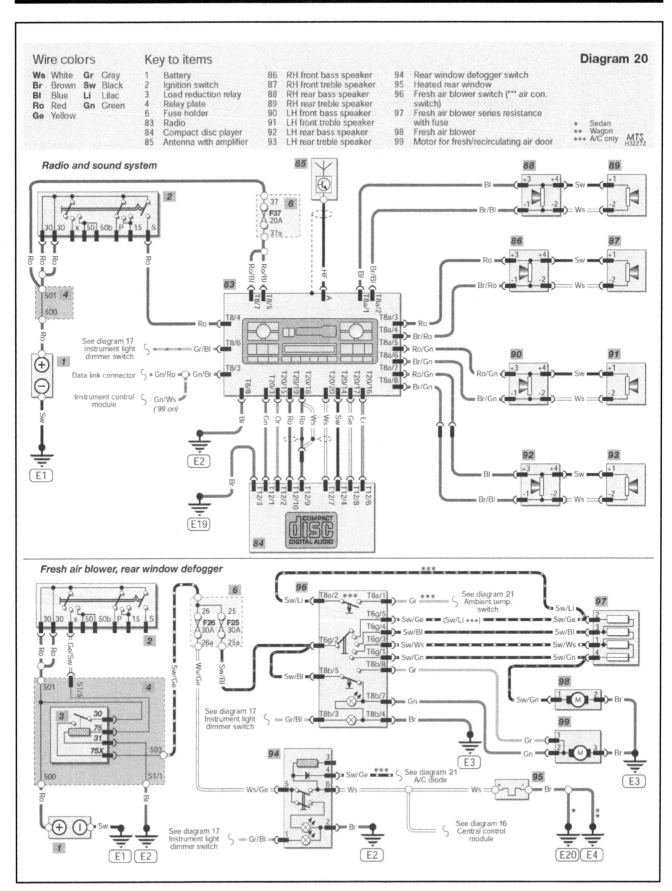

**Diagram 20: VW Models - Radio, fresh air blower and rear window defogger**

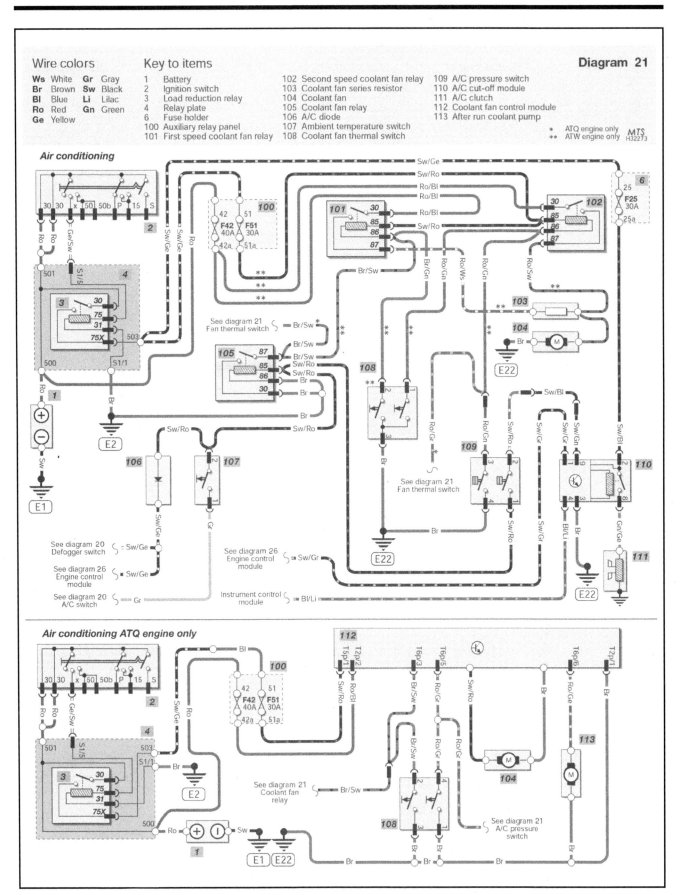

**Diagram 21: VW Models - Typical air conditioning**

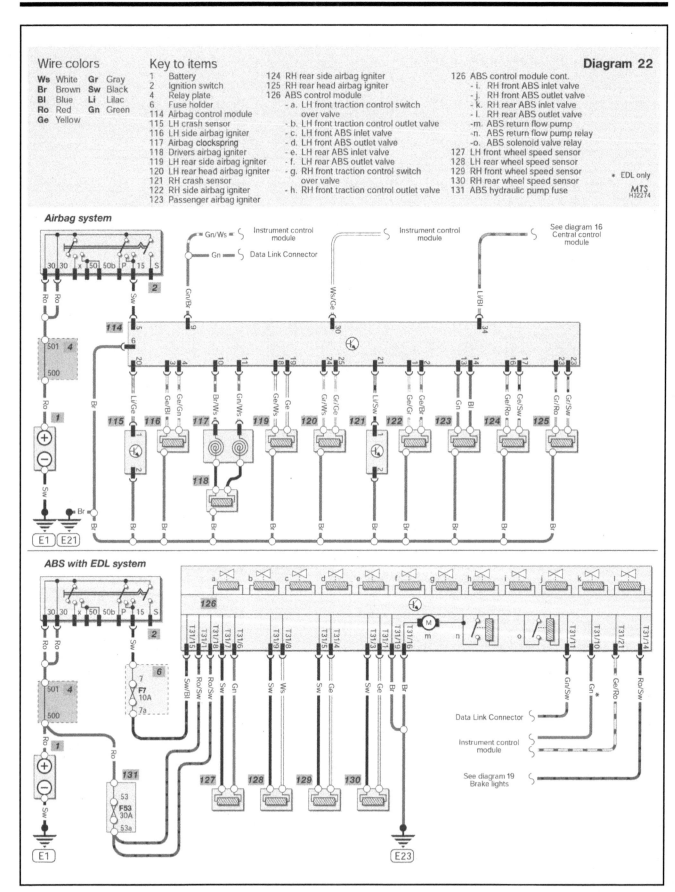

**Diagram 22**

**Wire colors**

| | | | |
|---|---|---|---|
| **Ws** | White | **Gr** | Gray |
| **Br** | Brown | **Sw** | Black |
| **Bl** | Blue | **Li** | Lilac |
| **Ro** | Red | **Gn** | Green |
| **Ge** | Yellow | | |

**Key to items**

1 Battery
2 Ignition switch
4 Relay plate
6 Fuse holder
114 Airbag control module
115 LH crash sensor
116 LH side airbag igniter
117 Airbag clockspring
118 Drivers airbag igniter
119 LH rear side airbag igniter
120 LH rear head airbag igniter
121 RH crash sensor
122 RH side airbag igniter
123 Passenger airbag igniter

124 RH rear side airbag igniter
125 RH rear head airbag igniter
126 ABS control module
 - a. LH front traction control switch over valve
 - b. LH front traction control outlet valve
 - c. LH front ABS inlet valve
 - d. LH front ABS outlet valve
 - e. LH rear ABS inlet valve
 - f. LH rear ABS outlet valve
 - g. RH front traction control switch over valve
 - h. RH front traction control outlet valve

126 ABS control module cont.
 - i. RH front ABS inlet valve
 - j. RH front ABS outlet valve
 - k. RH rear ABS inlet valve
 - l. RH rear ABS outlet valve
 -m. ABS return flow pump
 -n. ABS return flow pump relay
 -o. ABS solenoid valve relay
127 LH front wheel speed sensor
128 LH rear wheel speed sensor
129 RH front wheel speed sensor
130 RH rear wheel speed sensor
131 ABS hydraulic pump fuse

* EDL only

MTS
H32274

**Airbag system**

**ABS with EDL system**

**Diagram 22: VW Models - Airbag and ABS**

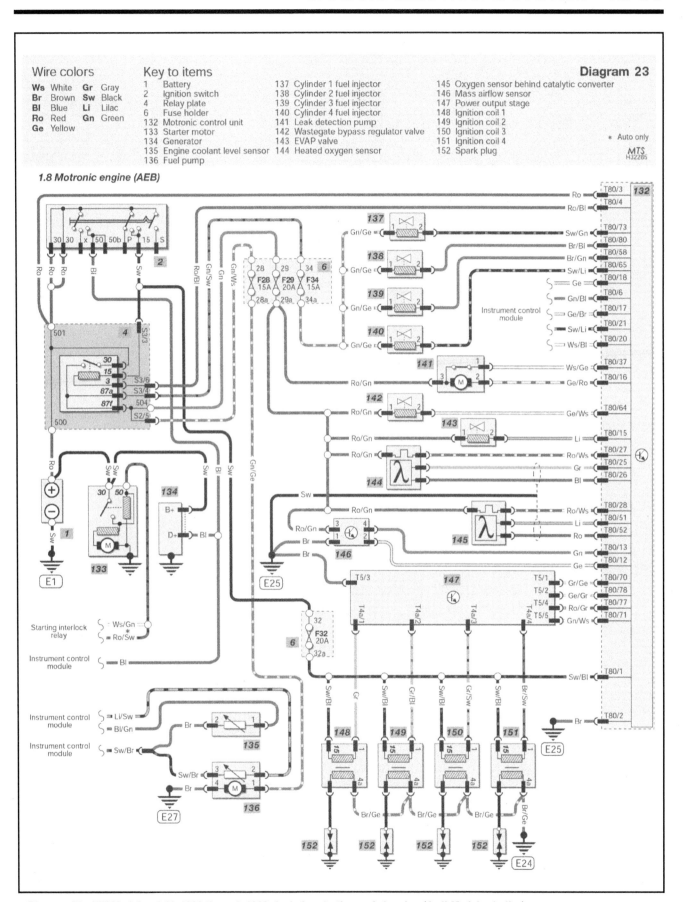

**Wire colors**

| | | | |
|---|---|---|---|
| **Ws** | White | **Gr** | Gray |
| **Br** | Brown | **Sw** | Black |
| **Bl** | Blue | **Li** | Lilac |
| **Ro** | Red | **Gn** | Green |
| **Ge** | Yellow | | |

**Key to items**

1 Battery
2 Ignition switch
4 Relay plate
6 Fuse holder
132 Motronic control unit
133 Starter motor
134 Generator
135 Engine coolant level sensor
136 Fuel pump

137 Cylinder 1 fuel injector
138 Cylinder 2 fuel injector
139 Cylinder 3 fuel injector
140 Cylinder 4 fuel injector
141 Leak detection pump
142 Wastegate bypass regulator valve
143 EVAP valve
144 Heated oxygen sensor

145 Oxygen sensor behind catalytic converter
146 Mass airflow sensor
147 Power output stage
148 Ignition coil 1
149 Ignition coil 2
150 Ignition coil 3
151 Ignition coil 4
152 Spark plug

**Diagram 23**

* Auto only

MTS
H32285

*1.8 Motronic engine (AEB)*

**Diagram 23:  VW Models - 1.8L 1996 through 1999, includes starting and charging (Audi Models similar)**

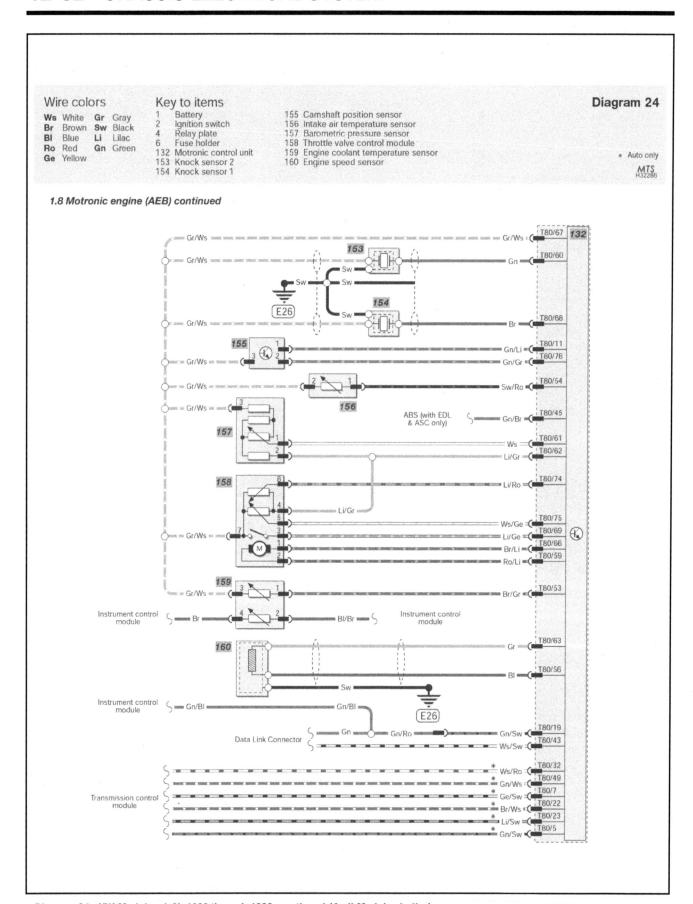

Wire colors

| | | | |
|---|---|---|---|
| **Ws** | White | **Gr** | Gray |
| **Br** | Brown | **Sw** | Black |
| **Bl** | Blue | **Li** | Lilac |
| **Ro** | Red | **Gn** | Green |
| **Ge** | Yellow | | |

Key to items

1 Battery
2 Ignition switch
4 Relay plate
6 Fuse holder
132 Motronic control unit
153 Knock sensor 2
154 Knock sensor 1
155 Camshaft position sensor
156 Intake air temperature sensor
157 Barometric pressure sensor
158 Throttle valve control module
159 Engine coolant temperature sensor
160 Engine speed sensor

**Diagram 24**

\* Auto only

MTS
H32286

*1.8 Motronic engine (AEB) continued*

**Diagram 24:  VW Models - 1.8L 1996 through 1999, continued (Audi Models similar)**

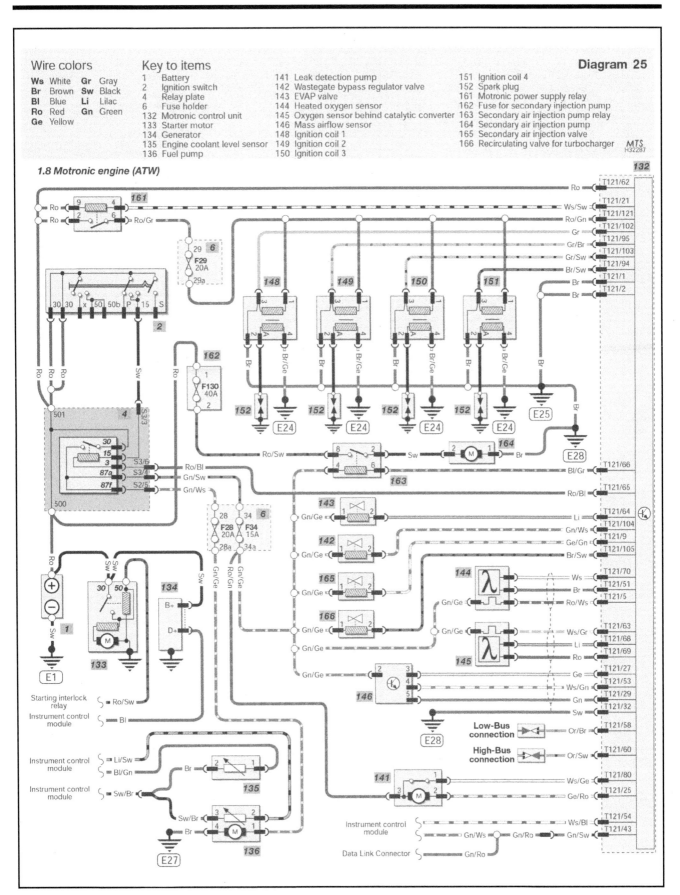

## Wire colors

**Ws** White **Gr** Gray
**Br** Brown **Sw** Black
**Bl** Blue **Li** Lilac
**Ro** Red **Gn** Green
**Ge** Yellow

## Key to items

1 Battery
2 Ignition switch
4 Relay plate
6 Fuse holder
132 Motronic control unit
133 Starter motor
134 Generator
135 Engine coolant level sensor
136 Fuel pump
141 Leak detection pump
142 Wastegate bypass regulator valve
143 EVAP valve
144 Heated oxygen sensor
145 Oxygen sensor behind catalytic converter
146 Mass airflow sensor
148 Ignition coil 1
149 Ignition coil 2
150 Ignition coil 3
151 Ignition coil 4
152 Spark plug
161 Motronic power supply relay
162 Fuse for secondary injection pump
163 Secondary air injection pump relay
164 Secondary air injection pump
165 Secondary air injection valve
166 Recirculating valve for turbocharger

**Diagram 25**

MTS
H32287

*1.8 Motronic engine (ATW)*

Diagram 25:  VW Models - 1.8L 2000 and later, includes starting and charging (Audi Models similar)

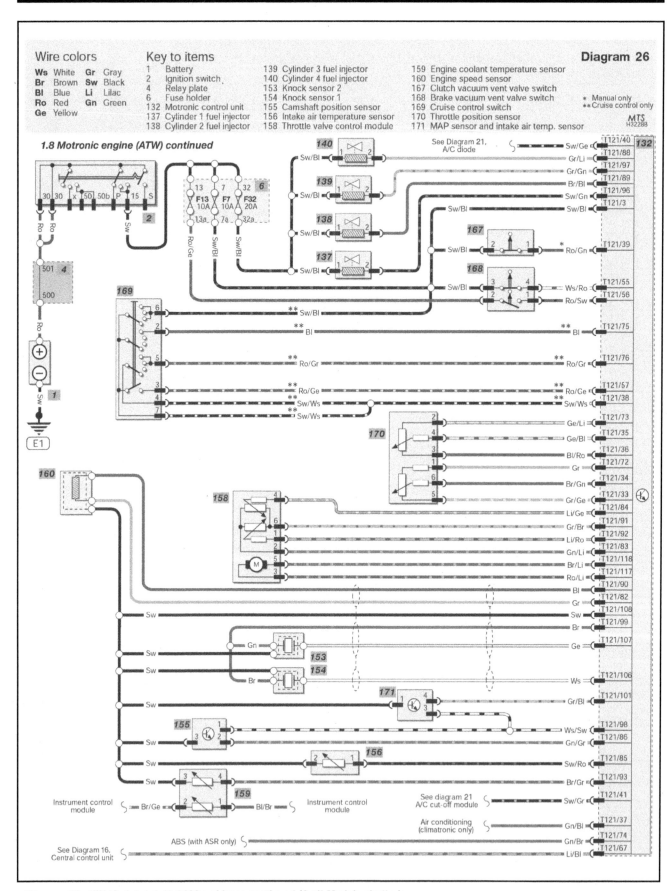

**Diagram 26: VW Models - 1.8L 2000 and later, continued (Audi Models similar)**

## Wire colors

**Ws** White  **Gr** Gray
**Br** Brown  **Sw** Black
**Bl** Blue  **Li** Lilac
**Ro** Red  **Gn** Green
**Ge** Yellow

## Key to items

1   Battery
2   Ignition switch
4   Relay plate
6   Fuse holder
132  Motronic control unit
133  Starter motor
134  Generator
135  Engine coolant level sensor
136  Fuel pump
137  Cylinder 1 fuel injector
138  Cylinder 2 fuel injector
139  Cylinder 3 fuel injector
140  Cylinder 4 fuel injector
141  Leak detection pump
143  EVAP valve
144  Heated oxygen sensor 1
145  Oxygen sensor behind
     catalytic converter 1
162  Fuse for secondary air injection pump
163  Secondary air injection pump relay
164  Secondary air injection pump
165  Secondary air injection valve
172  Cylinder 5 fuel injector
173  Cylinder 6 fuel injector
174  Intake manifold change over valve
175  Valve for camshaft adjustment 1
176  Valve for camshaft adjustment 2
177  Heated oxygen sensor 2
178  Oxygen sensor behind
     catalytic converter 2

**Diagram 27**

\* Auto only

MTS
H32289

*2.8 Motronic engine (AHA)*

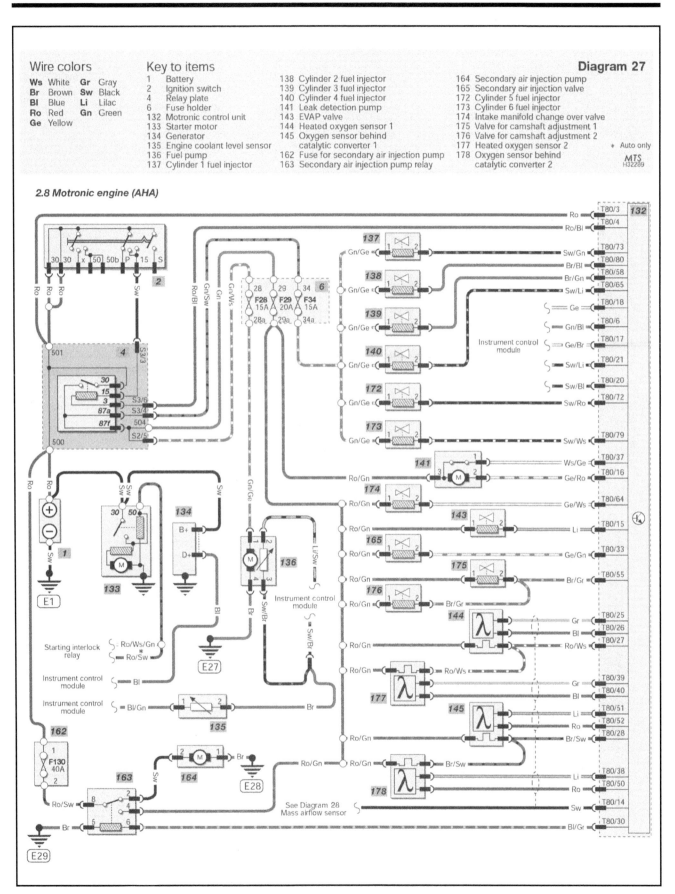

**Diagram 27:  VW Models - 2.8L 1998 and 1999, includes starting and charging (Audi Models similar)**

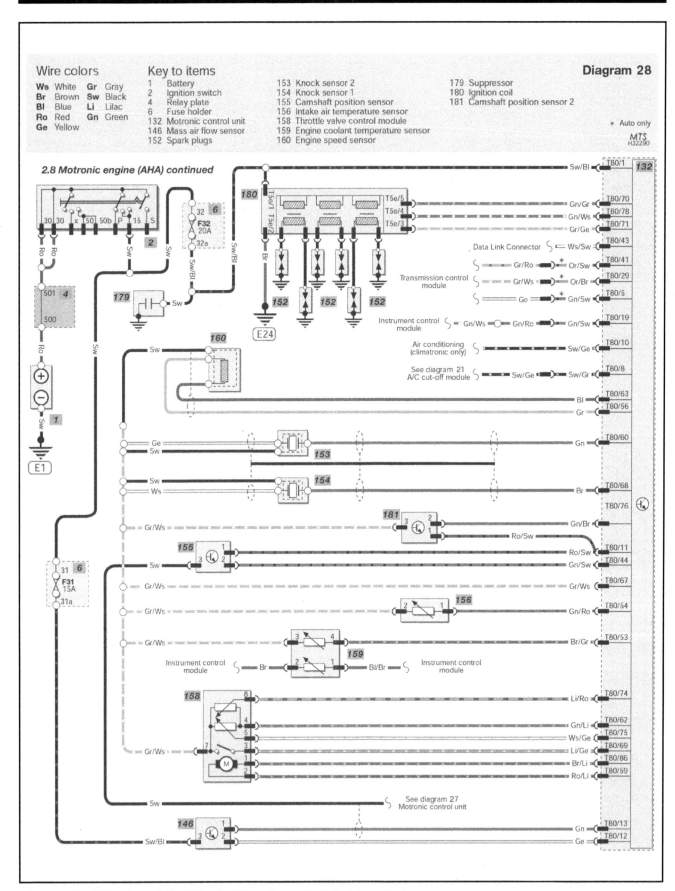

**Wire colors**

| | | | |
|---|---|---|---|
| **Ws** | White | **Gr** | Gray |
| **Br** | Brown | **Sw** | Black |
| **Bl** | Blue | **Li** | Lilac |
| **Ro** | Red | **Gn** | Green |
| **Ge** | Yellow | | |

**Key to items**

1 Battery
2 Ignition switch
4 Relay plate
6 Fuse holder
132 Motronic control unit
146 Mass air flow sensor
152 Spark plugs
153 Knock sensor 2
154 Knock sensor 1
155 Camshaft position sensor
156 Intake air temperature sensor
158 Throttle valve control module
159 Engine coolant temperature sensor
160 Engine speed sensor
179 Suppressor
180 Ignition coil
181 Camshaft position sensor 2

\* Auto only

**Diagram 28**

MTS
H32290

**2.8 Motronic engine (AHA) continued**

**Diagram 28: VW Models - 2.8L 1998 and 1999, continued (Audi Models similar)**

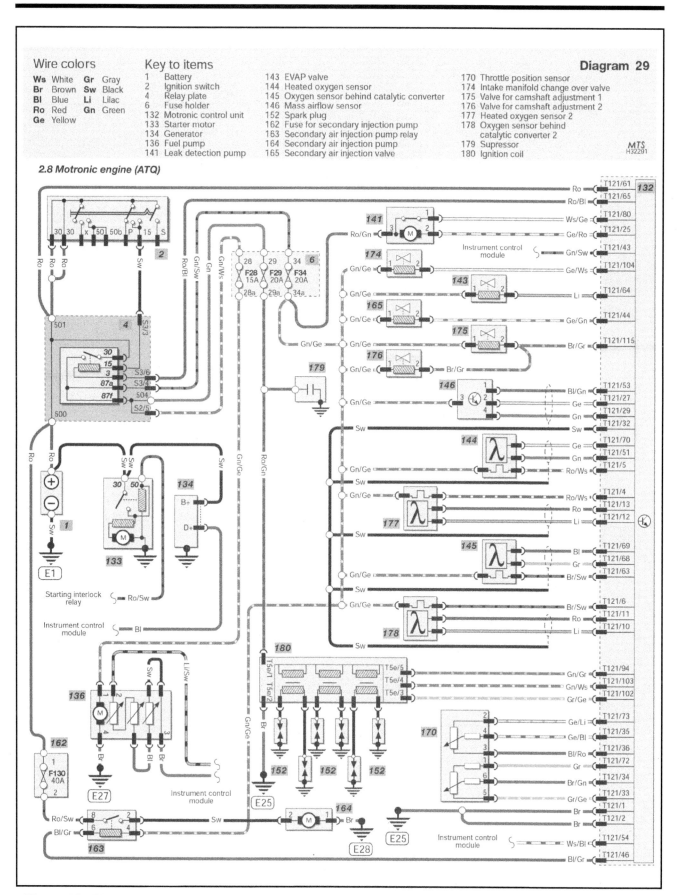

## Wire colors

**Ws** White    **Gr** Gray
**Br** Brown    **Sw** Black
**Bl** Blue     **Li** Lilac
**Ro** Red     **Gn** Green
**Ge** Yellow

## Key to items

1   Battery
2   Ignition switch
4   Relay plate
6   Fuse holder
132   Motronic control unit
133   Starter motor
134   Generator
136   Fuel pump
141   Leak detection pump
143   EVAP valve
144   Heated oxygen sensor
145   Oxygen sensor behind catalytic converter
146   Mass airflow sensor
152   Spark plug
162   Fuse for secondary injection pump
163   Secondary air injection pump relay
164   Secondary air injection pump
165   Secondary air injection valve
170   Throttle position sensor
174   Intake manifold change over valve
175   Valve for camshaft adjustment 1
176   Valve for camshaft adjustment 2
177   Heated oxygen sensor 2
178   Oxygen sensor behind catalytic converter 2
179   Supressor
180   Ignition coil

**Diagram 29**

MTS
H32291

### 2.8 Motronic engine (ATQ)

**Diagram 29:** VW Models - 2.8L 2000 and later, includes starting and charging (Audi Models similar)

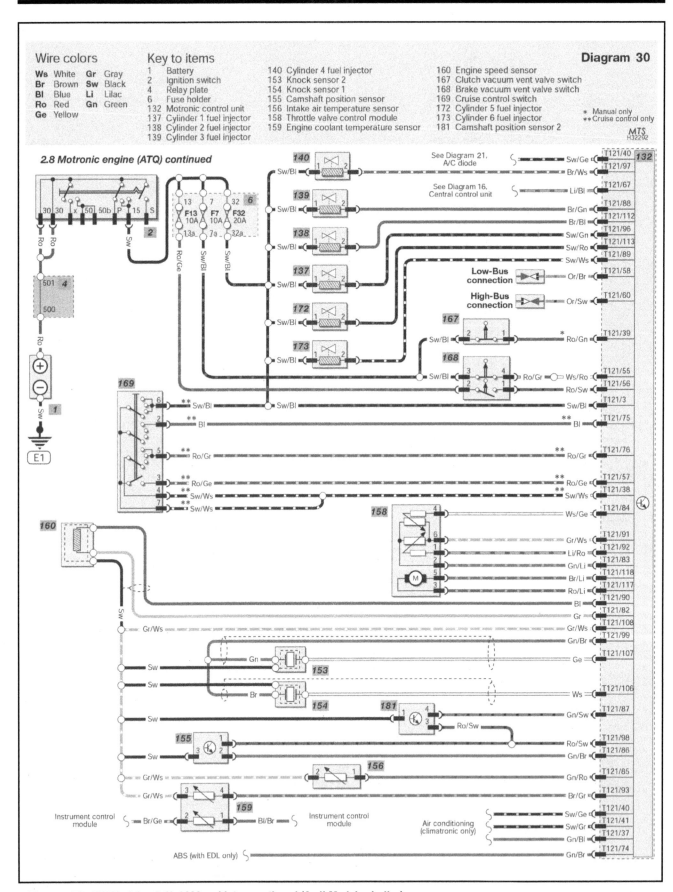

**Diagram 30**

### Wire colors

| | | | |
|---|---|---|---|
| **Ws** | White | **Gr** | Gray |
| **Br** | Brown | **Sw** | Black |
| **Bl** | Blue | **Li** | Lilac |
| **Ro** | Red | **Gn** | Green |
| **Ge** | Yellow | | |

### Key to items

1 Battery
2 Ignition switch
4 Relay plate
6 Fuse holder
132 Motronic control unit
137 Cylinder 1 fuel injector
138 Cylinder 2 fuel injector
139 Cylinder 3 fuel injector

140 Cylinder 4 fuel injector
153 Knock sensor 2
154 Knock sensor 1
155 Camshaft position sensor
156 Intake air temperature sensor
158 Throttle valve control module
159 Engine coolant temperature sensor

160 Engine speed sensor
167 Clutch vacuum vent valve switch
168 Brake vacuum vent valve switch
169 Cruise control switch
172 Cylinder 5 fuel injector
173 Cylinder 6 fuel injector
181 Camshaft position sensor 2

\* Manual only
\*\* Cruise control only

2.8 Motronic engine (ATQ) continued

**Diagram 30: VW Models - 2.8L 2000 and later, continued (Audi Models similar)**

## GLOSSARY

**AIR/FUEL RATIO:** The ratio of air-to-gasoline by weight in the fuel mixture drawn into the engine.

**AIR INJECTION:** One method of reducing harmful exhaust emissions by injecting air into each of the exhaust ports of an engine. The fresh air entering the hot exhaust manifold causes any remaining fuel to be burned before it can exit the tailpipe.

**ALTERNATOR:** A device used for converting mechanical energy into electrical energy.

**AMMETER:** An instrument, calibrated in amperes, used to measure the flow of an electrical current in a circuit. Ammeters are always connected in series with the circuit being tested.

**AMPERE:** The rate of flow of electrical current present when one volt of electrical pressure is applied against one ohm of electrical resistance.

**ANALOG COMPUTER:** Any microprocessor that uses similar (analogous) electrical signals to make its calculations.

**ARMATURE:** A laminated, soft iron core wrapped by a wire that converts electrical energy to mechanical energy as in a motor or relay. When rotated in a magnetic field, it changes mechanical energy into electrical energy as in a generator.

**ATMOSPHERIC PRESSURE:** The pressure on the Earth's surface caused by the weight of the air in the atmosphere. At sea level, this pressure is 14.7 psi at 32°F (101 kPa at 0°C).

**ATOMIZATION:** The breaking down of a liquid into a fine mist that can be suspended in air.

**AXIAL PLAY:** Movement parallel to a shaft or bearing bore.

**BACKFIRE:** The sudden combustion of gases in the intake or exhaust system that results in a loud explosion.

**BACKLASH:** The clearance or play between two parts, such as meshed gears.

**BACKPRESSURE:** Restrictions in the exhaust system that slow the exit of exhaust gases from the combustion chamber.

**BAKELITE:** A heat resistant, plastic insulator material commonly used in printed circuit boards and transistorized components.

**BALL BEARING:** A bearing made up of hardened inner and outer races between which hardened steel balls roll.

**BALLAST RESISTOR:** A resistor in the primary ignition circuit that lowers voltage after the engine is started to reduce wear on ignition components.

**BEARING:** A friction reducing, supportive device usually located between a stationary part and a moving part.

**BIMETAL TEMPERATURE SENSOR:** Any sensor or switch made of two dissimilar types of metal that bend when heated or cooled due to the different expansion rates of the alloys. These types of sensors usually function as an on/off switch.

**BLOWBY:** Combustion gases, composed of water vapor and unburned fuel, that leak past the piston rings into the crankcase during normal engine operation. These gases are removed by the PCV system to prevent the buildup of harmful acids in the crankcase.

**BRAKE PAD:** A brake shoe and lining assembly used with disc brakes.

**BRAKE SHOE:** The backing for the brake lining. The term is, however, usually applied to the assembly of the brake backing and lining.

**BUSHING:** A liner, usually removable, for a bearing; an anti-friction liner used in place of a bearing.

**CALIPER:** A hydraulically activated device in a disc brake system, which is mounted straddling the brake rotor (disc). The caliper contains at least one piston and two brake pads. Hydraulic pressure on the piston(s) forces the pads against the rotor.

**CAMSHAFT:** A shaft in the engine on which are the lobes (cams) which operate the valves. The camshaft is driven by the crankshaft, via a belt, chain or gears, at one half the crankshaft speed.

**CAPACITOR:** A device which stores an electrical charge.

**CARBON MONOXIDE (CO):** A colorless, odorless gas given off as a normal byproduct of combustion. It is poisonous and extremely dangerous in confined areas, building up slowly to toxic levels without warning if adequate ventilation is not available.

**CARBURETOR:** A device, usually mounted on the intake manifold of an engine, which mixes the air and fuel in the proper proportion to allow even combustion.

**CATALYTIC CONVERTER:** A device installed in the exhaust system, like a muffler, that converts harmful byproducts of combustion into carbon dioxide and water vapor by means of a heat-producing chemical reaction.

**CENTRIFUGAL ADVANCE:** A mechanical method of advancing the spark timing by using flyweights in the distributor that react to centrifugal force generated by the distributor shaft rotation.

**CHECK VALVE:** Any one-way valve installed to permit the flow of air, fuel or vacuum in one direction only.

**CHOKE:** A device, usually a moveable valve, placed in the intake path of a carburetor to restrict the flow of air.

**CIRCUIT:** Any unbroken path through which an electrical current can flow. Also used to describe fuel flow in some instances.

**CIRCUIT BREAKER:** A switch which protects an electrical circuit from overload by opening the circuit when the current flow exceeds a predetermined level. Some circuit breakers must be reset manually, while most reset automatically.

**COIL (IGNITION):** A transformer in the ignition circuit which steps up the voltage provided to the spark plugs.

**COMBINATION MANIFOLD:** An assembly which includes both the intake and exhaust manifolds in one casting.

**COMBINATION VALVE:** A device used in some fuel systems that routes fuel vapors to a charcoal storage canister instead of venting them into the atmosphere. The valve relieves fuel tank pressure and allows fresh air into the tank as the fuel level drops to prevent a vapor lock situation.

**COMPRESSION RATIO:** The comparison of the total volume of the cylinder and combustion chamber with the piston at BDC and the piston at TDC.

**CONDENSER:** 1. An electrical device which acts to store an electrical charge, preventing voltage surges. 2. A radiator-like device in the air conditioning system in which refrigerant gas condenses into a liquid, giving off heat.

**CONDUCTOR:** Any material through which an electrical current can be transmitted easily.

**CONTINUITY:** Continuous or complete circuit. Can be checked with an ohmmeter.

**COUNTERSHAFT:** An intermediate shaft which is rotated by a mainshaft and transmits, in turn, that rotation to a working part.

**CRANKCASE:** The lower part of an engine in which the crankshaft and related parts operate.

**CRANKSHAFT:** The main driving shaft of an engine which receives reciprocating motion from the pistons and converts it to rotary motion.

**CYLINDER:** In an engine, the round hole in the engine block in which the piston(s) ride.

**CYLINDER BLOCK:** The main structural member of an engine in which is found the cylinders, crankshaft and other principal parts.

**CYLINDER HEAD:** The detachable portion of the engine, usually fastened to the top of the cylinder block and containing all or most of the combustion chambers. On overhead valve engines, it contains the valves and their operating parts. On overhead cam engines, it contains the camshaft as well.

**DEAD CENTER:** The extreme top or bottom of the piston stroke.

**DETONATION:** An unwanted explosion of the air/fuel mixture in the combustion chamber caused by excess heat and compression, advanced timing, or an overly lean mixture. Also referred to as "ping".

**DIAPHRAGM:** A thin, flexible wall separating two cavities, such as in a vacuum advance unit.

**DIESELING:** A condition in which hot spots in the combustion chamber cause the engine to run on after the key is turned off.

**DIFFERENTIAL:** A geared assembly which allows the transmission of motion between drive axles, giving one axle the ability to turn faster than the other.

**DIODE:** An electrical device that will allow current to flow in one direction only.

**DISC BRAKE:** A hydraulic braking assembly consisting of a brake disc, or rotor, mounted on an axle, and a caliper assembly containing, usually two brake pads which are activated by hydraulic pressure. The pads are forced against the sides of the disc, creating friction which slows the vehicle.

**DISTRIBUTOR:** A mechanically driven device on an engine which is responsible for electrically firing the spark plug at a predetermined point of the piston stroke.

**DOWEL PIN:** A pin, inserted in mating holes in two different parts allowing those parts to maintain a fixed relationship.

**DRUM BRAKE:** A braking system which consists of two brake shoes and one or two wheel cylinders, mounted on a fixed backing plate, and a brake drum, mounted on an axle, which revolves around the assembly.

**DWELL:** The rate, measured in degrees of shaft rotation, at which an electrical circuit cycles on and off.

**ELECTRONIC CONTROL UNIT (ECU):** Ignition module, module, amplifier or igniter. See Module for definition.

**ELECTRONIC IGNITION:** A system in which the timing and firing of the spark plugs is controlled by an electronic control unit, usually called a module. These systems have no points or condenser.

**END-PLAY:** The measured amount of axial movement in a shaft.

**ENGINE:** A device that converts heat into mechanical energy.

**EXHAUST MANIFOLD:** A set of cast passages or pipes which conduct exhaust gases from the engine.

**FEELER GAUGE:** A blade, usually metal, or precisely predetermined thickness, used to measure the clearance between two parts.

**FIRING ORDER:** The order in which combustion occurs in the cylinders of an engine. Also the order in which spark is distributed to the plugs by the distributor.

**FLOODING:** The presence of too much fuel in the intake manifold and combustion chamber which prevents the air/fuel mixture from firing, thereby causing a no-start situation.

**FLYWHEEL:** A disc shaped part bolted to the rear end of the crankshaft. Around the outer perimeter is affixed the ring gear. The starter drive engages the ring gear, turning the flywheel, which rotates the crankshaft, imparting the initial starting motion to the engine.

**FOOT POUND (ft. lbs. or sometimes, ft.lb.):** The amount of energy or work needed to raise an item weighing one pound, a distance of one foot.

**FUSE:** A protective device in a circuit which prevents circuit overload by breaking the circuit when a specific amperage is present. The device is constructed around a strip or wire of a lower amperage rating than the circuit it is designed to protect. When an amperage higher than that stamped on the fuse is present in the circuit, the strip or wire melts, opening the circuit.

**GEAR RATIO:** The ratio between the number of teeth on meshing gears.

**GENERATOR:** A device which converts mechanical energy into electrical energy.

**HEAT RANGE:** The measure of a spark plug's ability to dissipate heat from its firing end. The higher the heat range, the hotter the plug fires.

**HUB:** The center part of a wheel or gear.

**HYDROCARBON (HC):** Any chemical compound made up of hydrogen and carbon. A major pollutant formed by the engine as a byproduct of combustion.

**HYDROMETER:** An instrument used to measure the specific gravity of a solution.

**INCH POUND (inch lbs.; sometimes in.lb. or in. lbs.):** One twelfth of a foot pound.

**INDUCTION:** A means of transferring electrical energy in the form of a magnetic field. Principle used in the ignition coil to increase voltage.

**INJECTOR:** A device which receives metered fuel under relatively low pressure and is activated to inject the fuel into the engine under relatively high pressure at a predetermined time.

**INPUT SHAFT:** The shaft to which torque is applied, usually carrying the driving gear or gears.

**INTAKE MANIFOLD:** A casting of passages or pipes used to conduct air or a fuel/air mixture to the cylinders.

**JOURNAL:** The bearing surface within which a shaft operates.

**KEY:** A small block usually fitted in a notch between a shaft and a hub to prevent slippage of the two parts.

**MANIFOLD:** A casting of passages or set of pipes which connect the cylinders to an inlet or outlet source.

**MANIFOLD VACUUM:** Low pressure in an engine intake manifold formed just below the throttle plates. Manifold vacuum is highest at idle and drops under acceleration.

**MASTER CYLINDER:** The primary fluid pressurizing device in a hydraulic system. In automotive use, it is found in brake and hydraulic clutch systems and is pedal activated, either directly or, in a power brake system, through the power booster.

**MODULE:** Electronic control unit, amplifier or igniter of solid state or integrated design which controls the current flow in the ignition primary circuit based on input from the pick-up coil. When the module opens the primary circuit, high secondary voltage is induced in the coil.

**NEEDLE BEARING:** A bearing which consists of a number (usually a large number) of long, thin rollers.

**OHM:** ($\Omega$) The unit used to measure the resistance of conductor-to-electrical flow. One ohm is the amount of resistance that limits current flow to one ampere in a circuit with one volt of pressure.

**OHMMETER:** An instrument used for measuring the resistance, in ohms, in an electrical circuit.

**OUTPUT SHAFT:** The shaft which transmits torque from a device, such as a transmission.

**OVERDRIVE:** A gear assembly which produces more shaft revolutions than that transmitted to it.

**OVERHEAD CAMSHAFT (OHC):** An engine configuration in which the camshaft is mounted on top of the cylinder head and operates the valve either directly or by means of rocker arms.

**OVERHEAD VALVE (OHV):** An engine configuration in which all of the valves are located in the cylinder head and the camshaft is located in the cylinder block. The camshaft operates the valves via lifters and pushrods.

**OXIDES OF NITROGEN (NOx):** Chemical compounds of nitrogen produced as a byproduct of combustion. They combine with hydrocarbons to produce smog.

**OXYGEN SENSOR:** Use with the feedback system to sense the presence of oxygen in the exhaust gas and signal the computer which can reference the voltage signal to an air/fuel ratio.

**PINION:** The smaller of two meshing gears.

**PISTON RING:** An open-ended ring with fits into a groove on the outer diameter of the piston. Its chief function is to form a seal between the piston and cylinder wall. Most automotive pistons have three rings: two for compression sealing; one for oil sealing.

**PRELOAD:** A predetermined load placed on a bearing during assembly or by adjustment.

**PRIMARY CIRCUIT:** the low voltage side of the ignition system which consists of the ignition switch, ballast resistor or resistance wire, bypass, coil, electronic control unit and pick-up coil as well as the connecting wires and harnesses.

**PRESS FIT:** The mating of two parts under pressure, due to the inner diameter of one being smaller than the outer diameter of the other, or vice versa; an interference fit.

**RACE:** The surface on the inner or outer ring of a bearing on which the balls, needles or rollers move.

**REGULATOR:** A device which maintains the amperage and/or voltage levels of a circuit at predetermined values.

**RELAY:** A switch which automatically opens and/or closes a circuit.

**RESISTANCE:** The opposition to the flow of current through a circuit or electrical device, and is measured in ohms. Resistance is equal to the voltage divided by the amperage.

**RESISTOR:** A device, usually made of wire, which offers a preset amount of resistance in an electrical circuit.

**RING GEAR:** The name given to a ring-shaped gear attached to a differential case, or affixed to a flywheel or as part of a planetary gear set.

**ROLLER BEARING:** A bearing made up of hardened inner and outer races between which hardened steel rollers move.

**ROTOR:** 1. The disc-shaped part of a disc brake assembly, upon which the brake pads bear; also called, brake disc. 2. The device mounted atop the distributor shaft, which passes current to the distributor cap tower contacts.

**SECONDARY CIRCUIT:** The high voltage side of the ignition system, usually above 20,000 volts. The secondary includes the ignition coil, coil wire, distributor cap and rotor, spark plug wires and spark plugs.

**SENDING UNIT:** A mechanical, electrical, hydraulic or electro-magnetic device which transmits information to a gauge.

**SENSOR:** Any device designed to measure engine operating conditions or ambient pressures and temperatures. Usually electronic in nature and designed to send a voltage signal to an on-board computer, some sensors may operate as a simple on/off switch or they may provide a variable voltage signal (like a potentiometer) as conditions or measured parameters change.

**SHIM:** Spacers of precise, predetermined thickness used between parts to establish a proper working relationship.

**SLAVE CYLINDER:** In automotive use, a device in the hydraulic clutch system which is activated by hydraulic force, disengaging the clutch.

**SOLENOID:** A coil used to produce a magnetic field, the effect of which is to produce work.

**SPARK PLUG:** A device screwed into the combustion chamber of a spark ignition engine. The basic construction is a conductive core inside of a ceramic insulator, mounted in an outer conductive base. An electrical charge from the spark plug wire travels along the conductive core and jumps a preset air gap to a grounding point or points at the end of the conductive base. The resultant spark ignites the fuel/air mixture in the combustion chamber.

**SPLINES:** Ridges machined or cast onto the outer diameter of a shaft or inner diameter of a bore to enable parts to mate without rotation.

**TACHOMETER:** A device used to measure the rotary speed of an engine, shaft, gear, etc., usually in rotations per minute.

**THERMOSTAT:** A valve, located in the cooling system of an engine, which is closed when cold and opens gradually in response to engine heating, controlling the temperature of the coolant and rate of coolant flow.

**TOP DEAD CENTER (TDC):** The point at which the piston reaches the top of its travel on the compression stroke.

**TORQUE:** The twisting force applied to an object.

**TORQUE CONVERTER:** A turbine used to transmit power from a driving member to a driven member via hydraulic action, providing changes in drive ratio and torque. In automotive use, it links the driveplate at the rear of the engine to the automatic transmission.

**TRANSDUCER:** A device used to change a force into an electrical signal.

**TRANSISTOR:** A semi-conductor component which can be actuated by a small voltage to perform an electrical switching function.

**TUNE-UP:** A regular maintenance function, usually associated with the replacement and adjustment of parts and components in the electrical and fuel systems of a vehicle for the purpose of attaining optimum performance.

**TURBOCHARGER:** An exhaust driven pump which compresses intake air and forces it into the combustion chambers at higher than atmospheric pressures. The increased air pressure allows more fuel to be burned and results in increased horsepower being produced.

**VACUUM ADVANCE:** A device which advances the ignition timing in response to increased engine vacuum.

**VACUUM GAUGE:** An instrument used to measure the presence of vacuum in a chamber.

**VALVE:** A device which control the pressure, direction of flow or rate of flow of a liquid or gas.

**VALVE CLEARANCE:** The measured gap between the end of the valve stem and the rocker arm, cam lobe or follower that activates the valve.

**VISCOSITY:** The rating of a liquid's internal resistance to flow.

**VOLTMETER:** An instrument used for measuring electrical force in units called volts. Voltmeters are always connected parallel with the circuit being tested.

**WHEEL CYLINDER:** Found in the automotive drum brake assembly, it is a device, actuated by hydraulic pressure, which, through internal pistons, pushes the brake shoes outward against the drums.

# MASTER INDEX